Lecture Notes in Business Information Processing

599

LNBIP reports state-of-the-art results in areas related to business information systems and industrial application software development – timely, at a high level, and in both printed and electronic form.

The type of material published includes

- Proceedings (published in time for the respective event)
- Postproceedings (consisting of thoroughly revised and/or extended final papers)
- Other edited monographs (such as, for example, project reports or invited volumes)
- Tutorials (coherently integrated collections of lectures given at advanced courses, seminars, schools, etc.)
- Award-winning or exceptional theses

LNBIP is abstracted/indexed in DBLP, EI and Scopus. LNBIP volumes are also submitted for the inclusion in ISI Proceedings.

Fabrizio Maria Maggi · Henrik Leopold ·
Flavia Santoro · Avigdor Gal
Editors

Business Process Management Forum

BPM 2026 Forum
Toronto, ON, Canada, September 27 – October 2, 2026
Proceedings

Springer

Editors
Fabrizio Maria Maggi
Free University of Bozen-Bolzano
Bozen-Bolzano, Italy

Henrik Leopold
Kühne Logistics University
Hamburg, Hamburg, Germany

Flavia Santoro
Universidade do Estado do Rio de Janeiro
Rio de Janeiro, Rio de Janeiro, Brazil

Avigdor Gal
Israel Institute of Technology
Haifa, Israel

ISSN 1865-1348 ISSN 1865-1356 (electronic)
Lecture Notes in Business Information Processing
ISBN 978-3-032-37876-7 ISBN 978-3-032-37877-4 (eBook)
https://doi.org/10.1007/978-3-032-37877-4

This Springer imprint is published by the registered company Springer Nature Switzerland AG
The registered company address is: Gewerbestrasse 11, 6330 Cham, Switzerland

Preface

This volume comprises all papers presented at the BPM Forum of the 24th International Conference on Business Process Management (BPM), held during September 2026 at York University in Toronto, Canada. The BPM Forum provides an opportunity for papers that, while not fully meeting the quality criteria for the main conference track, make meaningful and promising contributions to the BPM community. These contributions may include novel applications, early-stage results with strong potential, or creative perspectives on known problems. Forum papers were evaluated through the same rigorous review process as main track submissions and are presented in full length in this separate post-proceedings volume. All Forum papers were presented in a special session during the main conference, offering a platform for constructive feedback and engagement from the broader BPM community.

This year, the conference received a total of 169 submissions. After a rigorous double-blind review process involving at least three Program Committee members and one Senior Program Committee member per submission, and a discussion phase followed by meta-reviews, 42 papers were accepted for presentation at the main conference. In parallel, 24 papers were accepted for inclusion in the BPM Forum.

The Forum papers cover a diverse and timely set of topics, reflecting the evolving socio-technical and AI-enhanced landscape of BPM. These span the three thematic tracks: 4 in Foundations, 15 in Engineering, and 26 in Management. The 2026 Forum contributions explore themes such as AI-driven business processes, agentic BPM, automated workflow generation, automated diagnosis of process models, knowledge-intensive processes (KiPs), BPM capabilities, process governance, process model explainability, object-centric process management, process discovery and quality assessment, process complexity analysis, and object-centric process simulation with dynamic relationships.

Open Science aiming at reproducibility and replicability of research results remained a major principle for the BPM Forum. Authors were explicitly encouraged to provide supplementary materials, such as datasets, source code, and implemented prototypes. Papers that made such artifacts available are recognized with badges in this volume. We also promoted diversity, equity, and inclusion (DEI) across all conference activities. We extend our sincere thanks to the DEI Chairs and the entire BPM community for fostering a welcoming and representative environment.

We would like to thank all authors, both regular and senior members of the Program Committees, and all external reviewers. Their diligent reviews and effective collaboration made a rigorous and timely review process possible and enabled the high-quality research output reflected in the papers of this volume.

Finally, we express our heartfelt appreciation to the BPM 2026 General Chairs and the local organizers for creating a memorable event in the beautiful city of Toronto.

We hope that the papers presented at the BPM Forum 2026 and captured in this volume will provide fertile stimulation for BPM academics and professionals.

September 2026

Fabrizio Maria Maggi
Henrik Leopold
Flavia Santoro
Avigdor Gal

Organization

Steering Committee

Jan Mendling (Chair)	Humboldt Universität zu Berlin, Germany
Avigdor Gal	Technion – Israel Institute of Technology, Israel
Chiara Ghidini	University of Trento, Italy
Manfred Reichert	Ulm University, Germany
Hajo A. Reijers	Utrecht University, the Netherlands
Stefanie Rinderle-Ma	Technical University of Munich, Germany
Adela del Río Ortega	University of Seville, Spain
Michael Rosemann	Queensland University of Technology, Australia
Shazia Sadiq	University of Queensland, Australia
Flavia Santoro	Institute of Technology and Leadership, Brazil
Barbara Weber	University of St.Gallen, Switzerland
Matthias Weidlich	Humboldt Universität zu Berlin, Germany

Advisory Board

Wil van der Aalst	RWTH Aachen University, Germany
Mathias Weske	HPI, University of Potsdam, Germany
Marlon Dumas	University of Tartu, Estonia

Executive Committee

General Chairs

Arik Senderovich	York University, Canada
Ravi de Costa	York University, Canada

Main Conference Program Committee Chairs

Avigdor Gal	Technion - Israel Institute of Technology, Israel
Fabrizio Maria Maggi	Free University of Bozen-Bolzano, Italy
Henrik Leopold	Kühne Logistics University, Germany

Flavia Santoro	Institute of Technology and Leadership (Inteli), Brazil

Workshop Chairs

Xixi Lu	Utrecht University, The Netherlands
Jorge Muñoz-Gama	Pontificia Universidad Católica de Chile, Chile
Daniel Beverungen	Paderborn University, Germany

Demonstration and Resources Chairs

Iris Beerepoot	Utrecht University, The Netherlands
Martin Kabierski	University of Vienna, Austria
Andrea Delgado	Universidad de la República, Uruguay

Responsible BPM Forum Chairs

Marco Comuzzi	Ulsan National Institute of Science and Technology (UNIST), South Korea
Stephan A. Fahrenkrog-Petersen	University of Liechtenstein, Liechtenstein
Amy Van Looy	Ghent University, Belgium

Process Technology Forum

Manfred Reichert	University of Ulm, Germany
Joerg Evermann	Memorial University of Newfoundland, Canada
Barbara Re	University of Camerino, Italy

Industry Forum Chairs

Najah El-Gharib	University of Ottawa, Canada
Julian Lebherz	SAP, Singapore
Fredrik Milani	University of Tartu, Estonia

Education Forum Chairs

Lucineia Heloisa Thom	Federal University of Rio Grande do Sul, Brazil
Karolin Winter	Eindhoven University of Technology, The Netherlands
Krzysztof Kluza	AGH University of Krakow, Poland

Tutorial Chairs

Daniel Amyot	University of Ottawa, Canada
Adela del Río-Ortega	University of Seville, Spain
Thomas Grisold	University of St. Gallen, Switzerland

Doctoral Consortium Chairs

Andrea Marrella	Sapienza University of Rome, Italy
Jana-Rebecca Rehse	University of Mannheim, Germany
Pnina Soffer	University of Haifa, Israel

Journal-First Track Chairs

Sarah Winkler	Free University of Bozen-Bolzano, Italy
Sander Leemans	RWTH Aachen University, Germany
Carl Corea	University of Koblenz, Germany

Publicity Chairs

María Salas Urbano	University of Seville, Spain
Jari Peeperkorn	KU Leuven, Belgium

Proceeding Chairs

Johannes De Smedt	KU Leuven, Belgium
Alessandro Gianola	INESC-ID/Instituto Superior Técnico, Universidade de Lisboa, Portugal

Diversity, Equity and Inclusion Chairs

Shazia Sadiq	University of Queensland, Australia
Flavia Santoro	Institute of Technology and Leadership (Inteli), Brazil
Araham Jesus Martinez Lagunas	Concordia University, Canada

Track I: Foundations

Senior Program Committee

Johannes De Smedt	KU Leuven, Belgium
Claudio Di Ciccio	Utrecht University, Netherlands
Chiara Di Francescomarino	University of Trento, Italy
Thomas Hildebrandt	University of Copenhagen, Denmark
Sander J.J. Leemans	RWTH Aachen University, Germany
Xixi Lu	Utrecht University, Netherlands
Andrea Marrella	Sapienza University of Rome, Italy
Marco Montali	Free University of Bozen-Bolzano, Italy
Artem Polyvyanyy	University of Melbourne, Australia
Manfred Reichert	University of Ulm, Germany
Hagen Voelzer	University of St. Gallen, Switzerland
Matthias Weidlich	Humboldt-Universität zu Berlin, Germany
Mathias Weske	HPI, University of Potsdam, Germany
Wil van der Aalst	RWTH Aachen University, Germany
Jan Martijn van der Werf	Utrecht University, Netherlands

Program Committee

Simone Agostinelli	Universitas Mercatorum, Italy
Anti Alman	Free University of Bozen-Bolzano, Italy
Abel Armas Cervantes	University of Melbourne, Australia
Ahmed Awad	British University in Dubai, UAE
Mario Luca Bernardi	University of Sannio, Italy
Adam Burke	Queensland University of Technology, Australia
Izack Cohen	Bar-Ilan University, Israel
Marco Comuzzi	Ulsan National Institute of Science and Technology, South Korea
Patrick Delfmann	University of Koblenz, Germany
Jörg Desel	FernUniversität in Hagen, Germany
Ivan Donadello	Free University of Bozen-Bolzano, Italy
Rik Eshuis	Eindhoven University of Technology, Netherlands
Joerg Evermann	Memorial University of Newfoundland, Canada
Dirk Fahland	Eindhoven University of Technology, Netherlands
Peter Fettke	DFKI and Saarland University, Germany
Valeria Fionda	University of Calabria, Italy
Luciano García-Bañuelos	Tecnológico de Monterrey, Mexico
Laura Genga	Eindhoven University of Technology, Netherlands

Alessandro Gianola	University of Lisbon, Portugal
Georg Grossmann	University of South Australia, Australia
Maria Teresa Gómez López	University of Seville, Spain
Rick Hull	IBM, USA
Anna Kalenkova	University of Adelaide, Australia
Ekkart Kindler	Technical University of Denmark, Denmark
Akhil Kumar	Pennsylvania State University, USA
Francesco Leotta	Sapienza University of Rome, Italy
Hugo A. López	Technical University of Denmark, Denmark
Orlenys López-Pintado	University of Tartu, Estonia
Felix Mannhardt	Eindhoven University of Technology, Netherlands
Elisa Marengo	University of Turin, Italy
Chun Ouyang	Queensland University of Technology, Australia
Thomas Prinz	Friedrich Schiller University Jena, Germany
Barbara Re	University of Camerino, Italy
Daniel Ritter	SAP, Germany
Andrey Rivkin	Technical University of Denmark, Denmark
Massimiliano Ronzani	Fondazione Bruno Kessler, Italy
Lorenzo Rossi	University of Camerino, Italy
Tijs Slaats	University of Copenhagen, Denmark
Niek Tax	Meta, UK
Eric Verbeek	Eindhoven University of Technology, Netherlands
Sven Weinzierl	Friedrich-Alexander-Universität Erlangen-Nürnberg, Germany
Sarah Winkler	Free University of Bozen-Bolzano, Italy

Track II: Engineering

Senior Program Committee

Boualem Benatallah	Dublin City University, Ireland
Andrea Burattin	Technical University of Denmark, Denmark
Marco Comuzzi	Ulsan National Institute of Science and Technology, South Korea
Jochen De Weerdt	KU Leuven, Belgium
Benoît Depaire	Hasselt University, Belgium
Remco Dijkman	Eindhoven University of Technology, Netherlands
Agnes Koschmider	University of Bayreuth, Germany
Niels Martin	Hasselt University, Belgium
Massimo Mecella	Sapienza University of Rome, Italy
Jan Mendling	Humboldt-Universität zu Berlin, Germany

Jorge Munoz-Gama	Pontificia Universidad Católica de Chile, Chile
Oscar Pastor	Universitat Politècnica de València, Spain
Luise Pufahl	Technical University of Munich, Germany
Manuel Resinas	University of Seville, Spain
Stefanie Rinderle-Ma	Technical University of Munich, Germany
Shazia Sadiq	University of Queensland, Australia
Marcos Sepúlveda	Pontificia Universidad Católica de Chile, Chile
Pnina Soffer	University of Haifa, Israel
Minseok Song	Pohang University of Science and Technology, South Korea
Barbara Weber	University of St. Gallen, Switzerland
Ingo Weber	Technical University of Munich, Germany
Moe Wynn	Queensland University of Technology, Australia
Francesca Zerbato	Eindhoven University of Technology, Netherlands
Massimiliano de Leoni	University of Padua, Italy
Boudewijn van Dongen	Eindhoven University of Technology, Netherlands
Han van der Aa	University of Vienna, Austria

Program Committee

Lars Ackermann-Igl	University of Bayreuth, Germany
Daniel Amyot	University of Ottawa, Canada
Robert Andrews	Queensland University of Technology, Australia
Michael Arias	University of Costa Rica, Costa Rica
Bernhard Axmann	Technische Hochschule Ingolstadt, Germany
Saimir Bala	Humboldt-Universität zu Berlin, Germany
Andrea Delgado	Universidad de la República, Uruguay
Claudia Diamantini	Università Politecnica delle Marche, Italy
Joerg Evermann	Memorial University of Newfoundland, Canada
Stephan Fahrenkrog-Petersen	University of Liechtenstein, Liechtenstein
Marcelo Fantinato	University of São Paulo, Brazil
Walid Fdhila	University of Vienna, Austria
Carlos Fernandez-Llatas	Universitat Politècnica de València, Spain
Walid Gaaloul	Télécom SudParis, France
Luciano García-Bañuelos	Tecnológico de Monterrey, Mexico
Laura Genga	Eindhoven University of Technology, Netherlands
José González Enríquez	University of Seville, Spain
Oscar González-Rojas	Universidad de los Andes, Colombia
Daniela Grigori	Paris-Dauphine University - PSL, France
Georg Grossmann	University of South Australia, Australia
Joscha Grüger	University of Trier, Germany
Lucinéia Heloisa Thom	Federal University of Rio Grande do Sul, Brazil

Andrés Jiménez Ramírez	University of Seville, Spain
Martin Kabierski	University of Vienna, Austria
Anna Kalenkova	University of Adelaide, Australia
Krzysztof Kluza	AGH University of Krakow, Poland
Julius Köpke	University of Klagenfurt, Austria
Manuel Lama Penin	University of Santiago de Compostela, Spain
Francesco Leotta	Sapienza University of Rome, Italy
Orlenys López-Pintado	University of Tartu, Estonia
Elisa Marengo	University of Turin, Italy
Sarajane Marques Peres	University of São Paulo, Brazil
Alfonso Marquez-Chamorro	University of Seville, Spain
Laura Maruster	University of Groningen, Netherlands
Renata Medeiros de Carvalho	Eindhoven University of Technology, Netherlands
Giovanni Meroni	Technical University of Denmark, Denmark
Andrea Morichetta	University of Camerino, Italy
Julian Neuberger	University of Bayreuth, Germany
Helen Paik	University of New South Wales, Australia
Cesare Pautasso	University of Lugano, Switzerland
Jari Peeperkorn	KU Leuven, Belgium
Pierluigi Plebani	Politecnico di Milano, Italy
Pascal Poizat	Université Paris Nanterre, France
Simon Poon	University of Sydney, Australia
Pascal Ravesteijn	Utrecht University of Applied Sciences, Netherlands
Barbara Re	University of Camerino, Italy
Kate Revoredo	Humboldt-Universität zu Berlin, Germany
Sareh Sadeghianasl	Queensland University of Technology, Australia
Stefan Schönig	Universität Regensburg, Germany
Natalia Sidorova	Eindhoven University of Technology, Netherlands
Vinicius Stein Dani	Utrecht University, Netherlands
Victoria Torres	Universitat Politècnica de València, Spain
Maxim Vidgof	WU Vienna, Austria
Karolin Winter	Eindhoven University of Technology, Netherlands
Roy Jing Yang	Queensland University of Technology, Australia
Nicola Zannone	Eindhoven University of Technology, Netherlands
Nick van Beest	CSIRO Data61, Australia
Sebastiaan van Zelst	Celonis, Germany

Track III: Management

Senior Program Committee

Fernanda Baião	Pontifical Catholic University of Rio de Janeiro, Brazil
Daniel Beverungen	Paderborn University, Germany
Paul Grefen	Eindhoven University of Technology, Netherlands
Thomas Grisold	University of St. Gallen, Switzerland
Mojca Indihar Štemberger	University of Ljubljana, Slovenia
Marta Indulska	University of Queensland, Australia
Christian Janiesch	TU Dortmund University, Germany
Mieke Jans	Hasselt University, Belgium
Ralf Plattfaut	University of Duisburg-Essen, Germany
Maximilian Röglinger	University of Bayreuth, Germany
Michael Rosemann	Queensland University of Technology, Australia
Peter Trkman	University of Ljubljana, Slovenia
Oktay Turetken	Eindhoven University of Technology, Netherlands
Amy Van Looy	Ghent University, Belgium
Irene Vanderfeesten	KU Leuven, Belgium
Adela del Río Ortega	University of Seville, Spain

Program Committee

Amine Abbad-Andaloussi	University of St. Gallen, Switzerland
Banu Aysolmaz	Eindhoven University of Technology, Netherlands
Christian Bartelheimer	Paderborn University, Germany
Markus Becker	University of Southern Denmark, Denmark
Iris Beerepoot	Utrecht University, Netherlands
Elisabetta Benevento	University of Pisa, Italy
Paolo Bottoni	Sapienza University of Rome, Italy
Edyta Brzychczy	AGH University of Krakow, Poland
Silvia Dallavalle	University of São Paulo, Brazil
Djordje Djurica	WU Vienna, Austria
Mahendrawathi Er	Institut Teknologi Sepuluh Nopember, Indonesia
Irene Bedilia Estrada Torres	University of Seville, Spain
Joklan Imelda Camelia Goni	Ghent University, Belgium
Thomas Haskamp	University of Münster, Germany
Michael Leyer	Philipps-Universität Marburg, Germany
Andrea Magalhaes	Dheka BPM Consulting, Brazil
Azumah Mamudu	Queensland University of Technology, Australia

Martin Matzner	FAU Erlangen-Nürnberg, Germany
Nijat Mehdiyev	DFKI, Germany
Patrick Mikalef	Norwegian University of Science and Technology, Norway
Fredrik Milani	University of Tartu, Estonia
Geert Poels	Ghent University, Belgium
Pascal Ravesteijn	Utrecht University of Applied Sciences, Netherlands
Jana-Rebecca Rehse	University of Mannheim, Germany
Dennis M. Riehle	Universität Koblenz, Germany
Ben Roelens	Open Universiteit, Netherlands
Estefanía Serral	KU Leuven, Belgium
Rehan Syed	Queensland University of Technology, Australia
Konstantinos Tsilionis	Eindhoven University of Technology, Netherlands
Sven Weinzierl	Friedrich-Alexander-Universität Erlangen-Nürnberg, Germany
Bastian Wurm	LMU Munich, Germany
Inge van de Weerd	Utrecht University, Netherlands

Additional Reviewers

Acheli, Mehdi
Acitelli, Giacomo
Andreoni, Riccardo
Bahr, Charlotte
Banham, Adam
Basmer, Maike
Bein, Leon
Bergenthum, Robin
Biermanns, Matthias
Blatt, Jonas
Brabra, Hayet
Brennig, Katharina
Brunings, Mitchel
Budde, Carlos E.
Buliga, Andrei
Camargo, Manuel
Corea, Carl
Costa, Ana
El Haddad, Joyce
El-Gharib, Najah
Folz-Weinstein, Sabine
Fonger, Frederik
Galassi, Alessio
Gavric, Aleksandar
Grohs, Michael
Hambauer, Nico
He, Zhipeng
Häge, Marie-Christin
Kalukapuge, Savandi
Kirchmann, Henrik
Klievtsova, Nataliia
Kovář, Jakub
Käppel, Martin
Kölbel, Linda
König, Maximilian
Küpper, Sebastian
Lichtenstein, Tom
Liessmann, Annina
Lindner, Jonas
Lu, Yang
Maldonado, Andrea
Mangler, Juergen
Mele, Alessandro
Meneghello, Francesca

Min, Haeri
Mozzoni, Luca
Nagel, Sabine
Nour Eldine, Ali
Nunes, Izaque Dione
Pettinari, Sara
Poss, Leo
Safan, Aya
Slingerland, Erik
Valero-Ramon, Zoe
Wang, Weixin
Wei, Jia
Workman, Silas
Zilker, Sandra

Contents

Foundations

Engineering

Management

Track I: Foundations

Simulating Dynamic Relationships in Object-Centric Processes

Denise Cosolo[1(✉)], Chiara Di Francescomarino[2], Chiara Ghidini[1], Francesca Meneghello[3(✉)], Marco Montali[1], and Massimiliano Ronzani[4]

[1] Free University of Bozen Bolzano, Bolzano, Italy
{denise.cosolo,chiara.ghidini,montali}@unibz.it
[2] University of Trento, Trento, Italy
c.difrancescomarino@unitn.it
[3] Technical University of Denmark, Kgs. Lyngby, Denmark
frmen@dtu.dk
[4] Fondazione Bruno Kessler (FBK), Trento, Italy
mronzani@fbk.eu

Abstract. Several Business Process Simulation techniques have been proposed to generate case-based event logs. However, an accurate description of many real-world processes requires modeling the interdependent objects that participate in them, leading to the rapid growth of object-centric process mining. Despite this progress, approaches for simulating object-centric event logs remain limited, as state-of-the-art techniques fail to capture the dynamic evolution of objects and their interactions during process execution. This limitation has important practical consequences, as such models cannot reproduce activities that create, modify, or remove objects, nor capture how many objects are generated within a given simulation time window. In this paper, we address these issues by proposing an object-centric simulation model capable of reproducing multiple objects and their dynamic relationships: new objects can be generated, or destroyed, during the simulation, and their relationships evolve according to the stochastic behavior of the process.

Keywords: Object-centric Process Mining · Business Process Simulation · Digital Twin

1 Introduction

The growing interest in object-centric process mining [2] (OCPM) is driven by the limited expressiveness of traditional case-centric event logs in capturing interactions among multiple objects in real-world processes, where each object follows

This work was partially supported by the Innovation Fund Denmark (IFD) under the project P3AI4 - Predictive and Prescriptive Process Analytics for Industry 4.0 (Grant Nr. 4355-00010B).

F. M. Maggi et al. (Eds.): BPM 2026, LNBIP 599, pp. 3–20, 2027.
https://doi.org/10.1007/978-3-032-37877-4_1

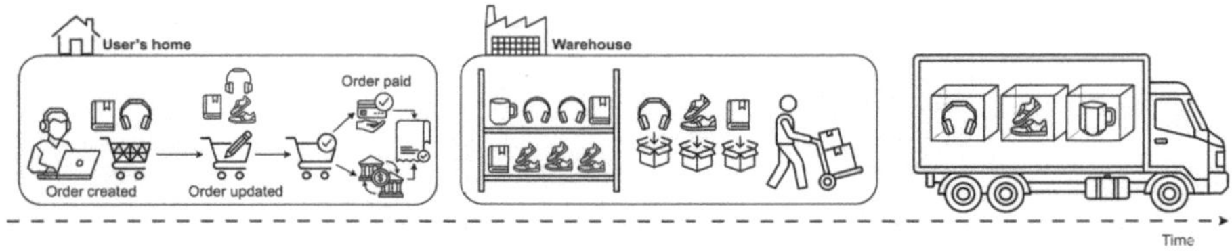

Fig. 1. An order-to-ship scenario

its own lifecycle and evolves in a partially (in)dependent flow. Despite significant advances in modeling and analyzing object-centric processes, and the recognized need for object-centric simulation [1], only a limited number of works have focused on developing Business Process Simulation (BPS) techniques for OCPM [9,11]. Additionally, the solutions proposed in these works exhibit significant limitations, particularly in the management of object relationships, failing to adequately capture the dynamic nature of such relations in realistic processes and their impact on process execution. For example, in an order-to-shipment process, they are unable to reproduce the dynamic creation and evolution of relationships resulting from the addition or removal of items from an order, or from the dynamic assignment of items to deliveries based on their arrival times at the shipment point. Moreover, they cannot capture object behavior driven by evolving data attributes of co-referenced objects, such as determining the payment method for an order based on the total cost after items have been added or removed.

In this work—inspired by the recent introduction of a formalism that addresses similar limitations in OC modeling, namely DOPID [7]—we present a novel simulation framework that overcomes several limitations of existing object-centric simulation techniques. In particular, our approach supports the definition, management and reproduction of dynamic relationships among objects during simulation, as well as the influence of object data attributes on their behavior and interactions. We formally present the components of our technique which model OC processes, along with a detailed description of their integration within a Discrete Event Simulation (DES) framework to generate process executions (Sect. 5) Moreover, we provide additional support for the soundness of the generated executions by showing how the building blocks of our simulation model can be related to the DOPID formalism, a rigorously defined modeling framework with well-established execution semantics (Sect. 6) Finally, we demonstrate the capabilities of our simulation method through two illustrative scenarios, highlighting its ability to address decision-support questions and dynamics that current object-centric simulation models are unable to capture (Sect. 7) The paper uses a running example illustrated in Sect. 2, while Sects. 3 and 4 reviews background notions the state of the art.

2 Running Example

Consider a typical order-to-ship process whose main steps are depicted in Fig. 1. A customer creates an order, specifying the delivery address. The order pertains a number of items. Each item has its own price, and represents a good to be delivered. Before checking out, the customer can modify the order by removing or adding items. Once all the items are confirmed, the order can proceed to check out and payment. The payment can be made by credit card or bank transfer: if the price of all items is lower than 1000 (EUR) the order is paid by Bank transfer, otherwise by Credit Card. As soon as the payment is confirmed the order proceeds with the packing and delivery by trucks.

Trucks are responsible for delivering items, which may belong to different orders; each truck has a capacity, namely the maximum number of items that it can deliver. After re-fueling and check-up, a truck is ready to load items that are ready to be shipped—namely, those that have completed packing—based on its capacity and/or the delivery addresses associated with the items. When the truck has terminated the delivery of all packages, it moves back to the warehouse and may load other ready items. The process terminates when all items belonging to the created orders have been delivered by truck.

This example contains several typical characteristics of object-centric processes. (1) *Coexistence of multiple objects:* no object type can be clearly identified as case. Orders, items, and trucks, flow through the process establishing mutual dependencies that influence how the process is executed. (2) *Activities with different focus:* several activities operate on dedicated objects, possibly influencing other objects as well. An order can be placed or paid, an item can be picked or removed, and a truck can be loaded and driven to a delivery address. (3) *Object relationships:* orders, items, and trucks are related one to another. Specifically, an order is related to at least one, possibly multiple, items, while each item has one and only one order. Similarly, a truck is related to all the items it has been loaded with. (4) *Dynamic relationships:* relationships can change over time. Items can be added or removed from an order, and trucks are loaded with different items every time they return to the warehouse. (5) *Existential dependencies*: objects can exist autonomously, or only because they depend on another object. For example: every item existentially depend on its containing order, while trucks exist independently of orders and items. Finally, (6) *synchronization points* are established between activities involving different objects. For example, an item can be packed only when its containing order has been paid.

3 Background

The section reports the background concepts used in the remainder of the paper.

Object Centric Event Log. We recall the standard definition object-centric event logs (see, e.g., [2,7]), where differently from the case-centric setting, each event can refer to one or more objects, possibly with different types.

Let Σ be the set of object types, $\mathcal{A}$ the set of activities, $\mathcal{R}$ the set of resources, $\mathcal{D}$ the set of attributes, and $\mathbb{T}$ the set of timestamps with a total order $\leq$. An event log is a tuple $\mathcal{L} = \langle E, \mathcal{O}, \pi_{\text{act}}, \pi_{\text{obj}}, \pi_{\text{time}}, \pi_{\text{res}}, \pi_{\text{attr}} \rangle$ where (i) E is a set of event identifiers; (ii) $\mathcal{O}$ is a set of object identifiers that are typed by the function $type : \mathcal{O} \to \Sigma$; (iii) π are projection functions, $\pi_{\text{act}} : E \to \mathcal{A}$, $\pi_{\text{obj}} : E \to P(\mathcal{O})$, $\pi_{\text{res}} : E \to \mathcal{R}$ that map each event $e \in E$ to an activity $a \in \mathcal{A}$, a set of affected objects identifiers $O \in P(\mathcal{O})$, and a resource $r \in \mathcal{R}$, respectively; (iv) $\pi_{\text{attr}} : \Sigma \to P(\mathcal{D})$ maps each object type to a set of attributes $D \in P(\mathcal{D})$; (v) $\pi_{\text{time}} : E \to \mathbb{T}$ maps each event to a timestamp.

Business Process Simulation (BPS). We recall the main characteristics of Business Process Simulation (BPS) models. We emphasize separation between two main components. First, a BPS contains a process component, providing the control-flow (and in case of object-centric processes, also the object) dependencies determining which activities are enabled at every point in time. Second, simulation parameters capture further perspectives related to time and resources.

Specifically, a BPS model is a tuple $\mathcal{M} = (\mathcal{N}, \mathcal{P})$, where $\mathcal{N}$ is process model, and $\mathcal{P} = \{\mathcal{P}_R, \mathcal{P}_T, \mathcal{P}_C, \mathcal{P}_S, \mathcal{P}_D\}$ a set of simulation parameters that describe different perspectives of the process: resource, time, control flow, inter-arrival time and data payload, respectively. In particular, $\mathcal{P}_R$ specifies the resources, often grouped in roles, involved in $\mathcal{M}$ along with allocation rules to execute the activities. $\mathcal{P}_T$ defines the processing time required to complete each activity in $\mathcal{N}$. In the presence of decision points in $\mathcal{N}$, $\mathcal{P}_C$ handles the selection among possible paths; otherwise, the simulation follows the semantics of $\mathcal{N}$. $\mathcal{P}_S$ defines the inter-arrival time of the new process instances. Finally, $\mathcal{P}_D$ defines possible attributes associated to process instances.

BPS models are discrete event simulation (DES) frameworks where elements run simultaneously based on their inter-arrival time, and compete for resources to execute activities and complete the process. Specifically, each next activity to execute is defined by $\mathcal{N}$ and can be executed if at least one resource according to $\mathcal{P}_R$ is available; otherwise, it waits until the necessary resource is available. Moreover, $\mathcal{P}_R$ may include calendars associated with the resources involved in the process, which define their availability. The output of the simulation $\mathcal{M}$ is an event log containing the generated traces.

4 Related Work

The section is structured around two classes of related literature: object-centric process models and object-centric simulation models.

4.1 Object Centric Process Models

Several works have proposed process models capable of handling object-centric processes, such as object-centric Petri Nets [2], synchronous proclets [5], Petri Nets with Identifiers (PNIDs) [12–14], and DOPIDs [7].

Object-centric nets [2] offer an implicit approach to object centricity where places and transitions have different object types. Simple arcs match with a single object at a time, while variable arcs handle arbitrarily many objects of the same type. However, the lack of explicit object relationships prevents the modeling of object synchronization and coreference, meaning that, after their creation, objects cannot refer to other objects through identifiers. For instance, after the creation of an order and its items, the information about which items belong to the order is missing.

Synchronous Proclets [5] proposes a modular and defined structure of the processes and can implicitly express the tracking of objects and their mutual relationships. In particular, each object is described by its own proclet, and the interactions between instances of different entities are performed via synchronous channels. The cardinality constraints on channels bound how many instances may participate in a given interaction. However, Synchronous Proclets [5] cannot properly handle modifications to existing relationships, as any update overwrites the previous one. For instance, adding a new item to an existing order results in the loss of the items initially associated with it.

Petri Nets with Identifiers (PNIDs) and their variants [12–14] are introduced to address the object synchronization and coreference issues limitations of Object-centric nets [2], by explicitly managing objects and their relationships through identifier tuples. Therefore, they provide mechanisms for creating, updating, and removing objects and relationships, although they lack variable arcs to handle multiple objects within a single transition.

Data object-centric Petri nets with identifiers (DOPIDs) [7] are the extended version of OPID [13] incorporating the structured data into the model, allowing object instances to carry typed attributes that evolve as the process unfolds, capturing both the control flow and the data perspective of each object's lifecycle. DOPIDs introduces exact synchronization semantics, formally specifying how multi-object events must coordinate, such that a transition fires only when all required object instances are available and meet the specified conditions.

Despite DOPIDs [7] offering significant expressive power, they are not well-suited to be used into simulation models. Indeed, their compact style of representation would require a complex specification of all simulation parameters to ensure compliance with DOPIDs semantics during execution. On the other hand, thanks to their modular structure, Synchronous Proclets [5] are adaptable for simulation models, but they still have the aforementioned limitations in handling relationships. The simulation model introduced in Sect. 5 exhibits similarities with Synchronous Proclets in the style of modeling, but it encompasses its limitations by handling dynamic relationships according to the DOPIDs semantics.

4.2 Object Centric Simulation Models

The majority of business process simulation approaches—namely, traditional approaches [4], agent-based approaches [8], and hybrid approaches [3,10]—follow the case-centric paradigm and cannot simulate object centric scenarios.

In [9], a data-driven Object-Centric Process Simulation Model is proposed, where, starting from an OCEL log, an object component with an object generator and an intertwined behavioral component are defined to route generated objects through a control-flow model. However, in this approach, objects and their relationships cannot be deleted once created. Moreover, relationships between objects are predefined and cannot be established dynamically during the simulation. These limitations prevent this simulation approach from capturing the dynamic nature of real-world processes. For example, it cannot simulate the removal of an item from a shopping cart, nor assign a delivery truck based on the actual arrival time of an item at the warehouse.

Renew [11] is a tool that supports the development and execution of object-oriented Petri nets, which include net instances, synchronous channels providing a flexible modelling approach. However, it does not allow the dynamic creation or deletion of relationships at runtime and requires a detailed specification of object types, relationships, and constraints.

5 Approach

In this section, we focus on the process component of an object centric BPS model showing all the complexities of our running example. As written in Sect. 4.1, we aim at a trade-off between the modularity of (synchronous) Proclets and the modelling features of DOPIDs. Proclets clearly encapsulate the behavior of each object type and where it needs to synchronize with that of other types, directly lending themselves to be realized in a simulator. DOPIDs provide the necessary modelling features to capture time-varying relationships and their usage in synchronization points, as well as conditions on data attributes.

We first introduce the proposed *Object-Centric Simulation Model (OCSM)* resulting from this trade-off, and then we describe its execution in a simulation environment. The execution semantics in terms of objects, their flows, and mutual relationships, is then given in Sect. 6 through an encoding into DOPIDs [7]. The framework captures object-centric processes as shown in Fig. 2.

5.1 Object-Centric Simulation Model Definition

We consider a Petri-net based simulation model for object-centric processes. As it will become clear in the technical exposition, the model can be seen as a refinement of the synchronous Proclet model, providing dynamic relationships (i.e., in the Proclet jargon, a dynamic version of *correlation sets*), more flexible forms of synchronization, and data-dependent conditions. These features require the full power of DOPIDs to be suitably formalized.

We consider a set Σ of object types, and an implicit universe of all the objects present in the domain, equipeed with a typing function *type* that, given an object o, maps it to its correpsonding type $type(o) \in \Sigma$. Every type can be further modelled using data attributes (such as the cost of an item), each mapping every object in that type to a corresponding value. Attributes may be static (such as the price of an item in our running example), or dynamic (such as the credit card number used to issue a payment). Due to space constraints, we do not enter into the formalization of attributes, but we keep the data-conditioned payment in our running example to witness the versatility of our approach.

Definition 1 (OCSM). Given a set Σ of object types, an OCSM $\mathcal{M}_\Sigma$ is a triple $(\mathbb{U}, \Phi, \mathbb{C})$, where: (i) $\mathbb{U}$ is a set of *object relationships* between object types from Σ; (ii) Φ is an object flow total function mapping each type in Σ to a corresponding *object flow* (captured as a Petri net, as detailed later); (iii) $\mathbb{C}$ is a set of *synchronization mechanisms* relating transitions from distinct object flows.

We next detail the three components of $\mathcal{M}_\Sigma$.

Object Relationships. Object relationships are defined at the type level, and then instantiated accordingly during the process execution among objects of the related types. We consider binary, one-to-many relationships. N-ary relationships, binary many-to-many relationships, can be indirectly represented through reification, which is needed in those cases to properly express synchronization mechanisms (as discussed in [5]). One-to-one binary relationships, or binary relationships with more sophisticated cardinality constraints can be directly accommodated, but are omitted for simplicity of presentation.

Given this premise, $\mathbb{U} \subseteq \Sigma \times \Sigma$ describes the one-to-many relationships existing among pairs of types in Σ. To allow for a clear specification of flows and their synchronization, we assume that for a pair of types σ_1 and σ_2, only a single one-to-many relationship can be expressed.

If $(\sigma_1, \sigma_2) \in \mathbb{U}$, then at runtime each "parent" object p with $type(p) = \sigma_1$ can be (dynamically) associated to zero or more "child objects" of type σ_2.

At a given time point, function π_{rel} is used to indicate, for every binary relationship $(\sigma_1, \sigma_2) \in \mathbb{U}$, the current link between each object of type σ_1 and a corresponding set of objects of type σ_2.

Example 1. In the example in Sect. 2, order may refer to multiple items. Thus, we have $\{order, item\} \subset \Sigma$, and $(order, item) \in \mathbb{U}$. If, at a given time point, order o_1 is related to items i_1 and i_2, than in that moment $\pi_{\mathsf{rel}}(o_1) = \{i_1, i_2\}$.

Object Flows. Every object type gets mapped by Φ to a local simulation model, with a Petri net-based process component and a corresponding set of simulation parameters. Simulation parameters are defined locally for each object type, governing temporal, resource, and object-flow aspects, similarly to what is done in case-centric simulation [4,10], with the addition of parameters regulating the interactions with other object types.

The process component is represented as a labeled Petri net. We keep the definition compact and adhering to the standard Petri net literature. However, as we will see in Sect. 5.2, since objects have also attributes, transitions can also be conditioned by the current attributes carried by an object.

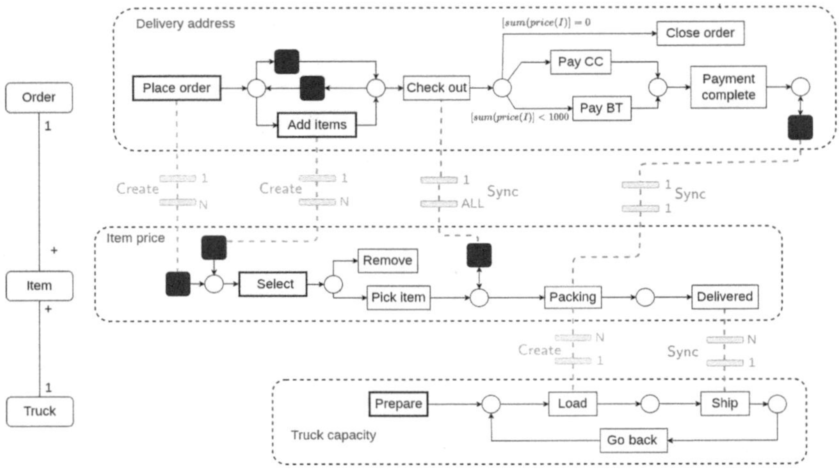

Fig. 2. Example of order-to-ship object centric process

Definition 2 (Labeled Petri net). A *labeled Petri net* is a tuple (P, T, F, L, l), where: (i) P is a finite set of places, (ii) T is a finite set of transitions with $P \cap T = \emptyset$, (iii) $F \subseteq (P \times T) \cup (T \times P)$ is a flow relation places to transitions and transitions to places, (iv) L is a set of activity labels, and (v) $l : T \to L \cup \{\tau\}$ is a labeling function mapping each transition to an activity label or to the special label $\tau \notin L$ denoting a silent move. A transition $t \in T$ is called an *origin* transition if it has no incoming arcs (i.e., $\forall p \in P, (p,t) \notin F$) and a *sink* transition if it has no outgoing arcs (i.e., $\forall p \in P, (t,p) \notin F$).

Definition 3 (Object Flow). Given an object type $\sigma \in \Sigma$, $\Phi(\sigma) = (\mathcal{N}^\sigma, \mathcal{P}^\sigma)$, where: (i) $\mathcal{N}^\sigma$ is a labeled Petri net; (ii) $\mathcal{P}^\sigma$ is a set of simulation parameters associated to $\mathcal{N}^\sigma$, as per Sect. 3, with the addition of a new parameter $\mathcal{P}^\sigma_X$ which defines the interaction with other objects types.

Example 2. Figure 2 illustrates the components of the simulation model based on the example in Sect. 2. The figure depicts for each of the three object types of the example (*order*, *item* and *truck*), the labeled Petri net $\mathcal{N}^\sigma$ of the related object flow $\Phi(\sigma)$. The top of the figure depicts the net $\mathcal{N}^{\text{order}}$ for to the *order* type, the middle one the *item* net, and the bottom one the *truck* net.

We assume here that transitions and places are assigned to a single object, i.e., for every two distinct types $\sigma_1, \sigma_2 \in \Sigma$, the nets $\Phi(\sigma_1)$ and $\Phi(\sigma_2)$ do not share places nor transitions. Events operating on different objects are in fact captured using channels, as described next.

Synchronization Mechanisms. We express synchronization mechanisms to link different types and their object flows using *channels*. Intuitively, channels indicate how relationships between objects are created, manipulated, and inspected. In the spirit of Proclets, this is done by linking transitions across different related object types, with decorations expressing the type of synchronization, and the involved cardinalities. However, there are clear distinctions with the Proclet channels, as here we deal with varying relationships.

A channel operates on a pair of transitions, belonging to the object flows of two types in a one-to-many relationship from $\mathbb{U}$. The overall set $\mathbb{C}$ of channels is partitioned into two distinct types of channels $\mathbb{C}_c$ and $\mathbb{C}_s$, respectively representing *create* channels, dealing with the creation of one or more instances of a given relationship, and *synch* channels, synchronizing the flow or one or more pair of interrelated objects through their respective transitions.

Definition 4 (Channel). Let σ_p and σ_c be two types from Σ with $(\sigma_p, \sigma_c) \in \mathbb{U}$, and let T_p and T_c denote the set of transitions in the two nets of $\Phi(\sigma_p)$ and $\Phi(\sigma_c)$ respectively. A channel $h \in \mathbb{C}_c$ is a tuple $(t_p, t_c, type, card)$ where: (i) $t_p \in T_p$ is the parent transition; (ii) $t_c \in T_c$ is the child transition; (iii) $type \in \{\texttt{Create}, \texttt{Sync}\}$ is the channel type indicating whether instances for relationship (σ_p, σ_c) are created, or accessed for synchronization; (iv) $card \in \{1{:}1, 1{:}\mathrm{N}, 1{:}\mathrm{ALL}\}$ is a cardinality creation constraint, indicating whether the channel operates on a single pair (1:1), or multiple pairs with the same parent object and distinct child objects (1:N), or multiple pairs with the same parent object and all active objects connected to that parent (1:ALL). The following well-formedness conditions are assumed: (i) for two transitions t_p and t_c, there is at most one channel connecting them; (ii) if the channel is of type `Create`, then *card* is either 1:1 or 1:N (*not* 1:ALL).

Let us comment on `Create` channels first. They handle the creation of one or multiple parent-children instances of some relationship. While this seems similar to init channels from synchronous Proclets [5], it is more sophisticated: while an init channel fixes a single moment of time where a parent object gets related to one or multiple child objects, in our case this set can be dynamically extended with further relationship instances, as the next example shows.

Example 3. Consider the *order* and *item* flows from Fig. 2. The Place Order transition in the net of $\Phi(order)$ is linked by a 1:N `Create` channel to a corresponding silent transition in the net of $\Phi(item)$. Since both such transitions are origin transitions, this represents the creation of a new order and a batch of multiple items that all get related to that order. That order-to-item relationship can be further extended if that order flows through the Add Items transition in the $\Phi(order)$ net. In fact, also that transition is linked to a dedicated silent transition in the $\Phi(item)$ net through a 1:N `Create` channel.

Consider now `Sync` channels. A 1:1 channel denotes that, in a given instant, the linked transitions are enabled for a parent and *a single* child object only if they are enabled according to the two local Petri nets, and the two objects are related. A 1:N channel extends this by indicating that *one or more* objects from the child side can simultaneously flow through transition t_c if their parent object simultaneously flows through transition t_p. A 1:ALL channel further strengthens this synchronization semantics by requiring that all active child objects with the same parent must flow (which implicitly requires that they are *all* enabling transition t_c in that instant). These constructs port the corresponding synchronization notions of synchronous Proclets to dynamic relationships as understood in our approach.

Example 4. Consider the Check out transition in Fig. 2. It is connected to a Sync 1:ALL channel to a corresponding silent transition in the *item* net. This enforces that an order can be checked out only when *all its active items* have been previously picked. We stress here the term *active*: items may have been dynamically added to the same order (through Place Order and/or Add Items executed on the same order object), but also dynamically removed from it in case they went through the Remove transition. Upon Payment complete, order objects are inserted in the downward place in the *order* net. The 1:1 Sync channel connected to the consequent read-write silent transition models that the items of an order can only be packed if that order has reached that place. Finally, the Ship transition in the *truck* net is linked via a 1:N Sync channel to the Delivered transition in the *item* net, capturing that a truck shipment simultaneously leads one or more of the items currently loaded in the truck (not necessarily all) to be delivered.

5.2 Execution of an OCSM

Given an object-centric simulation model $\mathcal{M}_\Sigma$ defined as before, we describe its simulation into a typical DES framework. The simulation starts by considering object types that have an origin transition that is either not connected to any channel, or if so it is on the parent side (i.e., on the one side of a one-to-many relationship). Objects for those types are instantiated: when, and how many, is determined by the $\mathcal{P}_S^\sigma$ parameters, which specify their arrival rate. The DES environment also sets the parameters $\mathcal{P}_D^\sigma$ determining the values of the attributes associated to those objects. Corresponding tokens are then emitted by firing as many copies of the origin transition as created objects. As in all models for object-centric processes, each resulting token refers to a dedicated object.

Subsequently, each object o follows the semantics of the associated Petri net $\mathcal{N}^\sigma$ where, like in synchronous Proclets, it is assumed that only tokens referring to the same object determine transition enablement, and are consumed and produced by its corresponding firing. In case multiple transitions are enabled for a given object o, the set P_C^σ defines the rules used to identify which transition is picked for firing. These rules can be specified as fixed probabilities or as custom logic based on the attributes of o and/or those of other objects related to it (see later how these are created and manipulated). An example is resolving the decision on how to pay an order: whenever a token referring to an order o is present in the downward place of Check out (see Fig. 2), then Pay CC is enabled. Depending on which items are currently associated to o, and in turn the value of the sum of their prices, then also Pay BT and/or Close order may be enabled as well for o.

How the simulator progresses the execution by picking and firing a transition actually also depends on the presence of channels. If a transition t does not belong to any channel, then its enablement and firing are determined by presence of tokens in its input places referring to the same object, and upon firing the corresponding activity is executed using its specific parameters for resource allocation and processing time estimation, namely $\mathcal{P}^{\sigma}_{R}$ and $\mathcal{P}^{\sigma}_{T}$.

Consider now a pair of transitions t_p and t_c in two nets for types σ_p and σ_c respectively, such that $(\sigma_p, \sigma_c) \in \mathbb{U}$ (i.e., σ_p is connected to σ_c by a one-to-many relationship). Firings for t_p and t_c are not unrelated, but depend on the internal distribution of tokens in their two nets (which we call next "local enablement"), as well as on the channel semantics, which establishes different forms of synchronization, in turn depending on which objects are related. The simulator then needs to keep a global memory to store which pairs of objects from σ_p and σ_c are related, ensuring that no two parent objects from σ_p are connected to the same child object from σ_c (this would violate the one-to-many nature of the relationship). We describe how this global memory is updated and influences the firing of t_p and t_c.

We first handle the case where t_p and t_c are connected by a `Create` channel, i.e., $(t_p, t_c, \texttt{Create}, \mathit{card}) \in \mathbb{C}$. Suppose that t_p is locally enabled for an object o, and t_c is enabled for one or more objects (denoting the overall set of such objects as Q).[1] Then, upon firing t_p for o, n copies of t_c simultaneously fire for a nonempty subset Q' of Q. Which objects are selected in Q', and how many, depends on *card* (if $\mathit{card} = 1{:}1$, then $|Q'| = 1$) and on the simulation parameters $\mathcal{P}^{\sigma_p}_{X}$ (see Definition 3). In parallel with the synchronous firings of all such transitions with their corresponding objects, the global memory is updated for relationship (σ_p, σ_c), indicating that for every $q \in Q'$, pair (o, q) provides a (new) instance for that relationship. Notice that existing pairs are maintained.

We now handle the case where t_p and t_c are connected by `Sync` channel of the form $(t_p, t_c, \texttt{Sync}, \mathit{card}) \in \mathbb{C}$. Suppose, as before, that t_p is locally enabled for some object o. The simulator needs to consider which instances of relationship (σ_p, σ_c) currently exist with o as first component. Denote by R the overall set of objects of type σ_c that are related to o in the current global memory for (σ_p, σ_c), and denote by Q the overall set of objects type σ_c for which transition t_c is currently enabled. If $\mathit{card} = 1{:}1$, then t_p is enabled for o if $|Q \cap R| \geq 1$, i.e., there is at least one object of type σ_c that is currently related to o and for which t_c is currently enabled. In that case, upon firing t_p then simultaneously also t_c fires for one object $q \in Q \cap R$. If $\mathit{card} = 1{:}\mathrm{N}$, then t_p is enabled for o if $|Q \cap R| \geq 1$, and upon firing t_p then the simulator selects a nonempty subset Q' of $Q \cap R$, and simultaneously fires t_c for every object $q \in Q'$—capturing the simultaneous firing of some objects of type σ_c related to o and for which t_c is enabled. If $\mathit{card} = 1{:}\mathrm{ALL}$, then t_p is enabled for o if $R \subseteq Q$, that is, *all* objects of type σ_c currently related to o are enabling t_c. In addition, upon firing t_p then the simulator simultaneously fires t_c for every such object $q \in Q$.

[1] Recall that, as a special case, an origin transition is always enabled for (infinitely many) *new* objects of its corresponding type, allowing for object creation.

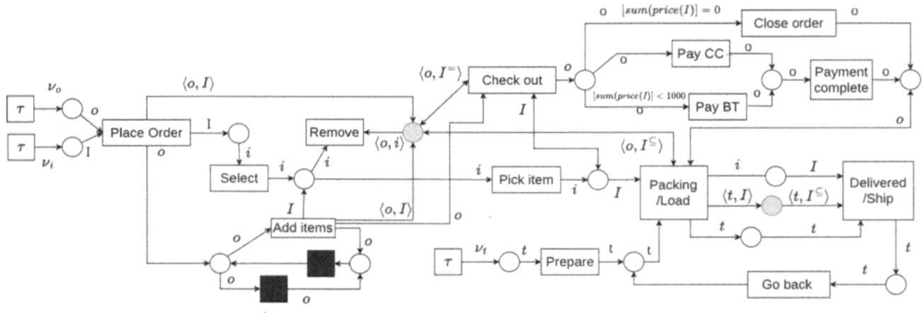

Fig. 3. Process representation using DOPID model

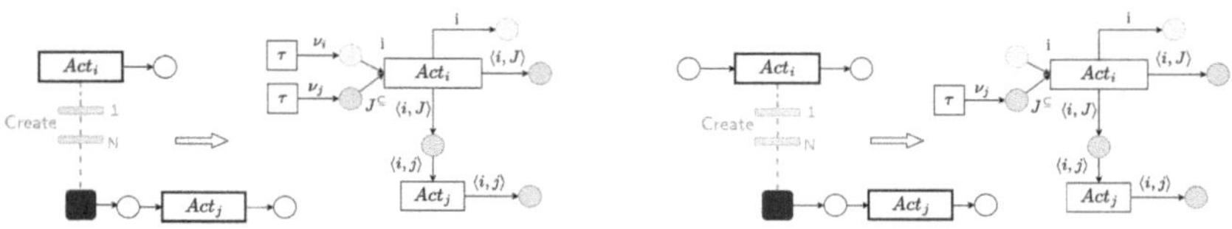

Fig. 4. DOPID encoding of `Create` channels

Finally, special consideration is given by the simulator to the firing of sink transitions, which have the effect of removing tokens from their local nets. In particular, the simulator enforces the following alignment with the global memory: when an enabled sink transition t fires consuming token(s) for an object q of type σ_c, and there are no other tokens left referring to the same object q (i.e., conceptually, object q is removed), then for every other type σ_p such that there is a relationship $(\sigma_p, \sigma_c) \in \mathbb{U}$, if there is an instance (o, q) currently present in the global memory for (σ_p, σ_c), then it is removed.[2]

The simulation ends either upon reaching a specified time window or when no further transitions are enabled due to constraints over the `Create` and `Sync` channels. For example, not all items generated during a simulation of the process in Fig. 2 may be delivered if the number of trucks is insufficient to complete the shipment. The output of the simulation of $\mathcal{M}_\Sigma$ is a simulated event log $\mathcal{L}$, conforming to the definition from Sect. 3.

6 Mapping OCSMs Into DOPIDs

In this section, we formally encode the process component of an OCSM into a corresponding DOPID, thereby unambiguously characterizing the execution semantics of an OCSM in terms of transition firings, object flows, channels, and the manipulation and inspection of object-to-object relationships. This backs up, formally, the description of transition firings from Sect. 5.2.

[2] Notice that since (σ_p, σ_c) is a one-to-many relationship, then at most one pair with second component q can exist in the working memory.

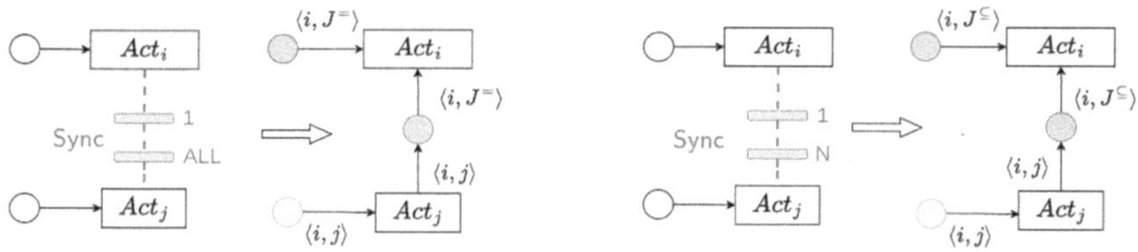

Fig. 5. DOPID encoding of `Sync` channels

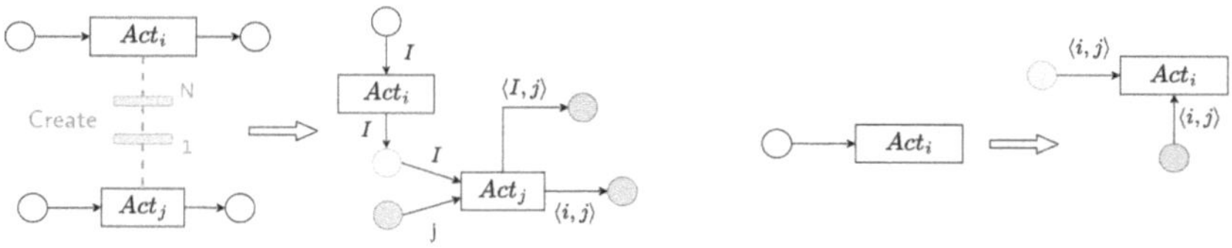

Fig. 6. From channel synchronization to DOPID: create and remove cases

The encoding extends the Proclet-to-DOPID encoding from [7], by reflecting the more sophisticated channels considered here. Essentially, the main difference with [7] is that here we have to introduce, for every relationship $(\sigma_p, \sigma_c) \in \mathbb{U}$, a dedicated, "global" place storing all pairs of objects that provide the current extension for that relationship, and manipulate its content according to the used channels. Figure 3 illustrates the DOPID obtained as the result of the encoding applied to the order-to-shipment OCSM from Fig. 2. Red places provide storage for the two one-to-many $(order, item)$ and $(truck, item)$ relationships.

For a general OCSM, the encoding works as follows. Every type translates into a corresponding DOPID fragment where inscriptions express that only tokens referring to the same object in that type can be consumed and produced. Transitions linked by a channel get fused into a single transition, which operates not only on objects of different types, but also insists on a further, dedicated place storing the pairs of the corresponding relationship. How this is done depends on the specific channel. We discuss the different cases next.

Figure 4 shows the encoding of two usages of `Create` Channel. In particular, the left side illustrates the case in which a new object is created from the beginning of a process, whereas the right side depicts the case in which a new object type is spawn from an already started process. In the mapping in the left hand side, the transition Act_i, identified as an originator by its bold border, is connected via a `Create` Channel to another emitter transition Act_j, and spawns n objects of type j. This pattern corresponds to the `Create` Channel between Place order and Select in Fig. 2. In the DOPID, the variables ν_i and ν_j denote freshly created objects of type i and j, respectively. Transition Act_i consumes a single parent object i along with some child objects J, and the output-transfer-template inscription $\langle i, J \rangle$ generates all the newly established pairs (i, j) (with $j \in J$) in the relationship storage place.

In the mapping on the right hand side, transitions Act_i and Act_j correspond to the Add items and Select transitions in Fig. 2. Here, the difference lies in the presence of a place preceding Act_i which is represented by the yellow place in the DOPID encoding. Since the process of the object i is already running, there is no need to inject fresh tokens into the net, as illustrated in Fig. 2, where the order process is already running and the activity updates the relationship between order and items by adding a new item.

Figure 5 illustrates the encoding of the different Sync Channels. The behavior of each channel varies according to its cardinality. On the left, Act_i and Act_j correspond to the Check out and Pick item transitions in Fig. 2, respectively. When all items related to an order have performed Pick item, the order can proceed with Check out. Additionally, the 1:ALL cardinality specifies that synchronization must involve all currently active J objects associated with a given i object, which is captured in the DOPID model by the use of inscription $\langle i, J^{=} \rangle$. This inscription consumes all pairs from the correlation set (red place) and the purple place, upon which the synchronization is performed.

The N:1 cardinality shown in the right hand side mapping captures scenarios in which only a subset of the active J objects associated with i is required to execute the transition. This is the case in the Sync Channel between Ship and Delivered in Fig. 2: the truck performs the shipment of the items it has loaded, and its completion triggers the firing of the Delivered transition in the item process. This behavior is expressed in the DOPID model through the inscriptions $\langle i, J^{\subseteq} \rangle$, which consumes a subset of pairs from both the correlation set and the purple place.

Figure 6 represents two additional behaviors. In the left hand side, we consider the case in which objects of types i and j already exist, and the channel establishes a new relation between them. In terms of Fig. 2, Act_i corresponds to the Packing transition and Act_j to the Load transition. Whenever an item is packed, it becomes available to be loaded onto the truck, which waits for N items to be packed before executing the Load transition. In the DOPID model, the synchronization is performed at the level of transition Act_j, and the correlation set is updated accordingly via the arc inscription $\langle I, j \rangle$.

In the right hand side, we include the removal of an object type, where the transition Act_i is encoded as a transition that consumes tokens from both the object place and the correlation set. This corresponds to the Remove transition in Fig. 2, where an item is removed from an order and the relation between the two objects is dissolved.

7 Simulation Scenario

In this section, we assess the capabilities of our framework. To this aim, we have defined a concrete scenario based on the running example discussed in Sect. 2 and illustrated in Fig. 2. Our simulator is developed with the SimPy[3] library, a

[3] https://simpy.readthedocs.io/en/latest/.

process-based Discrete Event Simulation framework in Python. The simulator takes as input a Petri net for each object type together with a set of simulation parameters defined for each type, governing temporal, resource, and object-flow aspects and synchronization with other object types. The implementation of the simulation model, together with the material to reproduce the simulations presented here, are available at[4].

The proposed simulation model comprises the three object types involved in the order-to-ship process. The model captures the lifecycle of each object independently as well as its relations with other objects. Synchronization between object types is implemented using the channels discussed in Sect. 5, ensuring that process steps are coordinated consistently across the simulation. In addition, each object type is characterized by a set of additional parameters, including the duration of each activity, modeled as a probability distribution, the roles and their assigned resources, and branching probabilities at decision points. For example, looking at the decision point after Select in the item process, the probability of executing Remove is 10% while that of executing Pick item is 90%. As discussed before, the model allows items to be dynamically added or removed from an order during the simulation. This possibility is captured through the `Create` Channel after Add item transition that triggers the creation of new item objects associated to the order with a probability of 10%. An additional element is the decision point after Check out, which depends on the total cost of the order. Each item is enriched with a price attribute, and the total cost of an order is computed as the sum of the prices of its associated items. This value determines the payment method to be applied and thus the respective activity to perform. Conversely, if all items of an order have been removed, the total cost is 0 and thus the order is closed. Looking at the truck process, the simulation model treats the truck as a reusable resource that remains continuously available for shipment. Once a delivery is completed, the truck returns to the warehouse to pick up the next batch of ready items (Go back).

We consider a simulation of 200 orders arriving over a one-week period. Order arrivals are modeled according to a uniform distribution between 10 min and 1 h. Each order initially includes between 2 and 3 items (Place Order), which can later be extended with additional items (Add Item) or reduced by removing items (Remove). Delivery is handled by three trucks, each with a capacity of 20 items and an operating schedule from 7 a.m. to 7 p.m.

In this experiment, we assess the impact of a shipment management rule on the delivery process, leveraging the expressive power of our simulation in managing dynamic relationships between objects. In particular we explore the effect of imposing four different minimum capacity thresholds, as a necessary condition for a truck to depart 50% full, 75% full, 90% full, and 100% (that is, completely) full. In our example we consider a total capacity of 20 items, and thus our thresholds correspond to a requirement that a truck departs with at least 10, 15, 18, and 20 items, respectively.

[4] https://github.com/francescameneghello/object_centric_simulator.

This comparison is relevant as it enables an investigation of the trade-off between delivery efficiency and truck costs. We expect that a lower threshold will increase the number of truck trips, leading to higher delivery costs, while also reducing the average delivery time of items.

Table 1. Results of the shipment threshold experiments

Min Load (%)	Exec. Time (days)	Truck Trips	Items Loaded			Del. Time (h)	Late del. (%)
			Min	Max	Avg		
50%	5.3	34.6	10	20	14.0	10.2	25.5
75%	5.3	29.1	15	20	16.8	10.7	28.9
90%	5.2	26.4	18	20	18.6	11.1	31.2
100%	5.2	24.2	20	20	20.0	11.0	31.2

Table 1 reports the results for each scenario, averaged over 10 simulation runs. As expected, increasing the truck load threshold reduces the total number of truck trips required to complete all deliveries, decreasing from a maximum of 35 trips at the 50% threshold to a minimum of 24 trips for fully loaded trucks. It is worth noting that in the 50% scenario, although the minimum number of items per truck is 10, some trucks depart fully loaded, resulting in an average of 14 items per trip. The higher number of trips at lower thresholds is offset by a reduction in the average item delivery time (column "Del. time" in Table 1), which decreases from 11 h for fully loaded trucks to just over 10 h for the 50% threshold. Although this difference in delivery time may seem small, it has a significant impact on the number of items delivered on a different day than that of the order payment (column "Late del. (%)"). Specifically, the percentage of items delivered late decreases from over 30% for fully loaded trucks to almost 25% at the 50% threshold, potentially leading to a better customer experience. Finally, note that increasing the threshold from 90% to full load is desirable, as it reduces the number of trips required to complete deliveries, lowering costs, without increasing the percentage of items delivered on a different day.

Although this example is relatively simple, it effectively showcases the potential of our simulation model in analyzing the impact of dynamically managing relationships between objects, highlighting how changes in object interactions can influence overall process performance.

8 Conclusions and Discussion

In this paper we have presented a novel simulation framework that overcomes several limitations of existing object-centric simulation techniques. In particular, our approach supports the definition, management and reproduction of dynamic relationships among objects during simulation, as well as the influence of object data attributes on their behavior and interactions. The capabilities of our simulation method are shown by means of two illustrative scenarios, highlighting its ability to address decision-support questions and dynamics that current object-centric simulation models are unable to capture.

This work provides an initial step towards simulating object-centric processes with dynamic relationships, with several open challenges ahead. We discuss in particular three of them. First, we do not focus in depth on the definition of simulation parameters (e.g., temporal and resource perspectives). Simulation parameters play a fundamental role in simulation dynamics. In fact, the stochastic and uncertain nature of temporal and resource-related behaviors is often one of the main drivers behind the need to manage dynamic relationships between objects. Therefore, aspects related to simulation parameters are included in our Running Example (Sect. 2) and evaluation (Sect. 7). However, with the notable exception of synchronization and relationship-management rules, the simulation parameters considered in this work are defined for individual object types, mirroring traditional case-centric simulation approaches [4,10]. Future work should investigate richer parameterization mechanisms that explicitly account for cross-object dependencies, allowing processing times, resource allocation policies, and routing decisions to depend on the attributes and states of interrelated objects. A second limitation of our work concerns the mapping of our OCSM framework to formal models for object-centric processes with dynamic relationships. This is essential to define a precise execution semantics and enable reasoning and analysis tasks. The mapping of OCSM components to DOPIDs [7] discussed in Sect. 6 shows a clear path in that direction. A third, connected future work concerns the relationship between formal analysis and simulation of OCSMs. On the one hand, it would be interesting to study how the presence or absence of formal guarantees (such as deadlock freedom, synchronization correctness, and soundness) impacts simulation, and whether fragments that guarantee soundness by design [14] cover relevant simulation scenarios. On the other hand, simulation could help detecting whether the input model violates such properties, considering that this task is highly undecidable in the object-centric setting [6].

References

1. van der Aalst, W.M.P.: Toward more realistic simulation models using object-centric process mining. In: Proceedings of ECMS. European Council for Modeling and Simulation (2023)
2. van der Aalst, W.M., Berti, A.: Discovering object-centric petri nets. Fundamenta Informaticae **175**(1-4) (2020)
3. Camargo, M., Dumas, M., Rojas, O.G.: Learning accurate business process simulation models from event logs via automated process discovery and deep learning. In: Proceedings of CAiSE. LNCS, vol. 13295, Springer (2022)
4. Chapela-Campa, D., López-Pintado, O., Suvorau, I., Dumas, M.: SIMOD: automated discovery of business process simulation models, vol. 30 (2025)
5. Fahland, D.: Describing behavior of processes with many-to-many interactions. In: Donatelli, S., Haar, S. (eds.) PETRI NETS 2019. LNCS, vol. 11522, pp. 3–24. Springer, Cham (2019). https://doi.org/10.1007/978-3-030-21571-2_1

6. Ghilardi, S., Gianola, A., Montali, M., Rivkin, A.: Petri net-based object-centric processes with read-only data. Inf. Syst. **107**, 102011 (2022)
7. Gianola, A., Montali, M., Winkler, S.: Object-centric processes with structured data and exact synchronization - formal modelling and conformance checking. In: Proceedings of CAiSE. LNCS, vol. 15702, Springer (2025)
8. Kirchdorfer, L., Blümel, R., Kampik, T., van der Aa, H., Stuckenschmidt, H.: Agentsimulator: An agent-based approach for data-driven business process simulation. In: Proceedings of ICPM. IEEE (2024)
9. Knopp, B., Pourbafrani, M., van der Aalst, W.M.P.: Discovering object-centric process simulation models. In: Proceedings of ICPM. IEEE (2023)
10. Meneghello, F., Di Francescomarino, C., Ghidini, C., Ronzani, M.: Runtime integration of machine learning and simulation for business processes: Time and decision mining predictions. Inf. Syst. **128** (2025)
11. Moldt, D., et al.: RENEW: modularized architecture and new features. In: Proceedings of PETRI NETS. LNCS, vol. 13929, Springer (2023)
12. Polyvyanyy, A., van der Werf, J.M.E.M., Overbeek, S., Brouwers, R.: Information systems modeling: language, verification, and tool support. In: Giorgini, P., Weber, B. (eds.) CAiSE 2019. LNCS, vol. 11483, pp. 194–212. Springer, Cham (2019). https://doi.org/10.1007/978-3-030-21290-2_13
13. Seidel, A., Winkler, S., Gianola, A., Montali, M., Weske, M.: To bind or not to bind? discovering stable relationships in object-centric processes. In: Proceedings of ER. LNCS, vol. 16189, Springer (2025)
14. van der Werf, J.M.E.M., Rivkin, A., Polyvyanyy, A., Montali, M.: Data and process resonance. In: Proceedings of PETRI NETS. Springer (2022)

Explaining Violations in Petri Nets

Valeria Fionda, Antonio Ielo(✉), and Francesco Ricca

University of Calabria, Rende, (CS), Italy
{valeria.fionda,antonio.ielo,francesco.ricca}@unical.it

Abstract. Conformance checking techniques quantify how much a trace deviates from a process model, but provide limited support for diagnosing *which* events cause a violation. We address this limitation by providing explanations in terms of subsequences of events that are sufficient to *make a violation unavoidable.* We formalize two complementary problems. First, we identify subsequences of events in a trace that make the trace un-replayable over a safe workflow net. Second, we study a complementary repair problem to identify the events to be removed in order to restore replayability. We exploit the notions of (minimal) *unsatisfiable cores* (MUCs) and *correction sets* (MCSes)— adapting well-established concepts from logics to the context of Process Mining. These concepts capture minimal explanations for violations and minimal repair sets, respectively. We implemented our approach using Answer Set Programming and evaluated it on real-world process models and event logs, showing its feasibility in providing explanations and repairs.

Keywords: Conformance Checking · Process Mining · Answer Set Programming

1 Introduction

Organizations rely on process models to understand, monitor, and improve the execution of business processes [1]. Process executions, captured in event logs, are continuously analyzed to check whether they conform to a reference model. In particular, *conformance checking* [9] techniques compare traces in an event log against a process model to determine whether the observed execution follows the expected process behavior. While existing conformance checking techniques can detect and quantify the extent of deviations, and possibly suggest how traces should be edited to restore conformance, their support for explaining *which* events in a trace lead to violating a model is limited. From the perspective of process diagnostics, identifying that a trace violates the model is only the first step of the analysis: one of the most interesting aspects is to have explanations at the level of individual events to support process analysts in diagnostic tasks, i.e., *which of the events observed in the trace cause the failure?*

F. M. Maggi et al. (Eds.): BPM 2026, LNBIP 599, pp. 21–38, 2027.
https://doi.org/10.1007/978-3-032-37877-4_2

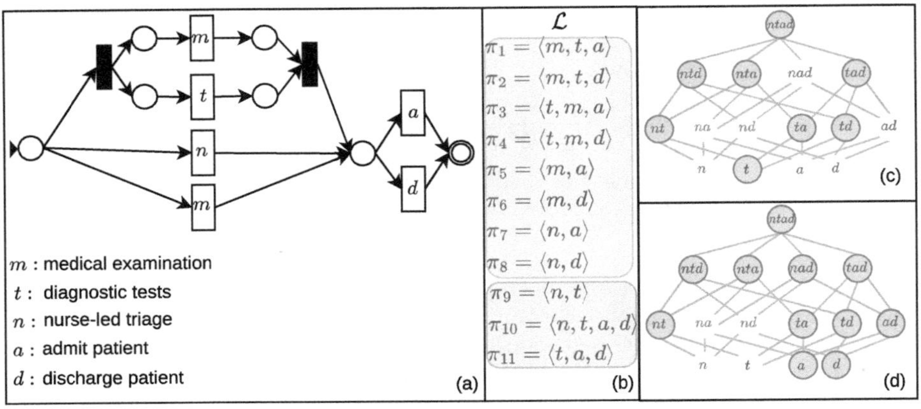

Fig. 1. (a) A simplified emergency department protocol modeled as a workflow net. (b) Example traces: in green, valid firing sequences; in red, those that violate the protocol. (cd) Subtrace lattice of the trace $\pi_{10}(Colorfigureonline) = \langle n, t, a, d \rangle$. Highlighted nodes (purple) represent invalid firing sequences on the model, illustrating the intuition behind minimal event-level explanations for non-conforming traces

Example 1. Figure 1(a) illustrates a simplified model for an emergency department. After arrival, a patient may follow different pathways based on symptom severity and diagnostic results. In the standard pathway, a medical examination (m) is performed. This may be combined with diagnostic tests (t), which can occur before or after the medical examination. In contrast, a nurse-led triage (n) may be conducted for low-severity cases. After the clinical assessment, the process concludes with either hospital admission (a) or patient discharge (d). The model admits traces such as $\pi_1 = \langle m, t, a \rangle$, $\pi_2 = \langle m, t, d \rangle$, $\pi_3 = \langle t, m, a \rangle$, $\pi_4 = \langle t, m, d \rangle$, as well as shorter variants like $\pi_5 = \langle m, a \rangle$ and $\pi_7 = \langle t, a \rangle$. However, not all observed event traces comply with the protocol (Fig. 1(b)). For example, the trace $\pi_{10} = \langle n, t, a, d \rangle$ is not valid, as diagnostic tests cannot be performed without a contextual medical examination, and as it includes incompatible clinical decisions. When such traces are rejected, natural questions arise: which part of the trace is responsible for the violation? How could it be "'*repaired*" so that it becomes conformant with the model? ■

In this paper, we address these issues by introducing event-level explanations for non-conforming traces. The core idea is to identify minimal subsequences of events whose presence ensures that the trace can never be accepted by the reference process model. Intuitively, such subsequences can be seen as minimal explanations for the violation. We formalize this intuition by studying two complementary problems. First, we consider subsequences of events in the trace that make it impossible to extend it into an accepted trace over the reference model. Second, we study a complementary *repair* problem by identifying which events should be removed from a trace to restore possibility of acceptance. In process mining, the term *repair* typically refers to process models [18]. In this paper, we use it to denote sets of events that can restore acceptance, in a purely diagnostic sense. It is not meant to suggest that past execution (i.e., completed traces) can be altered.

Example 2. For the trace $\pi_{10} = \langle n, t, a, d \rangle$, Fig. 1(c) and (d) illustrate two possible explanations for the violation: $\{t\}$ and $\{a, d\}$. The first one captures that the diagnostic test t cannot occur together with nurse-led triage n and without any medical examination m. The second one reflects the incompatibility between the two final decisions. Each node in the *subtrace lattice* (cfr. Fig. 2) represents a subtrace, and edges correspond to removing/adding events in the original trace. Green nodes correspond to valid firing sequences, while red nodes correspond to sequences that violate the protocol. Minimal sets of events that induce a sub-lattice containing only invalid sequences are minimal event-level explanations, i.e. the sets $\{t\}$ (Fig. 1(c)) and $\{a, d\}$ (Fig. 1(d)). These explanations naturally induce alternative repairs. Removing t together with either a or d yields the repaired traces $\pi'_{10} = \langle n, d \rangle$ or $\pi''_{10} = \langle n, a \rangle$. ■

One might wonder whether the problem studied here reduces to the classical alignment problem. Alignments [2] provide a way to compute *how much* and *in which terms* a trace violates a model, but cannot answer our questions. Consider the trace π_{10} and the model in Fig. 1, the optimal alignments are:

$$(n, n), (t, \gg), (a, a), (d, \gg) \qquad\qquad (n, n), (t, \gg), (a, \gg), (d, d)$$

where (x, x) represents a synchronous move over $x \in \mathcal{A}$, $(a, \gg)$ is a *log move* and $(\gg, a)$ is a *model move*. By comparing the two alignments, we observe that t is very likely a problem *on its own*, since both optimal alignments skip it. Indeed, from the Petri net structure, we can observe that to fire t, we also need to fire m, which is not in π_{10}. Similarly, we can observe that a and d are mutually exclusive as they share a place in the pre-set. Hence, the presence of both a and d in the same trace is enough to cause a violation. Indeed, the alignments show that skipping t and a (or, t and d) is *sufficient* to align behavior to the model, but is it *necessary*? Moreover, it is not immediately clear how one should act on these events (i.e., t, a, and d) to obtain an accepting firing sequence, nor whether t, a, and d are *indeed sufficient* to guarantee the non-existence of such a firing sequence.

This paper proposes a notion of *explanation* and *repair* for non-conforming traces that exactly characterize *(i)* why a set of events is a sufficient reason to cause (i.e., "explain") a violation and, *(ii)* which events are necessary to remove to solve (i.e., "repair") a violation. We aim to provide concise diagnostic information, and thus, we focus on minimal explanations and repairs, as minimality ensures that no redundant events are considered for explaining or repairing the inconsistency. In this respect, for trace π_{10} our approach will identify $\{a, d\}$ and $\{t\}$ as explanation and $\{t, d\}$ and $\{t, a\}$ as repairs. A (very!) keen observer would also notice that both $\{t, d\}$ and $\{t, a\}$ are *minimal hitting sets*[1] of $\{t\}$ and $\{a, d\}$ (and viceversa). This intuition reflects a duality: on the one hand, repairing a violation must "disable" every sufficient reason for the non-existence of an accepting firing sequence; on the other hand, any such sufficient reason is incompatible with the conditions necessary for the existence of an accepting firing sequence. This recalls a well-known duality relationship between *minimal correction sets* (MCSes) and *minimal unsatisfiable cores* (MUCs) in constraint systems [30,35]. These notions are the standard machinery for logic-based explainable intelligence techniques [31].

We implement a prototype in Answer Set Programming [7], and evaluate it on real-world logs. Results show that the framework can effectively compute explanations and repairs for non-conforming traces in real-world event logs.

The remainder of the paper is organized as follows. Section 2 reviews related literature. Section 3 introduces the necessary background. Section 4 presents the proposed framework for computing explanations and repairs. Section 5 discusses our ASP-based implementation. Section 6 reports the results of the experimental evaluation and Sect. 7 concludes the paper.

2 Related Work

The precise characterization of "failure explanations" proposed in this work has not been investigated before in the context of process mining. However, there are several adjacent research lines and techniques that are related to ours.

Alignments [2] are an edit distance-based optimization technique that quantifies how much a trace deviates from a reference process model. Computing alignments shows what "cost-optimal deviations" look like, and also provides information about "what to do to avoid violations". Our example in the introduction discusses how our notion of repair relates to alignments. Explanations are a different diagnostic question. Moreover, our approach can be extended to consider symbols that do not appear in the original trace but require fireable transitions (which makes it different from model moves). *Counterfactual explanations* [23] focus on how inputs ought to be changed to edit the outcome of an algorithm. In our context, the input is the trace violating a model, and the outcome is the violation itself: counterfactuals would be "how to change the trace", so that it becomes compliant with the process model. This has been explored in the context of declarative process models [8]. Our notion of repair can be understood as a counterfactual explanation to avoid trace violation in Petri net process models. Monitoring techniques in declarative process mining [14] detect, given a linear temporal logic over finite traces (LTL$_f$; [15]) specification, *unavoidable failure* at the earliest point in time. That is, the shortest prefix of a running trace that becomes permanently violated. This is similar to our definition, but

[1] Given a collection of sets $C = \{S_1, \ldots, S_k\}$ we say that H is an hitting set over C if $H \cap S_i$ is non-empty for all $S_i \in C$, and it is *minimal* if none of its proper subsets is an hitting set for C.

with two important differences. Our notion concerns subsequences that guarantee failure, and we analyze complete traces rather than finite prefixes of an "ongoing" trace. Moreover, we are also able to compute suggestions on how to restore coherence of the trace. Monitors do not support this, beyond suggesting to "undo" the action that leads to the violation. *Event structures* [32] have been introduced in conformance checking to compute *mismatches* between an event log and a Petri net [21], i.e. what is allowed by the model that does not occur in the log ("additional behavior") and what occurs in the log but the model does not admit ("unfitting behavior"). However, even if considering a log composed by a single-trace case, the method differs fundamentally from ours: it characterizes divergences in terms of pre-determined *behavioral relations* (causality, conflict, concurrency) rather than identifying minimal subsequences of events within the trace whose presence makes a violation unavoidable.

Our work is orthogonal, as we provide trace-level diagnostics (rather than a log-level summary) by means of (minimal) sets of events within the trace that *guarantee* violation of the process model regardless of other events that may occur in the trace.

Slicing techniques [34] for Petri nets study which subnets (i.e., which places and transitions) are sufficient to exhibit a given property. This can be helpful in debugging process models or aid the design process, providing a smaller network that supports desired/undesired behavior. Our approach does not focus on explaining violations in terms of *subnets*, but as subsequences of the input trace.

3 Preliminaries

In this paper, we assume that process models are safe Petri nets. We provide here the necessary definitions, and refer the reader to the standard literature [1] for a more comprehensive introduction in the context of Process Mining.

A *Petri net* is a tuple $N = (P, T, F)$ where P and T are finite disjoint sets of *places* and *transitions*, respectively, and $F \subseteq (P \times T) \cup (T \times P)$ is the *flow relation*. A *marking* M is a function $M : P \mapsto \mathbb{N}$ that assigns a number of tokens to each place. For each $t \in T$, we denote by $\bullet t = \{p \in P : (p,t) \in F\}$ its pre-set, that is the set of its *input places* and, by $t\bullet = \{p \in P : (t,p) \in F\}$ its post-set, that is the set of its *output places*. A transition t is *enabled* in a marking M if $M(p) \geq 1$ for all $p \in \bullet t$. *Firing* an enabled transition t in a marking M yields a marking M', where $M'(p) = M(p) - 1$ for $p \in \bullet t$, $M'(p) = M(p) + 1$ for $p \in t\bullet$ and $M'(p) = M(p)$ for $p \in P \setminus (\bullet t \cup t\bullet)$.

A *sequence* over a set Σ is a finite ordered tuple $\sigma = \langle s_1, \ldots, s_n \rangle$ where $s_i \in \Sigma$. A *firing sequence* is a sequence of transitions $\langle t_1, \ldots, t_n \rangle$ such that there exists a sequence of markings $M_0, \ldots, M_n$ where t_i is enabled in M_{i-1} and firing t_i in M_{i-1} yields the marking M_i. In this case, we say that M_0 *reaches* M_n via σ. A *labeled* Petri net associates to each transition a *label* in a set of activity names $\mathcal{A}$ via a labeling function $\Lambda : T \mapsto \mathcal{A} \cup \{\lambda\}$. Transitions labeled by λ are known as *silent* transitions. Given a firing sequence $\sigma = \langle t_1, \ldots, t_n \rangle$, we denote by $\Lambda(\sigma)$ the sequence $\langle \Lambda(t_1), \ldots, \Lambda(t_n) \rangle$.

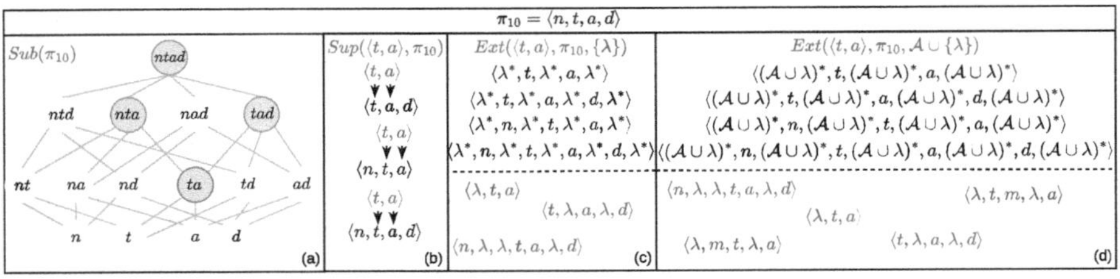

Fig. 2. (a) A graphical representation of $Sub(\pi)$, $Sup(\pi)$ and $Ext(\pi)$, for $\pi = \langle n, t, a, d\rangle$. (b) Embeddings of $\langle t, a\rangle$ into $\langle t, a, d\rangle$, $\langle n, t, a\rangle$ and $\langle n, t, a, d\rangle$. (c-d) Red and green sequences are, respectively, (examples of) firing sequences that correspond to accepted and rejected observables traces for the Petri net in Fig. 1, assuming $\mathcal{S} = \{\lambda\}$ (c) and $\mathcal{S} = \mathcal{A} \cup \{\lambda\}$ (d) (Color figure online)

A firing sequence is accepting if it reaches the final marking. The *observed trace* associated with a firing sequence $\sigma = \langle t_1, \ldots, t_m\rangle$ is the trace obtained from $\Lambda(\sigma)$ by removing all occurrences of λ. We denote this trace by $\Omega(\Lambda(\sigma))$. The language of a Petri net N, denoted by $L(N)$, is the set of *observed traces* associated with its accepting firing sequences. A trace π is accepted by N if $\pi \in L(N)$; otherwise, π is rejected by (or π violates) N. In the following, for convenience, we will sometimes identify an observed trace π with a labeled sequence s such that $\Omega(s) = \pi$. A Petri net is *safe* if every reachable marking contains at most one token per place, i.e., for each reachable marking M, and every place $p \in P$, it holds that $M(p) \leq 1$.

4 Framework

In the rest of the paper, we use the following notation. Let $n \in \mathbb{N}$. We denote by $[n]$ the set $\{1, \ldots, n\}$. Let N be a labeled Petri net over the activity alphabet $\mathcal{A}$ and let $\pi = \langle a_1, \ldots, a_n\rangle$ be a trace, that is a sequence over $\mathcal{A}$. We assume that silent transitions in N are labeled with a fresh symbol λ not appearing in $\mathcal{A}$. Firing sequences on N may contain symbols from $\mathcal{A} \cup \{\lambda\}$, where λ represents unobservable steps in the process model (and thus, in execution traces).

To reason about admissible completions of a subtrace, we need a notion that captures the preservation of the relative order of elements in two sequences. We formalize this through the notion of *embedding*.

Definition 1 (Embedding). *Let* $\pi = \langle \pi_1, \ldots, \pi_n\rangle$ *and* $\pi' = \langle \pi'_1, \ldots, \pi'_\ell\rangle$ *be sequences, with* $\ell \geq n$. *We say that* π embeds *into* π', *denoted as* $\pi \preceq \pi'$, *if there exists a strictly increasing function* $f : [n] \to [\ell]$ *such that* $\pi'_{f(i)} = \pi_i$ *for all* $i \in [n]$. π properly *embeds in* π' *if* $\ell > n$. ◀

Example 3. Let $\pi = \langle a, c\rangle$ and $\pi' = \langle b, a, d, c\rangle$. Then $\pi \preceq \pi'$ with the witness function f defined as $f(1) = 2$ and $f(2) = 4$. ■

Given a trace π of length n, each subset $I \subseteq [n]$ uniquely identifies a subtrace of π, which we denote by $\pi[I] = \langle a_i \mid i \in I, \text{ in increasing order}\rangle$. We denote by $Sub(\pi)$ the set of all subtraces of π, that is $Sub(\pi) = \{\pi[I] \mid I \subseteq [n]\}$. Given a trace π and a subtrace $\pi' \in Sub(\pi)$, we denote by $Sup(\pi', \pi)$ the set of subtraces of π in which π' embeds, formally:

$$Sup(\pi', \pi) = \{\pi'' \in Sub(\pi) \mid \pi' \prec \pi''\}$$

Given a trace π, a subtrace $\pi' \in Sub(\pi)$, and a set of symbols $\mathcal{S} \subseteq (\mathcal{A} \cup \{\lambda\})$ the extension of π' wrt π and $\mathcal{S}$, denoted by $Ext(\pi', \pi, \mathcal{S})$, is the set of sequences that can be obtained by: *(i)* reintroducing additional events from the original trace π, and *(ii)* adding extra symbols from $\mathcal{S}$ at arbitrary positions.

Definition 2 (Extension). *Let $\pi = \langle \pi_1, \ldots, \pi_n \rangle$ be a trace, $\pi' \in Sub(\pi)$, and $\mathcal{S} \subseteq (\mathcal{A} \cup \{\lambda\})$. We define the extension of π' relative to π over $\mathcal{S}$ as the set $Ext(\pi', \pi, \mathcal{S})$ of all sequences $s = \langle \mathcal{S}^*, \tau_1, \mathcal{S}^*, \tau_2, \ldots, \mathcal{S}^*, \tau_m, \mathcal{S}^* \rangle$ such that $\tau = \langle \tau_1, \ldots, \tau_m \rangle \in Sup(\pi', \pi)$, and * is the Kleene star operator.* ◀

An element $s \in Ext(\pi', \pi, \mathcal{S})$ represents a candidate labeled sequence that preserves the relative order of the events in π. Such a sequence corresponds to a candidate execution of the process model whenever there exists a firing sequence σ of N such that $\Lambda(\sigma) = s$. In this case, we say that s is executable in N. If σ is an accepting firing sequence then $\Omega(s)$ is accepted by N, i.e., $\Omega(s) \in L(N)$.

Intuitively, an extension alphabet $\mathcal{S}$ specifies which labeled sequences are allowed as completions of a given subtrace. Different alphabets correspond to different notions of admissible extensions. As we shall discuss in Sect. 4.1, the choice of the extension alphabet affects the notion of explanation based on it. Note that extensions can be ordered according to the inclusiveness of the alphabet on which they are defined. This ordering induces a monotonicity property. In particular, it holds that $Ext(\pi', \pi, \mathcal{S}) \subseteq Ext(\pi', \pi, \mathcal{S}')$ if $\mathcal{S} \subseteq \mathcal{S}'$. Figure 2 provides an example of the relationship between subsequences (*Sub*, Fig. 2(a)), supersequences (*Sup*, Fig. 2(b)) and extensions of a trace (Fig. 2(c)-(d)).

Example 4. Consider again the model in Fig. 1(a), the trace $\pi_{10} = \langle n, t, a, d \rangle$ and the subtrace $\pi'_{10} = \pi_{10}[\{2, 3\}] = \langle t, a \rangle$. If $\mathcal{S} = \{\lambda\}$ then $Ext(\pi'_{10}, \pi_{10}, \mathcal{S})$ is the set reported in Fig. 2(c). Note that no sequence s in the set is such that $\Omega(s) \in L(N)$, i.e., there exists no accepting firing sequence σ of N such that $\Lambda(\sigma) = s$. If instead we consider $\mathcal{S}' = \mathcal{A} \cup \{\lambda\} = \{a, d, m, n, t, \lambda\}$ then $Ext(\pi'_{10}, \pi_{10}, \mathcal{S}')$ (depicted in Fig. 2(d)) includes sequences for which there exists an accepting firing sequence. This is the case of $\langle \lambda, t, m, \lambda, a \rangle$ and the accepting firing sequence σ such that $\Lambda(\sigma) = \langle \lambda, t, m, \lambda, a \rangle$. ■

Definition 3 (Explanation). *Let π be a trace not accepted by N. A subtrace $\pi' \in Sub(\pi)$ is an* explanation *for π with respect to $\mathcal{S}$ if $Ext(\pi', \pi, \mathcal{S})$ does not contain any sequence s such that $\Omega(s) \in L(N)$. An explanation π' is* minimal *if no proper subtrace $\pi'' \in Sub(\pi')$ is an explanation.* ◀

Example 5. Consider again the Petri net N in Fig. 1(a) and the trace $\pi_{10} = \langle n, t, a, d\rangle$, that is not accepted by N. If $\mathcal{S} = \{\lambda\}$, then the subtrace $\pi'_{10} = \langle t, a\rangle$ is an explanation. Indeed, there exists no sequence $s \in Ext(\pi'_{10}, \pi_{10}, \{\lambda\})$ that is accepted by N (see Fig. 2(c)). However, π'_{10} is not a minimal explanation. In fact, the subtrace $\pi''_{10} = \langle t\rangle \in Sub(\pi')$ is also an explanation. On the other hand, $\pi'''_{10} = \langle n\rangle$ is not an explanation, since the sequence $s = \langle n, a\rangle \in Ext(\pi'''_{10}, \pi_{10}, \{\lambda\})$ is such that $\Omega(s) \in L(N)$. ■

Let π be a trace of length n, and $I \subseteq [n]$. We denote by π^I the trace we obtain by *removing* from π the events in positions $i \in I$, i.e. $\pi^I = \langle \pi_j : j \in [n] \setminus I\rangle$.

Definition 4 (Repair). *Let π be a trace that violates N. A set of positions $I \subseteq [n]$* repairs *π if there exists a sequence $s \in Ext(\pi^I, \pi, \mathcal{S})$ such that $\Omega(s) \in L(N)$. The repair is* minimal *if no proper subset of I repairs π. Whenever this does not cause ambiguities, with a slight abuse of notation, we will refer to π^I itself as the repair.* ◀

Example 6. Consider again the Petri net N in Fig. 1(a), the extension alphabet $\mathcal{S} = \{\lambda\}$, and the trace $\pi_{10} = \langle n, t, a, d\rangle$. The set $I = \{1, 4\}$, corresponding to removing the first and last events from π and obtaining the trace $\pi_{10}^{\{1,4\}} = \langle t, a\rangle$, is not a repair since $Ext(\pi_{10}^{\{1,4\}}, \pi, \{\lambda\})$ does not contain any sequence s that is accepted by N (see Fig. 2(c)). On the other hand, if we consider the extension alphabet $\mathcal{S}' = \mathcal{A} \cup \{\lambda\}$ the set $I = \{1, 4\}$ is a repair since the sequence $s = \langle \lambda, m, t, \lambda, d\rangle \in Ext(\pi_{10}^{\{1,4\}}, \pi, \mathcal{S}')$ is accepted by N (see Fig. 2(d)). In this case, I is also a minimal repair, since no proper subset of I is a repair. ■

4.1 Extension Alphabets and Explanation Semantics

The notion of extension (and consequently, explanation and repair) is parameterized by the extension alphabet $\mathcal{S}$, which specifies which activities can appear in the subtrace completions when looking for a valid execution on the process model.

A first natural choice is to restrict the subtrace completions to the activities already observed in the trace, consisting of setting $\mathcal{S} = \{\lambda\}$. This translates into the fact that, given a trace π, for each of its subtrace π', the extension $Ext(\pi', \pi, \{\lambda\})$ contains only sequences s such that $\Omega(s) \preceq \pi$. As a result, explanations correspond to subsequences that cannot be *extended* using only the behavior observed in the original trace.

Example 7. Consider the Petri net N in Fig. 1(a), the trace $\pi_{10} = \langle n, t, a, d\rangle$, whose $Sub(\pi_{10})$ is reported in Fig. 2(a). Let $\pi'_{10} = \langle t\rangle$ be a subtrace in $Sub(\pi_{10})$. Then $Sup(\pi'_{10}, \pi_{10}) = \{\langle n, t\rangle, \langle t, d\rangle, \langle t, a\rangle, \langle n, t, a\rangle, \langle n, t, d\rangle, \langle t, a, d\rangle, \langle n, t, a, d\rangle\}$. However, $Ext(\pi'_{10}, \pi_{10}, \{\lambda\})$ does not contain any labeled sequences s such that $\Omega(s) \in L(N)$. Thus, $\langle t\rangle$ is an explanation for π wrt $\{\lambda\}$. ■

A second natural choice is to consider the extension alphabet to be $\mathcal{A} \cup \{\lambda\}$. In this setting, subtraces can be extended with any activity appearing in the process model. This leads to a *stronger* notion of explanation, since for every trace $\pi' \in Sub(\pi)$ it holds that $Ext(\pi', \pi, \{\lambda\}) \subseteq Ext(\pi', \pi, \mathcal{A} \cup \{\lambda\})$. Consequently, explanations capture sequences that remain inconsistent with the model even when *arbitrary* activities are allowed to interleave with them.

Example 8. Consider again the Petri net N in Fig. 1(a), the trace $\pi_{10} = \langle n, t, a, d \rangle$, and the subsequence $\pi'_{10} = \langle t \rangle \in Sub(\pi)$. In this setting, $Ext(\pi'_{10}, \pi_{10}, \mathcal{A} \cup \{\lambda\})$ contains some labeled sequences s (e.g., $\langle \lambda, m, t, \lambda, a \rangle$) such that $\Omega(s) \in L(N)$. Thus, $\langle t \rangle$ is not an explanation for π_{10} wrt $\mathcal{A} \cup \{\lambda\}$. ■

The previous examples also illustrate how our notion of explanation differs from the diagnostic information provided by alignments. Consider the model in Fig. 1(a), the non-conformant trace $\pi_{10} = \langle n, t, a, d \rangle$ and the extension alphabet $\mathcal{A} \cup \{\lambda\}$. Our framework identifies two explanations, that are $\{n, t\}$ and $\{a, d\}$. As discussed in the introduction, alignments suggest that t appears to be problematic on its own, since it is skipped in both alignments. However, when considering both the trace and the process model, the violation can be more precisely explained by the *joint* presence of t and n. Indeed, t alone, together with a or d, can be extended to an accepting firing sequence on the model.

Moreover, our framework can also suggest repairs that make explicit how the trace could be corrected. For instance, the set $\{n, d\}$ is a valid repair. Removing these events corresponds to the subtrace $\langle t, a \rangle$, which admits the labeled sequence $s = \langle \lambda, m, t, \lambda, a \rangle \in Ext(\langle t, a \rangle, \pi_{10}, \mathcal{A} \cup \{\lambda\})$, such that $\Omega(s) \in L(N)$. Firing sequences "made available" by repairs suggested by our approach explicitly show *missing behavior* that leads to violations. While alignment provides one optimal way to edit the trace so that it fits the model, repairs in our framework provide a structural characterization of the cause of the violation.

5 Computing Bounded Explanations and Repairs

The notions of explanations and repairs proposed in this paper require reasoning about the existence of accepting firing sequences into which the input trace can embed. Intuitively, an explanation corresponds to any subsequence that cannot be embedded into any accepting firing sequence, whereas a repair is a set of events whose removal allows such embedding to exist. In both cases, the computational task involves reasoning about the (non-)existence of a subsequence of π that embeds into an accepting firing sequence. In general, deciding the existence of accepting firing sequences in Petri nets is a PSPACE-complete problem, even when restricted to safe nets [10]. In practice, it is common to assume an upper bound $k \in \mathbb{N}$ on the length of executions [20,26,29]. Under this assumption the complexity drops to NP, since one can non-deterministically guess a sequence of transitions and check it leads to the final marking. Accordingly, we consider k-bounded explanations and repairs by restricting the set of candidate extensions to sequences of length at most k. Formally, we substitute $Ext(\pi', \pi, \mathcal{S})$ with $Ext^k(\pi', \pi, \mathcal{S}) = \{x \in Ext(\pi', \pi, \mathcal{S}) : |x| \leq k\}$ in Definition 3 and Definition 4.

This observation enables us to tackle the problem using satisfiability-based techniques [6,33]. SAT encodings for similar problems have already been proposed in the literature, for instance, to compute alignments [6]. The core idea of our approach is to exploit satisfiability-based search from a *dual* perspective: given an incoherent problem instance, we compute unsatisfiable cores and correction sets of the corresponding *satisfiability encoding*, and translate them into meaningful explanations and repairs for the input trace [30]. In general, k-explanations are *monotone*: if a sequence is a k-explanation, it is also a k'-explanation for every $k' > k$. In practice, reasonable values for k can be estimated from the event log (i.e., by considering the length of typical firing sequences). Moreover, when computing explanations and repairs, one is interested in concise diagnostic information [31]. For this reason, we focus on *minimal* explanations and repairs, as minimality ensures that the analysts can look at essential, non-redundant sets of events.

Following this idea, we exploit unsatisfiability cores (UC) to identify explanations, and correction sets (CS) to identify repairs. To achieve this, the search for cores and correction sets must be restricted to the set of literals of the input trace. We implement a solution using Answer Set Programming (ASP) [7]. ASP is a declarative programming paradigm based on stable model semantics of logic programs, with many applications in both academic and industrial settings [17,19]. Its focus on knowledge representation and reasoning enables writing readable and uniform encodings compared to other modeling solutions. Moreover, state-of-the-art ASP systems exploit SAT solving technology, which yields efficient implementations, and provide very similar notions of unsatisfiable cores and correction sets [3,7]. In this paper, we use ASP for its high-level modeling capabilities and its support for easy prototyping of complex reasoning tasks. This choice is also supported by several recent applications of ASP to process mining and knowledge representation tasks [11–13,16,24,25].

Core idea. Let $N = (P, T, F)$ be a safe workflow net, with initial marking m_0 and final marking m_f, and $\pi \in \mathcal{A}^*$ a trace. The encoding searches for an accepting firing sequence of length at most k into which π embeds. Since, by assumption, $\pi \notin L(N)$, the logic program is incoherent. We therefore compute unsatisfiable cores (wrt the atoms encoding events of π) which represent subsequences π' that are sufficient to cause the incoherence (i.e., non-existence of such accepting firing sequence). In practical terms, satisfiability of the logic program captures the existence of a bounded length firing sequences that accept π. The computation of unsatisfiable cores determines subsequences $\pi' \prec \pi$ that guarantee unsatisfiability, i.e., such that $Ext(\pi', \pi, \mathcal{S})$ do not contain accepting firing sequences for π'. Thus, unsatisfiable cores map to explanations as by Definition 3. By the well-known duality between cores and correction sets, correction sets of the logic program map to repairs as by Definition 4. The computation and minimization of unsatisfiable cores and correction sets is a standard task, and it is implemented on top of ASP systems with standard algorithms [3]. The full ASP encoding is reported in Appendix 8.

6 Evaluation

In this section, we experimentally evaluate our approach. To the best of our knowledge, there are currently no other tools and techniques that compute objects comparable to the explanations and repairs proposed in this work. Thus, our evaluation focuses on assessing the feasibility of the proposed approach. In particular, we evaluate whether our approach can compute explanations and repairs on traces and process models extracted from real event logs. Experiments have been executed on an Intel Core i7-13700 CPU @ 5.2 GHz with 32 GB RAM and 16 physical cores, running up to 8 concurrent jobs via GNU Parallel. Each run was limited to a timeout of 60 CPU seconds. We report that the peak memory usage across all executions was below 500 MB. We used Python 3.12, `clingo` [22] 5.8 as the ASP solver, and the `pm4py` [5] library version 2.7.

Experimental Setting. We evaluated our approach on four event logs from the BPI Challenge 2020, which describe a reimbursement process in a Dutch institution. In particular, we considered the logs *Domestic Declarations*, *International Declarations*, *Request for Payment*, and *Prepaid Travel Costs*. For each log, we mined the corresponding Petri net by applying the *infrequent* Inductive Miner[2] (IMf; [27]). To obtain models with different levels of filtering, we used noise threshold $s \in \{.20, .30, .40\}$. Thus, for each input log L, and for each value of s we obtain a Petri net M_L^s.

We generated a dataset of 3294 problem instances of the form (M_L^s, π, k_L), where $\pi \in L$ is a trace that violates the model M_L^s, and $k_L \in \mathbb{N}$ is a bound on the length of labeled sequences that correspond to accepting firing sequences of the model. Since each transition in a firing sequence contributes exactly one label in the labeled sequence, this bound also limits the length of firing sequences. We compute k_L based on (log-wise) statistics about the average length of firing sequences for accepted traces in L. In particular, we estimate the average ratio between the length of accepting firing sequences and the length of their corresponding observed traces. Let $\hat{r}_L$ denote this ratio for log L, we then set $k_L = \lceil |\pi| \cdot \hat{r}_L \rceil$. Intuitively, $\hat{r}_L$ captures how much firing sequences tend to grow with respect to observed traces due to the presence of silent transitions. For example, if $\hat{r}_L = 1.5$ and $|\pi| = 5$, the bound is $k_L = 8$.

For traces that violate the model, we search for repairs whose firing sequence has length within the bound, and compute explanations that show that no firing sequence exists within that bound. For each triple (M_L^s, π, k_L), we compute both an explanation (i.e., a Minimal Unsatisfiable Core – MUC) and a repair (i.e., a Minimal Correction Set – MCS), under two extension alphabets: the strong setting $\mathcal{S} = \mathcal{A} \cup \{\lambda\}$ and the weak setting $\mathcal{S} = \{\lambda\}$.

Results. Table 1 reports summary statistics for the solved problem instances for the two settings (weak and strong). Overall, our approach was able to solve the majority of instances, for both explanations and repairs. Most instances are solved within a reasonable runtime. As expected, the strong setting is computa-

[2] We report all obtained models are safe.

Table 1. Results summary with solved and out-of-time (OOT) counts and runtime distribution statistics for each formulation and semantics combination. We report minimum runtime (min), maximum runtime (max) and the i-th quantiles (q_i). Runtime is reported in seconds

Target	Setting	Solved	OOT	min	$q_{0.25}$	$q_{0.50}$	$q_{0.75}$	max
Repair	weak	3284	10	0.001	0.040	0.097	0.270	60.536
	strong	2402	892	0.001	0.351	2.027	8.875	54.737
Explanation	weak	3290	4	0.001	0.049	0.156	0.435	47.902
	strong	2910	384	0.002	0.434	1.426	4.716	54.787

Table 2. Features correlation

	Repair		Explanation	
	weak	strong	weak	strong
k	0.88	0.92	0.91	0.77
$\|\pi\|$	0.87	0.90	0.89	0.76
$\|\mathcal{A}\|$	0.46	0.56	0.53	0.53
$\|P\|$	0.44	0.51	0.47	0.50
$\|T\|$	0.46	0.55	0.51	0.53

Table 3. Success counts stratified by $k \in [k_l, k_r]$

Target	$[k_l, k_r]$	Weak (#)	Strong (#)	$q_{0.5}$ (s)
Repair	0-20	626/626	626/626	0.026
	21-40	2225/2225	1754/2225	0.344
	41-60	417/427	22/427	0.896
	61+	16/16	0/16	6.869
Explanation	0-20	626/626	626/626	0.029
	21-40	2225/2225	2066/2225	0.536
	41-60	423/427	217/427	2.301
	61+	16/16	1/16	7.783

tionally harder than the weak one. In fact, as reported in the table, the number of solved instances decreases, and the runtime increases significantly. The memory consumption, instead, remains comparable. This behavior was expected since the strong setting allows a larger extension space, corresponding to a larger search space for the ASP solver. However, even in the strong setting, the approach is able to solve the majority of instances within the timeout.

To better understand the factors affecting runtime, we analyzed the correlation between instance features and the natural logarithm of runtime[3]. Table 2 shows linear correlation coefficients for several model- and trace-related features. Cell colors indicate the strength of correlation (faint to bright green). As trace-related features, we considered k (acceptance bound) and $|\pi|$ (trace length). As for model-related features we considered $|\mathcal{A}|$ (alphabet size), $|P|$ (number of model places), and $|T|$ (number of model transitions). The results show that k is the feature most strongly correlated with runtime, and the trace length has a similar but slightly weaker correlation. This was expected since increasing k exponentially increases the search space. Recall that k is also related to trace length since, for a given log L, $k_L = \lceil |\pi| \cdot \hat{r}_L \rceil$. Thus, increasing the length of the trace translates to a larger bound on the length. In the following analysis, we focus on k, as it is the parameter most strongly correlated with runtime.

[3] An exponential effect is expected, due to the nature of our solution.

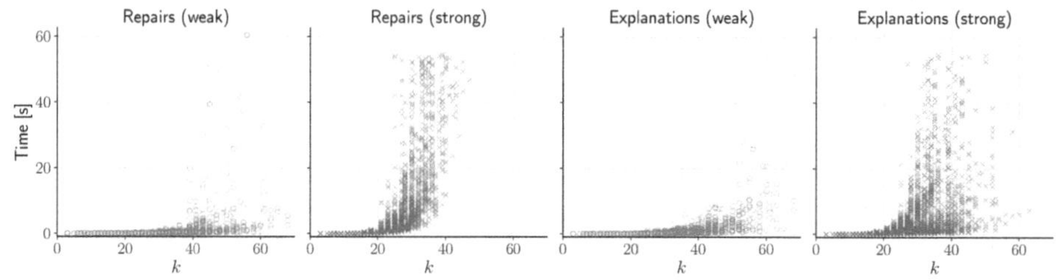

Fig. 3. Runtime vs. acceptance bound (k) for all formulations and settings. Each subplot shows successful instances only. Columns correspond to repairs and explanations for weak and strong settings

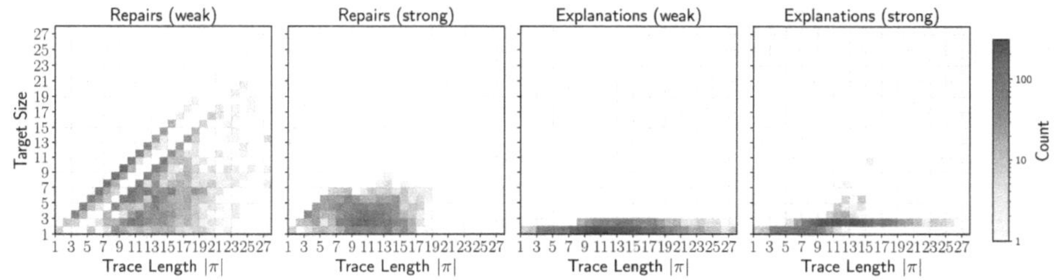

Fig. 4. Distribution of instances by trace length $|\pi|$ and target size (cardinality of repair or explanation)

Table 3 stratifies the results reported in Table 1 with respect to the length bound k. We observe that across both targets, the majority of instances under the strong setting can be solved for values of k up to 40, which covers the majority of control-flow variants in our logs. As expected, the difficulty of the problem increases as k grows, even if in the weak setting, our tool is able to solve almost all instances. The median runtime increases with k (last column in Table 3), confirming that the length bound has a significant impact on the computational time required. The scatterplot in Fig. 3 provides a fine-grained analysis of this trend, showing that runtime increases with k, and confirming the strong correlation reported in Table 2. This trend is particularly evident for the strong setting, where a fast increase in runtime can be observed even for a small increase of k. By comparing the behaviors of the weak and strong settings, one can note that the strong setting has a much larger variance, with some instances solved in a few seconds and some instances approaching the timeout, for the same value of k. This suggests that in the strong setting, our approach becomes more sensitive to the structure of the process model. Finally, the charts show that explanations and repairs have similar behavior. Indeed, in the weak setting, all instances remain tractable even for higher values of k, while in the strong setting, the computation of both becomes harder when $k > 40$.

Finally, Fig. 4 shows the relationship between the trace length ($|\pi|$, on the x-axis) and the size of the computed explanation/repair (on the y-axis). Each cell (i, j) represents the number of instances with trace length i and target size j. As expected, all points lie below the bisector, since the size of a repair or explanation cannot exceed the size of the trace, and it is typically smaller. From the chart, it can be noted that, in many cases, our approach is able to isolate very small targets, showing that violations can often be imputed to a limited number of events. These results suggest that the proposed method is able to produce concise explanations and repairs for non-conforming traces.

7 Conclusion

In this paper, we addressed the problem of explaining and repairing trace violations in Petri net process models. We formalized explanations as minimal subsequences of events of the trace that make a violation unavoidable, and repairs as minimal sets of events whose removal from the trace can, possibly, restore acceptance. We implemented our approach in Answer Set Programming by adapting the standard notions of minimal unsatisfiable cores and minimal correction sets of logic programs, and evaluated it on real-world event logs. The experimental results showed that the proposed framework is feasible and practical in realistic settings.

8 ASP Encoding

We assume the reader is familiar with ASP; otherwise, we refer to standard ASP textbooks [28]. We describe the logic program Π^k_{FS}, that jointly with facts $[N]$ (encoding the net) and $[\pi]$ (encoding the input trace) admits as answer sets accepting firing sequences of length at most k where π embeds into. Computation of unsatisfiable cores and correction sets (and their minimization) is a standard task, and is implemented on top of ASP systems with standard algorithms [3].

Encoding N and π. We encode N similarly to [4]. The predicates $ptarc/2$ and $tparc/2$ encode the flow relation F, with atoms $ptarc(p, t)$ for each $(p, t) \in P \times T \subseteq F$, and $tparc(t, p)$ for each $(t, p) \in T \times P \subseteq F$. If $\Lambda(t) = a$, we have the fact $trans(t, a)$, and we have the label *__lambda__* for silent transitions. We mark the source place p with the atom $im(p)$ and the sink place p' with the atom $fm(p)$. We encode that $a = \pi(t)$ by the atom $obs(t, a)$.

Defining the Search Space. The idea is to guess an accepting firing sequence where π embeds into. We proceed in steps. First, we guess the length of the firing sequence:

```
step(0..H-1) :- horizon(H).
{ time(T) } :- step(T).
time(T-1) :- time(T), T > 0.
last(T) :- time(T), not time(T+1).
```

The atom $time(t)$ means that t is an available time step. We backwardly propagate the existence of time-point $t - 1$ given t is a valid time-point, up to $t = 0$. $last(t)$ marks that t is the final time-point in the firing sequence. $horizon(h)$ is an input fact that denotes we are searching firing sequences of length at most h.

Then, we guess the embedding of π into the firing sequence, by guessing atoms $embed(i, j)$, meaning that the i-th state of π corresponds to the j-th transition in the firing sequence. The choice rules define the i–j mappings, and constraints enforce monotonicity as by Definition 1.

```
{ embed(I,J): time(J) } = 1 :- obs(I,A).
:- embed(I,J), embed(K,H), I < K, J >= H.
```

Since $k > |\pi|$, some time-points of the firing sequence would not be an image of states of π. We will assign them a symbol $s \in \mathcal{S}$, so that our search space represents $Ext(\pi, \pi, \mathcal{S})$. The predicate $trace/2$ will encode the guessed $s \in Ext(\pi, \pi, \mathcal{S})$, with $trace(t, a)$ meaning that $s(t) = a$.

```
trace(T,A) :- obs(I,A), embed(I,T).
{ trace(T,X): symbol(X) } = 1 :- time(T), not embed(_,T).
```

The first rule "copies" the input trace π to the correct time-steps (wrt $embed/2$), while the second choice rule fills in the empty states with symbols of $\mathcal{S}$. Over weak semantics, we will have $symbol(\_\_lambda\_\_)$, while with strong semantics we will have $symbol(a)$ for $a \in \mathcal{A}$.

Token Game. The following rules are an adaptation of [4] to encode a safe Petri net dynamics within ASP. The $holds(i, p)$ atom models that at the i-th step of the firing sequence, the place p contains a token. This is required to define the notion of *enabled transition*, as well as track the state of the Petri net. The initial state (marking) of the net is encoded as follows:

```
holds(0,P) :- im(P,1).
```

We encode the notion of transition ts being enabled at time t (that is, enough tokens are available in its preset) as follows:

```
-enabled(T,TS) :- trans(TS,_), time(T), not holds(T,P), ptarc(P,TS,_).
enabled(T,TS) :- trace(T,A), trans(TS,A), not -enabled(T,TS).
```

We define the search space of firing sequences, among enabled ones, over each time step. We use interleaved semantics, with exactly one transition firing per time step. At each time-point, we select exactly one transition that can fire among the enabled ones. The atom $fires(t, ts)$ denotes that at time t transition ts is fired. Notice that the definition of $enabled/2$ enforces that only transitions with a correct label (according to the guessed completion of π) can be fired.

```
{ fires(T,TS): enabled(T,TS) } = 1 :- time(T).
```

The following rules *update* the marking, by defining $holds(t+1, p)$ whenever we fire a transition ts and $p \in t\bullet$. Tokens are implicitly removed from $p \in \bullet t$, since the $holds(t+1, p)$ in this case is not supported. The last rule is an *inertia* rule, that maintains tokens unaffected by the fired transition in their current place:

```
holds(T+1,P) :- fires(T,TS), holds(T,P), not ptarc(P,TS,_).
holds(T+1,P) :- tparc(TS,P,_), fires(T,TS).
```

Finally, the following constraints enforce that the marking in the last time-step must be the final marking, i.e., one single token in the sink place:

```
:- last(T), holds(T+1,P), not fm(P,1).
:- fm(P,1), last(T), not holds(T+1,P).
```

Acknowledgments. This work has been partially funded by the Italian Ministry of Industrial Development (MISE) under project EI-TWIN n. F/310168/05/X56 CUP B29J24000680005 and also partially by projects SIGENERA (CUP J29I24001770005) and MOZART (J49I24001740005) selected within the framework of the PR FESR FSE Calabria 2021/2027 and implemented with the support of the Italian State and the Calabria Region.

References

1. van der Aalst, W.M.P.: Process Mining - Data Science in Action, Second Edition. Springer (2016). https://doi.org/10.1007/978-3-662-49851-4
2. Adriansyah, A., van Dongen, B.F., van der Aalst, W.M.P.: Conformance checking using cost-based fitness analysis. In: EDOC, pp. 55–64. IEEE Computer Society (2011)
3. Alviano, M., Dodaro, C., Fiorentino, S., Previti, A., Ricca, F.: ASP and subset minimality: enumeration, cautious reasoning and muses. Artif. Intell. **320**, 103931 (2023)
4. Anwar, S., Baral, C., Inoue, K.: Encoding petri nets in answer set programming for simulation based reasoning. Theory Pract. Log. Program. **13**(4-5-Online-Supplement) (2013)
5. Berti, A., van Zelst, S., Schuster, D.: Pm4py: A process mining library for python. Softw. Impacts **17**, 100556 (2023)
6. Boltenhagen, M., Chatain, T., Carmona, J.: Optimized SAT encoding of conformance checking artefacts. Computing **103**(1), 29–50 (2021)
7. Brewka, G., Eiter, T., Truszczynski, M.: Answer set programming at a glance. Commun. ACM **54**(12), 92–103 (2011)
8. Buliga, A., Francescomarino, C.D., Ghidini, C., Montali, M., Ronzani, M.: Generating counterfactual explanations under temporal constraints. In: AAAI, pp. 15622–15631. AAAI Press (2025)
9. Carmona, J., van Dongen, B.F., Weidlich, M.: Conformance checking: foundations, milestones and challenges. In: Process Mining Handbook, Lecture Notes in Business Information Processing, vol. 448, pp. 155–190. Springer (2022)
10. Cheng, A., Esparza, J., Palsberg, J.: Complexity results for 1-safe nets. Theor. Comput. Sci. **147**(1&2), 117–136 (1995)
11. Chesani, F.: Process discovery on deviant traces and other stranger things. IEEE Trans. Knowl. Data Eng. **35**(11), 11784–11800 (2023)
12. Chiariello, F., Ielo, A., Tarzariol, A.: An ilasp-based approach to repair petri nets. LPNMR. LNCS **15245**, 85–97 (2024) Springer
13. Chiariello, F., Maggi, F.M., Patrizi, F.: Asp-based declarative process mining. In: AAAI, pp. 5539–5547. AAAI Press (2022)
14. De Giacomo, G., Masellis, R.D., Maggi, F.M., Montali, M.: Monitoring constraints and metaconstraints with temporal logics on finite traces. ACM Trans. Softw. Eng. Methodol. **31**(4), 68:1–68:44 (2022)
15. De Giacomo, G., Vardi, M.Y.: Linear temporal logic and linear dynamic logic on finite traces. In: IJCAI, pp, 854–860 (2013). IJCAI/AAAI

16. Donadello, I., Maggi, F.M., Patrizi, F., Tessaris, S., Zorzi, M.: Flexible event log generation using answer set programming. Inform. Syst. 102714 (2026). https://doi.org/10.1016/j.is.2026.102714, https://www.sciencedirect.com/science/article/pii/S0306437926000281
17. Erdem, E., Gelfond, M., Leone, N.: Applications of answer set programming. AI Mag. **37**(3), 53–68 (2016)
18. Fahland, D., van der Aalst, W.M.P.: Model repair - aligning process models to reality. Inf. Syst. **47**, 220–243 (2015). https://doi.org/10.1016/J.IS.2013.12.007
19. Falkner, A.A., Friedrich, G., Schekotihin, K., Taupe, R., Teppan, E.C.: Industrial applications of answer set programming. Künstliche Intell. **32**(2–3), 165–176 (2018)
20. Fionda, V., Ielo, A., Ricca, F.: LTLF2asp: LTLF bounded satisfiability in ASP. LPNMR. LNCS **15245**, 373–386 (2024) Springer
21. García-Bañuelos, L., van Beest, N., Dumas, M., Rosa, M.L., Mertens, W.: Complete and interpretable conformance checking of business processes. IEEE Trans. Software Eng. **44**(3), 262–290 (2018). https://doi.org/10.1109/TSE.2017.2668418
22. Gebser, M., Kaminski, R., Kaufmann, B., Schaub, T.: Multi-shot ASP solving with clingo. Theory Pract. Log. Program. **19**(1), 27–82 (2019)
23. Guidotti, R.: Counterfactual explanations and how to find them: literature review and benchmarking. Data Min. Knowl. Discov. **38**(5), 2770–2824 (2024)
24. Huitzil, I., Mazzotta, G., Peñaloza, R., Ricca, F.: Asp-based axiom pinpointing for description logics. In: Kutz, O., Lutz, C., Ozaki, A. (eds.) Proceedings of the 36th International Workshop on Description Logics (DL 2023) co-located with the 20th International Conference on Principles of Knowledge Representation and Reasoning and the 21st International Workshop on Non-Monotonic Reasoning (KR 2023 and NMR 2023)., Rhodes, Greece, September 2-4, 2023. CEUR Workshop Proceedings, CEUR-WS.org (2023). https://ceur-ws.org/Vol-3515/paper-14.pdf
25. Ielo, A., Mazzotta, G., Ricca, F., Peñaloza, R.: Towards asp-based minimal unsatisfiable cores enumeration for ltlf. In: Porello, D., Vinci, C., Zavatteri, M. (eds.) Short Paper Proceedings of the 6th International Workshop on Artificial Intelligence and Formal Verification, Logic, Automata, and Synthesis, OVERLAY 2024, Bolzano, Italy, November 28-29, 2024, pp. 49–55. CEUR Workshop Proceedings, CEUR-WS.org (2024). https://ceur-ws.org/Vol-3904/paper6.pdf
26. Kuhlmann, I., Corea, C.: Inconsistency measurement in ltl_f based on minimal inconsistent sets and minimal correction sets. In: SUM, LNCS, vol. 15350, pp. 217–232. Springer (2024)
27. Leemans, S.J.J., Fahland, D., van der Aalst, W.M.P.: Discovering block-structured process models from event logs containing infrequent behaviour. In: Business Process Management Workshops. Lecture Notes in Business Information Processing, vol. 171, pp. 66–78. Springer (2013)
28. Answer Set Programming. Springer, Cham (2019). https://doi.org/10.1007/978-3-030-24658-7_9
29. Mannhardt, F., Koschmider, A., Baracaldo, N., Weidlich, M., Michael, J.: Privacy-preserving process mining - differential privacy for event logs. Bus. Inf. Syst. Eng. **61**(5), 595–614 (2019)
30. Marques-Silva, J.: Minimal unsatisfiability: Models, algorithms and applications (invited paper). In: ISMVL. pp. 9–14. IEEE Computer Society (2010)
31. Miller, T.: Explanation in artificial intelligence: Insights from the social sciences. Artif. Intell. **267**, 1–38 (2019)
32. Nielsen, M., Plotkin, G.D., Winskel, G.: Petri nets, event structures and domains, part I. Theor. Comput. Sci. **13**, 85–108 (1981). https://doi.org/10.1016/0304-3975(81)90112-2

33. Ogata, S., Tsuchiya, T., Kikuno, T.: SAT-Based Verification of Safe Petri Nets. In: Wang, F. (ed.) ATVA 2004. LNCS, vol. 3299, pp. 79–92. Springer, Heidelberg (2004). https://doi.org/10.1007/978-3-540-30476-0_11
34. Rakow, A.: Slicing petri nets with an application to workflow verification. In: Geffert, V., Karhumäki, J., Bertoni, A., Preneel, B., Návrat, P., Bieliková, M. (eds.) SOFSEM 2008. LNCS, vol. 4910, pp. 436–447. Springer, Heidelberg (2008). https://doi.org/10.1007/978-3-540-77566-9_38
35. Reiter, R.: A theory of diagnosis from first principles. Artif. Intell. **32**(1), 57–95 (1987). https://doi.org/10.1016/0004-3702(87)90062-2

Discovering the Lively World of Object-Centric Processes with Dynamic Relationships

Alessandro Gianola[1], Zeeshan Hameed[2], Marco Montali[2], Anjo Seidel[3], Mathias Weske[3], and Sarah Winkler[2](✉)

[1] INESC-ID/Instituto Superior Técnico, Universidade de Lisboa, Lisbon, Portugal
alessandro.gianola@tecnico.ulisboa.pt
[2] Free University of Bozen-Bolzano, Bolzano, Italy
{hameed,montali,winkler}@inf.unibz.it
[3] Hasso Plattner Institute, University of Potsdam, Potsdam, Germany
{anjo.seidel,mathias.weske}@hpi.de

Abstract. Object-centric process mining examines how processes interact with multiple co-evolving objects. After the introduction of the OCEL standard for object-centric event logs, several modeling approaches have emerged, including object-centric Petri nets (OCPNs) and (data-aware) object-centric Petri nets with identifiers (OPIDs). While OCPN discovery has been widely explored, OCPNs leave object relationships underspecified and cannot enforce synchronization. On the other hand, OPIDs address synchronization by tracking object identities and relations. However, existing discovery techniques are limited to the notion of stable relationships in OCELs, i.e., they do not change during process execution. In this paper, we present discovery techniques for OPIDs that support fine-grained dynamic relations between objects. Building on recent work that studies how dynamically changing one-to-many relationships can be detected from OCELs, we first identify patterns in OCELs that indicate when activities create, delete, and update links between objects pertaining to given relationships. Building on these patterns, we propose a novel discovery technique for OPIDs, capable of enforcing the detected dynamic relationships. This technique does not compromise fitness, as the original log can always be replayed. The approach is implemented and evaluated on available OCELs.

Keywords: Business Process Management · Process Discovery · Object-centric Petri Nets · Dynamically Changing Relations · Synchronization

1 Introduction

Process mining aims to extract actionable insights from the execution data recorded by information systems [27]. While traditional process mining assumes

F. M. Maggi et al. (Eds.): BPM 2026, LNBIP 599, pp. 39–57, 2027.
https://doi.org/10.1007/978-3-032-37877-4_3

that events refer to a single case notion, modern information systems store data about multiple interacting process instances that share co-evolving objects [22, 25]. This has motivated the development of object-centric process mining, where processes evolve several interrelated objects simultaneously. Executions are recorded using the standard format of object-centric event logs (OCEL) [8].

Various modeling formalisms have been proposed to represent processes with object-centric behavior. The most lightweight option are object-centric Petri nets (OCPNs) [28], which extend Petri nets to handle multiple object types and their token flows. Their structural conciseness makes OCPNs highly amenable to automated discovery, but OCPNs are known to be underspecified [26]: they do not keep track of object identity, and consequently they cannot model inter-object relationships or synchronization. This is problematic in scenarios where transitions must bind objects consistently, e.g., ensuring that an item stays linked to the same order along the entire execution. Recent works enhance OCPNs with synchronization mechanisms [31,32], but the kinds of synchronization are restricted to special cases of stable relationships in identified subprocesses. More expressive modeling formalisms capturing various forms of synchronizations exist in the Petri net literature, such as synchronous proclets [3], and Petri nets with identifiers (e.g., COA nets [9] and PNIDs [29]). A recent unifying proposal is object-centric Petri nets with identifiers (OPIDs) [12,13], which enhance OCPNs with explicit object identities and synchronization semantics.

In recent work, a postprocessing technique for OCPNs has been put forward that turns a discovered OCPN into an OPID that can maintain stable relationships throughout executions [23]. Here *stable* (or *rigid*) means that when an object of one type is linked to an object of another type, this association remains fixed throughout the execution. In fact this stability assumption is pervasive in discovery and log analysis techniques that consider relationships between objects [19,32]. This is surprising, given that *dynamic* relationships are ubiquitous in real life: the content of an order is often updated; the delivery address of a package may change during shipment; or in a logistics process, containers may be rescheduled and routes changed. Indeed, an evaluation showed that many publicly available event logs contain relationships between object types that appear as many-to-many when considered rigid, but many-to-one or even one-to-one when seen as dynamic [10].

In this paper, we address the problem of *discovering object-centric processes from event data, considering that relationships may dynamically change over time.* To this end, we provide a lightweight approach that enhances OCPNs into OPIDs that keep track of dynamic relationships, requiring a minimal amount of domain knowledge. More precisely, we provide a fourfold contribution. (1) Building on assumptions regarding how to represent dynamic relationships in an OCEL that are typically satisfied, and focusing on dynamically changing one-to-many relationships, we discover from a log the roles of each activity when it comes to the creation, deletion, and propagation of links in such relationships. These interaction patterns are reflected by a precise way to label activities that we introduce here (Sect. 3). (2) Building on these patterns, we propose a novel discovery technique capable of discovering dynamically changing relationships (Sect. 4). Due to their natural ability to combine the main modelling constructs for object-centric processes, we use OPIDs as a target formalism. Our approach first applies off-the-shelf existing OCPN discovery techniques (extensively validated in terms of performance [28]), and afterwards transforms the obtained OCPN into an OPID that suitably manipulates dynamic relationships. (3) We prove that the resulting OPID, despite being more precise, does not compromise fitness, as the initial log can always be replayed. Moreover, we prove a bound on the size of the resulting OPID. (4) This procedure is implemented and evaluated on publicly available OCELs, showing that the procedure is indeed lightweight and efficient (Sect. 5). In Sect. 6 we conclude with a discussion and a comparison to related work. Some supporting material can be found in an extended version [11].

2 Background

We build on the standard notion of object-centric event logs [28]:

Definition 1. *Given a set of object types* Σ*, a set of activities* $\mathcal{A}$*, and a set of timestamps* $\mathcal{T}$ *totally ordered by* $<$*, an* object-centric event log *(OCEL) is a tuple* $L = (E, \mathcal{O}, \pi_{act}, \pi_{obj}, \pi_{time})$ *where: (i)* E *is a set of events; (ii)* $\mathcal{O}$ *is a set of object identifiers, each typed by the function* $ot\colon \mathcal{O} \to \Sigma$*; (iii) the functions* $\pi_{act}\colon E \to \mathcal{A}$*,* $\pi_{obj}\colon E \to \mathcal{P}(\mathcal{O})$*, and* $\pi_{time}\colon E \to \mathcal{T}$ *map each event to an activity, a set of objects, and a timestamp, respectively.*

Example 1. Consider an order management process that keeps tracks of orders along with the items they contain. To obtain a simple example, we only consider the activities $\mathcal{A} = \{\textsf{create order}, \textsf{change order}, \textsf{ship order}\}$ that indicate the creation of a new order, an update of the items in an existing order, and shipment; and object types $\Sigma = \{order, item\}$. An event log may look as follows:

$$\langle \#_1, \textsf{create order}, \{o_1, i_1, i_2\}\rangle, \langle \#_2, \textsf{change order}, \{o_1, i_1, i_3\}\rangle, \langle \#_3, \textsf{ship order}, \{o_1, i_1, i_3\}\rangle, \langle \#_4, \textsf{create order}, \{o_2, i_4\}\rangle, \langle \#_5, \textsf{change order}, \{o_2, i_5\}\rangle, \langle \#_6, \textsf{ship order}, \{o_2, i_5\}\rangle$$

Here the objects o_1, o_2 are of type *order*, while $i_1, \ldots, i_5$ are of type *item*.

2.1 Object-Centric Petri Nets

Next, we recall the notion of OCPNs by van der Aalst and Berti [28]:

Definition 2. *Given a set of object types* Σ*, and a set of activities* $\mathcal{A}$*, an* object-centric Petri net *(OCPN) is defined as a tuple* $N = (P, T, F, pt, F_{var}, \ell)$*, where:*

1. P *and* T *are finite sets of places and transitions such that* $P \cap T = \emptyset$*,*
2. $F \subseteq (P \times T) \cup (T \times P)$ *is the flow relation between places and transitions,*
3. $pt\colon P \to \Sigma$ *maps every place to an object type,*
4. $F_{var} \subseteq F$ *is the set of variable flow relations,*

5. $\ell\colon T \to \mathcal{A} \cup \{\tau\}$ *is the transition labelling where* τ *marks an invisible activity.*

The pre-places of a transition t are all places with ingoing flow relations $\bullet t = \{p \in P \mid (p,t) \in F\}$, while the post-places of t are all those connected with outgoing flow relations $t\bullet = \{p \in P \mid (t,p) \in F\}$. For a transition t, the function $tpl(t)$ returns the types of all its corresponding places $p \in \bullet t \cup t\bullet$. $tpl_{var}(t)$ returns all place types connected via variable flow, and $tpl_{nv}(t)$ all place types connected via non-variable flow. An OCPN is well-formed if for each transition no object type is connected with both a variable and non-variable flow.

All possible tokens and their location in a place are defined as $Q = \{(p,o) \in P \times \mathcal{O} \mid pt(p) = type(o)\}$. We assume that a marking M is a subset of Q.[1] Furthermore, each object type has one initial place and one final place.

A binding b is a function $\Sigma \to \mathcal{P}(\mathcal{O})$ that maps each object type $\sigma \in \Sigma$ to a set of objects of appropriate type. The set of binding executions B contains pairs of a transition and a binding s.t. $B = \{(t,b) \mid dom(b) = tpl(t) \wedge \forall \sigma \in tpl_{nv}(t) : |b(\sigma)| = 1\}$. We denote by $codom(b) = \bigcup_{\sigma \in \Sigma} b(\sigma)$ the codomain of b. The execution of a binding consumes a set of object tokens from the transition's pre-places, $cons(t,b) = \{(p,o) \in Q \mid p \in \bullet t \wedge o \in b(pt(p))\}$. Similarly, it produces object tokens $prod(t,b) = \{(p,o) \in Q \mid p \in t\bullet \wedge o \in b(pt(p))\}$. This binding is enabled in a marking M_i if $cons(t,b) \leq M_i$, and the resulting marking is $M_j = M_i - cons(t,b) + prod(t,b)$. The firing is denoted as $M_i \xrightarrow{(t,b)} M_j$.

An *accepting OCPN* combines an OCPN N with an initial marking M_{init} and a final marking M_{final}. An accepted run is a sequence of binding executions from initial to final marking $M_{init} \xrightarrow{*} M_{final}$. The accepting OCPN can be used to *replay* the log: The recorded objects in the initial places of their type form the initial marking, and each event in the log and its recorded objects are interpreted as a binding execution of the corresponding transition. Finally, the OCPN accepts the log if a final marking M_{final} is reached. If an event allows no corresponding binding or M_{final} is not reached, the log is rejected by the OCPN.

Example 2. The following OCPN can be obtained by the discovery technique [28] from the event log in Example 1, and can also replay it. Places to be populated by initial resp. final markings are marked by the symbol ▶ resp. ■ resp.

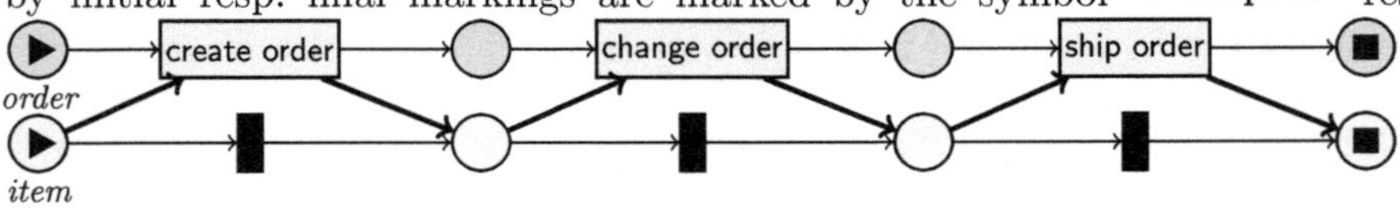

.

2.2 Object-Centric Petri Nets with Identifiers

We summarize the definitions from [12]. Every object type $\sigma \in \Sigma$ is assumed to have a domain $dom(\sigma) \subseteq \mathcal{O}$, given by all objects in $\mathcal{O}$ of type σ. We also need *list* types with a base type in Σ, such a type is denoted as $[\sigma]$ for $\sigma \in \Sigma$. OPIDs are

[1] A marking is generally defined as a multiset of token-object pairs [28], but contemporary discovery algorithms [28,30] ensure an object multiplicity of at most 1.

a variant of Petri nets where each place has a *color* that is a cartesian product of data types from Σ; i.e., the set of colors $\mathcal{C}$ is the set of all $\sigma_1 \times \cdots \times \sigma_m$ such that $m \geq 1$ and $\sigma_i \in \Sigma$ for all $1 \leq i \leq m$. We fix a set of variables $\mathcal{X} = X \uplus X_{list} \uplus \Upsilon$ that is the disjoint union of a set X of "normal" variables that refer to single objects with type in Σ, denoted by lower-case letters like x; a set X_{list} of list variables that refer to a list of objects of the same type, denoted by upper case letters like U, with type$(U) = [\sigma]$ for some $\sigma \in \Sigma$, and a set Υ of variables that refer to fresh objects, with a type in Σ, denoted by ν with some subscript.

In OPIDs [12], tokens are object tuples, and each object tuple naturally has a color in $\mathcal{C}$. To define relationships between objects in consumed and produced tokens when firing a transition, *arc inscriptions* are used, which are labellings of arcs by tuples of variables. Arc inscriptions can be *template* or *simple*, corresponding to variable and non-variable arcs in OCPNs: a template inscription uses one variable in X_{list} and the respective arc can carry an arbitrary number of tokens of the same color; while a simple inscription uses only variables in X or Υ, and a corresponding arc can carry only a single token. The color of an inscription ι coincides with the color of the tokens it carries: if $\iota = \langle x_1, \ldots, x_m \rangle$ then color$(\iota) = \langle \sigma_1, \ldots, \sigma_m \rangle$ where $\sigma_i = \text{type}(x_i)$ if $x_i \in X \cup \Upsilon$, and $\sigma_i = \sigma'$ if x_i has type $[\sigma']$. We set vars$(\iota) = \{x_1, \ldots, x_m\}$. The set of all inscriptions is denoted Ω. Based on this, an OPID is formally defined as follows [12].

Definition 3. *An* object-centric Petri net with identifiers *(OPID) is defined as a tuple* $N = (\Sigma, P, T, F_{in}, F_{out}, \text{color}, \ell)$, *where:*

1. *P and T are finite sets of places and transitions such that $P \cap T = \emptyset$,*
2. *color*$: P \to \mathcal{C}$ *maps every place to a color,*
3. *$\ell: T \to \mathcal{A} \cup \{\tau\}$ is the transition labelling where τ marks an invisible activity,*
4. *$F_{in} \subseteq P \times T \times \Omega$ is a relation that designates the* input flow, *such that* color$(\iota) =$ color(p) *for every $(p, t, \iota) \in F_{in}$, had to modify the def of OPID to allow arrows with multiple inscriptions, please . Ale: do you mean that now multiple arrows instead of a single arrow are allowed?*
5. *$F_{out} \subseteq T \times P \times \Omega$ is a relation that designates the* output flow, *such that* color$(\iota) =$ color(p) *for every $(t, p, \iota) \in F_{out}$,*

To ensure that all resulting tokens of a transition firing are well-defined, we set $\text{vars}_{in}(t) = \bigcup_{(p,t,\iota) \in F_{in}} \text{vars}(\iota)$ *and* $\text{vars}_{out}(t) = \bigcup_{(p,t,\iota) \in F_{out}} \text{vars}(\iota)$ *and require that* $\text{vars}_{in}(t) \cap \Upsilon = \emptyset$ *and* $\text{vars}_{out}(t) \subseteq \text{vars}_{in}(t) \cup \Upsilon$, *for all $t \in T$.*

A tuple $(p, t, \iota) \in F_{in}$ is called a *variable* flow if ι is a template inscription, and *non-variable* flow otherwise; and similar for output flows.[2]

Example 3. The following simple OPID can be obtained from the OCPN in Example 2
by simply labelling edges with singleton inscriptions and adding emitter and

[2] We slightly deviate from [12], allowing multiple flows between the same place and transition. Thus F_{in} and F_{out} are defined as relations rather than partial functions.

consumer transitions, as formalised by the transformation $\mathbb{T}_1$ in [23], which simply adds singleton inscriptions to all arcs and emitter and consumer transitions.

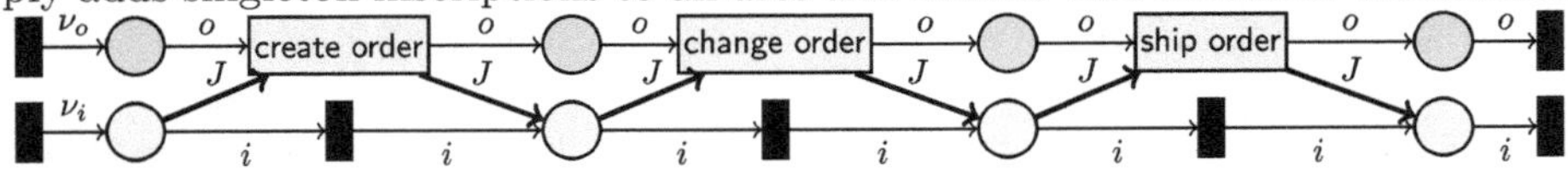

In examples, we write singleton inscriptions without parantheses, e.g. o instead of $\langle o \rangle$. Arcs from the emitter transitions on the left have ν-inscriptions as they generate fresh objects, variable arcs get a list variable as inscription, indicated by the upper case name, and all other transitions have normal inscriptions.

Example 4. The following OPID represents instead a simple order process with synchronization: the bottom (green) place stores (*item*,*order*) pairs, and ensures that an order is always shipped with the same item set it was created with.

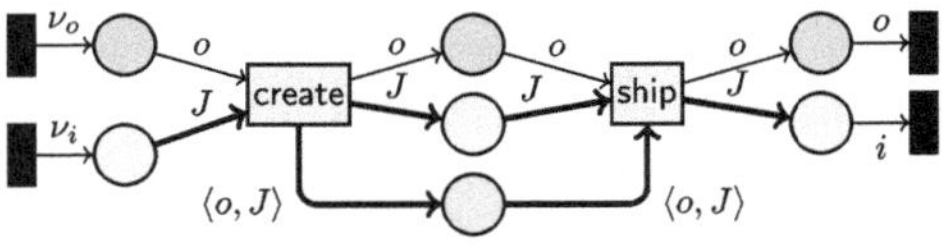

We describe here the semantics of OPIDs informally, details can be found in [11, Sec. A]. Given the set of tokens $\boldsymbol{\mathcal{O}}$ as the set of object tuples, a *marking* is a function $M: P \to 2^{\boldsymbol{\mathcal{O}}}$ mapping each place to a set of tokens of appropriate color. Next, *bindings* are defined to fix which objects are involved in a transition firing: A *binding* for a transition t is a type-preserving function β that maps each variable in $\text{vars}_{in}(t) \cup \text{vars}_{out}(t)$ to a single object (for variables in $X \cup \Upsilon$) or a non-empty list of objects (for variables in X_{list}) of appropriate type. Variables in Υ must be mapped to fresh and distinct objects. A binding β for t is extended to inscriptions of in- and out-flows of t, denoted $\boldsymbol{\beta}$, and defined such that $\beta(\iota)$ is the set of object tuples obtained by combining all objects in the binding of list variables with the binding of normal variables. E.g., if J is a list variable of type *item* and o is a simple variable of type *order*, a binding can set $\beta(o) = \mathsf{o}_1$ and $\beta(o) = o_1$ and $\beta(J) = [i_1, i_2, i_3]$, assuming $o_1, i_i, i_2, i_3 \in \mathcal{O}$. Then for an inscription $\langle o, I \rangle$, we have $\boldsymbol{\beta}(\langle o, I \rangle) = \{\langle o_1, i_1 \rangle, \langle o_1, i_2 \rangle, \langle o_1, i_3 \rangle\}$.

Then, as is common, a binding β is enabled for a transition t, if all object tuples resulting from $\boldsymbol{\beta}$ occur in the current marking, for all places that have a flow to t. Accordingly, firing t with β yields a new marking where all object tuples resulting from $\boldsymbol{\beta}$ are transported from the pre- to the post-places of t. We write $M \xrightarrow{t,\beta} M'$ to denote that t is enabled for binding β in M, and its firing yields M'. An *accepting* OPID N is an OPID together with a set of initial markings M_{init} and a set of final markings M_{final}. A run is accepted if it starts in the initial and ends in the final marking $M_{init} \xrightarrow{*} M_{final}$.

2.3 Detection of Dynamic Relationships in OCELs

In [10], assumptions were identifed under which dynamic relationships can be unambiguously detected from OCELs. In this work, we build on these assumptions, hence recall them here for completeness. Let L be an OCEL with object types Σ and activities $\mathcal{A}$. For clarity, we use the term *relationship* at the type level (e.g., *item–order*), and *link* at the instance level (e.g., a link between i_1 and o_1).

(1) *Snapshot semantics.* The OCEL L is recorded under the snapshot semantics, i.e., each event captures the links of the involved objects after its execution, rather than reporting the parameters of an operation. E.g., if an event create order$\{o_1, i_1, i_2\}$ associates order o_1 with items i_1, i_2, and afterwards an event update order$\{o_1, i_1\}$ is recorded, the latter event is interpreted as the deletion of the link between o_1 and i_2, rather than an update of the link between o_1 and i_1. Thus, events represent a snapshot of the current state.
(2) *Knowledge of relationships.* The many-to-one and one-to-one relationships on Σ are known, with information which object type is the *one* resp. *many* side, and L agrees with these relationships. E.g., if type *order* is in a one-to-many relationship with *item*, each order can contain multiple items (child objects) but each item belongs to a single order (parent object). Then each event in L referring to both orders and items must refer to only one order.
(3) *Reference types.* Every activity occurs with a fixed set of object types,[3] and has one object type as a reference type, and accordingly every event in L has exactly one object of the reference type of its activity, called the reference object. E.g., all activities in Example 1 may have *order* as reference type.
(4) *Perspectives.* The snapshot semantics is refined to distinguish between parent and child perspectives. Let e be an event with reference object o, and suppose there is a one-to-many relationship between σ and μ.
   (i) *Parent reference object:* if o has type σ and e mentions objects $x_1, \dots, x_k$ of type μ, then $x_1, \dots, x_k$ are *all* objects of type μ related to o after e.
   (ii) *Child reference object:* if o has type μ and e also mentions an object x of type σ, then x is related to o after e, but x may have other children.

   E.g., suppose that besides the relationship *item–order* there is also a many-to-one relationship *order–employee*, and an event with activity create order also mentions the handling employee. For instance, let e be create order$\{o_1, i_1, i_2, p_1\}$ where p_1 is an employee (the reference type being still *order*). Then wrt. the relationship *item-order*, case (*i*) applies, and i_1 and i_2 are the only items related to o_1 after e. However, wrt. the relationship *order–employee*, case (*ii*) applies, and p_1 is related to o_1 after e, but p_1 may be related to other orders as well, and if e is followed by an event create order$\{o_2, i_3, p_1\}$ then this does not affect the link between p_1 and o_1.
(5) *Locality.* Each event only modifies links that involve its reference object. E.g., if an event assign to employee$\{o_1, p_2\}$ with reference type *employee* and reference object p_2 would occur after the two events above, then in order to maintain the relationship *order-employee* as many-to-one, the link between p_1 and o_1 would have to be deleted, which would violate locality. However, no violation would occur if assign to employee had reference type *order*.

[3] This is also a prerequisite to OCPN discovery [28], which we use in our approach.

By Assumption (5), if an event refers to objects of two related types but neither is the reference type, the implied links must have already existed beforehand. This implies that in Assumption (4), the case of a reference type different from σ and μ need not be considered because in this case the links remain the same.

An evaluation of publicly available event logs [10] showed that, when assuming relationships between any two types that co-occur in some event, for many logs suitable reference types can be found that satisfy Assumptions (1)–(5).

3 Detection of Relationship Patterns

We next present a technique to detect from an OCEL the roles that activities play in the dynamic manipulation of links for a given relationship. Below we restrict to many-to-one relationships, but it is straightforward to adapt the technique to the case of one-to-one, as detailed in [11, Sec. C]. Many-to-many relationships are known to make it difficult to properly handle synchronization, but can be reified into a pair of many-to-one relationships with a newly introduced object type [3], hence we do not treat many-to-many relationships explicitly.

To adopt the assumptions from Sect. 2.3, we assume that the object types connected by relationships and the reference types of activities are provided as a form of minimal domain knowledge, captured by the next definition.

Definition 4. *For a set of object types* Σ *and a set of activities* $\mathcal{A}$*, a* relationship summary *is a tuple* $\mathcal{R} = (m2o, reftype)$ *such that* $reftype\colon \mathcal{A} \to \Sigma$ *maps each activity to a reference type, and* $m2o \subseteq \Sigma^2$ *is a set of many-to-one relations such that for all* $(\mu, \sigma) \in m2o$*,* μ *is the many-type,* σ *is the one-type, and* $\mu \neq \sigma$*.*

Note that $m2o$ need not contain all pairs of object types that co-occur in events of L, but only those for which the tracking of the dynamic relationship is of interest. In the sequel, we assume that an event log L and a relationship summary $\mathcal{R} = (m2o, reftype)$ are given, both defined over the same set of object types Σ and activities $\mathcal{A}$. Moreover, we assume that L respects Assumptions (1)–(5) wrt. the relationships in $m2o$ and the reference types indicated by *reftype*. For instance, the log from Example 1 respects the assumptions wrt. the many-to-one relationship *item–order* if *order* is fixed as reference type for all activities.

Below, we focus on a single many-to-one relationship $(\mu, \sigma) \in m2o$, where σ is the one-type and μ is the many-type; for multiple relationships, the procedure can be reiterated. Let $E_{\mu,\sigma}$ be a list of all events in L that refer to objects of both types σ and μ, ordered by timestamp; and $Act_{\mu,\sigma}$ the set of all activities of events in $E_{\mu,\sigma}$. Our aim is to label first events, and in a second step activities according to whether they create, maintain, update, or delete links of the relationship. To this end, let $RL = \{\mathsf{create}, \mathsf{maintain}, \mathsf{update}, \mathsf{delete}\}$ be the set of labels.

Our labelling procedure is presented in Algorithm 1. It takes as input the event log L, the relationship summary $\mathcal{R} = (m2o, reftype)$, and a pair of types $(\mu, \sigma) \in m2o$. It outputs a labelling Lab that assigns to each event $e \in E_{\mu,\sigma}$ a set of labels $Lab(e) \subseteq RL$. The procedure maintains a set R that contains all

Algorithm 1. Event classification for a many-to-one relationship

1: **Input:** log L, relationship summary $\mathcal{R}$, pair of types $(\mu, \sigma) \in m2o$
2: **Output:** labelling function $Lab \colon E_{\mu,\sigma} \to \mathcal{P}(RL)$
3: $R = \emptyset$ ▷ set of current links of type (μ, σ)
4: compute $E_{\mu,\sigma}$ from L
5: **for** $e \in E_{\mu,\sigma}$ **do**
6: let u be the object of type σ in e
7: $Lab(e) = \emptyset$
8: **if** e has reference type σ **then**
9: let $O_e = \{o_1, \dots, o_k\}$ be all objects of type μ in e
10: **for** each object $o_i \in O_e$ **do**
11: **if** $(o_i, u) \in R$ **then**
12: add maintain to $Lab(e)$
13: **else**
14: add (o_i, u) to R, and add create to $Lab(e)$
15: $D = \{o \mid (o, u) \in R \text{ but } o \notin O_e\}$
16: **if** $D \neq \emptyset$ **then**
17: remove (o, u) from R, for all $o \in D$
18: add delete to $Lab(e)$
19: **if** e has reference type μ **then**
20: let o be the single object of type μ in e
21: **if** $(o, u) \in R$ **then**
22: add maintain to $Lab(e)$
23: **else if** $(o, u') \in R$ for some $u \neq u'$ **then**
24: remove (o, u') from R and add (u, o)
25: add update to $Lab(e)$
26: **else**
27: add (o_i, u) to R, and add create to $Lab(e)$
28: **return** Lab

current links, i.e., all pairs of objects currently in the $\mu - \sigma$ relationship. At the same time, labels are assigned to events as follows:

- The procedure loops over all events that involve objects of both types. Reflecting Assumption (4), a case distinction is performed according to whether the reference type of e is σ or μ (as in other cases, links are not modified).
- In the former case, if there is some object o_i of type μ mentioned in e such that $(o_i, u) \in R$, i.e., the link is already in the relation, we add label maintain to e (Line 12). If there is some o_i such that (o_i, u) does currently not occur in R, i.e., a new link, we add create to the labels of e (Line 14) and (o_i, u) to R. In addition, if there are objects o of type μ that were up to now linked to u (i.e., $(o, u) \in R$) but do not occur in e, we remove (o, u) from R and add label delete for e (Line 18).
- If the reference type of e is μ, only a single object o of type μ occurs in e by Assumption (3). Similar as above, we add label maintain resp. create if (o, u) is already in R resp. if o has no binding in R (Lines 22 and 27). If instead o already has a binding (o, u') in R for some $u' \neq u$, we perform an *implicit*

deletion to maintain the relationship many-to-one, i.e., remove (o, u') from R, add (o, u) instead, and remember this case with label update (Line 25).

We illustrate the labeling on two examples.

Example 5. Consider the event log from Example 1, with a many-to-one relationship *item—order* where the items in an order are dynamically updated. In the next table, we give for every event e the state of R when processing e according to Algorithm 1, and the labels of e. The reference type is always assumed to be *order*.

	activity	order	item	R	event labels
$\#_1$	create order	o_1	i_1, i_2	$\{(i_1, o_1), (i_2, o_1)\}$	{create}
$\#_2$	change order	o_1	i_1, i_3	$R_2 := \{(i_1, o_1), (i_3, o_1)\}$	{create, maintain, delete}
$\#_3$	ship order	o_1	i_1, i_3	R_2	{maintain}
$\#_4$	create order	o_2	i_4	$R_2 \cup \{(i_4, o_2)\}$	{create}
$\#_5$	change order	o_2	i_5	$R_5 := R_2 \cup \{(i_5, o_2)\}$	{create, delete}
$\#_6$	ship order	o_2	i_5	R_5	{maintain}

Note that e.g. $\#_2$ has multiple labels, because it creates a new link (i_3, o_1), maintains an existing link (i_1, o_1), and deletes an existing link (i_2, o_1).

Example 6. Consider the following event log describing the formation of teams, with a many-to-one relationship *person—team* (i.e., each person can only be a member of one team). The reference types are underlined in all activity names.

	activity	team	person	R	event labels
$\#_1$	create <u>team</u>	t_1	p_1, p_2	R_1	{create}
$\#_2$	add <u>person</u> to team	t_1	p_3	R_2	{create}
$\#_3$	create <u>team</u>	t_2	p_4, p_5	R_3	{create}
$\#_4$	add <u>person</u> to team	t_2	p_3	R_4	{update}
$\#_5$	notify <u>team</u>	t_1	p_1, p_2	R_4	{maintain}

Here $R_1 := \{(p_1, t_1), (p_2, t_1)\}$, $R_2 := R_1 \cup \{(p_3, t_1)\}$, $R_3 := R_2 \cup \{(p_4, t_2), (p_5, t_2)\}$, and $R_4 := (R_3 \setminus \{(p_3, t_1)\}) \cup \{(p_3, t_2)\}$. Note how in $\#_4$ person p_3 is added to t_2, so since p_3 can only be a member of one team, she is removed from t_1, which is reflected in R_4. Consequently, when all team members of t_1 are notified in $\#_5$ and p_3 is not included, no deletion happens.

To show that Algorithm 1 correctly interprets the $\mu - \sigma$ relationship as a dynamic many-to-one relationship, one can prove the invariant that at any point in time, for every object o of type μ, there is at most one u of type σ such that $(u, o) \in R$:

Lemma 1. *If L satisfies Ass. (5), after every loop iteration in Algorithm 1, for every object o of type μ, there is at most one object u of type σ such that $(o, u) \in R$.*

Proof. The proof is by induction on the number of loop iterations. It suffices to show that for any object o of type μ, whenever (u, o) is added in R, there is no $(u', o) \in R$ with $u' \neq u$. This is clear in Line 27, because it is only executed if there is no $(u', o) \in R$. It also holds for Line 24, because (u', o) is removed, and by the induction hypothesis there is at most one such $(u', o) \in R$. Finally, suppose there was some $(u', o) \in R$ before Line 14. However, then event e would, in order to guarantee the many-to-one relation, need to implicitly delete the parent of object o, which is not the reference object, contradicting Assumption (5). Thus we can exclude this situation. proof could be moved to appendix. If Marco expands the intro, yes. Otherwise we could keep it, since I don't think we have a space issue at the moment □

Finally, the *relationship labels* of an activity $a \in Act_{\mu,\sigma}$ are simply the set of combinations of creation, maintenance, update, and deletion that occur in events with this activity.

Definition 5. *Given $Lab\colon E_{\mu,\sigma} \rightarrow \mathcal{P}(RL)$, let the* relationship labelling *be defined as $RL_{\mu,\sigma}(a) = \{Lab(e) \mid e \text{ occurs in } L \text{ with activity } a\}$ for all $a \in Act_{\mu,\sigma}$.*

For instance, for Example 5 we obtain the relationship labelling $RL_{item,order}$ that assigns $\{\{\mathsf{create}\}\}$ to create order, but $\{\{\mathsf{create}, \mathsf{maintain}, \mathsf{delete}\}, \{\mathsf{create}, \mathsf{delete}\}\}$ to change order, and $\{\{\mathsf{maintain}\}\}$ to ship order. This reflects the fact that in create order, links are always created. On the other hand, in change order, sometimes new links are created and old ones removed, but sometimes in addition links are also maintained, so two different patterns occur. For Example 6 the labelling $RL_{person,team}$ assigns $\{\{\mathsf{create}\}\}$ to create team but $\{\{\mathsf{create}\}, \{\mathsf{update}\}\}$ to add person to team and $\{\{\mathsf{maintain}\}\}$ to notify team.

4 Discovery of OPIDs with Dynamic Relations

As a starting point of our discovery approach, we assume that from the given event log L that satisfies the assumptions wrt. $\mathcal{R} = (m2o, \mathit{reftype})$, an OCPN N_0 was already discovered, e.g., using [28], and an equivalent OPID $N_1 = \mathbb{T}_1(N_0)$ was obtained by applying translation $\mathbb{T}_1$ from [23]. Below, we assume a pair (μ, σ) of object types as input, for which a dynamic many-to-one relationship should be tracked in the OPID. If multiple relationships are to be tracked, the following approach to augment the OPID is applied to each pair of types.

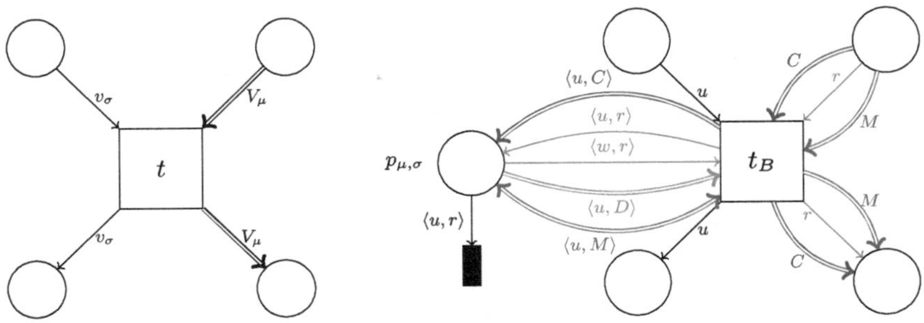

Fig. 1. Extending a transition t with capabilities to track a dynamic relationship.

Transformation. Informally, for each $(\mu, \sigma) \in m2o$, we proceed as follows. Let $N_1 = (\Sigma, P_1, T_1, F_{in,1}, F_{out,1}, color_1, \ell_1)$ be an OPID that was either obtained directly as $N_1 = \mathbb{T}_1(N_0)$ or from an earlier transformation step for another pair of types. We assume that the relationship labelling $RL_{\mu,\sigma}$ was computed using Algorith 1. Our transformation, which we will call $\mathbb{T}_D$, changes N_1 as follows:
(1) Relationship place. We add a *relationship place* $p_{\mu,\sigma}$ that will keep track of all current links of type (μ, σ).
(2) Copies of transitions. For each transition t with activity $a \in Act_{\mu,\sigma}$, we remove t itself, and add *for every label set* $B \in RL_{\mu,\sigma}(a)$ a transition t_B, that implements the transition gadget in Fig. 1:(b). However, the arrows included depend on the labels in B, as follows:

1. if create $\in B$, then the arcs labeled $\langle C\rangle$ resp. $\langle u, C\rangle$ are included;
2. if delete $\in B$ then the arc labeled $\langle u, D\rangle$ is included; and
3. if update $\in B$, then the arcs labeled $\langle r\rangle$, $\langle w, r\rangle$ and $\langle u, r\rangle$ are included;
4. if maintain $\in B$, then the arcs labeled $\langle M\rangle$ resp. $\langle u, M\rangle$ are included.

Any additional incoming/outgoing arcs to places with types different from σ and μ are maintained. Note that in step (2) for each label set $B \in RL_{\mu,\sigma}(a)$, a separate copy t_B is created. The relationship place $p_{\mu,\sigma}$ is shared, though. In addition, we add a single consuming transition for each relationship place.
Formally, $N_2 = \mathbb{T}_D(N_1) = (\Sigma, P_2, T_2, F_{in,2}, F_{out,2}, color_2, \ell_2)$ is as follows: we could move this formal description to the appendix, too

- The set of places is $P_2 = P_1 \cup \{p_{\mu,\sigma}\}$ and $color_2$ is like $color_1$ except that $color_2(p_{\mu,\sigma}) = \langle \mu, \sigma\rangle$.
- Let $T_{\mu,\sigma} = \{t \mid t \in T_1 \text{ and } \ell(t) \in Act_{\mu,\sigma}\}$ be all transitions with an activity in $Act_{\mu,\sigma}$. Then, $T_2 = T_1 \setminus T_{\mu,\sigma} \cup \{t_B \mid t \in T_{\mu,\sigma} \text{ and } B \in RL_{\mu,\sigma}(\ell_1(t))\} \cup \{t^c_{\mu,\sigma}\}$, and ℓ_2 extends ℓ_1 by setting $\ell_2(t_B) = \ell_1(t)$ for all $t \in T_{\mu,\sigma}$, and $\ell_2(t^c_{\mu,\sigma}) = \tau$.
- For all $t \in T_1 \setminus T_{\mu,\sigma}$, the flows $F_{in,2}$ and $F_{out,2}$ coincide with $F_{in,1}$ and $F_{out,1}$. For $t \in T_{\mu,\sigma}$ and $B \in RL_{\mu,\sigma}(\ell_1(t))$:
  - For all places $p \in P_1$ with $color_1(p) \neq \mu$ we set $(p, t_B, \iota) \in F_{in,2}$ iff $(p, t, \iota) \in F_{in,1}$ and $(t_B, p, \iota) \in F_{out,2}$ iff $(t, p, \iota) \in F_{out,1}$.

  - There must be a single place p_i of color $\langle\mu\rangle$ with $(p_i, t, \iota) \in F_{in,1}$ for some ι, and a single place p_o of color μ with $(t, p_o, \iota) \in F_{out,1}$. We set $(p_i, t, \iota) \in F_{in,2}$ and $(t, p_o, \iota) \in F_{out,2}$ for all $\iota \in I$ where I is the smallest set that satisfies all of the following: if create $\in B$ then it contains $\langle C\rangle$, if maintain $\in B$ it contains $\langle M\rangle$, and if update $\in B$ it contains $\langle r\rangle$.
  - There must be a single place p_σ of color $\langle\sigma\rangle$ with $(p_\sigma, t, \iota) \in F_{in,1}$ for some ι. Since σ is the one-type in the relationship, this flow cannot be a variable flow, so we assume it has inscription $\langle u\rangle$. Then the flow $(t, p_{\mu,\sigma}, \iota) \in F_{out,2}$ for all $\iota \in I$ where I is the smallest set that satisfies the following: if create $\in B$ then it contains $\langle u, C\rangle$, if maintain $\in B$ then it contains $\langle u, M\rangle$, and if update $\in B$ then it contains $\langle u, r\rangle$.
- We set $(p_{\mu,\sigma}, t^c_{\mu,\sigma}, \langle u, r\rangle) \in F_{in,1}$ for the silent consumer transition.

Next, we illustrate our approach on an example.

Example 7. Consider again the log from Example 1 and Example 5, for which Algorithm 1 yields the activity labelling RL(change order) $=$ {{create, maintain, delete}, {create, delete}}, RL(create order) $=$ {{create}}, and RL(ship order) $=$ {{maintain}}. Starting from the OCPN N_0 in Example 2 discovered from this log, we obtain the OPID $N_1 = \mathbb{T}_1(N_0)$ shown in Example 3. Finally, Fig. 2 shows the OPID $N = \mathbb{T}_1(N_0)$ obtained after our transformation. Note that there are two change order transitions because RL(change order) contains two sets, reflecting the fact that in some cases a change order event only deletes and creates links, while in others it also maintains links.

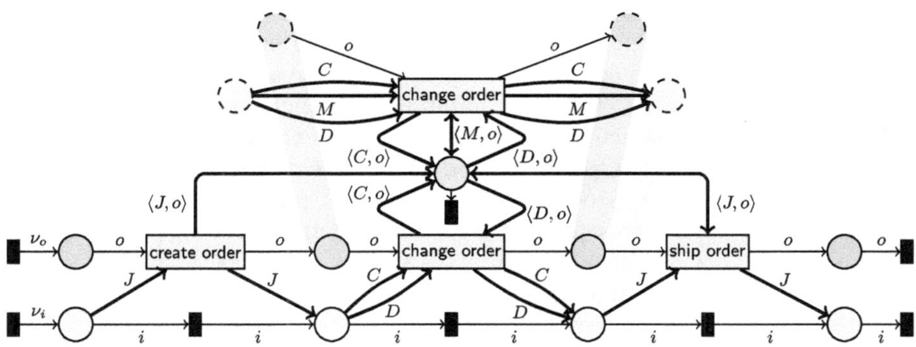

Fig. 2. OPID tracking dynamic relations obtained for the OCEL in Example 7

One can show that the OPID N obtained by $\mathbb{T}_D$ has two key properties:

(i) even though the OPID restricts the behaviour of the original OCPN with additional arcs, if the OCPN can replay the log then so can N, and
(ii) the tuples in the link places reconstruct the dynamic relations in L:

Theorem 1. *Let* $N = \mathbb{T}_D(\mathbb{T}_1(N_0))$ *an OPID obtained from an OCPN* N_0*,* L *an event log that is well-formed wrt. the relationship summary* $\mathcal{R}$*, and* $N = \mathbb{T}_D(N_0, \mathcal{R})$*. If* N_0 *can replay the log* L*, then* N *can do so as well, reconstructing the links in* L*.*

Note that Algorithm 1 requires only a single pass on the OCEL, i.e., it is linear in the size of the log. In fact a single pass also suffices for multiple relationships, if for each relationship a separate set of links R is used and different labellings are constructed. Finally, we provide a bound on the size of the resulting OPID, compared to the input OCPN.

Proposition 1. *Let* N_0 *be an OCPN with set of places* P_0 *and set of transitions* T_0*, and* $N_2 = \mathbb{T}_D(\mathbb{T}_1(N_0))$ *an OPID with places* P_2 *and transitions* T_2*, both over object types* Σ*, and let* $k = |m2o|$*. pls check if statement is understandable and sounds positive enough despite the exponential Then* $|P_2| \leq |P_0| + k$ *and* $|T_2| \leq 7^k(|T_0| + 2|\Sigma|)$*. Moreover, if the activities modifying the different relationships are disjoint then* $|T_2| \leq 7(|T_0| + 2|\Sigma|)$ *so that the size of* N_2 *is linear in the size of* N_0*.*

5 Evaluation

Our technique has been fully implemented[4] in Python, for both one-to-one and many-to-one relationships. As input, the tool takes an OCEL together with a small file in json format, which represents the relationship summary $\mathcal{R}$ (cf. Def. 4). The tool first discovers an OCPN from the given OCEL using `pm4py` [28], and applies afterwards the transformation $\mathbb{T}_1$ from [23] to obtain an OPID N_1. By applying the pattern recognition approach in Algorithm 1, activity labellings are obtained, which are then used to augment the OPID N_1 with relationship-manipulation capabilities (cf. Sect. 4). The approach thus combines the validated technique in [28] with computationally light post-processing.

We applied our approach to examples taken from the literature and various public OCELs[5], where reference types could be detected according to [10] (cf. Assumption (3)). The one-to-one and many-to-one relationships with respect to which the approach was applied were inferred from the logs by checking which pairs of types occur together in events. This is in general a considerable *overapproximation* of the relationships that should be tracked: e.g., if there are many-to-one relationships between *products* and *packages*, *packages* and *employees*, and *products* and *employees* then the last one is redundant. To illustrate efficiency for different configurations of relationships, we counted the number of one-to-one and many-to-one relationships in each OCEL, and applied our algorithm once for many-to-one relationships only, and once for both many-to-one and one-to-one relationships. For instance the container logistics log was tested with both with only one many-to-one relationship, as well as one many-to-one relationship together with six one-to-one relationships. The results are

[4] https://doi.org/10.5281/zenodo.20425098.

[5] https://www.ocel-standard.org/.

Table 1. The evaluation of dynamic relationship discovery in terms of time and size

OCEL	#Σ	oo/mo/mm	used oo/mo	time (s)	OCPN size	OPID size
bicycle assembly [23]	3	0/2/0	0/2	0.1	12/4	14/12
container logistics	7	6/1/3	0/1	1.2	46/50	47/65
			6/1	4.3		53/77
order management	6	3/7/3	0/7	4.8	44/46	51/72
			3/7	4.9		54/87
procure-to-pay	7	3/5/1	0/5	1.0	30/27	35/47
			3/5	1.1		38/53
hinge production	12	29/3/0	0/3	16.1	65/59	68/86
			29/3	26.6		97/137
LRMS p2p	6	7/1/0	0/1	0.6	34/38	35/53
	6		7/1	0.7		42/74
LRMS hiring	6	7/1/0	0/1	0.8	46/52	47/65
			7/1	1.5		54/76
LRMS hospital	8	9/0/0	9/0	0.9	42/36	51/70
inventory management	7	17/0/0	17/0	0.6	68/65	85/149

shown in Table 1, where $|\Sigma|$ indicates the number of object types; *oo/mo/mm* the number of one-to-one/many-to-one/many-to-many relationships discovered from the log as described above; *used oo/mo* the number of respective relationships in $\mathcal{R}$ to which the transformation was applied; the *time* in seconds taken to apply our transformation (OCPN discovery excluded); and *OCPN/OPID size* the size of the resulting OCPN and OPID, respectively, in terms of number of places/transitions. This comparison of the OCPN and OPID sizes shows that in most cases the size grows only moderately, and suggests that the general exponential upper bound from Prop. 1 is actually never reached.

6 Related Work and Discussion

Several works discuss the semantics of object type relationships in OCEL. Both the OCEL 2.0 standard [2] and the OCED core model [4] assume that relationships are timeless, i.e., that relationships are stable and do not change, which is discussed in [23]. However, the need to represent changing relationships has already been highlighted [14]. As the existing log formats [2,4] do not provide a semantics of dynamic relationships, we used the interpretation from [10] to derive an update semantics for activities. We note that this is not the only possibility: e.g. works from the domain of planning rely instead on parameter semantics instead of snapshot semantics [16]. However, once the update semantics is fixed, our discovery approach (Sect. 4) can be applied in any case, regardless of these assumptions. In the future, the existing data formats [2,4] should be extended to allow the explicit tracking of dynamic relationships. Our approach in Sect. 3 can serve as a starting point to ensure backward compatibility.

The discovery of dynamic relationships has been investigated to only a limited extent. In [19], temporal object type models (TOTeM) are mined from OCELs, but relationships are supposed to be governed by object lifespan, i.e., a link is supposed to exist as long as both objects exist. However, this is under-specific and does not consider any changes within the relationships during the co-existence. In [31], relationship patterns between object types can be discovered to identify *temporal implications between two types*. By embedding these temporal implications into object-centric process trees, it is possible to identify sub-trees, in which objects are related accordingly. However, this only allows to identify specific subprocesses, in which object relationships are stable, similar to the notion in [23]. Moreover, as argued in [10], implicit assumptions like snapshot semantics are also needed in this approach. As our approach is based on discovery of OCPNs [28], we are limited by the assumptions underlying this technique, e.g. that event types in an OCEL always operate on the same set of object types. However, our approach is in fact also compatible with other OCPN discovery techniques. As proven, our extension preserves the fitness of a discovered OCPN. However, we are not able to quantify its precision, because current object-centric conformance metrics [1,21] do not consider object relationships at all. To alleviate this shortcoming seems a worthwhile topic for future work.

Our approach allows to bridge the gap between object-centric process models whose scope is purely replay-centered [28,32] to their use as prescriptive models to generate new traces. While existing high-level modeling approaches such as fragment-based Case Management [15], Philharmonic Flows [18], or MERODE [24] already define dynamic relationships, OPIDs can now serve as a formal basis between modeling and mining. In future work, we plan to extend our approach to *data-aware* OPIDs (DOPIDs) [13], which extend OPIDs with rich data support for a variety of data types, such as arithmetic types and uninterpreted functions, as customary in data-aware process mining [5–7,20]. We also plan to generalise it to handle n-ary relationships with richer cardinality constraints. The formal analysis of OPIDs with changing relationships is non-trivial. Therefore, the correctness notions for PNIDs [29] should be extended to OPIDs. Furthermore, discovered object-centric simulation models also assume stable relationships [17] but more realistic simulation models are needed [26], where we provide a starting point.

7 Conclusion

This paper introduces the first rigorous discovery technique for dynamic relationships in object-centric process mining, i.e., relationships between objects that may change over time. We presuppose a minimal amount of domain knowledge that allows to interpret the dynamic relationships represented in an OCEL in a unique way. On this basis, we show how relationship changes triggered by activities can be discovered from an OCEL. We then develop an enhancement of a discovered OCPN into an OPID that effectively enforces these relationship manipulations. Our technique is linear in the size of the log, and we prove that it preserves the original OCPN's fitness, while being more precise. It is also minimally intrusive in that the basic net structure is preserved and the original log can be replayed. The approach is fully implemented for many-to-one and one-to-one relations, and our evaluation shows that it is indeed very efficient, and for a reasonable number of relationships the size increase of the OPID is moderate.

In summary, this paper opens up the novel perspective that makes dynamic relationships in object-centric process mining first-order citizens. This approach bridges process mining and process modeling, and open a way for many lines of future research: the discovery techniques can be extended to data-aware OPIDs [13], conformance checking techniques that take relationships into account can be developed, and process analysis techniques that can analyze properties with respect to relationships could be studied. All these process mining techniques could also be approached for semantics differing from the snapshot view, e.g. parameter semantics or a semantics that is explicitly provided by an action model.

Acknowledgments. This work was partially supported by the NextGenerationEU FAIR PE0000013 project MAIPM (CUP C63C22000770006). This work was also partly supported by Portuguese national funds through Fundação para a Ciência e a Tecnologia, I.P. (FCT) under projects UID/50021/2025 (DOI: https://doi.org/10.54499/UID/50021/2025) and UID/PRR/50021/2025 (DOI: https://doi.org/10.54499/UID/PRR/50021/2025).

References

1. Adams, J.N., van der Aalst, W.M.P.: Precision and fitness in object-centric process mining. In: Proceedings of the 3rd ICPM, pp. 128–135 (2021)
2. Berti, A.: OCEL (Object-Centric Event Log) 2.0 Specification (2024)
3. Fahland, D.: Describing behavior of processes with many-to-many interactions. In: Donatelli, S., Haar, S. (eds.) PETRI NETS 2019. LNCS, vol. 11522, pp. 3–24. Springer, Cham (2019). https://doi.org/10.1007/978-3-030-21571-2_1
4. Fahland, D., et al.: Towards a simple and extensible standard for object-centric event data (OCED) - core model, design space, and lessons learned. CoRR **abs/2410.14495** (2024)
5. Felli, P., Gianola, A., Montali, M., Rivkin, A., Winkler, S.: Conformance checking with uncertainty via SMT. In: Proceedings of BPM 2022, pp. 199–216. Lecture Notes in Computer Science, Springer (2022)
6. Felli, P., Gianola, A., Montali, M., Rivkin, A., Winkler, S.: Data-aware conformance checking with SMT. Inf. Syst. **117**, 102230 (2023)
7. Felli, P., Gianola, A., Montali, M., Rivkin, A., Winkler, S.: Multi-perspective conformance checking of uncertain process traces: An SMT-based approach. Eng. Appl. Artif. Intell. **126**, 106895 (2023)
8. Ghahfarokhi, A.F., Park, G., Berti, A., van der Aalst, W.M.P.: OCEL: a standard for object-centric event logs. In: Bellatreche, L., Dumas, M., Karras, P., Matulevičius, R., Awad, A., Weidlich, M., Ivanović, M., Hartig, O. (eds.) ADBIS 2021. CCIS, vol. 1450, pp. 169–175. Springer, Cham (2021). https://doi.org/10.1007/978-3-030-85082-1_16

9. Ghilardi, S., Gianola, A., Montali, M., Rivkin, A.: Petri net-based object-centric processes with read-only data. Inf. Syst. **107**, 102011 (2022)
10. Gianola, A., Hameed, Z., Montali, M., Seidel, A., Weske, M., Winkler, S.: Detecting dynamic relationships in object-centric event logs (2026)
11. Gianola, A., Hameed, Z., Montali, M., Seidel, A., Weske, M., Winkler, S.: Discovering the lively world of object-centric processes with dynamic relationships (extended version) (2026), available from https://doi.org/10.5281/zenodo.20425097
12. Gianola, A., Montali, M., Winkler, S.: Object-centric conformance alignments with synchronization. In: Proceedings of the 36th CAiSE. LNCS, vol. 14663, pp. 3–19 (2024)
13. Gianola, A., Montali, M., Winkler, S.: Object-centric processes with structured data and exact synchronization - formal modelling and conformance checking. In: Proceedings of the 37th CAiSE. LNCS, vol. 15702, pp. 185–202. Springer (2025)
14. Goossens, A., De Smedt, J., Vanthienen, J.: Object-centric event logs: characteristics, comparative analysis and road map. In: Business Process Management Forum, pp. 37–54. Springer (2024)
15. Haarmann, S., Montali, M., Weske, M.: Refining case models using cardinality constraints. In: International Conference on Advanced Information Systems Engineering, pp. 296–310. Springer (2021)
16. Jansen, N., Gösgens, J., Geffner, H.: Learning lifted action models from traces of incomplete actions and states. In: Proceedings of the 22nd KR, pp. 832–842 (10 2025)
17. Knopp, B., Pourbafrani, M., van der Aalst, W.M.P.: Discovering object-centric process simulation models. In: Proceedings of the 5th ICPM. pp. 81–88 (2023)
18. Künzle, V., Reichert, M.: PHILharmonicFlows: towards a framework for object-aware process management. J. Softw. Maint. Evol. Res. Pract. **23**(4), 205–244 (2011)
19. Liss, L., Adams, J.N., van der Aalst, W.M.P.: TOTeM: temporal object type model for object-centric process mining. In: Proceedings of the BPM Forum, pp. 107–123. Springer (2024)
20. Mannhardt, F., de Leoni, M., Reijers, H.A., van der Aalst, W.M.P.: Balanced multi-perspective checking of process conformance. Computing **98**(4), 407–437 (2016)
21. Park, G., Adams, J.N., van der Aalst, W.M.P.: Conformance checking and performance analysis using object-centric directly-follows graphs. In: Proceedings of the Business Process Management Forum, pp. 179–196. Springer (2024)
22. Seidel, A., Weske, M., Montali, M., et al.: Object-centric process management: a research manifesto. Inf. Syst. **141**, 102728 (2026)
23. Seidel, A., Winkler, S., Gianola, A., Montali, M., Weske, M.: To bind or not to bind? Discovering stable relationships in object-centric processes. In: Proceedigns of the 44th ER. LNCS, vol. 16189, pp. 223–241. (2025)
24. Snoeck, M., Verbruggen, C., De Smedt, J., De Weerdt, J.: Supporting data-aware processes with MERODE. Softw. Syst. Model. **22**(6), 1779–1802 (2023)
25. van der Aalst, W.M.P.: Object-centric process mining: dealing with divergence and convergence in event data. In: Software Engineering and Formal Methods, pp. 3–25. Springer (2019)
26. van der Aalst, W.M.P.: Toward more realistic simulation models using object-centric process mining. In: Proceedings of the 37th ECMS, pp. 5–13 (2023)
27. van der Aalst, W.M.P., et al.: Process mining manifesto. In: Proceedings of the BPM Workshops. Springer (2012)

28. van der Aalst, W.M.P., Berti, A.: Discovering object-centric petri nets. Fund. Inform. **175**(1–4), 1–40 (2020)
29. van der Werf, J.M.E.M., Rivkin, A., Montali, M., Polyvyanyy, A.: Correctness notions for petri nets with identifiers. Fund. Inform. **190**(2–4), 159–207 (2024)
30. van Detten, J.N., Schumacher, P., Leemans, S.: Discovering compact, live and identifier-sound object-centric process models. In: Proceedings of the 6th ICPM, pp. 113–120 (2024)
31. van Detten, J.N., Schumacher, P., Leemans, S.: Modeling and discovering dynamic identity relations in object-centric process mining. In: Proceedings of the 7th ICPM (2025)
32. van Detten, J.N., Schumacher, P., Leemans, S.J.J.: Object synchronizations and specializations with silent objects in object-centric petri nets. In: Proceedings of the 22nd BPM. LNCS, vol. 14940, pp. 57–74. Springer (2024)

Process-Aware Design of Just-In-Time Adaptive Interventions: An Object-Centric Perspective

Maria Inês Ribeiro[1(✉)], Laura Genga[1], Monique Simons[2], and Pieter Van Gorp[1]

[1] Eindhoven University of Technology, Eindhoven 5612 AZ, Netherlands
{m.i.da.graca.jorge.da.silva.ribeiro, l.genga,p.m.e.v.gorp}@tue.nl
[2] Wageningen University & Research, 6706 KN Wageningen, Netherlands
monique.simons@wur.nl

Abstract. Just-in-time adaptive interventions (JITAIs) support health behavior change by using smartphone and wearable data to detect the best moment to deliver tailored support based on predefined decision rules. Micro-randomized trials (MRT) support the causal estimates of intervention effects and moderation analyses that inform JITAI design (e.g., tailoring variables and decision rules). However, existing analytical methods fail to capture the partially structured nature of the lifestyle and health behaviors in which JITAIs are embedded. This paper proposes JITAI mining (JITAIm), a conceptual framework that integrates causal estimation with process-aware analysis of human behavior to inform JITAI (re-)design. We (i) operationalize a sensor-augmented object-centric event log (OCEL) extending the OCEL 2.0 standard with a sensor dimension and (ii) specialize the object-centric paradigm for the JITAI domain, by mapping object-centric entities to JITAI components and introducing a human-centric process execution notion. A simulated proof-of-concept demonstrates the feasibility and potential of our JITAIm framework.

Keywords: Just-in-Time Adaptive Intervention · Object-Centric Process Mining · Digital Health · Sensor Data

1 Introduction

Non-communicable diseases, such as diabetes and cardiovascular conditions, are leading causes of death worldwide and are frequently driven or exacerbated by unhealthy modern lifestyles [17]. Pervasive digital health tools, including mobile apps and wearables, enable innovative behavior change interventions promoting healthy lifestyles, using continuously collected digital trace data.

F. M. Maggi et al. (Eds.): BPM 2026, LNBIP 599, pp. 58–74, 2027.
https://doi.org/10.1007/978-3-032-37877-4_4

A prominent framework in this area is the just-in-time adaptive intervention (JITAI), which tailors the intervention type, dose, and timing to an individual's dynamic states of opportunity and receptivity to change [20]. Digital trace data are continuously collected and used to infer tailoring variables (e.g., location, recent activity) that capture these dynamic states. Decision rules then map these variables to intervention options, enabling interventions to be delivered when individuals are most likely to benefit from them.

Traditional behavioral theories guide intervention design by identifying behavioral determinants (i.e., factors influencing health behavior) and specifying mechanisms of behavior change (i.e., causal pathways through which interventions produce change). Yet, they do not explicitly represent how behavior and its determinants evolve over time [25]. JITAIs operationalize traditional theoretical insights by capturing rapidly changing determinants as tailoring variables through digital trace data and dynamically adapting intervention delivery over time [20].

However, JITAI techniques overlook the *process structure* inherent in human behavior across lifestyle routines. Research on habits, time-use, and socioecological systems suggests that human behavior unfolds in a partially ordered temporal structure, with activities like sleep, work, and leisure recurring in temporal patterns shaped by physiological rhythms, social obligations, and environmental constraints [8,12]. These sequences create predictable windows for intervention. For example, sedentary periods can reliably follow a prolonged work block or precede a meeting, yielding distinct contexts that affect opportunity and receptivity to a physical activity (PA) prompt. Therefore, experiments to evaluate and redesign JITAIs should explicitly model these temporal patterns to understand when and why adaptive interventions fail, yet current designs lack such process awareness.

The gold standard experimental design for JITAIs is the micro-randomized trial (MRT), in which intervention options are repeatedly randomized at decision points to estimate causal effects (i.e., the change attributable to delivering vs. not delivering an option) and to analyze candidate effect moderators (i.e., whether these effects vary with observable variables) [22]. MRT digital trace data are a time series of decision points, indicating whether an intervention was delivered, candidate effect moderators (e.g., location, recent activity), and a binary proximal outcome (i.e., whether the target behavior occurred during the time window when the intervention should be effective). Although this enables rigorous causal estimation, it can miss behavioral dynamics, structured temporal patterns, and how intervention events, health behaviors, and lifestyle routines evolve interdependently over time [11].

We argue that designing effective JITAIs can significantly benefit from interpretable and process-aware representations that (i) capture the order, duration, and temporal relations of behavioral events across these interdependent dimensions without collapsing them into a single process notion and (ii) reveal temporal patterns in lifestyle routines when opportunity and receptivity to change emerge. This enables the analysis of the moderation effect of such heterogeneous and local patterns, which may serve as tailoring variables in future JITAI designs or as features for AI-driven personalization.

Building on this motivation, we investigate: *"How can process-aware analysis of digital trace data inform JITAI design?"*. We frame JITAI design as a process-aware analysis problem at the intersection of Business Process Management (BPM) and digital health, and apply process mining (PM) to JITAI digital traces to model the partially structured lifestyle routines in which interventions and health behaviors emerge [1]. Specifically, we (i) introduce a framework for process-aware analysis to inform JITAI design (Sect. 3), (ii) propose a technical approach to implement it using sensor-augmented object-centric PM (Sect. 4), and (iii) provide a proof-of-concept using a simulated dataset (Sect. 5).

2 Related Work

Prior research to evaluate and refine a JITAI analyzed digital trace data as a time series of decision points, overlooking the process structure of human behavior [22]. In this section, we review process-aware approaches to understanding human behavior and position our contribution.

PM is a family of data-driven techniques to discover, monitor, and improve processes from event data recorded in information systems [1]. Events labeled with a type and timestamp are stored in event logs and grouped by case identifiers (i.e., process instances). PM is increasingly used to study human behavior from digital trace data and reconstruct routines, habits, and user journeys in domains such as e-commerce, smart environments, software usage, and mobile health [5,10,16,23,27,28]. Yet, existing PM frameworks offer little guidance for process-aware analysis in MRTs to inform JITAI design. The user behavior mining framework reconstructs software usage behavior from digital trace data, limiting the scope to the participants interactions with the system itself [23]. Authors in [7] propose a conceptual model to incorporate IoT in BPM to model and mine human-centric processes. The human behavior mining framework [11] shows how to leverage digital trace data to formulate and test hypotheses about dynamic behavior. They do not address how to make these insights actionable for guiding the design of JITAIs [11].

More fundamentally, prior PM applications for studying human behavior use traditional *case-centric* analysis describing the structured lifecycle of a single object (e.g., an individual's behavior during one mobile app session). However, in JITAI analysis, multiple behavioral dimensions are inherently interdependent, unfold at different time scales, and constrain one another. Intervention components are repeatedly randomized, lifestyle routines represent a partially ordered context that shapes how health-related behaviors unfold, and individual or contextual determinants shape opportunity and receptivity. The object-centric PM (OCPM) paradigm addresses this limitation by allowing behavioral events to relate to multiple interdependent objects and by representing their relationships in an object-centric event log (OCEL) [9]. For JITAI analysis, such a representation is relevant because it preserves the interleaving of interventions, routines, and health behaviors, and helps avoid convergence and divergence distortions caused by flattening the log to single case notions. However, state-of-the-art techniques applying PM for human behavior analysis do not leverage the OCPM paradigm.

In addition, prior PM applications for studying human behavior overlook or adopt a simplified approach to deal with *sensor data*, which are a fundamental source of data in JITAIs to observe behavior. PM traditionally relies on event logs in which semantically meaningful events have already been recorded. In recent IoT applications, these events often must be inferred from raw sensor streams during log construction [6,7]. Prior work has proposed preserving raw sensor streams alongside derived events, to keep logs reproducible, retain the full context for semantic insights, and enable additional inference during post-hoc analysis [13]. In JITAI analysis, passive sensing captures observable lifestyle and health behaviors (e.g., PA, mobility) and individual and contextual behavioral determinants (e.g., location, social context, weather proxies) that shape an individual's opportunity and receptivity to behavior change. Converting these sensor streams into behavioral insights requires domain-specific inference, especially from research areas like human activity recognition [3]. Different intervention components and proximal outcomes require different behavioral abstractions, and moderation analyses often demand additional post-hoc inference from raw streams (e.g., deriving PA intensity bands such as light vs. vigorous), which, hence, must be preserved. Still, previously proposed sensor-rich logging approaches remain largely case-centric rather than object-centric.

Altogether, the literature highlights the potential of PM methods for process-aware analysis of JITAIs but lacks end-to-end guidance on: (i) representing interdependent intervention, lifestyle, and health-behavior episodes in an object-centric log that preserves sensor evidence for inference, and (ii) using the resulting process models to derive actionable diagnostics that improve standard causal estimates and inform JITAI redesign.

3 Just-in-Time Adaptive Intervention Mining Framework

We propose the JITAIs Mining (JITAIm) framework to help behavioral scientists evaluate and redesign JITAIs by accounting for participants' partially structured lifestyles and health-related behaviors through process-aware analysis (Fig. 1).

Behavioral Theories. The framework starts with a selected behavioral theory that (i) specifies behavioral determinants as measurable tailoring variables for capturing states of opportunity and receptivity, (ii) links mechanisms to intervention options via decision rules at event-based decision points, (iii) defines proximal outcomes and their causal detection temporal window, and (iv) guides which lifestyle events should be captured for process-aware analysis. Behavioral scientists should choose the theory whose constructs best explain the proximal outcome and can be measured with available data; for example, the COM-B

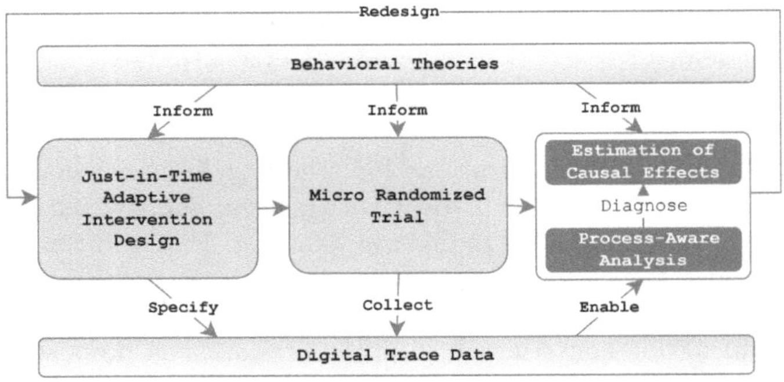

Fig. 1. Just-in-time adaptive intervention mining framework

model (Capability-Opportunity-Motivation-Behaviour) fits adaptations targeting momentary capability, opportunity, or motivation [18].

JITAI Design. The theory-informed JITAI design specifies what intervention options will be delivered and when, and what data will be recorded to support adaptation and analysis. For clarity, we provide brief definitions of the core JITAI components: (i) *outcomes*: distal (ultimate target) and proximal (short-term mediators) behavioral goals; (ii) *opportunity/receptivity states*: conditions when support is expected to help (opportunity) and when an individual is likely to engage with it (receptivity); (iii) *decision points*: times when the system decides whether to intervene and, if so, which option to deliver; (iv) *tailoring variables*: time-varying information about the individual's states of opportunity/receptivity used in decision rules; (v) *intervention components and options*: behavior change support components (e.g., prompts) and the array of deliverable options at decision points, including *provide nothing*; (vi) *decision rules*: protocolized mappings from tailoring variables to intervention options (if–then statements); and (vii) *additional observations*: data not used as tailoring variables, but useful for post-hoc analysis (e.g., effect moderators, engagement). In our JITAIm framework, we propose collecting lifestyle events as additional observations to identify local behavioral patterns at decision points and later treat them as candidate effect moderators.

Micro Randomized Trials. An MRT sets up the theory-informed JITAI design into an experimental design to evaluate its effectiveness. An MRT repeatedly randomizes intervention options at decision points with a pre-defined probability, generating digital trace data for post-hoc analysis [22].

Digital Trace Data. The MRT collects heterogeneous digital trace data capturing timestamped observations of the intervention events, lifestyle and health-related behavior events, and their individual or contextual determinants. This data includes: (i) *active* events created by the participant (e.g., self-reports, or manual logs) or by the JITAI system itself (e.g., decision points, or intervention delivery); and (ii) *passive* sensor data streams (e.g., accelerometer, GPS) that continuously observe behavior and its determinants at high temporal resolution [19].

Process-Aware Analysis and Estimation of Causal Effects. After an MRT, state-of-the-art regression-based estimators use digital trace data to quantify JITAI causal effects and test effect moderators [22]. Excursion causal effects compare behavior when an option is delivered versus not delivered. Moderation analysis tests whether other observed variables alter the size or direction of this effect.

In state-of-the-art MRT analysis, effect moderators are additional observed variables constructed as point-wise or aggregated covariates of the decision-point time series. While these variables are not part of the evaluated JITAI design for decision rules, those that meaningfully shift the effect size/direction can be promoted to tailoring variables in subsequent redesign cycles. However, because moderators are typically part of the decision point time series, they typically fail to capture where decision points occur within partially structured lifestyle routines. We propose a process-aware approach to help address this gap by (i) using process discovery to surface recurrent, local behavioral patterns in lifestyle routines from available digital trace data and (ii) for each decision point, encoding an indicator of whether that decision point occurs within one of the discovered patterns (e.g., decision points following prolonged sedentary bouts during evening leisure). The output of i) provides behavioral scientists with explicit, visual hypotheses about the routine patterns in which decision points arise and proximal outcomes are (un)achieved, supporting theory-driven reasoning about the determinants that generate these patterns. The resulting process-derived indicators are interpretable, routine-grounded variables for moderation analysis that directly test whether intervention effects depend on the surrounding behavioral context. The significant ones can be promoted to a tailoring variable for iterative JITAI refinement. Identifying such tailoring variables is often non-trivial from digital trace data and may require multiple MRT cycles of ad hoc exploration; process models make these patterns traceable and actionable.

4 Sensor-Augmented and Object-Centric Process Mining

In JITAIs, digital health tools (e.g. smartphones, mobile apps, wearables) continuously collect digital trace data tracking intervention logs, lifestyle or health-related events, together with their contextual determinants. This section explains how to map these data to a sensor-augmented OCEL for process-aware analysis. We (i) operationalize a sensor-augmented OCEL extending the OCEL 2.0 standard with a sensor dimension and (ii) specialize the object-centric paradigm for the JITAI domain by mapping object-centric entities to JITAI components and introducing a human-centric notion of process execution. Figure 2 conceptualizes the sensor-augmented OCEL in the four dimensions operationalized below: time, sensor events, behavior events, and objects.

Objects. An object is a first-class entity with a defined type and dynamic attributes, which participates in or relates to events through many-to-many rela-

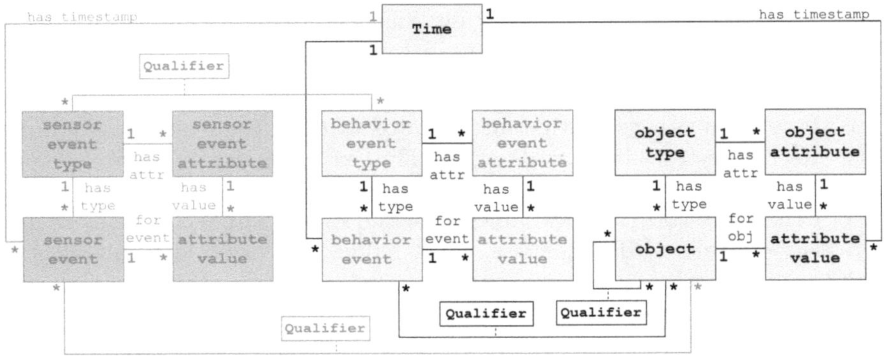

Fig. 2. Sensor-augmented object-centric event data

tions. From the literature, we derive the following object types to map the core JITAI components:

- **Participant**: the individual targeted by the intervention.
- **Individual and contextual determinants**: determinants of the dynamic states of opportunity and receptivity of an individual toward an intervention option, serving as tailoring variables or effect moderators.
- **Intervention component**: a concrete intervention component with its options, tailoring variables, decision rules, and randomization probability in the MRT.
- **Health-related behaviors**: observable health behavior episodes with an explicit lifecycle that are targeted by the JITAI and represent proximal outcomes (e.g. PA bouts).
- **Lifestyle behaviors**: observable lifestyle routine instances that provide context and potential triggers for intervention decision points and are expected to mediate JITAI proximal outcomes (e.g. daily lifestyle routines). Lifestyle routines can be specified at multiple time scales (e.g. daily, weekly), which typically have a lower granularity than the health behavior episodes and intervention component lifecycles.
- **Temporal**: relevant timescales, such as days, weeks, or intervention periods, defining the granularity at which the progress towards the outcomes is expected to occur. Temporal objects operationalize timescales for querying process executions, and they enable bottom-up aggregation and top-down constraints across timescales, e.g., with dynamic attributes that keep relevant information from other timescales (e.g. weekly hybrid work schedule).

Object types are related to each other through interleaved processes. For example, contextual and individual determinants define opportunity and receptivity states, which trigger intervention decision points events. These events link routine to intervention objects, which in turn causally affect health-behavior objects representing proximal outcomes. As such, relations between object types capture both causal dependencies and contextual embedding.

Behavior Events. Behavior events are semantic activities recorded in or inferred from digital trace data, each linked to a single behavior event type with structured attributes. Non-atomic events are represented as paired instances with a lifecycle attribute (start/end). In a JITAI, behavior refers to any observable action or inaction taken by individuals or a JITAI system, as captured in digital trace data, including: (i) health- and life-related events passively inferred from sensor data (e.g. PA bouts, commutes), (ii) events actively logged by the participant (e.g. work calendar events), and (iii) events generated by the JITAI system (e.g. decision points, intervention delivered, user response, proximal outcome).

Sensor Events. The proposed sensor dimension adds to the event log atomic events in a high-frequency time series, each tied to one sensor event type (e.g., accelerometer) and defined by structured attributes (e.g., axes x, y, z). In JITAIs, sensors are embedded in digital health tools through smartphones or wearables. Modeling the sensor dimension separately from semantic dimensions, and preserving raw sensor data, allows multiple behavioral views to be constructed in the JITAI analysis after an MRT.

Time. In PM, time is the primary ordering and synchronization dimension. It provides a single timestamp for sensor events, behavior events, and dynamic object attributes. Digital trace data in JITAI systems is similarly timestamped. In JITAIs, time is the dimension through which behavior unfolds and across which changes in an individual's states trigger intervention delivery.

Relations and Inference Enrichment. Instance-level relations formalize the links between sensor events, behavior events, and objects. Concretely, we record relations from (i) a sensor event instance to one or more object instances (sensor-event-to-object, $SE2O$; e.g., an accelerometer reading linked to a participant object), (ii) a behavior event instance to one or more object instances (behavior-event-to-object, $BE2O$: e.g., PA event linked to a participant and a daily routine object), and (iii) between object instances (object-to-object, $O2O$; e.g., a daily routine object linked to a day and a participant). These relations may include a semantic qualifier specifying the object's role.

These relations are created directly from the source digital trace data while building the log. Afterwards, the log can be enriched, online or post-hoc, by applying inference operators. First, *type-level relations* ($SET2BET$, sensor-event-type-to-behavior-event-type) specify which sensor event types may infer which behavior event and object types under qualified semantics. These relations define the admissible abstraction space but do not yet instantiate events. Second, inference operators transform raw sensor streams into semantically meaningful behavioral constructs for PM. We define *inference* as the abstraction of multivariate sensor event time series into (i) semantic behavior events (e.g., PA bouts) or (ii) dynamic object attribute updates (e.g., location). Time series are obtained by querying sensor events linked to an object (e.g., a participant) with existing methods and ordering them in time. Instead of introducing new inference techniques, we build on established methods from fields like human activity recognition, where algorithms routinely infer high-level behavior from low-level sensor data [3]. The inferred entities are inserted into the log and linked via the same instance-level relations (e.g., *BE2O*, *O2O*).

Process Executions Extraction. In OCPM, defining a process execution means extracting an OCEL subgraph by selecting relevant objects and retaining their related events [2]. Process execution extraction is a required preliminary step that constructs a unit of analysis for follow-up analysis, identifying coherent object-centric process instances. For JITAI analysis, we propose extracting *human-centric executions* by applying a query that (i) anchors the subgraph in a single participant, (ii) selects the relevant object types for analysis (e.g., intervention, daily routine, and health behavior), and (iii) filters events through explicit semantic predicates.

Process Discovery. We propose implementing process-aware analysis through object-centric process discovery to uncover recurrent, theory-relevant local behavioral patterns. Process discovery identifies regularities across extracted process execution instances. We use object-centric directly-follows graphs (OC-DFGs) annotated with frequency, which place each JITAI decision point in its surrounding sequence of lifestyle and/or health events, making explicit the local behavioral patterns in which decision points and proximal outcomes occur. Participants are modeled as first-class objects to anchor human-centric execution extraction, but they are not used in process discovery because they do not constitute meaningful behavioral execution units. Discovery focuses on dynamic objects with explicit lifecycles (e.g., daily routine, intervention, and health-behavior instances).

5 Proof-of-Concept

To demonstrate the feasibility and potential of the JITAIm framework, we offer a proof-of-concept using simulated MRT digital trace data, with tutorial-style code in an open-source repository [24]. The focus of the evaluation is to demonstrate the capability of the framework to model human behaviors and reveal patterns in participants' routine that can be used to inform JITAI design. We evaluate our approach on synthetic data in a realistic yet controlled setting that provides behavioral variation and multi-object interactions for validation. The simulation explicitly models interleaved object lifecycles, where daily routines, JITAI interventions, and PA episodes evolve and interact over time. We plan to apply the framework to real-world data in future work.

JITAI Design and Micro-Randomized Trial. The first steps in our framework are to design the JITAIs and the corresponding MRT. Our design choices are inspired by previous JITAI studies promoting PA (e.g., [15]) and on recent research on sedentary behavior among office workers in onsite, remote, and hybrid work models [14]. The distal outcome of our JITAI is to reduce prolonged sedentary behavior, while the proximal outcome is engaging in a light PA (LPA) break after a prompt. Decision points occur after 60 min of sedentary time as opportune conditions and when receptivity constraints are met ($\leq$3 prompts/day; not during meetings or PA; only 08:00–21:00; $\geq$2 h since the last prompt). These represent tailoring variables to decide between two intervention options: deliver a prompt vs. provide nothing. The JITAI design was theory-informed by the COM-B model: we target *physical opportunity* by identifying feasible time windows in the work schedule (e.g. short gaps between calendar events) and prompting an active break, making the option salient and easy to carry out [18].

In the MRT setup illustrated in Fig. 3, intervention options are randomized at 50% probability. The primary research question is: *"What is the overall effect of prompting active breaks after prolonged sedentary periods (> 60 min) to promote LPA bouts (>1 min) in the next 15 min, compared to no prompt?"*. The MRT includes 20 participants across 8 weeks and 40 weekdays.

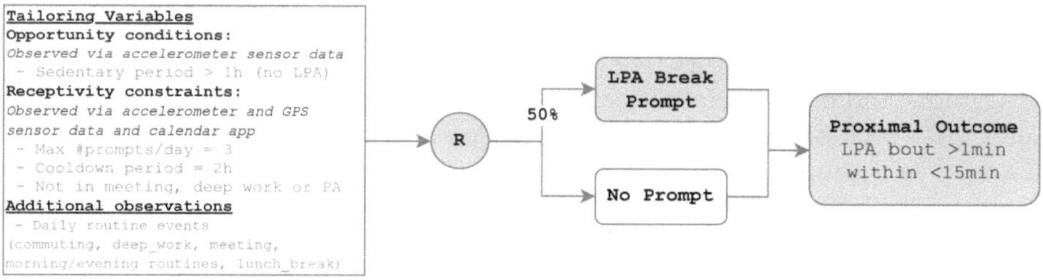

Fig. 3. Micro randomized trial experimental setup

Digital Trace Data Simulation. We simulated digital trace data that would be collected in a real-world MRT to evaluate the JITAI design above, including: (i) intervention events (including decision points, delivery, user response, proximal outcome); (ii) lifestyle events logged in a calendar app (e.g. deep work, meetings, lunch breaks) or passively observed (including commuting events related to GPS and accelerometer sensor data, and morning/evening routine events created between waking up and the first calendar event or the last work event and falling asleep); (iii) PA behavior events related to GPS and accelerometer sensor data. Availability of these traces is realistic and observed in prior MRTs: intervention events are logged by the JITAI system, and PA is sensed by wearables and smartphones [15]. Calendar apps have also been used in prior work to extract interpretable routines from schedules and activity traces, supporting the feasibility of their integration [26].

We simulated an MRT dataset for 20 participants using four *personas*, i.e. virtual archetypes that model individual health behaviors, routines, and environmental interactions. Personas represent hybrid office workers (a population characterized by high sedentary exposure) whose reported daily routines alternate between working from home and in the office within a weekly structure [14]. Daily schedules were generated from probabilistic models encoding chronotype and routine structure (e.g., normal sleeping hours, variability of meetings/ deep work timing) with temporal jitter (e.g., ±20 min meeting duration). The baseline LPA probability during work breaks was set at 20%, increased by +50% when prompts are accepted, realistic within canonical MRT effect sizes [15]. Prompt acceptance/compliance followed a stage-like trajectory (action → maintenance → relapse; e.g., 60%→90%→30%) motivated by the Transtheoretical Model and engagement decay reported in prior MRTs [15,21]. To reflect plausible context-dependent receptivity, we configured that prompt acceptance increases by 20% at decision points immediately before meetings (reflecting anticipated transitions) and decreases by 20% at decision points immediately after deep work blocks (reflecting mental fatigue) because these transitions often occur after prolonged sedentary periods and trigger decision points.

Starting from each persona configuration, we executed an 8-step simulation pipeline to (1–2) instantiate participants, temporal objects, intervention components, and contextual factors (e.g., weather); (3–4) generate daily routine events with timing and order variability and derive eligible JITAI decision points; (5–6) apply MRT randomization to simulate intervention delivery/response/proximal outcomes and propagate outcomes back into routine events (e.g. add an active break); and (7–8) output proximal PA events (LPA > 1 min) with corresponding synthetic accelerometer and GPS streams. A more detailed explanation is provided in the companion repository [24].

Sensor-Augmented Object-Centric Event Log. We mapped the simulated digital trace data into the sensor-augmented OCEL (see Sect. 3). To demonstrate the value of sensor augmentation, we infer PA behavior events from one participant-day of accelerometer data using two parameterizations: an LPA threshold with a 1-min minimum bout and a moderate-vigorous PA threshold with a 10-min minimum bout. In both cases, SET2BET relations link the inferred PA behavior event types to the same sensor event type. For brevity, Fig. 4 plots only the Euclidean Norm Minus One (ENMO) signal along with inferred LPA bouts. ENMO is an accelerometer-derived magnitude feature commonly used for threshold-based PA intensity classification. Note that retaining sensor events enables *post-hoc* re-inference: thresholds, bout definitions, or target behaviors can be revised retrospectively to address new questions and generate different inferred behavioral activities (e.g., recovering short LPA bouts that would be discarded by real-time pipelines). This is illustrated in the repository [24].

Extraction of Process Executions. We instantiate the execution query intuitively introduced in Sect. 4 over the sensor-augmented OCEL as follows. We (i) anchored the extraction at each *participant* object; (ii) retain the object types needed to represent the component within its temporal and outcome context (*daily routine*, the focal *intervention-component*, and *LPA* objects); and (iii) selected behavior events related simultaneously to both the relevant *daily routine* and the focal *intervention component*, whereas lifestyle events (e.g. meeting) and LPA events were included when related to any retained object. This extraction

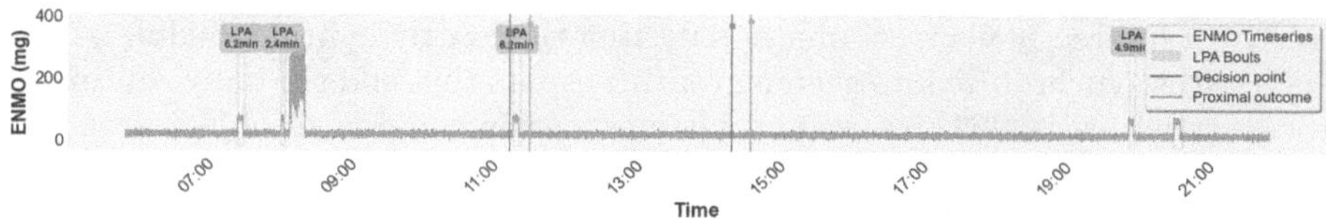

Fig. 4. Light physical activity events inference from accelerometer time series

operationalized the MRT research question as intervention component-specific executions that are semantically coherent, avoiding contamination from intervention events of other components occurring within the same day. The resulting executions provide the analysis unit for intervention analysis.

For the sake of space, we report only key OCEL summary statistics. The log contains 17,642 events and 8,287 objects across 12 activities and 3 object types, with 21612 event–object relations. Event-type frequencies are: *lpa1_bout* (5,909), *meeting* (1,990), *deep_work* (1,776), *intervention_{decision_point (1,604), proximal_outcome (1,581), delivered (791), response (791)}*, *morning_routine* (800), *lunch_break* (800), *evening_routine* (800), and *commuting_{to_work (400), from_work (400)}*. Object-type counts are: *lpa1* (5,883), *intervention_component_lpa_break* (1,604), and *daily_routine* (800).

Process-Aware Analysis. We used PM4Py to discover DFGs with frequency annotations [4]. Figure 5 shows the output process model, integrating the behavioral objects of the intervention component (in purple), daily routine (in blue), and LPA (in yellow); for readability, we only show the directly-follows edges observed at least 100 times. Note that participants are not included in discovery, because it focuses on the objects with explicit lifecycles.

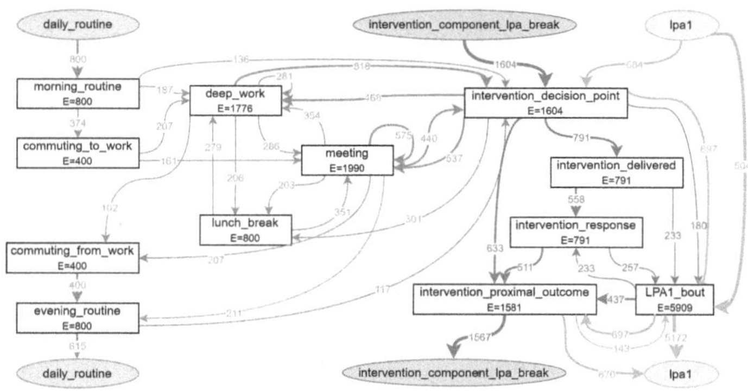

Fig. 5. Object-centric directly follows graph obtained, with edge threshold of 100

The daily routines exhibit the partial structure configured in the simulator: morning routine events transition to a variable sequence of work events (either directly or indirectly via a commute, reflecting the weekly hybrid schedule), after which paths synchronize into evening routine events that end the daily routine. A process structure is also observed in the intervention component object: from the decision point to randomized delivery or no delivery (i.e., event is skipped), participant response, and proximal outcome detection, within which PA is enacted or not. Similarly, the health behavior object regarding PA (*lpa*1) shows how bouts may occur after an intervention decision point or not.

The discovered DFG underscores why our framework relies on object-centric process mining to model behavior in JITAIs. The intervention decision point event, linked to daily routine and intervention object types, captures a multi-dimensional relationship: it occurs at an opportune moment in the routine and either triggers or does not trigger intervention delivery as these elements evolve interleaved over time. By keeping object instances of distinct types (e.g., a daily routine instance and multiple intervention instances) and their relations, the model avoids convergence and divergence issues. For example, if we flatten the log using the intervention instance as the case identifier, daily-routine events (e.g., lunch breaks) are duplicated across the many intervention traces occurring on the same day (convergence distortion). Conversely, if we use the daily routine as the case identifier, the intervention events within that day must be interleaved with routine events; this can impose spurious directly-follows relations, even though routine and intervention activity actually unfold in parallel after the intervention decision event (divergence distortion).

The discovered DFG demonstrates that our process-aware JITAI framework makes recurrent, local lifestyle-routine patterns around intervention decision points explicit, in a way that can be operationalized for effect moderation analysis. Using the frequency annotations, a behavioral scientist can identify behavioral contexts in which intervention decision points occur by inspecting the incoming and outgoing of the decision point event in the DFG (e.g., here, the most frequent relations are after deep work, $n = 818$; before meetings, $n = 537$). This supports theory-grounded hypotheses about potential effect heterogeneity (e.g., whether prompts right before meetings are more or less effective). Note that more complex paths can also be considered as candidate moderators. We plan to explore this opportunity in future work.

Table 1. Summary of causal excursion effects estimated

Model	Term	Estimate	95% CI	SE	p-value	RR (95% CI)
No Moderator	Intercept	1.052	[0.883, 1.221]	0.080	<0.001	2.86 [2.42, 3.39]
Meeting	Intercept	0.804	[0.614, 0.995]	0.090	<0.001	2.24 [1.85, 2.71]
	meeting	0.894	[0.607, 1.181]	0.136	<0.001	
Deep Work	Intercept	1.190	[0.955, 1.425]	0.111	<0.001	3.29 [2.60, 4.16]
	deep_work	-0.246	[-0.584, 0.091]	0.160	0.142	

After extracting the process model, we leverage this knowledge to support the analysis of the causal estimators. We filtered the process executions in which the intervention decision points emerged for each of the discovered patterns, and obtained new DFGs, available in the companion repository for the sake of space. These visualizations allow the analyst to check within each process-defined context, whether the assumption of 50% randomization holds. In our data, the observed allocation differs across contexts (before meetings $\hat{r}_{meeting} = 240/537 \approx 44.6\%$; after deep work $\hat{r}_{deepwork} = 420/818 \approx 51.3\%$). Under a binomial randomization, the expected random fluctuation is small ($SD(\hat{r}_{meeting}) \approx \pm 2.1\%$; $SD(\hat{r}_{deepwork}) \approx \pm 1.7\%$), so the before-meeting deviation indicates context-specific imbalance and threatens internal validity of the estimation of causal effects. Therefore, when estimating moderation by this process-derived context, we use the observed randomization rate.

Estimation of Causal Effects. Following the proposed framework, we encoded the discovered routine patterns (before a meeting and after deep work) as binary indicators and included them as candidate effect moderators in regression-based MRT estimation [22]. Table 1 reports the excursion effect of the primary intervention component and moderation by the two patterns, estimated using the MRTAnalysis R package. Overall, the intervention component demonstrated a consistent positive causal excursion effect. In the baseline model without moderators, treatment increased the outcome probability substantially (RR = 2.86, 95% CI: 2.42–3.39). When including meeting as a moderator, both the intercept and the meeting term were significant, indicating that the treatment effect remained strong overall (RR = 2.24, 95% CI: 1.85–2.71) and was higher before meeting periods. In contrast, adding deep work as a moderator did not reveal a significant moderating effect (p-value = 0.142), although the overall treatment effect in this model remained large (RR = 3.29, 95% CI: 2.60–4.16).

These results align with the simulation design. The significant positive moderation by the meeting pattern shows that the injected higher receptivity before meetings was recovered in the causal estimates. The deep work pattern had the expected negative direction but was not significant, as designed.

Overall, these findings show how process-derived routine patterns can be used as candidate effect moderators to identify when in the lifestyle routine the JITAI is most effective. Combining process-level behavioral reconstruction with causal moderation analysis revealed that JITAI effectiveness is not uniform but shaped by routine-dependent states. Concretely, a behavioral scientist may redesign the JITAI by prioritizing meeting-preceding moments, and deprioritizing moments following deep work, to preserve the daily nudge budget and minimize burden.

6 Discussion

This paper makes a conceptual contribution at the intersection of BPM and digital health by framing JITAI (re-)design as a process-aware problem. State-of-the-art JITAI approaches do not explicitly represent the partial temporal structure in lifestyle routines within which intervention decision points emerge and health-related behaviors unfold [19]. Our proposed JITAIm framework addresses this gap by treating interventions as embedded in recurring behavioral contexts. We introduce the first sensor-augmented object-oriented framework to model human behavior and guide the (re-)design of MRT interventions. The OCPM paradigm is well-suited to explicitly represent the interleaved structure of multiple object types, linking process-aware behavior representations with causal inference in JITAIs. By jointly modeling context (lifestyle routines), mechanism (intervention), and outcomes (health behavior), the framework clarifies how and when interventions work. In addition, preserving raw sensor streams alongside inferred behavioral events supports post-hoc reinterpretation and contextualization of behavior.

The results from the proof-of-concept showcase the capability of the framework in supporting the discovery of local routine patterns that signal opportunity and receptivity for a JITAI, making it possible to encode such patterns for moderator analysis. In this way, process-aware analysis can actionably support the improvement of JITAI delivery.

This study offers a feasibility proof-of-concept on simulated data. Although the simulation was designed considering available literature to reflect realistic human behavior and JITAI settings, validation on real-world digital trace data is still needed to determine whether lifestyle routines and health-related behaviors indeed exhibit recoverable process structure at useful levels of granularity (e.g. days, weeks), and what preprocessing is required to detect this structure robustly in noisy and missing data. A second limitation concerns observability: the framework can only recover routine structure that is recorded in, or inferable from, the digital trace data. Although several data collection devices and apps have been recently introduced in JITAIs, there may be ethical or practical barriers limiting what can be observed in a MRT experiment [26]. Third, we demonstrate the framework on a single lifestyle routine time scale, the daily routine. Future work should investigate cross-timescale interactions (e.g., how weekly work patterns shape daily intervention opportunities), and further examine how dynamic determinants (e.g., weather or social context) can be incorporated [20]. Furthermore, more complex process patterns can be considered when determining the intervention context. Finally, qualitative work is needed to understand what representations are most interpretable for behavioral scientists and how process insights can best be translated into JITAI redesign.

Acknowledgments. This publication is part of the project "LIFTS: Healthy Lifestyle for low liTerate teenagerS" with file number KICH1.GZ03.21.011 of the research program "Zorg in eigen leefomgeving", which is (partly) financed by the Dutch Research Council (NWO) under the grant "Kennis- en Innovatieconvenant (KIC), 2020–2023". AI tools were used to suggest language improvements within the manuscript and partially generate code used in the research process.

Data Availability. Code and data are available for reproducibility at [24].

References

1. van der Aalst, W.: Process mining: Discovery, Conformance and Enhancement of Business Processes. Springer, Berlin (2011). https://doi.org/10.1007/978-3-642-19345-3
2. Adams, J.N., Schuster, D., Schmitz, S., Schuh, G., van der Aalst, W.: Defining cases and variants for object-centric event data. In: 2022 4th International Conference on Process Mining (ICPM), pp. 128–135 (2022). https://doi.org/10.1109/ICPM57379.2022.9980730
3. Arshad, M.H., Bilal, M., Gani, A.: Human Activity Recognition: Review, Taxonomy and Open Challenges. Sensors **22**(17), 6463 (2022). https://doi.org/10.3390/s22176463
4. Berti, A., van Zelst, S., Schuster, D.: PM4Py: a process mining library for Python. Softw. Impacts **17**, 100556 (2023). https://doi.org/10.1016/j.simpa.2023.100556
5. Bertrand, Y., Van den Abbeele, B., Veneruso, S., Leotta, F., Mecella, M., Serral, E.: A survey on the application of process discovery techniques to smart spaces data. Eng. Appl. Artif. Intell. **126**, 106748 (2023). https://doi.org/10.1016/j.engappai.2023.106748
6. Brzychczy, E., Aleknonytė-Resch, M., Janssen, D., Koschmider, A.: Process mining on sensor data: a review of related works. Knowl. Inf. Syst. **67**(6), 4915–4948 (2025). https://doi.org/10.1007/s10115-024-02297-y
7. Burattin, A., Meroni, G., Federico, G.D., López, H.A., Rivkin, A., Kindler, E.: Human-centric processes: where IoT and people meet. In: Internet of Things Meets Business Process Management: A Synergistic Integration, pp. 63–86. Springer (2026). https://doi.org/10.1007/978-3-031-90746-3_3
8. Carden, L., Wood, W.: Habit formation and change. Curr. Opin. Behav. Sci. **20**, 117–122 (2018). https://doi.org/10.1016/j.cobeha.2017.12.009
9. van Der Aalst, W.: Object-centric process mining: unraveling the fabric of real processes. Math. **11**(12), 2691 (2023). https://doi.org/10.3390/math11122691
10. van Eck, M.L., Sidorova, N., van der Aalst, W.: Enabling process mining on sensor data from smart products. In: 2016 IEEE Tenth International Conference on Research Challenges in Information Science (RCIS), pp. 1–12 (2016). https://doi.org/10.1109/RCIS.2016.7549355,iSSN: 2151-1357
11. Fallon, M., Rehse, J.R., Heinzl, A.: Human behavior mining: a framework for theorizing about mHealth behavior using digital trace data. J. Assoc. Inf. Syst. **26**(3), 633–667 (2025). https://doi.org/10.17705/1jais.00938
12. Gershuny, J., Sullivan, O.: What We Really Do All Day: Insights from the Centre for Time Use Research. Penguin, UK (2019)
13. Grüger, J., et al.: SensorStream: An XES Extension for Enriching Event Logs with IoT-Sensor Data (2022). https://doi.org/10.48550/arXiv.2206.11392
14. Jaudzeme, M., Jeņkova, J., Eimane, A., Pavlovska, I., Reste, J.: Motives for interrupting sedentary behaviour among office workers in onsite, remote, and hybrid work models. J. Public Health (2026). https://doi.org/10.1007/s10389-026-02710-6
15. Klasnja, P., Smith, S., Seewald, N.J., Lee, A., Hall, K., Luers, B., Hekler, E.B., Murphy, S.A.: Efficacy of contextually tailored suggestions for physical activity: a micro-randomized optimization trial of HeartSteps. Ann. Behav. Med. **53**(6), 573–582 (2019). https://doi.org/10.1093/abm/kay067

16. Leotta, F., Mecella, M., Veneruso, S.: Discovering Human Habits Through Process Mining: State of the Art and Research Challenges. In: Activity Recognition and Prediction for Smart IoT Environments, pp. 1–18. Springer Nature, Switzerland (2024). https://doi.org/10.1007/978-3-031-60027-2_1
17. Li, J., Pandian, V., Davidson, P.M., Song, Y., Chen, N., Fong, D.Y.T.: Burden and attributable risk factors of non-communicable diseases and subtypes in 204 countries and territories, 1990–2021: a systematic analysis for the global burden of disease study 2021. Int. J. Surg. **111**(3), 2385–2397 (2025). https://doi.org/10.1097/JS9.0000000000002260
18. Michie, S., van Stralen, M.M., West, R.: The behaviour change wheel: A new method for characterising and designing behaviour change interventions. Implement. Sci. **6**(1), 42 (2011). https://doi.org/10.1186/1748-5908-6-42
19. Nahum-Shani, I., Murphy, S.A.: Just-in-time adaptive interventions: Where are we now and what is next? Annu. Rev. Psychol. **77**(1), 679–703 (2026). https://doi.org/10.1146/annurev-psych-121024-044244
20. Nahum-Shani, I., et al.: Just-in-Time Adaptive Interventions (JITAIs) in Mobile Health: Key Components and Design Principles for Ongoing Health Behavior Support. Ann. Behav. Med., 1–17 (2016). https://doi.org/10.1007/s12160-016-9830-8
21. Prochaska, J.O., Velicer, W.F.: The transtheoretical model of health behavior change. Am. J. Health Promot. **12**(1), 38–48 (1997). https://doi.org/10.4278/0890-1171-12.1.38
22. Qian, T., Walton, A.E., Collins, L.M., Klasnja, P., Lanza, S.T., Nahum-Shani, I., Rabbi, M., Russell, M.A., Walton, M.A., Yoo, H., Murphy, S.A.: The micro-randomized trial for developing digital interventions: experimental design and data analysis considerations. Psychol. Methods **27**(5), 874–894 (2022). https://doi.org/10.1037/met0000283
23. Rehse, J.R., Abb, L., Berg, G., Bormann, C., Kampik, T., Warmuth, C.: User behavior mining. Bus. Inf. Syst. Eng. **66**(6), 799–816 (2025)
24. Ribeiro, M.I., Genga, L., Simons, M., Van Gorp, P.: JITAI-mining: v1.0.0 (2026). https://doi.org/10.5281/zenodo.20441960
25. Scholz, U.: It's time to think about time in health psychology. Appl. Psychol. Health Well Being **11**(2), 173–186 (2019). https://doi.org/10.1111/aphw.12156
26. Stein Dani, V., Lu, X., Beerepoot, I.: Integrating wearable data into process mining: event, case and activity enrichment. In: 1st International Workshop on Personal and Human-Centric Process Mining (PHPM) (2025). https://doi.org/10.48550/arXiv.2512.05203
27. Topaloglu, B., Oztaysi, B., Dogan, O.: Cart-state-aware discovery of E-commerce visitor journeys with process mining. J. Theor. Appl. Electron. Commer. Res. **19**(4), 2851–2879 (2024). https://doi.org/10.3390/jtaer19040138
28. Winter, M., Langguth, B., Schlee, W., Pryss, R.: Process mining in mHealth data analysis. npj Digit. Med. **7**(1), 1–10 (2024). https://doi.org/10.1038/s41746-024-01297-0

Track II: Engineering

Simulation-Based Assessment of Business Process Robustness

Robert Blümel[1,2(✉)], Alexander Kraus[3], Timotheus Kampik[2,4], and Han van der Aa[5]

[1] Doctoral School Computer Science, University of Vienna, Vienna, Austria
robert.bluemel@sap.com
[2] SAP Signavio, Walldorf, Germany
timotheus.kampik@sap.com
[3] Data and Web Science Group, University of Mannheim, Mannheim, Germany
alexander.kraus@uni-mannheim.de
[4] Department of Computing Science, Umeå University, Umeå, Sweden
[5] Faculty of Computer Science, University of Vienna, Vienna, Austria
han.van.der.aa@univie.ac.at

Abstract. Business process robustness describes a process's ability to meet performance targets despite variations in parameters such as arrival rates or resource availability. Assessing business process robustness quantitatively is challenging because dependencies between process parameters and performance are complex and stochastic. Business process simulation avoids the need to specify those dependencies by simulating the process and observing the resulting behavior. However, the space of possible parameter combinations grows quickly, which makes it challenging to identify parameters that still meet the required performance targets. In this paper, we address this challenge by proposing a simulation-based approach that systematically explores parameter configurations to efficiently identify those that maintain acceptable performance. Our evaluation demonstrates that the proposed approach substantially reduces assessment complexity compared to our baselines while accurately identifying parameter combinations under which a process remains robust. The approach provides users with insights into how parameter variations influence performance, enabling a step toward a quantitative foundation for assessing and improving business process robustness.

Keywords: Business process robustness · Robustness assessment · Business process simulation · Process mining

1 Introduction

Organizations operate in environments characterized by constant change. These changes influence business processes and their parameters, such as arrival rates or resource availability [8]. Consequently, process performance measures, such

F. M. Maggi et al. (Eds.): BPM 2026, LNBIP 599, pp. 77–94, 2027.
https://doi.org/10.1007/978-3-032-37877-4_5

as the average cycle time or resource cost, may fluctuate over time. When processes are expected to achieve specific performance targets, variations in process parameters can cause these targets to be missed. Hence, it is essential to understand the extent to which process parameters can vary while maintaining desired performance levels. This relates to the concept of robustness [3,9,27], which captures the ability of a business process to sustain its performance targets despite changes in underlying parameters.

Despite the importance of business process robustness, its quantitative assessment remains a challenging and largely unexplored task in process mining. This challenge arises from the need to understand how specific parameter configurations influence process performance, a relationship that is often difficult to determine due to its inherent complexity. For example, while higher arrival rates generally increase cycle times, the precise impact is not straightforward to predict. The stochastic nature of processes [24] further adds variability to performance outcomes. Consequently, developing robustness assessment approaches is difficult, since they need to account for complex and stochastic dependencies between process parameters and performance outcomes.

One way of dealing with process complexity and stochasticity is to use business process simulation (BPS). Recent advances in BPS have enabled quantitative assessment of business process robustness. By modeling processes and simulating them under varying parameter configurations, BPS makes it possible to examine how parameter changes affect overall process performance [1]. However, existing BPS tools are mainly designed for simple what-if analyses and typically used for identifying parameters that optimize selected performance measures [19,20,22,23], rather than for determining the complete set of configurations that maintain acceptable performance. As a result, current approaches cannot efficiently identify all parameter settings under which a process meets its performance targets. This gap highlights the need for a dedicated solution for robustness assessment.

To address this gap, we propose a simulation-based approach for assessing business process robustness. Our approach quantifies how much variation a process can withstand while maintaining acceptable levels of process performance. Our contribution is twofold. First, we formalize business process robustness as an algorithmic assessment problem. Second, we develop an approach that systematically explores parameter variations to efficiently identify configurations that ensure acceptable performance using information recorded about the process in an event log. Conducted evaluation experiments demonstrate that our proposed approach substantially reduces the assessment complexity compared to other search baselines, while accurately identifying the parameter configurations under which the process remains robust. As a result, our approach gives users clear insights into the parameter configurations under which a process meets its performance targets. This marks a first step toward a quantitative foundation for assessing and improving business process robustness.

The remainder of the paper is organized as follows. In Sect. 2, we discuss the problem of business process robustness assessment. Then, we describe our simulation-based approach for its assessment in Sect. 3 and evaluate it in Sect. 4. In Sect. 5 we discuss related work, before we conclude in Sect. 6.

2 Problem Statement

This section formalizes the problem of assessing business process robustness by defining its core elements and the related assessment challenges.

Event Logs. An event log L contains data recorded by a process-aware information system during process execution [2]. Each event $e \in L$ includes at least a case ID, an activity, and a timestamp. Events with the same case ID ordered by time form a *trace*. An event log may also contain further attributes relevant for robustness analysis, such as resource information, costs, and lifecycle data.

Process Parameters. Process parameters are factors that influence the behavior and performance of a business process. They are represented as a vector $X = (x_1, x_2, \ldots, x_n) \in \mathbb{R}^n$, where each x_i denotes a specific parameter, such as the number of available resources or the average processing time of an activity.

Each process parameter x_i takes values from a defined range $\mathbb{I}^*_{x_i} = [a_i, b_i]$, for example, the number of available resources between 1 and 20. The *parameter domain* D is the Cartesian product of these ranges, so that $X \in D \subset \mathbb{R}^n$. D thus represents all parameter configurations considered for robustness assessment.

Process Performance Indicators. Process performance indicators (PPIs) are quantifiable measures of process effectiveness or efficiency [25]. They are represented as a vector $P = (p_1, p_2, \ldots, p_m) \in \mathbb{R}^m$, where each p_i denotes a performance metric such as time, cost, quality, or resource usage. Common PPIs include average lead time, process execution cost, and resource utilization.

Each PPI p_i has an acceptable performance range $\mathbb{I}^*_{p_i} = [l_i, u_i]$, where l_i and u_i are the lower and upper bounds. If only one bound is specified, the other is set to $+\infty$, $-\infty$, or zero, depending on the context. The *acceptable performance space* is defined as $P^* = \mathbb{I}^*_{p_1} \times \mathbb{I}^*_{p_2} \times \cdots \times \mathbb{I}^*_{p_m}$.

Performance Function. We represent the relationship between process parameters and outcomes through a process's *performance function*. This function maps the parameter domain to PPIs and can be viewed as a sample function of a stochastic process [15], producing one realization of performance for a given parameter vector. Formally, it is defined as $f(X) = P$, where $X = (x_1, \ldots, x_n)$ is the parameter vector and $P = (p_1, \ldots, p_m)$ the resulting vector of PPIs.

Monotonicity. We assume that each performance function is strictly monotonic with respect to each parameter. Formally, let $f_j(X)$ denote the j-th PPI for configuration X. For any parameter x_i, and for any two configurations X and X' that differ only in the i-th coordinate, strict monotonicity requires

$(x_i - x_i')(f_j(X) - f_j(X')) > 0$ whenever $x_i \neq x_i'$. That is, changing one parameter while keeping all other parameters fixed changes the corresponding PPI in a consistent direction. Note that monotonicity does not require a linear or otherwise specific functional relationship between parameters and PPIs, only that each parameter affects a given PPI in a consistent direction. This assumption is reasonable in our context, since parameter effects often follow predictable monotonic trends. For example, adding additional resources generally reduces the average lead time, while more case arrivals increase it (with everything else remaining equal).

Robustness Assessment Challenges. Given a parameter domain D, an acceptable performance space P^*, and a process performance function f, the assessment of business process robustness is characterized by two main challenges:

Process Stochasticity. Business processes are complex socio-technical systems [4] with many interacting elements, including human actors. Consequently, business processes are stochastic, and the same parameters may yield different PPI outcomes. For example, task durations depend on contextual factors such as worker experience. The performance function f is therefore stochastic and unknown, and must be approximated from an event log L.

State-Space Explosion. Even if an unbiased estimation of the performance function f is available, identifying the robustness space remains computationally challenging due to state-space explosion when the process parameter domain $D \subseteq \mathbb{R}^n$ is multidimensional. This makes an efficient search strategy necessary to determine the robustness space within reasonable computational limits.

Robustness Space. Given these challenges, the objective of this work is to approximate business process robustness efficiently and with high accuracy. Therefore, we define a robustness space $S^{\alpha,\beta}$ as the set of parameter configurations that lead to acceptable process performance:

$$S^{\alpha,\beta} = \{\, X \in D \mid f(X) = P \in P^* \,\},$$

where:

- $\alpha \in (0,1)$ defines the *confidence level*, i.e., the likelihood that parameter configurations within the robustness space yield acceptable performance. Formally, $\mathbb{P}(f(X) \in P^* \mid X \in S^{\alpha,\beta}) \geq \alpha$. This condition accounts for the stochastic nature of a process by ensuring that configurations in $S^{\alpha,\beta}$ meet the performance requirements with a confidence level of at least α, e.g., 0.90.
- $\beta \in (0,1)$ defines the *accepted uncertainty fraction* of the parameter domain that may contain configurations that also yield acceptable performance but lie outside the estimated robustness space, i.e., $\mathbb{P}(X \in D \setminus S^{\alpha,\beta} \mid f(X) \in P^*) \leq \beta$. This condition addresses state-space explosion by allowing a part of

the parameter domain, e.g., up to 10%, to remain uncertain, avoiding costly searches by trading off search efficiency and accuracy.

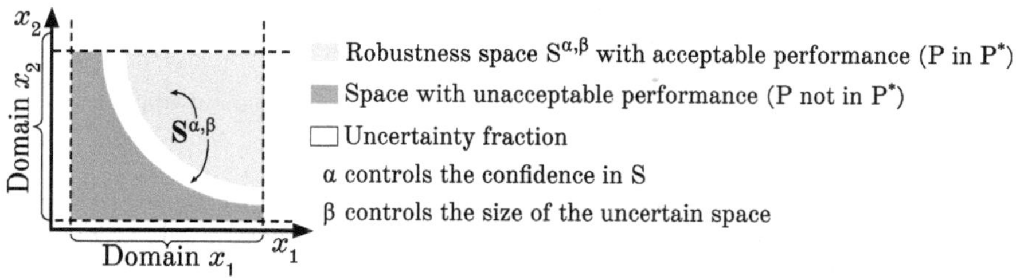

Fig. 1. Robustness space $S^{\alpha,\beta}$ (light green) with confidence level α and accepted uncertainty fraction β (white) within the parameter domain D

Figure 1 illustrates such a space $S^{\alpha,\beta}$. It contains all parameters $X \in D$ for which the resulting performance $f(X) = P$ lies within P^*. The confidence level α specifies the confidence of points in the robustness space (light green), while β limits the size of the accepted uncertainty space in the parameter domain (white).

Finally, we formalize the problem of business process robustness assessment: **Problem Statement.** Given an event log L, a parameter domain D, an acceptable performance range P^*, a desired confidence level α, and an accepted uncertainty fraction β, the task of business process robustness assessment is to determine the approximated robustness space $S^{\alpha,\beta}$ such that every parameter configuration $X \in S^{\alpha,\beta} \subseteq D$ satisfies $f(X) = P \in P^*$, with $\mathbb{P}(f(X) \in P^* \mid X \in S^{\alpha,\beta}) \geq \alpha$ and $\mathbb{P}(X \in D \setminus S^{\alpha,\beta} \mid f(X) \in P^*) \leq \beta$.

3 Business Process Robustness Assessment Approach

This section presents our approach for approximating the robustness space of a business process. As shown in Fig. 2, the approach takes as input an event log and a configuration that specifies the parameters to vary and the acceptable PPI ranges. It then iteratively simulates selected parameters and refines the search space through an adaptive search strategy until a termination condition is met. The result is an approximation of the robustness space, that is, the set of parameter configurations under which the process meets its performance targets.

3.1 Step 1: Simulate and Assess

In the first step, we simulate the process for a selected parameter configuration and assess its resulting performance. The result indicates whether the process meets the robustness requirements under this configuration. This information guides the search in Step 2 by showing which spaces of the parameter domain

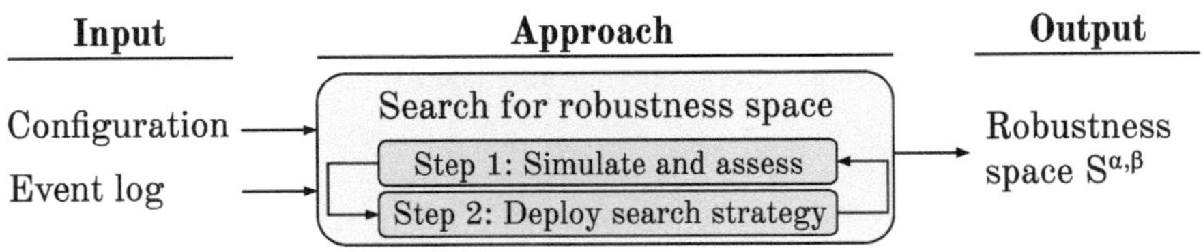

Fig. 2. Overview of the business process robustness assessment approach

require further analysis. In the following, we describe the discovery of the BPS model, performed once, and the process simulation and performance assessment procedure, applied to each evaluated parameter configuration.

BPS Model Discovery. To simulate a given process parameter configuration, we first discover a BPS model. This model can be discovered automatically from an event log L by using data-driven BPS techniques, such as Simod [6], AgentSimulator [12], or DeepSimulator [5]. Alternatively, a BPS model can be constructed manually and provided to the approach.

Process Simulation and Performance Assessment. Once the BPS model is available, we simulate the process under a parameter configuration X and assess the resulting performance. This combination of simulation and performance assessment yields an estimate of the performance function. To capture the stochastic nature of the process, each configuration is simulated N times. The resulting performance values are aggregated into an empirical distribution. We compute the fraction of outcomes that fall within the acceptable performance range for each PPI. A configuration X is classified as part of the robustness space $S^{\alpha,\beta}$ when this fraction reaches at least α for all PPIs; otherwise, X is outside of the robustness space.

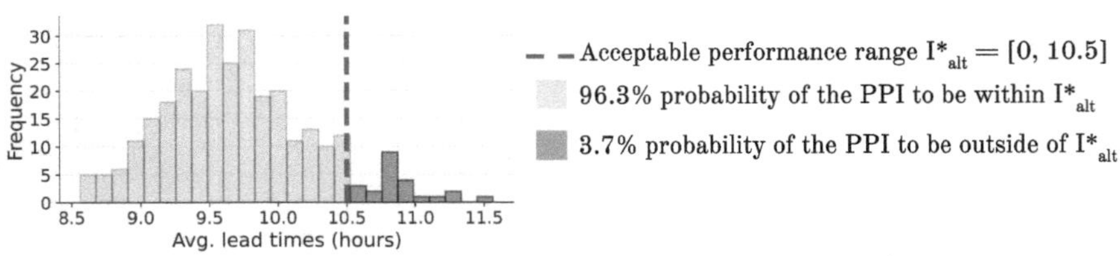

Fig. 3. Performance distribution after 300 process executions

Example. Figure 3 illustrates the idea behind the performance assessment step, using the average lead time (*alt*) as the PPI of interest. To decide whether a parameter configuration X leads to acceptable process performance, we simulate the process $N = 300$ times and obtain the corresponding performance values. This results in the frequency plot shown in the figure and the associated empirical distribution of the target PPI. Assuming that the acceptable performance range is $\mathbb{I}^*_{alt} = [0, 10.5]$ (the upper bound is indicated by the dashed blue line), we observe that 96.3% of the simulated outcomes fall within this range. Therefore, for any confidence level α below 96.3%, configuration X leads to acceptable performance and is classified as part of the robustness space.

3.2 Step 2: Deploy Search Strategy

In this step, we deploy our search strategy to approximate the robustness space. We use a quadtree-based [26] search that adaptively explores the parameter domain and avoids exhaustive evaluation of all configurations. Since we apply this to a d-dimensional parameter domain, we will refer to this as a hyperquadtree (HQT) [16]. In the following, we provide an overview of the deployed search strategy and then detail its key steps.

Overview of the Search Strategy. The goal of the search strategy is to approximate the set of parameter configurations that yield acceptable process performance while keeping the number of performance evaluations low. Although the exact performance function of a business process is unknown and stochastic, we assume that it is strictly monotonic (see Sect. 2). This assumption allows us to infer the behavior of a parameter subdomain from a limited number of simulated and assessed configurations using the hyperquadtree-based search strategy. This strategy represents the parameter domain as a hierarchical tree of nested hyperrectangular subspaces, referred to as nodes. Each node is defined by its hyperrectangular corners corresponding to specific parameter configurations in the domain. In an d-dimensional parameter domain, each node contains 2^d corner configurations. By evaluating the performance at these corner points, we can determine whether all configurations in the node are likely to meet or violate the performance requirements without sampling its interior. Nodes whose corner points all yield acceptable or unacceptable performance are classified accordingly. Nodes with a mix of acceptable and unacceptable outcomes remain uncertain and are recursively subdivided. This focuses simulations on the most informative parts of the domain, while large spaces can be classified with few evaluations.

Key Steps of the Search Strategy. Our search consists of the following key steps: initialization, node classification, node subdivision, and termination. Figure 4 illustrates the first three iterations of these steps in a 2-dimensional domain.

Initialization. We begin the search by considering the entire parameter domain as a *root node* with an unknown robustness status, as shown in Fig. 4a. This root node serves as the basis for subsequent recursive classification and subdivision.

Node Classification. We classify a node and, with that, the subspace it represents by assessing whether each of its corners meets the performance targets, using Step 1. Based on the obtained results, a node is classified as:

- *robust*, i.e., part of $S^{\alpha,\beta}$, if all corners meet the performance targets.
- *non-robust*, i.e., outside of $S^{\alpha,\beta}$, if no corners meet the performance targets.
- *uncertain*, if some corners meet the performance targets while others do not.

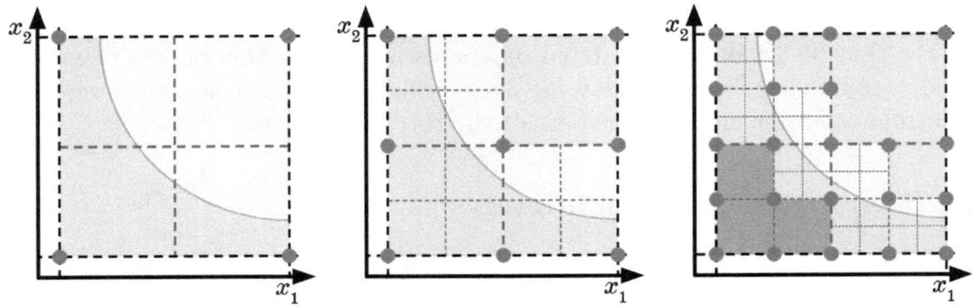

(a) Iteration 1. (b) Iteration 2. (c) Iteration 3.

Fig. 4. Illustration of the hyperquadtree-based search strategy in 2 dimensions

After classification, the parameter configurations of the robust nodes are added to $S^{\alpha,\beta}$. The non-robust nodes are excluded from further consideration. The uncertain nodes are marked for recursive refinement in subsequent iterations. As a result, this strategy focuses computational effort on nodes where the boundary between robust and non-robust configurations is unclear.

For Iteration 1, shown in Fig. 4a, we evaluate the root node using its four corner points, marked by the blue dots. The light green area represents the ground-truth robustness space, while the red area contains configurations that do not meet the performance requirements. Based on this ground truth, the root node is classified as uncertain (indicated by the white semi-transparent overlay) because the corner evaluations yield a mix of accepted and non-accepted performance outcomes. As a result, the root node is subdivided into four child nodes for the next iteration, shown in Fig. 4.

Subdivision of Uncertain Nodes. In the next step, we subdivide all uncertain nodes. Specifically, each uncertain node is split at the midpoint of every parameter dimension, which produces 2^d child nodes that are considered for classification in the next iteration. Through this recursive refinement, the hierarchical structure adapts dynamically: nodes classified as robust or non-robust are not explored further, while uncertain nodes are examined in increasing detail. Figure 4 illustrates this refinement process across the first three iterations.

Search Termination. The recursive search stops when two conditions are satisfied. First, a minimum refinement depth must be reached to avoid stopping at an overly coarse partition. Second, the fraction of the parameter domain covered by the uncertain nodes compared to the entire parameter domain must drop below the accepted tolerance threshold β. The recursive search is performed until both termination conditions hold. Then, no further simulations are performed, and the search returns the final approximation of the robustness space $S^{\alpha,\beta}$.

In our example, none of the termination conditions are met after Iteration 1 (Fig. 4a), so the algorithm proceeds with Iteration 2 (Fig. 4b). In this iteration, the node in the top right area of the domain shows acceptable performance for all of its corner points. This node is therefore classified as robust, and its parameter configurations become part of the robustness space $S^{\alpha,\beta}$. The remaining nodes show a mix of accepted and non-accepted performance outcomes at their corners and are thus classified as uncertain. These nodes are subdivided and evaluated in Iteration 3 (Fig. 4c). In this third iteration, the search classifies three nodes as non-robust, three nodes as robust, and seven nodes as uncertain. In the next iteration, the search will explore these seven uncertain nodes, further reducing the uncertain space of the parameter domain and providing a more accurate approximation of the robustness space.

4 Evaluation

This section presents experiments conducted to assess the accuracy and efficiency of our approach in comparison to four baselines. We first describe the evaluation setup and then discuss the results. To support reproducibility, we provide our data, implementation, and results in our repository[1].

4.1 Setup

In this section, we discuss the relevant setup of our experiments.

Simulated Process. We evaluated our approach on an established loan application process (cf. [8, Chapter 10.8]) that is represented by the BPMN model shown in Fig. 5. The model comprises 12 activities, a loop, a parallel structure, and several exclusive gateways, resulting in broad behavioral variation. The configurations for the simulation model, such as the distributions underlying task durations and branching probabilities, are discovered from an event log of the process (cf. [7]) using Simod [6].

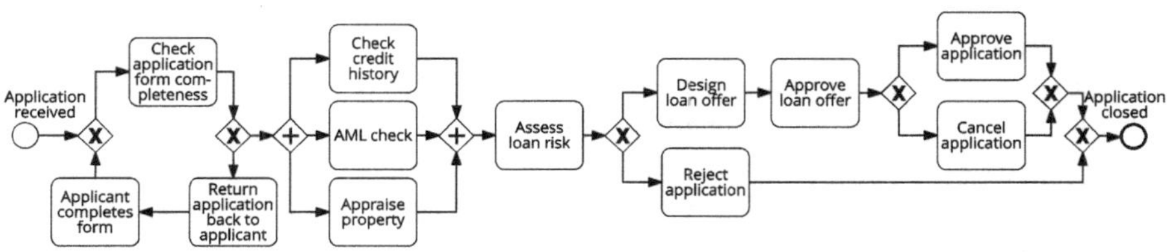

Fig. 5. BPMN model of the loan application process

Input Configurations. We consider the following input configurations for the robustness assessment of the loan application process:

Process Parameters. We assess robustness with respect to parameters related to different process perspectives:

[1] https://github.com/robert-l-b/robustness.

- *Arrival times (at)—(case perspective):* We model the arrival time by adjusting the mean of the discovered case inter-arrival time distribution. In our setup, this distribution was found to be a constant inter-arrival time that we vary from 0 to 125 min, that is, $\mathbb{I}^*_{at} = [0, 125]$.
- *Resources (res)—(resource perspective):* We consider a single resource type, where all resources follow the same seven-to-four schedule and are cross-trained, meaning that they can execute any activity when they are available. We vary the number of resources from 1 to 20, i.e., $\mathbb{I}^*_{res} = [1, 20]$.
- *Branching probability (bp)—(control-flow perspective):* We vary the branching probability at the XOR gateway following *Assess Loan Risk* from 0 to 1, that is, $\mathbb{I}^*_{bp} = [0, 1]$. This setting allows us to examine how varying strictness in risk assessment affects the robustness of the process.

Process Performance Indicators. We assess the robustness of the process with respect to two PPIs that cover the two remaining process perspectives:

- *Average lead time (alt)—(time perspective):* The average lead time is defined as the mean time between the first and the last event across all simulated cases. The accepted performance range $\mathbb{I}^*_{alt}$ is the interval $[0, 11]$ hours.
- *Cost (cost)—(data perspective):* The total cost is calculated as the working hours of all resources multiplied by a cost factor of 20 per hour. The corresponding accepted performance range $\mathbb{I}^*_{cost}$ is given by the interval $[0, 60,000]$.

Evaluation Scenarios. We define four evaluation scenarios (S1–S4) that cover increasing search space sizes, as well as single- and multi-objective performance constraints. Scenarios 1–3 use the same two-dimensional parameter domain, combining arrival times and number of resources, but differ in the PPIs: S1 evaluates average lead time, S2 evaluates cost, and S3 evaluates both. Scenario S4 increases the dimensionality by adding the branching probability as a third parameter and assesses both average lead time and cost.

Approach Parameters. Across all scenarios, each configuration is simulated $N = 30$ times. Each run produces a log with 300 traces, a size determined using Welch's method [17] to remove warm-up effects and ensure stable estimates. We set the confidence level to $\alpha = 0.90$. For Scenarios 1–2, the accepted uncertainty fraction is $\beta = 0.15$, whereas for Scenarios 3–4, we set $\beta = 0.25$ to account for the increased complexity due to multi-objective constraints in both scenarios and the higher-dimensional parameter space. The minimum refinement depth is set to 3, so the search performs at least three rounds of subdivision before termination is checked. This limits the influence of any single corner evaluation on the resulting classification. Additionally, we conduct a sensitivity analysis, in which we vary $\alpha \in \{0.75, 0.80, 0.85, 0.90, 0.95, 0.99\}$ and $\beta \in \{0.10, 0.15, 0.20, 0.25\}$ to study their effect on the robustness assessment.
Baselines. We compare our approach against four baselines that represent alternative strategies for approximating the robustness space.

Grid Search (GS). GS is a simple baseline that evaluates all parameter combinations on a predefined grid, providing a clear reference without relying on model assumptions or sampling strategies. The grid covers all parameter configurations defined by the domain step sizes: 50 options for arrival times, 20 for resources, and 11 for branching probability. Scenarios 1–3 therefore require 1,000 evaluations, and Scenario S4 requires 11,000. Since the true robustness space is unknown, GS provides a reliable, though costly, reference for evaluating the accuracy and efficiency of all other techniques.

Fixed Space Partitioning (FSP). FSP partitions the parameter domain into a fixed set of equally sized subspaces and represents a non-adaptive variant of the hyperquadtree approach. Each subspace is classified by evaluating its corner configurations, following the node classification logic described in Sect. 3.2. The classification of the subspaces continues until all subspaces are labeled or the uncertain space falls below β. To ensure comparability, we iteratively decrease the grid size until the resulting uncertainty fraction meets the target β, and report the number of evaluations at that grid size.

K-Nearest Neighbors (KNN). We consider KNN as a simple classification-based baseline. Parameter configurations are randomly sampled and labeled as described in Sect. 3.1, and the model is trained to distinguish between robust and non-robust configurations. The hyperparameter k is selected via a five-fold cross-validation. KNN is then applied to classify all configurations in the grid search reference, where a configuration is classified only if the predicted probability exceeds the threshold α. We report the setting that requires the fewest training points while satisfying the uncertainty criterion.

Regression Tree (RT). We evaluate regression trees as a more expressive classification baseline, using the same setup as for KNN. Hyperparameters are selected via grid search with five-fold cross-validation on the sampled training set, with each fold containing an equal number of randomly assigned configurations.

Evaluation Measures. We consider four measures that capture the efficiency (*Evals* and *Time*) and accuracy ($\hat{\beta}$ and *MCC*) for each scenario:
Number of Evaluated Configurations (Evals). Measures how many parameter configurations the approach simulates.
Computation Time in Minutes (Time). Measures the total runtime of the robustness assessment.
Uncertainty Fraction ($\hat{\beta}$). Measures the proportion of unclassified evaluations.
Matthews Correlation Coefficient (MCC). Measures how well our approach classifies the configurations of the grid search based on $S^{\alpha,\beta}$. The MCC considers all entries of the confusion matrix and is defined as [21]:

$$MCC = \frac{(TP \times TN) - (FP \times FN)}{\sqrt{(TP + FP)(TP + FN)(TN + FP)(TN + FN)}},$$

where TP and TN are correct robust and non-robust classifications, and FP and FN are the corresponding misclassifications. Unclassified configurations in the uncertain space (fraction β) are excluded. MCC ranges from -1 to $+1$, where $+1$ means perfect classification, 0 random guessing, and -1 full disagreement.

Note that for KNN and RT, we repeat each experiment five times with different random samples and report the average results to account for variability.

Implementation. The evaluation of our approach and the baselines was carried out in the same technical environment. We implemented all search strategies in Python and executed them on an Apple MacBook Pro (macOS version 15) equipped with a 16-core M2 Pro processor and 16 GB of RAM. To avoid effects related to parallel execution, all experiments were run on a single core.

4.2 Results

In this section, we discuss the evaluation results of our experiment. We first examine the efficiency and accuracy of our approach in comparison to the four baselines and then conduct a sensitivity analysis to assess how the parameters α and β influence the performance of our approach.

Table 1. Experiment results of our approach for Scenarios 1–4

Tech-niques	Scenario S1				Scenario S2				Scenario S3				Scenario S4			
	Evals	Time	$\hat{\beta}$	MCC	Evals	Time	$\hat{\beta}$	MCC	Evals	Time	$\hat{\beta}$	MCC	Evals	Time	$\hat{\beta}$	MCC
GS	1,000	88.0	–	–	1,000	79.8	–	–	1,000	92.3	–	–	11,000	975.5	–	–
FSP	220	18.2	0.14	**1.00**	180	13.4	**0.12**	**1.00**	300	31.1	0.20	**1.00**	6,730	561.7	0.24	**0.71**
KNN	150	24.9	0.14	0.99	100	16.6	0.15	0.99	200	37.7	0.20	0.98	4,000	738.8	0.24	0.68
RT	**75**	**12.7**	0.13	0.87	100	15.1	0.15	0.87	200	70.3	0.23	0.69	5,000	990.0	0.26	0.62
HQT	115	15.2	**0.08**	**1.00**	**54**	**8.5**	0.15	**1.00**	**176**	**25.9**	**0.14**	0.99	**2,874**	**243.6**	**0.14**	0.67

Note: The highlighted values indicate the best result in each column

Efficiency and Accuracy. Table 1 reports the evaluation results for all scenarios, comparing our approach, hyperquadtree (HQT), with the four baselines. In the following, we first discuss these results for the two-dimensional Scenarios S1–S3, and then look at the three-dimensional Scenario S4.

Scenarios S1–S3. Figure 6 presents the HQT results for the first three scenarios. All three scenarios share the same two-dimensional parameter domain but differ in the PPIs considered: average lead time (S1), cost (S2), and both measures (S3). The green and red dots show the configurations evaluated by GS, while the blue dots indicate the configurations evaluated by HQT. The red area marks configurations with unacceptable performance, and the green area represents the robustness space estimated by HQT. The figure shows that the robustness space identified by HQT closely matches that of GS and that the evaluation effort is efficiently concentrated along the boundary between the two areas.

Quantitatively, as reported in Table 1, HQT achieves near-perfect classification with MCC values of 1.00 or 0.99 across all three scenarios relative to GS. This confirms that grid points in the green and red areas are correctly classified.

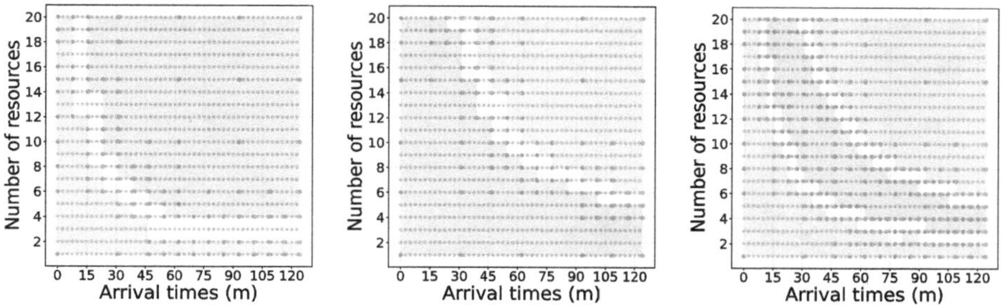

(a) Scenario S1. (b) Scenario S2. (c) Scenario S3.

Fig. 6. HQT vs. GS comparison showing the robustness space and boundary-focused evaluation

For each scenario, HQT's unclassified fraction of the domain (shown in white) accounts for about 8%–15%. These results should be interpreted in light of the uncertainty inherent in GS at the boundary between robust and non-robust configurations, which amounts to 7.2%, 5.6%, and 12.6% for S1–S3, respectively. Moreover, GS exhibits locally mixed behavior. For example, in S1, along the row with three resources, green and red GS points alternate within the white space. This indicates that GS itself is an uncertain approximation of the ground truth.

In comparison to all baselines, HQT provides the most favorable efficiency–accuracy trade-off. In terms of evaluation effort, HQT requires only 5–12% of the GS evaluations in S1 and S2 (115 and 54 vs. 1,000), whereas the other baselines require substantially more, with RT needing around 75–100 evaluations, KNN 100–150, and FSP up to 220. In S3, HQT uses 18% of GS evaluations (176 vs. 1,000), whereas the other baselines require 200–300. In terms of runtime, S1 is the only scenario where HQT is slightly slower than RT, but in S2 and S3, it achieves speedups of a factor of 1.5–3 over all other baselines. In terms of accuracy, HQT maintains near-perfect MCC (1.00 in S1–S2, 0.99 in S3), whereas FSP achieves 1.00 but with substantially more evaluations, KNN reaches 0.98–0.99, and RT drops to 0.69–0.87. Furthermore, HQT consistently yields the smallest unclassified fraction $\hat{\beta}$ and reaches these results in a single adaptive run. In contrast, the classification-based baselines KNN and RT achieve competitive accuracy, but the reported results correspond to the smallest number of training points for which the resulting classification satisfies the uncertainty criterion. Finding this number required testing several values for the number of training points, and the simulation calls used in those preceding trials are not included in Table 1. The same holds for the partitioning depth required by FSP. Unlike the baselines, HQT does not require any prior choice of the number of training points or the partitioning depth. Instead, it adaptively concentrates evaluation effort on uncertain parts of the domain and terminates once this uncertainty falls below the accepted uncertainty fraction β.

Scenario S4. Figure 7 presents the results for S4, which considers a three-dimensional parameter domain and two PPIs. The red cubes denote parts of the domain with unacceptable performance, the white cubes represent the uncertain space, and the green cubes mark the detected robustness space.

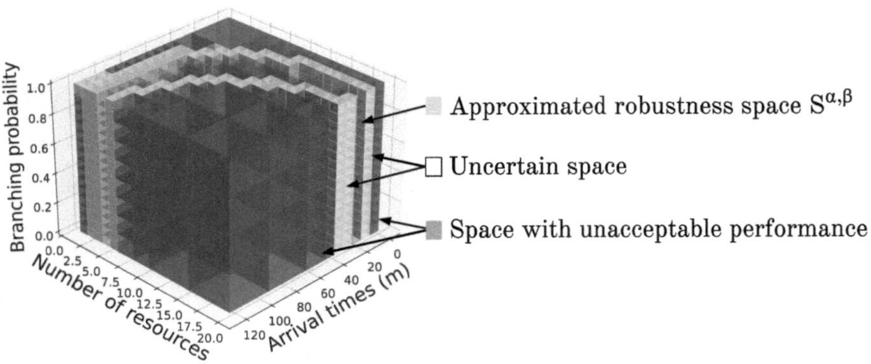

Fig. 7. Search results of HQT for Scenario S4

Consistent with the previous scenarios, HQT maintains a favorable efficiency–accuracy balance in S4 despite the increased dimensionality of the parameter space. As reported in Table 1, HQT evaluates 2,874 configurations–about 26% of GS–and reduces runtime to 243 min, a factor of 3.8 faster. The MCC is 0.67, with 14% of the domain unclassified. The decrease in accuracy is primarily due to configurations near the uncertain boundary of the three-dimensional space. Compared to that, FSP reaches a slightly higher MCC of 0.71 but requires 6,730 evaluations–more than twice the effort of HQT. KNN and RT require 4,000 and 5,000 evaluations, respectively, to achieve similar coverage, with RT still slightly failing to meet the target unclassified fraction $\hat{\beta} = 0.25$. Their MCC values (0.68 for KNN and 0.62 for RT) remain at or below that of HQT. Overall, these results confirm that HQT concentrates evaluations on uncertain regions, achieving higher efficiency while maintaining comparable accuracy.

Table 2. Sensitivity analysis results for varying α and β (Scenario S3)

	$\beta_{0.25}$			$\beta_{0.20}$			$\beta_{0.15}$			$\beta_{0.10}$		
	Evals	Time	MCC	Evals	Time	MCC	Evals	Time	MCC	Evals	Time	MCC
$\alpha_{0.75}$	66	17.7	1.00	176	29.5	0.99	176	29.2	0.99	438	44.7	1.00
$\alpha_{0.80}$	66	17.2	1.00	176	26.1	0.99	176	29.4	1.00	436	38.4	1.00
$\alpha_{0.85}$	66	17.8	1.00	176	30.5	0.99	176	29.9	1.00	436	45.7	1.00
$\alpha_{0.90}$	66	17.2	1.00	176	25.9	1.00	176	30.0	0.99	436	38.6	1.00
$\alpha_{0.95}$	66	16.3	1.00	176	26.3	1.00	176	27.3	0.99	434	40.3	0.98
$\alpha_{0.99}$	66	16.3	1.00	176	28.9	0.99	176	27.3	0.99	438	45.7	0.97

Sensitivity Analysis. Table 2 shows the sensitivity of efficiency and accuracy to the parameters α and β for Scenario S3. Our results are only slightly affected by these parameters. A higher β reduces the number of evaluations and computation time, while accuracy stays high with MCC values near 1.00. This is expected, since a larger uncertainty fraction reduces the need to classify points near the transition between robust and non-robust spaces. Changes in α have a minor impact: the number of evaluations remains nearly constant, and the MCC stays close to 1.00. Only for high confidence levels (0.95 and 0.99) do we observe a small rise in computation time and a slight drop in MCC of 1 to 2% points.

Overall, our evaluation shows that HQT can accurately approximate the robustness space of a business process across several scenarios when compared with naive GS and provides a favorable efficiency-accuracy trade-off against other baselines. It offers an efficient and effective method for assessing business process robustness and represents an initial step toward more robust business processes.

5 Related Work

In this section, we discuss work relevant to business process robustness assessment from three perspectives: robustness in other domains, related process characteristics, and approaches in BPM addressing similar search challenges.
Robustness in Other Domains. Studying robustness has a long tradition in science, yet "is rich in still insufficiently studied methodological and philosophical implications" [28]. In domains where it has been studied–software systems [10], biological [3] and pharmaceutical processes [9], and production systems [27]–robustness is typically understood as the ability of a system to continue functioning properly under varying conditions. In business process management (BPM), robustness has largely been considered only as a property of algorithms, denoting their ability to produce outcomes with guarantees despite imperfect input, such as noise [18]. As a property of the process itself, however, it has not been assessed.
Related Process Characteristics. Business process robustness relates to two process characteristics, process flexibility and resilience, both of which concern how processes respond to changes. *Flexibility* refers to "the ability to react to changes" [8], emphasizing active adaptation to changes. Robustness, by contrast, refers to the extent to which a process can withstand changes without requiring adaptation. *Resilience*, in turn, describes a process's ability to restore performance after a disruption [13]. While resilience addresses short-term disruption, robustness emphasizes long-term stability and requires continuous performance within acceptable bounds rather than returning to them after a deviation.
Related Approaches in BPM. Several approaches in BPM relate to our work, either by addressing a closely connected problem using BPS or by employing a similar search strategy but addressing a different problem. The closest line of related approaches concerns *process optimization.* These approaches use BPS to evaluate different parameter configurations, exploring varying activity batching or resource allocation policies, to identify settings that produce

performance-optimal outcomes [20,22,23]. Although methodologically related, these approaches aim to locate an optimal configuration rather than to search for the full space of configurations that satisfy given performance requirements. Another related approach applies *tree-based adaptive search* to find parameter configurations of Markov models [11], which are also commonly used in process prediction [14]. While the search strategy aligns with ours, it is applied to formally verify properties of a given model, whereas our work uses it to assess the robustness of a business process based on simulation. Together, these studies illustrate approaches that are related, but none of them address the problem of identifying the set of parameter configurations under which a process consistently meets its performance requirements.

6 Conclusion

In this paper, we introduced the problem of business process robustness assessment and proposed an approach for its simulation-based approximation that leverages information recorded about the process in an event log. Our approach efficiently identifies the robustness space, i.e., the range of parameter configurations under which a process maintains acceptable performance. The evaluation demonstrated that our approach substantially reduces assessment complexity compared to baseline techniques while accurately identifying parameter combinations under which a process meets its performance targets. The approach provides clear, decision-oriented insights into how parameter variations affect performance and offers a quantitative basis for managing processes under change.

Our approach and its evaluation have some limitations. First, the approach assumes that simulation tools are accurate, which may not always be the case. Second, it also assumes that process performance is monotonic in the process parameters. While this is generally reasonable, there may be cases where the assumption does not hold. For example, adding an extra resource can reduce team productivity for certain process types due to increased communication and coordination effort. Third, the evaluation of our approach is based on a single process, with an illustrative set of parameters and performance indicators. In future work, we aim to relax the monotonicity assumption by exploring search strategies that do not rely on it and remain effective in more general settings. We also seek further optimization by adjusting the number of simulation runs and simulated cases based on the stability of the performance targets over time. Building on these extensions, we plan to evaluate the approach on a broader set of real-life processes and parameters. Despite these limitations, our work provides a foundation for a systematic assessment of business process robustness.

References

1. Van der Aalst, W.: Business process simulation survival guide. In: Handbook on Business Process Management 1: Introduction, Methods, and Information Systems, pp. 337–370. Springer (2014)
2. Van der Aalst, W.: Process Mining: Data Science in Action. Springer, Heidelberg (2016)
3. Becker, L., Sturm, J., Eiden, F., Holtmann, D.: Analyzing and understanding the robustness of bioprocesses. Trends Biotechnol. **41**(8), 1013–1026 (2023)
4. Brocke, J.v., van der Aalst, W., et al.: Process science: the interdisciplinary study of socio-technical change. Process Sci. **1**, 1 (2024)
5. Camargo, M., Báron, D., Dumas, M., González-Rojas, O.: Learning business process simulation models: a hybrid process mining and deep learning approach. Inf. Syst. **117**, 102248 (2023)
6. Camargo, M., Dumas, M., González-Rojas, O.: Automated discovery of business process simulation models from event logs. Decis. Support Syst. **134**, 113284 (2020)
7. Chapela-Campa, D., Benchekroun, I., Baron, O., Dumas, M., Krass, D., Senderovich, A.: A framework for measuring the quality of business process simulation models. Inf. Syst. **127**, 102447 (2025)
8. Dumas, M., Rosa, M., Mendling, J., Reijers, H.: Fundamentals of business process management. Springer (2018)
9. Glodek, M.: Process robustness-a pqri white paper. Pharm. Eng. **26**(6), 1–11 (2006)
10. Gribble, S.: Robustness in complex systems. In: Proceedings Eighth Workshop on Hot Topics in Operating Systems, pp. 21–26 (2001)
11. Junges, S.: Parameter synthesis for Markov models: covering the parameter space. Formal Methods Syst. Des. **62**(1), 181–259 (2024)
12. Kirchdorfer, L., Blümel, R., Kampik, T., van der Aa, H., Stuckenschmidt, H.: Discovering multi-agent systems for resource-centric business process simulation. Process Sci. **2**(1), 4 (2025)
13. Kraus, A., Rehse, J., van der Aa, H.: Data-driven assessment of business process resilience. Process Sci. **1**(1), 4 (2024)
14. Lakshmanan, G.T., Shamsi, D., Doganata, Y.N., Unuvar, M., Khalaf, R.: A Markov prediction model for data-driven semi-structured business processes. Knowl. Inf. Syst. **42**(1), 97–126 (2015)
15. Lamperti, J.: Stochastic processes: a survey of the mathematical theory, vol. 23. Springer Science & Business Media (2012)
16. Laszlo, M., Mukherjee, S.: A genetic algorithm using hyper-quadtrees for low-dimensional k-means clustering. IEEE Trans. Pattern Anal. Mach. Intell. **28**(4), 533–543 (2006)
17. Law, A., Kelton, D.: Simulation modeling and analysis. Mcgraw-hill, NY (2007)
18. Leemans, S.J.J.: Process Mining. Presented at the (2022). https://doi.org/10.1007/978-3-030-96655-3_3
19. López-Pintado, O., Rosenbaum, J., Berx, J., Dumas, M.: Optimos: a tool for simulation-driven business process optimization. In: BPM (Demos/Resources Forum), pp. 126–130 (2024)
20. López-Pintado, O., Rosenbaum, J., Dumas, M.: Optimization of activity batching policies in business processes. In: BPM, pp. 397–414. Springer (2025)
21. Matthews, B.: Comparison of the predicted and observed secondary structure of t4 phage lysozyme. Biochimica et Biophysica Acta (BBA) - Protein Structure **405**(2), 442–451 (1975)

22. Meneghello, F., et al.: Optimizing resource allocation policies in real-world business processes using hybrid process simulation and deep reinforcement learning. In: BPM, pp. 167–184. Springer (2024)
23. Middelhuis, J., Bianco, R., Sherzer, E., Bukhsh, Z., Adan, I., Dijkman, R.: Learning policies for resource allocation in business processes. Inf. Syst. **128**, 102492 (2025)
24. Polyvyanyy, A., Moffat, A., García-Bañuelos, L.: An entropic relevance measure for stochastic conformance checking in process mining. In: ICPM, pp. 97–104 (2020)
25. del Río-Ortega, A., Resinas, M., Cabanillas, C., Ruiz-Cortés, A.: On the definition and design-time analysis of process performance indicators. Inf. Syst. **38**(4), 470–490 (2013)
26. Samet, H.: The quadtree and related hierarchical data structures. ACM Comput. Surv. (CSUR) **16**(2), 187–260 (1984)
27. Stricker, N., Lanza, G.: The concept of robustness in production systems and its correlation to disturbances. Procedia CIRP **19**, 87–92 (2014)
28. Wimsatt, W.: Robustness, Reliability, and Overdetermination (1981), pp. 61–87. Springer Netherlands, Dordrecht (2012)

From Event Logs to Process Performance Indicators via Declarative Process Mining

Alfonso Bravo[1(✉)], Simone Agostinelli[2], Adela del-Río-Ortega[1], Andrea Marrella[3], Manuel Resinas[1], and Jacopo Rossi[3]

[1] SCORE Lab, I3US, Universidad de Sevilla, Seville, Spain
{abllanos,adeladelrio,resinas}@us.es
[2] Universitas Mercatorum, Rome, Italy
simone.agostinelli@unimercatorum.it
[3] Sapienza Universitá di Roma, Rome, Italy
{marrella,j.rossi}@diag.uniroma1.it

Abstract. Organizations increasingly rely on event logs to monitor and improve business process performance, yet identifying meaningful and measurable Process Performance Indicators (PPIs) remains a challenging and largely manual task. Recent work has shown that Large Language Models (LLMs) can support this process by suggesting, operationalizing and computing PPIs against an event log based on process descriptions and business goals. However, such approaches depend heavily on user input quality and are not explicitly grounded in observed process behavior. In this paper, we present Declare2PPI, a fully automated pipeline for generating PPIs directly from event logs. Our approach follows a bottom-up strategy in which declarative process mining extracts behavioral constraints that guide LLM-based generation. The resulting PPIs reflect process-wide execution patterns and are automatically translated into executable definitions and computed to ensure measurability. The approach also supports the exploration of multiple PPI sets and prioritizes indicators based on their relevance and variability. We evaluate the approach through a qualitative comparison with a state-of-the-art top-down method and a quantitative analysis of correctness and robustness. The results show that grounding PPI generation in observed behavior leads to more relevant, non-redundant, and consistently computable PPIs, supporting data-driven performance analysis with minimal effort.

Keywords: process performance indicators · declarative process mining · process mining · large language models

1 Introduction

Measuring the performance of business processes is essential for organizations seeking to assess how well operational execution supports strategic objectives. To this end, Process Performance Indicators (PPIs) provide quantifiable metrics that evaluate the efficiency and effectiveness of one or more processes and

F. M. Maggi et al. (Eds.): BPM 2026, LNBIP 599, pp. 95–111, 2027.
https://doi.org/10.1007/978-3-032-37877-4_6

can be measured from execution data, support decision-making and continuous improvement [21].

The increasing availability of event logs has enabled the automatic computation of PPIs directly from recorded process executions. Existing approaches [17, 18, 24] show that, once a PPI is properly defined and linked to event data, its value can be computed reliably. However, a key challenge remains largely unresolved: identifying which PPIs should be measured in the first place [10]. In practice, PPI definition is often a manual and time-consuming activity that depends on domain expertise [14], while their operationalization requires detailed knowledge of the underlying event data and its structure. As a result, defining and implementing PPIs typically requires close collaboration between business stakeholders, who specify performance needs, and technical experts, who translate these needs into computable definitions. This gap complicates and slows down the adoption of data-driven performance analysis.

A recent approach [1] explores the use of Large Language Models (LLMs) to automatically suggest candidate PPIs and translate them into executable definitions that can be computed from event logs. It follows a top-down strategy, relying on process descriptions, activity and attribute names, and explicitly provided business goals to guide the generation. While promising, this strategy presents two key limitations: the generated PPIs are sensitive to how these inputs are formulated, and they are not explicitly grounded in the behavior observed in the event log. Consequently, the approach may return PPIs that are computable but not necessarily the most informative for the actual process execution, since frequent paths, behavioral regularities, and other execution-specific patterns are not adequately reflected in the inputs.

In this paper, we propose a bottom-up, behavior-aware approach for the automatic generation of PPIs directly from event logs. Instead of relying on external descriptions or predefined goals, our approach derives guidance from the data itself by extracting declarative constraints that summarize the observed process behavior. These constraints, combined with attribute information, are used to steer an LLM in generating candidate PPIs that are inherently grounded in the execution semantics of the process.

Building on this idea, we present Declare2PPI, an approach that takes an event log as its sole input and produces a set of measurable PPIs together with their executable definitions and their computed values. The approach supports the generation of multiple candidate PPI sets under different configurations and includes a selection step to identify balanced, non-redundant, and informative indicators. In addition, it provides mechanisms to support interpretation and analysis, such as grouping PPIs by business perspective, and prioritizing them based on their variability across process executions. To mitigate hallucination effects, only PPIs that can be successfully translated into executable models and computed against the event log are retained, ensuring that all reported indicators are measurable in practice.

We evaluate our proposal combining a qualitative comparison with a state-of-the-art top-down approach and a quantitative analysis that assesses (i) the executability of generated PPIs, (ii) their stability across LLM runs, and (iii) the impact of discovery parameters on the resulting PPI sets.

The rest of the paper is organized as follows. Section 2 introduces the background on declarative process mining. Section 3 discusses related work on PPI definition and computation and LLM-assisted performance analysis. Section 4 presents the Declare2PPI approach and tool. Section 5 reports the evaluation and results. Finally, Sect. 6 concludes the paper with limitations and future work.

2 Background on Declarative Process Mining

To support the bottom-up, behavior-aware generation of PPIs, our approach relies on declarative process mining to extract compact representations of the observed process behavior. Instead of prescribing a fixed control-flow, declarative models describe process behavior through constraints that executions should satisfy, typically defined using templates with formal semantics (e.g., LTL-based) [16]. A widely used formalism is the *DECLARE* language, which expresses temporal relations between activities. For instance, the response constraint Resp(A, B) specifies that whenever activity A occurs, activity B must eventually follow. Such constraints provide a compact and interpretable representation of recurring behavioral patterns in event logs.

Declarative process mining techniques enable the automatic discovery of these constraints from event logs by instantiating templates over observed activities and evaluating them using quantitative measures. In particular, we rely on *support*, defined as the proportion of cases in which a constraint is satisfied [5], to assess how representative a behavioral pattern is. Constraints whose support exceeds a user-defined threshold are retained. This threshold controls the granularity of the resulting model: lower values yield richer but potentially noisier constraint sets, while higher values emphasize more stable regularities.

Declarative constraints are particularly suitable for performance analysis because they are closely aligned with the types of properties typically measured by PPIs, such as ordering relations, temporal dependencies, and compliance conditions. Moreover, their associated support values quantify how representative each pattern is in the event log, providing an additional signal of relevance. This combination of expressiveness and quantification makes declarative constraints a suitable abstraction for linking observed behavior with measurable performance indicators. In our approach, they serve as a bridge between event data and performance indicators. We use the Declare4Py library [8] to perform constraint discovery and evaluation.

3 Related Work

Defining appropriate PPIs for business processes is widely recognized as a complex and resource-intensive task [14]. It requires both identifying relevant indicators from a business perspective and ensuring their computability from event data, which creates a gap between business stakeholders and technical experts [21].

Several approaches support PPI identification at a conceptual level. Frameworks such as Balanced Scorecard [13] and Devil's Quadrangle [10], as well as domain-specific catalogs [23], provide general dimensions and reusable indicators. However, they offer limited guidance on how to derive process-specific PPIs or operationalize them using event data. Guidelines such as SMART [4] and semantic or collaborative frameworks [7] improve the structuring of PPIs, but still rely heavily on manual input and do not ensure alignment with available data.

Complementary work has explored data-driven support for PPI definition and computation. Some approaches use data mining techniques to identify candidate indicators [11,15], while process mining tools and languages such as Celonis PQL [24] and ppinot4py [18] enable their computation. Recent work also combines natural language processing with event log analysis to map textual descriptions to computable indicators [3,17]. However, these approaches still assume that PPIs or their descriptions are externally provided and require technical expertise, thus not addressing the identification problem.

Recent research has investigated the use of LLMs to support performance analysis and indicator generation [25]. In BPM, Large Process Models [12] propose integrating LLMs with process knowledge and traditional tools to translate high-level goals into executable queries over event logs. Closest to our work, the approach in [1] uses LLMs to suggest PPIs and translate them into computable definitions based on process descriptions and business goals. While promising, these approaches follow a top-down strategy that depends on user-provided inputs and does not explicitly leverage behavioral patterns observed in event logs.

In contrast, our work adopts a bottom-up approach in which declarative constraints extracted from the event log guide PPI generation, ensuring that the resulting PPIs are both behaviorally grounded and measurable by construction.

4 Proposal

Declare2PPI is a fully automated approach for deriving PPIs directly from event logs. Next, we detail the guiding principles behind our proposal and the activities of the process, which are summarized in Fig. 1.

4.1 Guiding Principles

The proposal is designed around a set of guiding principles that define its objectives and methodological choices.

Behavioral grounding. PPIs should be grounded in observed process behavior. In Declare2PPI, this grounding is achieved through Declare constraints discovered from the event log, which provide the evidence from which the LLM derives indicators.

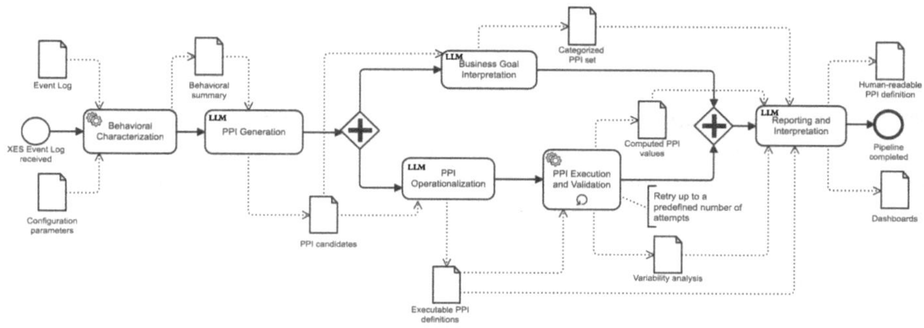

Fig. 1. Overview of Declare2PPI process

Automatic operationalization. The approach should go beyond suggesting indicators by ensuring their measurability. Each generated PPI is automatically translated into an executable definition and evaluated on the event log so that the output is composed of computable indicators rather than only conceptual suggestions.

Alignment with PPI design best practices. PPIs should follow established good practices in PPI definition, aligning with SMART principles. *Specificity* and measurability are ensured by construction, while relevance is supported through behavioral grounding. Time-boundedness is achieved by allowing users to configure the temporal interval for computation. The *achievable* property is out of scope, as it relates to target setting rather than metric definition [1], though it could be addressed by complementary approaches [19]. Moreover, beyond individual indicators, the approach should promote a limited, balanced, and non-redundant set of PPIs covering different perspectives.

Minimal user interaction. The event log is treated as the primary input, and the knowledge required for indicator generation is extracted automatically. This supports exploratory scenarios where users may lack clear objectives or detailed guidance. At the same time, the generated PPIs are intended as a starting point that users can refine, extend, or filter iteratively.

Justification and reliability. PPIs should be accompanied by explanations of their rationale, and only those that can be successfully operationalized and computed are retained. Following [1], reported PPIs are derived from their executable definitions rather than initial descriptions, ensuring alignment between computation and interpretation.

4.2 Declare2PPI Process

Declare2PPI is realized as a process that starts from an event log to derive a set of behaviorally grounded and computable PPIs (see Fig. 1). The process involves six activities that are described next. The artifact is publicly available at [2], including the full process, and the prompts used in the LLMs calls.

Behavioral Characterization of the Process. The process starts by extracting a behavioral characterization of the event log, received as input, consisting of two elements: (i) a set of Declare constraints, discovered with Declare4Py [8], together with their support, and (ii) a summary of activity occurrences, capturing which activities appear and how frequently. The granularity of the characterization, in particular the minimum support threshold, can be configured by the user. This parameter is important because it determines how strict or permissive the behavioral characterization will be. Higher thresholds yield more stable but conservative summaries, while lower thresholds capture more varied patterns.

Generation of Candidate PPIs. Once process behavior has been characterized, the resulting summary is provided to an LLM to generate an initial set of candidate PPIs. The model is guided to propose indicators grounded in the discovered constraints and observed activities, thus capturing process execution patterns rather than arbitrary or purely generic performance concerns. Optionally, event attributes can be included by augmenting the prompt with an attribute profile extracted from the log (including attribute names, data types, representative values, and basic statistics). This enables the generation of indicators that incorporate both control-flow and data perspectives. The prompt does not prescribe specific indicator definitions, but constrains the model to produce a limited and diverse set of indicators, avoiding elements not present in the log-derived summary. The objective is to produce a balanced, non-redundant set of candidate PPIs rather than an exhaustive list.

Business-Goal Interpretation. In this activity, an LLM assigns the generated indicators to higher-level business-goal categories. Rather than requiring goals as input, the model interprets each indicator in terms of broader concerns (e.g., timeliness, compliance, e!ciency), producing a categorized PPI set with an associated rationale as its output. This step adds a semantic organization layer that improves interpretability. Although the approach is bottom-up and log-driven, grouping indicators by business perspective makes the results easier to inspect and compare from a managerial viewpoint.

Operationalization into Executable PPI Definitions. In this activity, an LLM transforms the generated indicators into executable PPI definitions relying on PPINOT [20], thus moving from identification to operationalization. PPINOT provides a formal framework for defining process performance indicators in a computable form, while ppinot4py [18] enables their automatic evaluation over event logs. Rather than producing only descriptions, the model generates Python code based on ppinot4py that can be directly executed on the event log, leveraging the strong capabilities of LLMs for code generation [6]. Using directly ppinot4py instead of an intermediate model like in [1] allows us to compute more complex PPIs. The operationalization is grounded in the data context by explicitly incorporating activity and attribute names from the log in the prompt, ensuring that the resulting definitions refer only to available elements and avoiding invented concepts. The prompt provides the documentation of the ppinot4py library and instructs the model to produce executable code that defines the final metric (e.g., in `kpi_metric`), eliminating the need for intermediate specifications or additional translation steps.

Execution and Validation. In this activity, the ppinot4py-based definitions generated in the previous activity are executed against the event log to compute the corresponding PPI values. If a definition fails to execute, the LLM is prompted again to regenerate the corresponding code, and this process is repeated up to a predefined number of attempts. This enforces measurability by construction: only indicators whose definitions can be successfully executed and produce a valid result are retained. The rest are discarded. In addition, the approach supports variability analysis of computed PPIs over time. Given a user-defined temporal grouping (e.g., weeks or months), variability is assessed using the coefficient of variation, defined as $CV = \frac{\sigma}{\mu}$, where μ and σ denote the mean and standard deviation of the PPI values. This enables the identification of potentially uninformative PPIs due to low variation.

Reporting and Interpretation Support. This activity bridges the gap between executable results and their practical use, presenting PPIs in a form that is both analyzable and interpretable. It consolidates the validated PPIs, their business-goal categorization, the rationale for such categorization, executable definitions, and computed results into an integrated report. Since LLM-based generation may produce different but plausible indicators across executions, the implementation also allows the same configuration to be run multiple times and the successfully computed PPIs to be inspected jointly as an aggregated pool. To improve interpretability, each PPI is accompanied by both its textual ppinot4py description and a human-readable version generated by an LLM. Since ppinot4py definitions can become complex, involving multiple metrics, the LLM rewrites these textual descriptions into more accessible explanations. This rewriting is based solely on the ppinot4py-generated description, not on the original candidate PPI, which is no longer used at this stage to avoid propagating potential hallucinations from earlier steps, and is clearly marked as AI-generated.

Mitigating Hallucinations and Errors. Declare2PPI follows the same strategies as in [1] to mitigate hallucinations and errors that can be introduced by LLMs during the generation of candidate indicators, and their operationalization into executable definitions: (i) All PPIs whose computation leads to an error, either because the computable PPI definition is not well-formed or because it refers to activities or attributes that do not belong to the event log, are discarded, and (ii) the PPI description returned by Declare2PPI is the one obtained from the operationalization, thus we ensure that it accurately matches what is being computed.

4.3 Running Example

We illustrate the Declare2PPI process with the request for reimbursement process of [9]. In this process, an employee submits a declaration, a supervisor reviews and final-approves it, a payment request is then issued, and the case ends when the reimbursement is handled. While this is the expected main path, real cases may also contain delayed payments, missing approvals, or other deviations, which makes the process suitable for monitoring. Table 1 shows an example of how this process is handled by Declare2PPI across its different steps. Crucially, the discovered declarative constraints are not merely reported as a model, but are used as behavioral evidence to guide indicator synthesis. For example, *Init[Declaration SUBMITTED by EMPLOYEE]* directly suggests a *submission-first compliance* indicator (percentage of cases satisfying the constraint). Similarly, *Exactly1[Payment Handled]* grounds a *payment integrity* indicator capturing single-payment consistency. This running example highlights the central idea of Declare2PPI: instead of generating indicators directly from a textual process description alone, the proposal grounds PPIs on behavioral regularities discovered from the log itself, thereby producing indicators that are not only meaningful with respect to the process and its businness goal, but also grounded in the actual observed behavior.

Table 1. Example of a Declare2PPI execution on request for reimbursement log.

Artifact	**Example in the running process**
Declarative constraints	Discovered declarative evidence constraints with support like *Init[Declaration SUBMITTED by EMPLOYEE]* (support ≈ 98.7%), *Response[SUBMITTED, FINAL_APPROVED]* (support ≈ 95.9%), *Chain Response[Request Payment, Payment Handled]* (support ≈ 95.6%), *Exactly1[Payment Handled]* (support ≈ 95.7%).
Candidate PPIs set	Generated indicators such as *Submission-first compliance* (% of cases satisfying *Init[SUBMITTED]*), *Approval completion* (% of submitted cases followed by final approval), and *Straight-through reimbursement* (% of cases where *Request Payment* is immediately followed by *Payment Handled*).
Categorized PPI set with rationale	Category Compliance&integrity. *Payment integrity* is included because *Exactly1[Payment Handled]* has high support & violations indicate duplicate reimbursements or logging inconsistencies.

Executable PPI definitions set	*Straight-through reimbursement* (ppinot4py sketch):
	`count_payment_handled = model.CountMeasure('`concept:name` == "Payment Handled"')`
	`exactly_once = model.DerivedMeasure('c == 1', {'c': count_payment_handled})`
	`kpi_metric = model.AggregatedMeasure(exactly_once, 'AVG')`

Computed PPI values for each PPI	Exemplary computed value for one PPI: *Straight-through reimbursement* = 95.6% of cases where *Request Payment* is immediately followed by *Payment Handled*.
Human-readable PPI definition	Exemplary textual description: "*Percentage of reimbursement cases in which payment handling immediately follows payment request*".

5 Evaluation

We evaluate Declare2PPI with the goal of assessing its effectiveness in generating relevant, measurable, and robust process performance indicators from event logs. The implementation is conducted using our implemented prototype, supporting the full Declare2PPI process. Given the multiple aspects involved in the proposal, we adopt a mixed evaluation combining quantitative and qualitative analyses.

5.1 Tool Implementation

The prototype code of the prototype is available at [2]. The backend is built with FastAPI and exposes an endpoint that accepts an uploaded XES event log together with the main configuration parameters, including the LLM provider and model, the use of attributes, the minimum support for Declare discovery, and the optional time-grouping settings for variability analysis. It then executes the full Declare2PPI process and returns the resulting PPIs, executable PPI definitions, execution outcomes, and variability information in a normalized format.

5.2 Quantitative Analysis

The goal of the quantitative analysis is to assess the following five aspects of the Declare2PPI process: (Q1) stability of the generated PPIs across repeated runs, (Q2) correctness and robustness of operationalization, (Q3) generalizability across LLMs, (Q4) sensitivity of the results to the declarative discovery threshold, and (Q5) semantic alignment between PPI formulations and their executable PPINOT/PPINot4Py translations.

Experimental Setup. The full code, execution results, and analysis can be consulted at [2]. The experiments were repeated over three event logs, with 10 runs per condition to account for the inherent variability of LLM-based generation. Following the experimental philosophy of [1], we use a recent OpenAI GPT model as the default LLM configuration, since these models are widely used for LLM-based generation. In particular, we use OpenAI GPT-5.2 by default. To assess whether the results are dependent on a single LLM provider, Q3 additionally compares this configuration with Anthropic Opus 4.6.

The selected logs correspond to an IT incident management process [22], a request for reimbursement of domestic trips process [9], and a manuscript review management process. These are the same logs used in the evaluation of [1], which allows us to compare Declare2PPI with the prior top-down approach under comparable experimental conditions. Finally, we use a minimum support threshold of 0.8 for Declare discovery as it is a typical configuration in Declare discovery tools and provides a good trade-off between robustness and completeness. However, since this parameter may affect the granularity of the behavioral evidence provided to the LLM, Q4 explicitly analyzes its impact by repeating the experiments with four different support values.

Q1. Declare2PPI consistently produces around ten PPIs per run across all three logs (between 9.0-10.8, with standard deviation between 0.97-1.41), indicating stable output size under repeated executions and a manageable number of PPIs. To assess overlap between runs, we rely on the textual ppinot4py descriptions because they are deterministically generated from the metric that is actually computed. We define two metrics: *exact match*, counting PPIs with identical descriptions, and *approx match*, capturing semantically equivalent PPIs based on identical values and a textual similarity above 0.8. The latter accounts for minor variants in metric definition such as naming differences. Results show low overlap across runs (*exact match*: 0.13-1.33; *approx match*: 1.06-2.53), confirming substantial variability, consistent with prior findings [1]. Rather than a limitation, variability enables the exploration of diverse but valid indicator sets, motivating the use of multi-run strategies to obtain richer and more robust PPI collections.

Q2. To assess the correctness and robustness of the operationalization, we analyze the *failure rate*, defined as the proportion of generated PPIs that cannot be successfully executed (i.e., result in an error or an empty value). We report this metric both before and after applying the retry mechanism. The baseline average failure rate is already low (0.01-0.08), and remains similarly low after retries (0-0.06), indicating that most PPIs are correctly operationalized. In some cases, retries resolve execution errors but lead to empty values, due to conditions not being satisfied in the log. Overall, results indicate that the pipeline reliably produces executable PPI definitions, with the retry mechanism further improving robustness while maintaining transparency.

Q3. To assess the generalizability of the approach across different models, we repeated the experiment in Q2 using another state-of-the-art LLM (Anthropic Opus 4.6). The results are consistent with those obtained with OpenAI models, with a *baseline failure rate* of 0.01 and a final *failure rate* of 0 after retries. These findings suggest that Declare2PPI is robust across different LLMs and not tied to a specific provider, although further experiments would be required to systematically compare their performance across scenarios.

Q4. We repeat the experiment in Q1 with different values for the minimum support threshold $\{0.2, 0.4, 0.6, 0.8\}$. Varying the declarative discovery threshold has limited impact on both the volume and the exact stability of the generated PPI sets. Across the three logs, the pipeline consistently generates approximately the same number of PPIs per run (about 9–11 on average) for all tested support values. Across the three logs, the cross-support *approx match rate* ranges between 0.41 and 0.84, and the average number of heuristic matches per comparison remains in the same order (roughly between 1.66 and 2.65). Notably, these values are consistent with the observation from Q1 that exact match is low (for the three logs, the heuristic match rates were 0.232, 0.310, and 0.129, respectively). This indicates that changing the support threshold does not introduce an additional drop in exact overlap beyond the intrinsic variability of the generation step. Qualitative inspection suggests that different support values mainly lead to alternative formulations and operationalizations of similar monitoring intents (e.g., different aggregations, scope filters, or equivalent ways of encoding ordering/time constraints), rather than to systematically different PPIs. Hence, in our setting, minimum support changes the evidence granularity supplied to the LLM but it does not meaningfully affect the size or exact-set stability of the resulting PPI portfolio.

Q5. Two authors manually reviewed five randomly sampled executions of Declare2PPI (42 PPIs) to evaluate the semantic alignment between (i) candidate PPIs and executable definitions, and (ii) executable definitions and human-readable definitions. Only two mismatches were found in the former and none in the latter, supporting the robustness of the translation steps. The analysis also revealed recurring patterns: the LLM often introduces additional filters during operationalization (e.g., when computing durations) to ensure that referenced events occur, which is sometimes reflected in the human-readable description through clauses such as *"considering only cases where..."*, although this clause is not always surfaced even when the filter is applied. Additionally, ordering constraints are frequently encoded using time measures as a proxy, e.g., by checking that the duration between events is non-negative, indicating that the translation process preserves the intended semantics while revealing minor inconsistencies in how these are communicated in the final descriptions.

Discussion. The evaluation highlights several strengths of Declare2PPI. Across logs, runs, and LLMs, the approach consistently produces a manageable number of indicators, achieves high executability through the repair loop, and maintains strong semantic alignment between generated indicators, executable definitions, and human-readable descriptions. In addition, the approach provides mechanisms to support the interpretation and prioritization of indicators, including business-goal categorization and the assessment of temporal variability through the coefficient of variation. The results also reveal some limitations. The exact set of PPIs generated may vary across executions and LLMs, despite being grounded in the same event log. Furthermore, some declarative semantics are more difficult to express directly in ppinot4py than others, occasionally requiring approximations during operationalization.

Taken together, these observations suggest that the goal of Declare2PPI is not to identify a single definitive set of PPIs for a process. Rather, it provides analysts with a structured and behaviorally grounded starting point for performance analysis, combining automatic generation, operationalization, and interpretation support. In this sense, Declare2PPI should be understood as a guided indicator engineering pipeline rather than as a deterministic PPI discovery mechanism.

5.3 Qualitative Analysis

We conducted a qualitative study to compare the quality of PPIs generated by Declare2PPI with those produced by a state-of-the-art top-down approach, PPIPilot [1]. To the best of our knowledge, PPIPilot is the only existing approach that supports the automatic suggestion and computation of PPIs from event logs, making it the closest comparable baseline. The goal of the study is to assess the perceived usefulness of such PPIs for process monitoring and improvement.

To this end, we designed an online questionnaire composed of two parts. The first part collected participant background information regarding prior experience with business processes and PPIs. The second part focused on the evaluation of the generated PPIs and was structured around two different processes. For each process, participants were presented with two sets of PPIs and asked to evaluate them according to the selected quality criteria using a 5-point Likert scale. The criteria include *relevance*, *coverage*, *actionability*, *non-redundancy*, *usefulness*, and *understandability*. In addition, participants were asked to indicate which set they would prefer to use as a starting point for process performance monitoring and to justify their choice. To reduce potential ordering bias, the presentation of the approaches was alternated across processes. For the first process, Set A corresponded to PPIPilot and Set B to Declare2PPI, while for the second process the order was reversed.

The study was conducted using two event logs with different characteristics in order to assess the robustness of the approach across domains and to engage participants with diverse backgrounds. The selected logs correspond to an IT incident management process (BPI Challenge 2013 [22]) and a request for reimbursement process (BPI Challenge 2020 [9]). The questionnaire was distributed to participants from both academia and industry, including industry experts, researchers, and master's students. In total, 126 invitations were sent and 22 responses were collected from participants based in Italy, Spain, the Netherlands, and Switzerland, corresponding to a response rate of 17.5%. The respondents include five industry experts, 15 academics, and two master's students. All participants joined the study voluntarily and anonymously, with all data treated with strict confidentiality post-collection. The complete results and the analysis performed are available at [2].

To support the qualitative comparison, we conducted statistical analyses using the Wilcoxon signed-rank test, which is appropriate for paired Likert-scale evaluations. Moreover, the criteria were measured using 5-point Likert scales and the sample size was relatively small, so a non-parametric test was more appropriate than tests assuming normally distributed interval data. All the analysis is publicly available at [2]. As depicted in Fig. 2, for *Request for Reimbursement*, Declare2PPI obtained higher mean scores than PPIPilot in five out of the six evaluated dimensions: relevance, coverage, actionability, non-redundancy, and usefulness. PPIPilot only obtained a higher score in understandability. The Wilcoxon analysis confirmed this trend, with statistically significant differences after multiple-comparison correction in all six dimensions favoring Declare2PPI and moderate-to-large effect sizes. At the aggregated level, Declare2PPI achieved a significantly higher overall score than PPIPilot (3.88 vs. 3.33; $p = 0.022$), and 72.7% of respondents preferred the Declare2PPI set.

A similar pattern was observed for *Incident Management*. Declare2PPI obtained higher mean scores in all six dimensions, with the largest differences appearing in coverage and non-redundancy. Although the criterion-level differences did not remain statistically significant after Holm correction, the corresponding effect sizes were again moderate to large, particularly for coverage and non-redundancy. Aggregating all dimensions, Declare2PPI achieved a higher overall score than PPIPilot (3.89 vs. 3.38), with the difference almost approaching statistical significance ($p = 0.054$), while 66.7% of respondents preferred the Declare2PPI indicators. Overall, the statistical analysis reveals a consistent pattern across both processes, systematically favoring Declare2PPI and providing quantitative support for the observed participant preferences.

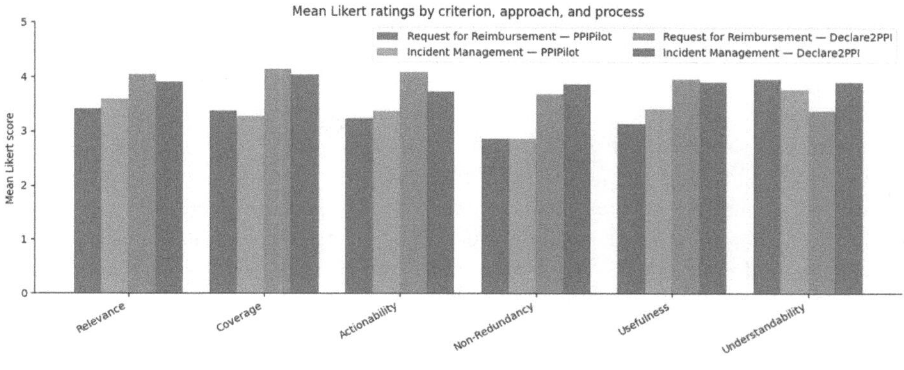

Fig. 2. Likert 1-5 means for the six measures for both evaluated event logs.

Discussion. Taken together, the qualitative results indicate that Declare2PPI is generally perceived as producing *better PPI sets than PPIPilot*, especially in terms of relevance, coverage, actionability, and non-redundancy. The strongest evidence appears in the *Request for Reimbursement* process, where Declare2PPI significantly outperformed PPIPilot in the aggregated evaluation and was preferred by a clear majority of respondents. For *Incident Management*, the pattern remains favorable to Declare2PPI, but the evidence is weaker, suggesting that the relative advantage of the proposal may depend on the characteristics of the process and the event log.

The open-ended responses help explain these results. Several participants described the preferred sets as more *comprehensive*, *concrete*, and better aligned with the critical stages of the process, particularly when they captured end-to-end behavior, bottlenecks, timing between key activities, and exception or rework situations. In this sense, respondents seemed to value PPI sets that provide a more holistic and process-aware view, rather than isolated activity-level counts. For instance, for the first process, one participant noted that: "Set B would be preferable as a starting point because it covers more relevant aspects of the reimbursement process, such as timing between key activities, rejection rates, sequence compliance, and payment handling.". This observation is consistent with the higher scores obtained by Declare2PPI in criteria such as coverage, actionability, and usefulness, and suggests that grounding indicator generation on declarative behavioral evidence can lead to PPI portfolios that are perceived as more informative and decision-orientedm, as illustrated with the comment of another user for the second case: "I find proposal C clearer and more concrete, where the pain points of the process can be clearly seen, where each instance may be deviating, etc.".

At the same time, the comments also reveal an interesting trade-off between *expressiveness* and *understandability*. PPIPilot was perceived as more understandable in the reimbursement case, despite being outperformed by Declare2PPI in most of the remaining criteria. This suggests that the stronger grounding of Declare2PPI on declarative behavioral evidence may lead to PPI sets that are more comprehensive and actionable, but sometimes slightly less straightforward to interpret at first glance. Several respondents noted that some indicators were too long, too tightly coupled to raw log labels, or too complex to understand without additional context. Participants also suggested that a process representation (e.g., a BPMN model or simple trace examples), clearer naming, or grouping indicators by theme (which is implemented in our approach and tool) could improve interpretability. This trade-off is consistent with the nature of the proposal, which prioritizes measurability and behavioral faithfulness over purely surface-level simplicity.

Overall, the qualitative evaluation reinforces the quantitative findings. While the quantitative analysis showed that Declare2PPI produces executable and semantically aligned PPIs, the qualitative results suggest that these indicators are also perceived as more useful for process monitoring than those generated by the baseline. At the same time, the open-ended feedback highlights a key direction for improvement: preserving the richer, more behaviorally grounded nature of the generated indicators while improving how they are presented and explained to end users. Hence, the evidence supports the claim that grounding PPI synthesis on declarative regularities not only improves computability, but also improves the perceived quality of the resulting indicators.

5.4 Threats to Validity

Regarding internal validity, the behavior of LLM-based systems is inherently variable. Differences may arise both from the non-deterministic nature of LLMs and from the use of different models, potentially leading to different sets of PPIs being generated from the same event log. This variability is explicitly addressed in the evaluation. In particular, Q1 analyzes the variability of generated PPIs across repeated executions, while Q3 evaluates the impact of using different LLMs. Moreover, the assessments of operationalization correctness (Q2) and semantic alignment (Q5) are based on multiple executions rather than a single run, reducing the influence of any particular generated PPI set.

A second threat to validity concerns the assessment of semantic alignment (Q5). Determining whether a generated indicator, its executable definition, and its human-readable description preserve the same meaning requires expert interpretation and therefore involves a degree of subjectivity. To mitigate this threat, the analysis was conducted independently by two authors using explicit comparison criteria. While this does not eliminate all subjectivity, it provides a systematic assessment of semantic consistency across the different representations of a PPI.

Regarding external validity, the qualitative evaluation was conducted with a limited but heterogeneous group of participants and focused on two representative processes. As a result, the observed preferences and perceptions should be interpreted as indicative rather than universally generalizable. Additional studies involving a larger number of participants, processes, and application domains would be needed to further assess the generalizability of these findings.

6 Conclusions and Future Work

In this paper, we presented Declare2PPI, a bottom-up approach for the automatic generation of PPIs from event logs. By leveraging declarative process mining, the approach extracts behavioral constraints that serve as grounding evidence to guide LLM-based PPI synthesis. The resulting indicators are not only derived from observed process behavior, but also automatically translated into executable definitions and computed, ensuring measurability in practice. The evaluation combined quantitative and qualitative analyses to assess the properties of the generated PPIs. The results show that Declare2PPI produces indicators that are consistently executable, semantically aligned with their intended meaning, and robust across repeated runs and parameter configurations. In addition, the qualitative comparison with a state-of-the-art top-down approach indicates that behaviorally grounded PPIs are perceived as more relevant, less redundant, and overall more useful for process monitoring and improvement.

These results support the idea that grounding PPI generation in declarative representations of process behavior provides an effective bridge between event data and performance measurement, reducing reliance on user-provided context and enabling the systematic derivation of process-wide indicators. Future work includes improving the presentation and interpretability of the generated PPIs, extending the support for complex declarative patterns, and conducting larger-scale evaluations across additional processes and user groups. In addition, future work should further strengthen the validation of semantic alignment between generated indicators, executable definitions, and human-readable descriptions. This could be addressed through a more systematic evaluation protocol involving independent evaluators, as well as through automatic or semi-automatic LLM-assisted assessment based on explicit alignment criteria, while keeping human supervision for ambiguous cases.

Acknowledgment. This paper is part of the projects PID2021-126227NB-C21, PID2022-140221NB-I00 and PID2024-156482NB-I00 funded by MICIU/AEI/10.13039/501100011033 and by ERDF/EU, and by the Junta de Andalucía (Consejería de Universidad, Investigación e Innovación) and co-funded by the EU through the FEDER Andalucía 2021-2027 Programme under project DGP_PIDI_2024_01609. A. Bravo is supported by PRE2022-102929 financed by MICIU/AEI/10.13039/501100011033 and by ERDF/EU. The research of S. Agostinelli was partially funded by the Ministry of University and Research (MUR) of Italy, grant number FIS-2023-00545 (CUP B53C24009470001). The work of A. Marrella has been supported by the Sapienza projects CAI4PPI and FOND-AIBPM, and the EU Horizon Europe Research and Innovation Programme under Grant 101299128 (euroFMX).

References

1. Agostinelli, S., del-Río-Ortega, A., Goñi-Medina, R., Marrella, A., Resinas, M., Rossi, J.: Automating performance insights: suggesting and computing process performance indicators from event logs. In: CAiSE 2025, pp. 221–237 (2025)

2. Bravo, A., Agostinelli, S., del-Río-Ortega, A., Marrella, A., Resinas, M., Rossi, J.: Supplementary material for Declare2PPI (2026). https://doi.org/10.5281/zenodo.20338704
3. Barbieri, L., Madeira, E.R.M., Stroeh, K., van der Aalst, W.M.: A natural language querying interface for process mining. J. Intell. Inf. Syst. (2022)
4. Cappiello, C., Comuzzi, M., Plebani, P., Fim, M.: Assessing and improving measurability of process performance indicators based on quality of logs. Inf. Syst. **103**, 101874 (2022)
5. Cecconi, A., De Giacomo, G., Di Ciccio, C., Maggi, F.M., Mendling, J.: Measuring the interestingness of temporal logic behavioral specifications in process mining. Inf. Syst. **107**, 101920 (2022)
6. Chen, M., Tworek, J., Jun, H., et al.: Evaluating large language models trained on code. CoRR abs/2107.03374 (2021)
7. Diamantini, C., Potena, D., Storti, E.: SemPI: a semantic framework for the collaborative construction and maintenance of a shared dictionary of performance indicators. Futur. Gener. Comput. Syst. **54**, 352–365 (2016)
8. Donadello, I., Riva, F., Maggi, F.M., Shikhizada, A.: Declare4py: A python library for declarative process mining. BPM (PhD/Demos) **3216**, 117–121 (2022)
9. van Dongen, B.: BPI challenge 2020 domestic declarations. https://doi.org/10.4121/uuid:52fb97d4-4588-43c9-9d04-3604d4613b51 (2020)
10. Dumas, M., La Rosa, M., Mendling, J., Reijers, H.A.: Fundamentals of Business Process Management. Springer, Heidelberg (2018). https://doi.org/10.1007/978-3-662-56509-4_9
11. Estrada-Torres, B., Richetti, P.H.P., del Río-Ortega, A., Baião, F.A., Resinas, M., Santoro, F.M., Ruiz-Cortés, A.: Measuring Performance in Knowledge-intensive Processes. ACM Trans. Internet Technol. **19**(1), 15:1–15:26 (2019)
12. Kampik, T., Warmuth, C., Rebmann, A., Agam, R., Egger, L.N., Gerber, A., Hoffart, J., Kolk, J., Herzig, P., Decker, G., et al.: Large process models: a vision for business process management in the age of generative ai. KI-Künstliche Intelligenz **39**(2), 81–95 (2025)
13. Kaplan, R.S., Norton, D.P.: The Balanced Scorecard: Translating Strategy into Action. Harvard Business School Press, Boston (1996)
14. Neely, A., et al.: Performance Measurement System Design: Developing and Testing a Process-based Approach. Int. J. Oper. Prod. Manage. **20**(10), 1119–1145 (2000), publisher: MCB UP Ltd
15. Peng, W., Sun, T., Rose, P., Li, T.: A semi-automatic system with an iterative learning method for discovering the leading indicators in business processes. In: Proceedings of the 2007 DDDM International Workshop, pp. 33–42. ACM (2007)
16. Pesic, M., Schonenberg, H., Van der Aalst, W.M.: Declare: full support for loosely-structured processes. In: 11th IEEE International Enterprise Distributed Object Computing Conference (EDOC 2007), pp. 287–287. IEEE (2007)
17. Resinas, M., del Río-Ortega, A., van der Aa, H.: From text to performance measurement: automatically computing process performance using textual descriptions and event logs. In: BPM, pp. 266–283. Springer, Cham (2023)
18. Resinas, M., del Río-Ortega, A., Ruiz-Cortés, A.: PPINOT computer and ppinot4py: two libraries to compute process performance indicators. In: ICPM (2021)
19. del Río-Ortega, A., García, F., Resinas, M., Weber, E., Ruiz, F., Ruiz-Cortés, A.: Enriching decision making with data-based thresholds of process-related KPIs. In: Advanced Information Systems Engineering, pp. 193–209. Springer (2017)

20. del Río-Ortega, A., Resinas, M., Cabanillas, C., Ruiz-Cortés, A.: On the definition and design-time analysis of process performance indicators. Inf. Syst. **38**(4), 470–490 (2013)
21. del Río-Ortega, A., Resinas, M., Durán, A., Bernárdez, B., Ruiz-Cortés, A., Toro, M.: Visual Ppinot: a graphical notation for process performance indicators. Bus. Inf. Syst. Eng. **61**(2), 137–161 (2019)
22. Steeman, W.: BPI challenge 2013, incidents (2013). https://data.4tu.nl/articles/dataset/BPI_Challenge_2013_incidents/12693914/1
23. Van Looy, A., Shafagatova, A.: Business process performance measurement: a structured literature review of indicators, measures and metrics. SpringerPlus **5**(1), 1797 (2016)
24. Vogelgesang, T., Ambrosy, J., Becher, D., Seilbeck, R., Geyer-Klingeberg, J., Klenk, M.: Celonis PQL: a query language for process mining. In: Process Querying Methods, pp. 377–408. Springer (2022)
25. Xu, X., Yin, J., Gu, C., Mar, J., Zhang, S., E., Dow, J.L., S.P.: Jamplate: exploring LLM-enhanced templates for idea reflection. In: ICIUI, pp. 907–921 (2024)

GNN4PPM: Multi-Target Predictive Process Monitoring with Relational Graph Convolutional Networks

Ana Costa(✉), Johannes Mäkelburg, and Luise Pufahl

School of Computation, Information and Technology, Technical University of Munich, Heilbronn, Germany
{a.costa,johannes.maekelburg, luise.pufahl}@tum.de

Abstract. Predictive Process Monitoring (PPM) aims at predicting at runtime and as early as possible the future states of a process execution. Common tasks include predicting the next event, the time to completion of a trace, and outcomes. Existing approaches typically consider an event from the perspective of the executed activities along with their timestamps and case identifiers. This leads to the disadvantage that in real-life settings, there is much more information recorded in the event log that is not captured or completely ignored when performing prediction tasks. We introduce GNN4PPM, an approach that predicts all next events along with their complete data payload at once. We represent event information in a heterogeneous knowledge graph that captures the event log as an RDF semantics, and train the embeddings with a Relational Graph Convolutional Network (R-GCN). Our approach is promising in comparison to existing solutions, and experiments with state-of-the-art solutions prove the accuracy and applicability of GNN4PPM in complex settings.

Keywords: Predictive Process Monitoring · Heterogeneous Graph Neural Networks · Relational Graph Convolutional Networks · Next Event Prediction

1 Introduction

Predictive Process Monitoring (PPM) is a branch of process mining that focuses on forecasting the future of an ongoing business process case [8]. Information from historical event data for each case is recorded, such as the order of executed activities, their timestamp, and additional attributes [26]. Instead of looking only at what has happened, predictive monitoring helps in understanding what will happen next, allowing process stakeholders to intervene before issues occur. PPM uses as input data an event log, a structured collection of event traces that record how individual process cases evolve [18]. Each event refers to an executed activity

F. M. Maggi et al. (Eds.): BPM 2026, LNBIP 599, pp. 112–130, 2027.
https://doi.org/10.1007/978-3-032-37877-4_7

and includes at least a case identifier and a timestamp, optionally enriched with event attributes while traces themselves may contain case-level attributes.

One of the tasks in PPM, the next event prediction, focuses on predicting sequences of future activities and related data payloads [8]. These data payloads refer to attributes associated with events, such as the timestamp or the resource performing the next activity. A naive approach is to consider only the recorded activity within its case, along with its timestamps [18]. This leads to the disadvantage that in real-life settings, there is much extra *information recorded in the event log that is not captured or is completely ignored when performing prediction tasks*, such as the resource of the event, its role, the characteristics of the case, and other relevant attributes [14]. For example, in the loan application process at a financial institution, loan acceptance depends on a combination of application characteristics, such as application type, loan goal, requested amount, and other institutional context [15]. Considering only the activity along with the timestamp ordering for prediction provides poor information on a loan acceptance or rejection.

Even though business processes are traditionally represented and discovered in process models as graph structures with their activities, execution relations, and case information [21], predictive approaches do *not properly capture the interrelations between events, case attributes, and event attributes*. As a result, contextual information available in logs is not fully exploited during the prediction. Many deep learning approaches do not take advantage of the dependencies between events since the prefix is embedded into a single vector. Therefore, features of each execution can be overwritten [38].

In predictive process monitoring, the prediction target can vary across different event attributes (e.g., activity, resource, loan risk value), yet most approaches *focus on a single target attribute* [7]. As a result, separate models must be trained from scratch for each attribute on the same event log [43], leading to increased manual effort and training time. When in reality, prediction models are so target-specific that even when using the same event log with a different target, the model needs to be completely retrained [27].

We consider that event and case attributes should participate in triggering the future events and their attributes. We predict, therefore, in our multi-target model, all the next events along with the complete data payload of the event log at once. This enables the model to capture inter-attribute dependencies that are ignored by single-target approaches while reducing the retraining effort required for different prediction targets. To incorporate the contextual relations among event attributes into the learning process [47] and the sequence aspects of the traces [48], our prediction model combines sequential information and graph embeddings into the Graph Neural Network (GNN) training. In principle, neighbours can be aggregated for more informative node embeddings [2] by using a Relational Graph Convolutional Network (R-GCN) architecture, which is designed to handle such heterogeneous graphs [42]. The strong performance of this type of neural network on graph-structured data [9,42] and their ability to handle highly multi-relational data characteristics motivate our interest in exploring the architecture. Our contributions include, therefore (1) the design of a heterogeneous knowledge graph capable of including the complete data payload of the event log and its relations, (2) the combination of sequential information and RDF2Vec embeddings for representation of trace characteristics, and (3) the prediction of all next event attributes using an R-GCN architecture with focus on the attributes neighborhood relations.

The paper is organized as follows. Section 2 and 3 describe recent achievements in literature, preliminaries, and problem definition. Section 4 describes our approach, Sect. 5 describes the experimental setup and evaluation results, and finally, Sect. 6 presents the conclusion and future work.

2 Related Work

Although previous studies in PPM focus on exploring different model architectures for next event prediction [6,39], few approaches consider the prediction of more than one event attribute in the same model. [17] addresses the prediction of the next event, showing that Recurrent Neural Networks (RNNs) can capture sequential dependencies in event logs, but focusing on one event attribute. [43] investigates Long Short-Term Memory (LSTMs) for the prediction of the next event and its timestamp, and [7] studies how to train RNNs with LSTMs to predict the next activity, its timestamp, and the associated resource as separate events. While [44] focuses on predicting one attribute of the event by enhancing a discovered Petri net model with time decay functions, [12,32] propose Convolutional Neural Network (CNN) architectures specifically for Next Activity Prediction (NAP). This architecture is further explored by other authors, who predict the next activity by enriching instance graphs [10] or by adding an explainable layer after the prediction [5].

Recent studies focus on suffix prediction mainly using activity as the event attribute [3]. While [48] addresses full-context-aware (i.e., considering also event attributes) suffix prediction, [35] proposes an LLM method to generate the sequence of future activities from a given prefix. Using a graph-based representation, [36] introduces an explainable approach that handles the next activity prediction, but [14] integrates activities, timestamps, resources, and shows superiority in terms of accuracy compared to the previous method with a similar architecture. The few approaches that extend next event prediction to other attributes within the same model include, therefore, the timestamp of the activity or its resource [14]. The reasoning behind this lack of event attribute consideration emphasizes that current deep learning architectures, such as RNNs and LSTMs, represent the events as sequential data rather than with the multi-target view that GNNs can capture.

Some authors, such as [33,34], recognize the problem by presenting multiple process perspectives in a multi-view setting, but target again, only the prediction of one event at a time. Other authors, such as [25], extend next activity prediction to incorporate other attributes, but disregard time or numerical attributes. Since graph-based approaches facilitate the identification of the behavioral patterns

[38], some authors have investigated GNNs for both next activity prediction [10,46,47] and remaining time [4,16]. Therefore, approaches are still limited to having as a target attribute one or a few attributes.

Several graph-based approaches target next-event prediction using homogeneous graphs [10,46,47]. Two approaches extend this architecture to heterogeneous graphs, which encode multiple node and edge types [14,36]. The advantage of using heterogeneous graphs is that they directly embrace the multi-target representation of event data [36], which is what an event log with multiple event attributes contains. However, these previous approaches opt for Graph Attention Networks (GAT) instead of R-GCN. Since R-GCNs are developed specifically to handle the highly multi-relational data characteristic of realistic knowledge bases [42], we scale the relational data of our graphs by applying this neural architecture, which, to our knowledge, has not been explored before for the multi-target prediction problem within PPM.

3 Preliminaries

This section describes the essential definitions used in the approach, as well as the problem statement. We first introduce the formal definitions of the key data constructs of PPM, such as events, traces, prefixes, and logs, drawing on established literature [31].

Definition 1 (Event). *An* event *is a tuple* $e = (a, cid, t, v_1, \ldots, v_n)$ *where* $a \in A$ *represents the activity label,* $cid \in C$ *is the unique identifier for the case, also called* case ID, $t \in \mathcal{T}$ *represents the event's timestamp and* $v_1, \ldots, v_n$ *are the event-specific attributes, where* $\forall 1 \leq k \leq n : v_k \in \mathcal{V}_k$ *is the domain of the* k^{th} *attribute. These variables create a multi-dimensional space for the universe of events* $\mathcal{E}$.

Definition 2 (Trace). *A* trace $\sigma \in \mathcal{E}^*$ *is a finite sequence of unique events* $\sigma = \langle e_1, e_2, \ldots, e_{|\sigma|} \rangle$, *where* $|\sigma|$ *indicates the number of events in the trace, also called the* trace length. *The events are ordered chronologically and share the case identifier cid.*

Similarly to the *event-specific attributes*, each trace contains a set of *trace-specific attributes*, which remain constant throughout the entire trace. Every trace mandatorily includes the *case ID*, and may optionally include trace-specific attributes $w_1, \ldots, w_m$, where each $w_j \in \mathcal{W}_j$ belongs to the domain of the jth attribute. We define the set of all traces by $\mathcal{S} \subseteq \mathcal{E}^*$, with each trace $\sigma \in \mathcal{S}$.

Definition 3 (Prefix and Suffix). *The* prefix *and* suffix *are specific types of partial traces, obtained by employing the* $hd^i(\sigma)$ *and* $tl^i(\sigma)$ *functions, respectively. This is realized with a selection operator* $(\cdot) : \sigma(i) = e_i, \forall i \in [1, |\sigma|] \subset \mathbb{N}$, *such that* $hd^i(\sigma) = \langle e_1, e_2, \ldots, e_{\min(i,|\sigma|)} \rangle$ *and* $tl^i(\sigma) = \langle e_y, e_{y+1}, \ldots, e_{|\sigma|} \rangle$, *where* $y = \max(1, |\sigma| - i + 1)$.

Definition 4 (Event Log). *An* event log *L is a set of traces denoted by $L = \{\sigma_1, \sigma_2, \ldots, \sigma_z\}$ and $\sigma_i \in \mathcal{S}$ for $1 \leq i \leq z$, $z \in \mathbb{N}^+$. Each σ_i is a trace as previously defined.*

In this work, we make use of knowledge graphs derived from an event log. Following the Resource Description Framework (RDF) standard [40], a knowledge graph (KG) [23] is a collection of *(subject, predicate, object)* triples. Subjects and predicates are Internationalized Resource Identifiers (IRIs), while objects can either be IRIs representing entities or literals with a specific datatype (e.g., strings or integers).

Definition 5 (Knowledge Graph). *A knowledge graph is a set of triples $\mathcal{KG} \subseteq (IRI \times IRI \times (IRI \cup Literal))$, where each triple $(s, p, o) \in \mathcal{KG}$ consists of a subject s, a predicate p, and an object o.*

In the context of logs, events, cases, and categorical attributes are represented as entities, whereas numerical values, such as timestamps, are represented as literals.

Problem Definition. Given an event log $L = \{\sigma_1, \sigma_2, \ldots, \sigma_z\}$ where each trace $\sigma = \langle e_1, e_2, \ldots, e_{|\sigma|}\rangle$ consists of events $e_i = (a_i, cid_i, t_i, v_{1_i}, \ldots, v_{n_i})$, and an event prefix $hd^i(\sigma)$ with $i < |\sigma|$. The task is to learn a predictive function $f : hd^i(\sigma) \rightarrow \hat{e}_{i+1}$ that predicts the next event $\hat{e}_{i+1}$, such that $\hat{e}_{i+1}$ approximates the ground-truth e_{i+1}. The prediction target includes the activity label, timestamp, and all event-specific attributes.

4 GNN4PPM

In this section, we present our graph-based approach, GNN4PPM. The solution predicts the next events along with the complete data payload of the event log by following the steps presented in Fig. 1. In the offline phase, an event log is first loaded, preprocessed, and transformed into a knowledge graph using an event log ontology and mapping rules. RDF2Vec embeddings are generated from the knowledge graph structure to capture semantic relationships between entities. These embeddings serve as initial node features for an R-GCN, which learns to predict event attributes by propagating information through the graph's type edges. The trained R-GCN model is applied to validation cases in the online phase to generate predictions.

4.1 Event Log Knowledge Graph Construction

To enable relational learning over event logs, we transform the tabular event log representation into a multi-relational knowledge graph. This representation preserves the structural dependencies among cases, events, and attributes and enables relational message parsing in the R-GCN model. In contrast to purely

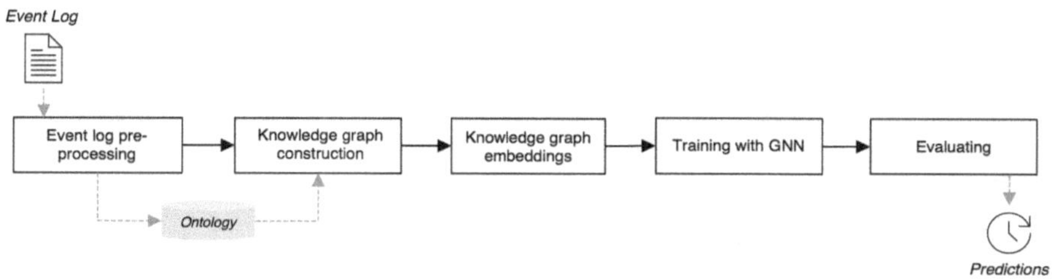

Fig. 1. Pipeline of GNN4PPM

sequential encoding, the graph structure allows explicit modeling of temporal dependencies between attribute values and cross-attribute interactions.

The first step of the pipeline transforms the event log into a structured knowledge graph while preserving its semantics and temporal relations. During preprocessing, we extend the classical definition of an event e by introducing a unique event identifier $eid \in \mathbb{N}^+$, reflecting the execution order within a case. Modeling eid as an explicit node in the KG, as shown in Fig. 2, prevents identical activity labels across different cases from being merged and ensures that relational dependencies are captured at the level of concrete event instances. An event is thus represented as $e = (eid, cid, a, t, v_1, \ldots, v_n)$.

All other components of the tuple follow Definition 1. Furthermore, the *event attributes* $v_1, \ldots, v_n$, and *trace attributes* $w_1, \ldots, w_m$, are classified into *numerical* or *categorical* to facilitate the embedding and training.

Event Log Knowledge Graph. Given an event log $L = \{\sigma_1, \ldots, \sigma_z\}$, we construct an event log knowledge graph $\mathcal{KG}_L$ consisting of RDF triples. The graph represents the structural relations between cases, events, and attribute values contained in the event log.

Each trace $\sigma = \langle e_1, e_2, \ldots, e_{|\sigma|} \rangle$ consists of events $e = (eid, cid, a, t, v_1, \ldots, v_n)$ and trace-level attributes $(cid, w_1, \ldots, w_m)$, as defined in Sect. 3. The knowledge graph is constructed by generating RDF triples describing event attributes, trace relations, and temporal dependencies between events. For each event e, we create triples linking the event identifier to its activity label, timestamp, and attribute values:

$$T_e = \{(eid, r_a, a), (eid, r_t, t)\} \cup \{(eid, r_{v_k}, v_k) \mid 1 \leq k \leq n\}, \tag{1}$$

These triples connect the event node to the nodes representing its activity label and attribute values. Trace-level relations connect events to their corresponding case ID and represent trace attributes:

$$T_\sigma = \{(cid, r_{w_j}, w_j) \mid 1 \leq j \leq m\} \cup \{(cid, r_{\text{contains}}, eid_i) \mid e_i \in \sigma,\ 1 \leq i \leq |\sigma|\}, \tag{2}$$

These triples associate each case with its trace attributes and the events belonging to that trace. Temporal ordering within a trace is represented through a *directly-follows* relation. For each pair of consecutive events e_i and e_{i+1} in a trace σ, we add triples that connect the values of the same attribute types across both events: For every attribute $x \in \{a, t, v_1, \ldots, v_n\}$ we denote by $x(e_i)$ the value of attribute x in event e_i. The set of directly-follows triples is then formally defined as:

$$T_{df} = \{(x(e_i), r_{\mathrm{dF}}, x(e_{i+1})) \mid e_i, e_{i+1} \in \sigma,\ i < |\sigma|,\ \sigma \in L,\ x \in \{a, t, v_1, \ldots, v_n\}\}. \tag{3}$$

This relation captures how attribute values evolve from one event to the next within the same trace. Modeling directly-follows at the attribute level, enables capturing temporal dependencies not only between events but also between evolving attribute values. Unlike sequential models that receive a prefix as input, GNN4PPM encodes prefix context implicitly. The knowledge graph contains, e.g., *directlyFollows* edges linking consecutive attribute values across events, so the R-GCN aggregates multi-hop neighborhood information from earlier events into the embedding during message passing. The prediction head then operates on the resulting embedding of the node, which carries the accumulated context of the preceding trace. This design allows learning how specific attribute combinations influence subsequent attribute values. Additionally, type triples are added to associate each instance with its corresponding domain class. The resulting event log knowledge graph is given by:

$$\mathcal{KG}_L = T_e \cup T_\sigma \cup T_{df}. \tag{4}$$

The knowledge graph can be interpreted as a directed labeled graph $G_L = (V_L, E_L)$ where the nodes V_L correspond to the entities and literals appearing in the triples and the edges E_L correspond to the triples themselves.

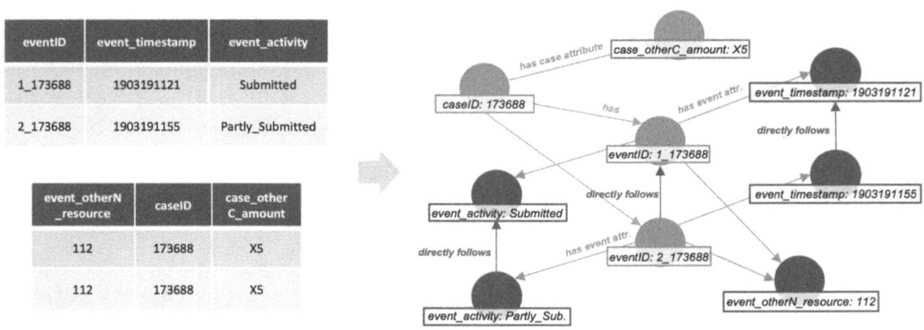

Fig. 2. Knowledge graph representation of the entity relations of an event log.

Besides the entity relations shown in Fig. 2, the class relations in the event log on the left indicate that *eid* contains its event attributes, and *cid* contains *eids* and their case attributes. In the figure, the edges indicate the direction, and the edge labels indicate the relation type. Additionally, all other attributes of the log were classified into *other numerical* or *other categorical.* Although their instances are unique, this classification was fundamental for the embeddings and loss calculation in the neural network. The class hierarchy and conceptual relations of the event log are defined through an ontology inspired by the XES standard [20]. The ontology defines node types *CaseID*, *EventID*, categorical attribute nodes, and numerical literals, connected by structural and temporal predicates. Instance-level triples are generated by YARRRML-Mappings [45] and RML-Mappings [13], which transfer each combination of instances of the event log into RDF triples, allowing the event data to be represented as a knowledge graph. Since a detailed explanation of the ontology is not the purpose of this work, a formalization of the ontological structure is left for future work.

The resulting knowledge graph serves as the structured input for embedding generation. Moreover, the constructed graph can be related to object-centric settings, in which object types correspond to attribute types with their own relational structure. Both approaches make use of heterogeneous graphs, which are well-suited for expressing inter-dependencies between objects.

4.2 Knowledge Graph Embeddings

Knowledge graph embeddings are created using RDF2Vec, which learns low-dimensional representations of entities by performing biased random walks over the RDF graph and training a Word2Vec model on the resulting sequences [41]. This embedding technique was chosen because it captures multiple paths from each entity, encoding rich contextual information from the neighbouring nodes and predicates. Since the learned embeddings preserve similar entities close to each other [29], event attributes for each event instance can be kept closer in the vector space, making RDF2Vec a suitable embedding for the given prediction task.

Performing biased walks instead of random walks allows us to prioritize event attributes and directly follow relations according to breadth-first and depth-first parameters [19]. This is important when representing our event logs since uniform random walks would pick any neighbor with equal probability without following the temporal sequence of traces. After the biased walks are performed on the graph, we train the resulting sequences with Continuous Bag of Words (CBOW) to predict the center token from its surrounding context window [28]. For example, in a walk such as [*caseID:173688, has, eventID:1, directly follows, eventID:2*], CBOW learns *eventID:1* from the context of both its case and its next event simultaneously (see Fig. 2).

Since the embedding quality depends on walk and CBOW settings (e.g., walk length, window size, embedding size), we perform unsupervised model selection by clustering the embeddings with K-Means and choosing hyperparameters that maximize the silhouette score on the training set, so that similar entities land close together in the embedding space. The silhouette-based selection is performed exclusively on training-set entities, as both the random walks and the

Word2Vec training are restricted to the training graph, ensuring that no information from validation cases influences the embedding hyperparameter choice. These pre-trained embeddings are stored with a corresponding *id* mapping and serve as initial node features for training with the R-GCN.

4.3 Training with R-GCN

This section describes the R-GCN architecture used for representation learning and the multi-target objective employed for model training.

Model Architecture. The RDF2Vec entity embeddings serve as initial node features for the R-GCN encoder. During training, R-GCN receives these embeddings along with the graph structure (edge index and edge types). We use R-GCN because the event log knowledge graph contains multiple semantic relation types, and R-GCN explicitly models relation-specific message passing in heterogeneous graphs. Unlike conventional heterogeneous GNN frameworks that define distinct aggregation operators (e.g., GraphSAGE or GAT) per relation type, our model applies a uniform aggregation function across all edge types. Relation-type heterogeneity is captured through the relation-specific transformation matrices, which are parameterized via basis decomposition to keep the number of learnable parameters controlled across the large number of RDF predicate types in the knowledge graph. Architectures without explicit relation modeling would treat all edges uniformly and cannot distinguish between temporal and contextual dependencies.

The encoder consists of three R-GCN layers with layer normalization, residual connections, and ReLU activations, ϕ. Three layers were chosen to allow sufficient neighborhood aggregation depth, capturing event, case, and attribute-level context. A residual connection from the first to the second layer, combined with layer normalization after each of the first two layers, preserves node-specific information and stabilizes training, mitigating the risk of over-smoothing. In the forward pass, each R-GCN layer aggregates messages from neighboring nodes, with separate transformation matrices for each relation type. To handle the multiple relation types in our event log knowledge graphs, we employ basis decomposition, which represents relation-specific weight matrices as linear combinations of a smaller set of basis matrices. Formally, the forward propagation of each R-GCN layer follows the update rule proposed by [42]. For a node i at layer l, the updated representation $h_i^{(l+1)}$ is computed as:

$$h_i^{(l+1)} = \phi\left(\sum_{r\in\mathcal{R}} \sum_{j\in\mathcal{N}_i^r} \frac{1}{c_{i,r}} W_r^{(l)} h_j^{(l)} + W_0^{(l)} h_i^{(l)}\right), \tag{5}$$

where $\mathcal{N}_i^r$ is the set of neighbors of i under relation r, $W_r^{(l)}$ is the relation-specific transformation matrix, $W_0^{(l)}$ is the self-loop weight, $\phi(\cdot)$ is the ReLU activation function, and $c_{i,r} = |\mathcal{N}_i^r|$ is a degree-based normalization constant equal to the number of neighbors connected to node i through relation r. This formulation enables the model to learn different aggregation patterns for different relation types, making relations such as *directly-follows* or *has-resource* more meaningful during information propagation.

Training Objective. The model is trained in a multi-target setting, predicting all categorical and numerical attributes of the next event. Since these attributes differ fundamentally in their type and scale, a unified loss function would not adequately capture their prediction objectives; therefore, multiple loss functions are required.

For categorical attributes, we use cross-entropy loss with inverse frequency class weighting to address class imbalance [22]. Cross-entropy is well-suited for multi-class prediction, as it directly optimizes the likelihood of the correct class while producing calibrated probability estimates. Event logs typically have an unbalanced attribute value distribution. Inverse frequency weighting compensates these imbalances and prevents the model from biased predictions. For a categorical attribute k, the loss, $\mathcal{L}_{\mathrm{CE}}$, is defined as:

$$\mathcal{L}_{\mathrm{CE}}^{(k)} = \mathrm{CE}\Big(\hat{y}^{(k)}, y^{(k)}, w_k\Big)\,, \tag{6}$$

where w_k denotes the inverse-frequency class weights. For the timestamp, we apply the mean squared error (MSE) on log-transformed and normalized values. Time differences in real-world processes are often unbalanced. Therefore, applying a logarithmic transformation compresses large time gaps and prevents them from dominating during training [30]. After normalization, we define the timestamp loss $\mathcal{L}_t$ using MSE, which provides a simple and effective regression objective for continuous time prediction, since the values are now distributed more balanced.

For other numerical attributes, we employ Huber Loss [24] to improve robustness against outliers. For a numerical attribute k, the corresponding loss $\mathcal{L}_{\mathrm{Huber}}^{(k)}$ is defined as:

$$\mathcal{L}_t = \mathrm{MSE}\Big(\hat{y}^{(t)}, y^{(t)}\Big)\,, \qquad \mathcal{L}_{\mathrm{Huber}}^{(k)} = \mathrm{Huber}\Big(\hat{y}^{(k)}, y^{(k)}\Big)\,. \tag{7}$$

Unknown or missing target values are masked and excluded from gradient computation. The total loss is computed as a weighted aggregation of the individual task losses:

$$\mathcal{L}_{\mathrm{total}} = \mathcal{L}_{\mathrm{CE}}^{(\mathrm{act})} + \lambda_t\,\mathcal{L}_t + \lambda_c \sum_{k\in\mathcal{C}} \mathcal{L}_{\mathrm{CE}}^{(k)} + \lambda_n \sum_{k\in\mathcal{N}} \mathcal{L}_{\mathrm{Huber}}^{(k)}, \tag{8}$$

where $\mathcal{C}$ and $\mathcal{N}$ denote the sets of categorical and numerical attributes, respectively, and λ_t, λ_c, and λ_n are task-specific balancing coefficients optimized via hyperparameter search.

5 Evaluation

In this section, we evaluate the performance of GNN on several real-world event logs. Our evaluation addresses: (Q1) How does GNN4PPM perform on the next activity prediction task compared to the state-of-the-art methods; (Q2) How accurately can the model predict the additional attributes of the next event; and (Q3) How does each event attribute influence the prediction performance. We first describe the datasets and experimental setup, then compare with state-of-the-art approaches. We then analyze the results of our defined multi-target task, and lastly, investigate the influence of individual event attributes on the prediction performance. The source code is available online[1] [11].

5.1 Datasets

We used six real-life event logs to evaluate our solution (see Table 1). All used logs are publicly available at 4TU Centre for Research [1] and contain processes performed in different domains. The datasets BPIC12A, BPIC12W, and BPIC12WC describe the application process for a personal loan or overdraft at a Dutch Financial Institute. The first contains the complete lifecycle traces of the subprocess 'Application', the second contains the subprocess 'Work', and the third contains the complete lifecycle traces of the subprocess 'Work'. The event log BPIC13O consists of the open problems in the problem management process at VolvoIT Belgium, which includes the activities required to diagnose the root cause of incidents and to secure the resolution of those problems. While BPIC20P contains events pertaining to two years of travel expense claims, BPIC20R contains events related to the request for payments.

Table 1. Structural characteristics of the event logs and constructed knowledge graph information.

Dataset	#Traces	#Events	#Variants	#Case attr.	#Event attr.	Triples	Entities	Predicates
BPIC12A	13,087	60,849	17	1	3	597,071	144,870	10
BPIC12W	9,658	170,107	2921	1	3	1,554,389	360,122	10
BPIC12WC	9,658	72,413	2963	1	3	685,039	164,741	10
BPIC13O	819	2,351	108	3	6	32,925	5,985	14
BPIC20P	10,500	56,437	896	8	5	254,401	59,081	16
BPIC20R	6,886	36,796	76	3	4	368,206	110,655	7

5.2 Experimental Setup

This section describes the experimental configuration, including data handling, training procedure, hyperparameter optimization and evaluation. All experiments were conducted on a compute node using one NVIDIA H100 GPU with 96 GB of memory.

[1] https://github.com/ana-luisa-costa/GNN4PPM.git.

Data Handling and Splits. The prediction task is formulated as event-to-event prediction. For each consecutive pair (e_i, e_{i+1}) within a trace, the model uses the graph representation at event e_i to predict the attributes of e_{i+1}. Data were split at the case level (80% training, 20% validation) using a fixed random seed, following the random case-level split convention used in benchmark studies in this domain [37]. All embeddings and graph structures were constructed exclusively from training cases.

Hyperparameter Optimization. Model hyperparameters were selected using automated Bayesian optimization with 15 trials and early pruning to discard underperforming configurations. The search space included: hidden dimension $\{64, 128, 256\}$, prediction head hidden dimension $\{64, 128, 256\}$, number of basis functions $[5, 15]$, dropout rate $[0.1, 0.5]$, learning rate $[10^{-4}, 10^{-2}]$ (log scale), and task-specific loss weights for multi-task balancing. Each trial was trained for 100 epochs, and the configuration with the lowest validation loss was selected for final training. The task-specific loss balancing coefficients were optimized over the following search ranges: $\lambda_t \in [0.1, 1.0]$, $\lambda_c \in [0.3, 1.0]$, and $\lambda_n \in [0.1, 0.5]$.

RDF2Vec Configuration. The walk length, CBOW context window size, and embedding dimensionality were selected through unsupervised Bayesian optimization, maximizing the silhouette score on K-Means training embeddings. The corresponding search spaces were $[6, 12]$ for walk length, $[3, 8]$ for context window size, and $\{64, 128, 256\}$ for embedding dimensionality.

Training Configuration. Training was performed using the Adam optimizer with cosine annealing. We applied early stopping with a patience of 30 epochs, gradient clipping with a maximum norm of 1.0. Models were trained for up to 100 epochs.

Evaluation Metrics. Categorical attributes are evaluated using macro-averaged precision, recall, F1-score, and accuracy. Numerical attributes are evaluated using Mean Absolute Error (MAE) in normalized log-space. All reported metrics were computed exclusively on the validation set.

5.3 Comparison Against State-of-the-Art

To put our performance into perspective and address (Q1), we compare our approach against a set of state-of-the-art methods. All compared approaches focus on the NAP task since they do not allow for the prediction of the other attributes in the same training. In contrast, our approach predicts the next event together with its associated attributes. However, to ensure comparability with existing work, we report NAP accuracy in this section.

The most recent approaches [14,35,36,39] were implemented and executed by us for all datasets. The older approaches [7,17,43,44] were run by us for datasets BPIC13O, BPIC20P, and BPIC20R. For the other three datasets, we rely on the results of [37], who performed an extensive benchmark on these logs. For each method, we followed the training configuration and evaluation protocol recommended in the respective publication, including the specified data splits, cross-validation procedures, and, where available, hyperparameter settings. For datasets on which certain approaches were not originally evaluated, we applied the same configuration strategy proposed by the authors and conducted additional experiments accordingly. In these cases, hyperparameters were selected according to published guidelines or the default settings provided in the official implementations. Table 2 summarizes the accuracy results across all benchmark datasets. Other measures such as F1 score, precision, and recall, are available in the data file online[2].

Table 2. Accuracy of the Next Activity Prediction Task compared to different approaches. Best result per dataset highlighted in bold; second-best is underlined.

Approach	BPIC12A	BPIC12W	BPIC12WC	BPIC13O	BPIC20P	BPIC20R
SEPHIGRAPH [14]	67.8	78.6	46.6	20.2	73.8	76.5
LUPIN [35]	19.3	35.7	21.2	24.4	39.7	51.1
PROPHET [36]	76.7	**89.2**	<u>80.3</u>	62.0	83.8	87.8
BEST [39]	76.1	69.0	40.0	48.8	<u>86.9</u>	89.0
CAMARGO [7]	75.9	76.4	68.9	<u>77.2</u>	84.4	**91.9**
EVERMANN [17]	75.8	75.4	67.5	65.1	**88.9**	<u>90.5</u>
TAX [43]	<u>79.5</u>	85.3	69.8	60.9	84.4	89.0
THEIS [44]	65.5	<u>86.2</u>	80.1	59.3	83.2	89.7
Our approach						
GNN4PPM	**88.8**	80.8	**83.2**	**78.1**	80.4	87.8

Overall, GNN4PPM achieves the best performance on BPIC12A, BPIC12WC, and BPIC13O, and remains competitive on the remaining three. BPIC13O is the smallest dataset in terms of traces and events, but it contains a relatively high number of case and event attributes relative to its size. On this dataset, GNN4PPM outperforms all competing methods. Both BPIC12A and BPIC12WC, while larger, exhibit a moderate number of event attributes and a high number of traces. GNN4PPM also achieves the strongest performance on these datasets, indicating that the approach performs particularly well on small to medium-sized event logs. In contrast, BPIC12W contains more events and a low number of event attributes. On this dataset, PROPHET achieves the best performance. This suggests that while certain approaches may benefit from very large logs, our approach maintains stable performance across different scales. BPIC20P and BPIC20R contain a larger number of activities and more case and event attributes. On these datasets, Evermann and Camargo achieve the strongest results, while GNN4PPM remains competitive, especially on BPIC20R.

The results show that GNN4PPM performs well across different log sizes with multiple event attributes, while remaining competitive on larger, more complex datasets. These results demonstrate that predicting the complete data pay-

[2] https://figshare.com/s/cde2c35c6ab3f7ca5422.

load positively affects activity prediction performance while achieving competitive results across all datasets.

5.4 Multi-target Prediction

Table 3. Multi-target prediction results across datasets given as accuracy for event attributes and as normalized MAE for timestamp.

Dataset	activity	resourcecountry	org_resource	lifecycle	org_group	org_role	timestamp (MAE)
BPIC12A	88.75	64.20	100	-	-	-	0.3184
BPIC12W	80.34	56.76	83.46	-	-	-	0.5476
BPIC12WC	83.19	14.83	100	-	-	-	0.7667
BPIC13O	78.15	63.16	67.04	97.78	87.14	91.85	0.7100
BPIC20P	80.44	99.38	-	-	88.67	-	0.3785
BPIC20R	87.79	99.58	-	-	90.53	-	0.6400

To address the question (Q2) and the problem definition in Sect. 3, our approach predicts the next event together with its complete data payload. Figure 3 shows the number of distinct attributes per dataset and Table 3 summarizes the prediction results across all datasets. Since not all datasets contain the same event attributes, some entries are not applicable.

For the `timestamp` prediction results the prediction with lowest errors are observed for BPIC12A (0.3184) and BPIC20P (0.3785), while BPIC13O (0.7100) and BPIC12WC (0.7667) have relatively large errors. This can be due to how regular the intervals between timestamp is in each process. While BPIC12A and BPIC20P have more consistent intervals between events, BPIC13O and BPIC12WC have times that range from seconds to days, making them harder to predict. Another influencing factor is the number of targets in the prediction task, as BPIC13O has the most categorical prediction heads, so the encoder must balance multiple classification gradients.

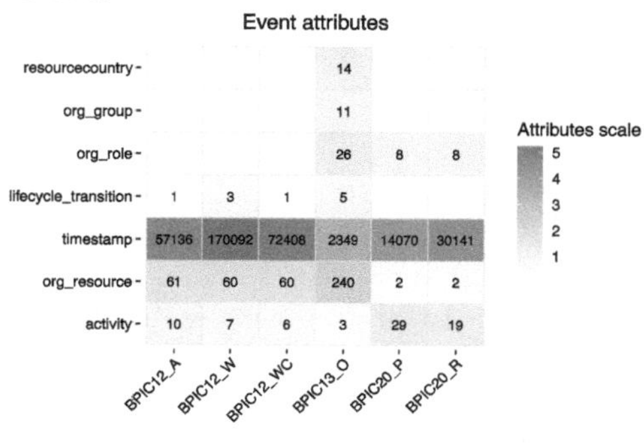

Fig. 3. Number of distinct event attributes

For predictions in which attributes are classes, the number of distinct values highly influences the results. In datasets like BPIC20P and BPIC20R, where the number of distinct `org_resource`-values is low, the accuracy is very high (99.38% and 99.56% respectively), in comparison to datasets with many distinct resources. The same effect is observed in the prediction of the `lifecycle` and `org_role` attributes. BPIC13O shows the lowest accuracy among the datasets, but also has the most distinct values for these attributes. Overall, the results indicate that the prediction difficulty of individual attributes is influenced by attribute cardinality, the number of distinct values per class, and consistency on intervals between events, indicating that datasets with lower concept drift would perform better in the multi-target task. Attributes with few distinct identifiers can be predicted reliably, whereas those with very large numbers, such as resources or role, remain challenging.

5.5 Attribute Influence

To address the question (Q3), we analyze how individual event attributes influence the prediction performance. Figure 4 illustrates the influence of individual event attributes on the activity prediction accuracy across the datasets. The values (Δ) represent the difference in accuracy when a specific attribute is not in the model. For the comparison, the attribute was removed from the raw dataset, and the model was retrained with the new setting. Negative values, therefore, indicate that the attribute contributes positively to prediction performance, while positive values indicate that removing the attribute improves accuracy.

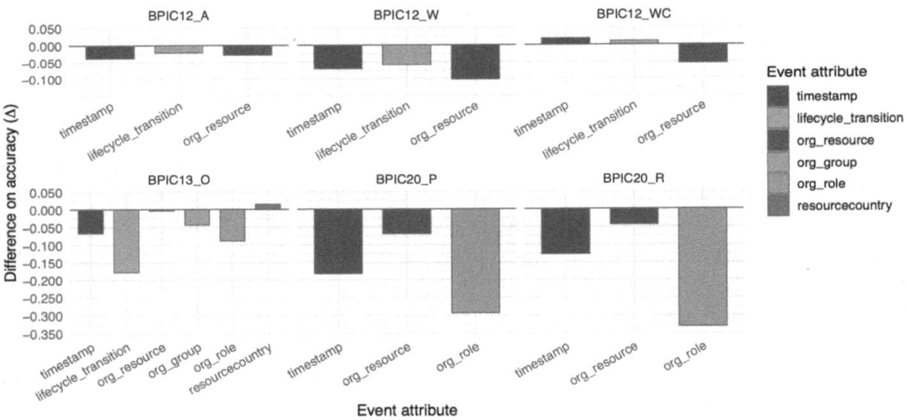

Fig. 4. Influence of each event attribute on activity prediction accuracy when removed from the model

Across most datasets, the `timestamp` attribute positively influences prediction accuracy. Removing it leads to a decrease in performance, particularly in BPIC20P ($\Delta = -0.1821$) and BPIC20R ($\Delta = -0.1272$). The `org_resource` and `org_role` attributes also improves prediction performance. For example, removing the first reduces accuracy in BPIC12W ($\Delta = -0.1027$) and BPIC20P ($\Delta = -0.0694$), while removing the second results in large decreases in accuracy ($\Delta = -0.2937$ and $\Delta = -0.3314$). At the same time, some attributes show small positive values, meaning that removing them slightly improves prediction performance. This effect appears, for example, in BPIC12WC, where removing `timestamp` ($\Delta = 0.0182$) or `lifecycle_transition` ($\Delta = 0.0105$) slightly increases accuracy. These attributes either have high cardinality or weak correlations with the control flow, which may introduce noise into the model.

The results indicate that temporal information, resources, and organizational roles provide the strongest predictive targets, while some higher-level attributes may contribute little information. Therefore, including case and event attributes as context in the knowledge graph improves the prediction performance of the next activity prediction task, achieving better accuracy across many datasets when compared with an architecture that considers fewer attributes.

6 Conclusion

We introduce GNN4PPM, an approach to multi-target predictions that captures contextual relations among event and case attributes that sequential models do not preserve. Despite these promising results, several limitations must be acknowledged. Experiments showed that GNN4PPM performs well on small to medium-sized event logs but is less competitive on larger logs that contain fewer attributes, such as the BPIC12W. Furthermore, prediction accuracy is directly related to attribute cardinality, since attributes with many distinct values, such as resources and roles, remain difficult to predict regardless of the model architecture. The timestamp prediction is also sensitive to the process irregularity of time intervals. Additionally, a direct baseline comparison for the multi-target task is not possible since other approaches do not predict the complete data payload of the next event. Finally, the evaluation is further limited by the absence of an independent test set and by being conducted on six publicly available event logs, all drawn from a limited set of domains, which restricts the generalization of the findings to other processes where attribute structures differ.

Future work includes first extending GNN4PPM from next-event to suffix prediction, which allows predicting complete sequences of future events with their full data payloads. Second, there is a potential for applying the approach to the prediction of objects in object-centric logs. Our heterogeneous graph structure, which is better suited for expressing inter-dependencies between attributes, indicates the applicability of GNN4PPM to object-centric event logs, where event attributes are objects with different type relations in the knowledge graph. Third, formalizing in detail the event log knowledge graph ontology and schema and automating the graph construction. Additionally, incorporating explainability into the R-GCN predictions, for example, through graph attention or post-hoc explanation methods, would allow identifying which attribute relations influence individual predictions. Lastly, evaluating GNN4PPM on a broader set of domains, which include multiple targets to be predicted, would strengthen the generalization of the findings.

References

1. 4TU.ResearchData: 4tu.researchdata: an international research data repository (2026). https://data.4tu.nl/

2. Agarwal, C., Queen, O., Lakkaraju, H., Zitnik, M.: Evaluating explainability for graph neural networks. Sci. Data **10**(1), 144 (2023)
3. Ali, M.A., Dumas, M., Milani, F.: Enhancing the accuracy of predictors of activity sequences of business processes. In: Araújo, J., de la Vara, J.L., Santos, M.Y., Assar, S. (eds.) International Conference on Research Challenges in Information Science, pp. 149–165. Springer (2024). https://doi.org/10.1007/978-3-031-59465-6_10
4. Amiri Elyasi, K., van der Aa, H., Stuckenschmidt, H.: PGTNET: a process graph transformer network for remaining time prediction of business process instances. In: Guizzardi, G., Santoro, F., Mouratidis, H., Soffer, P. (eds.) International Conference on Advanced Information Systems Engineering. Springer (2024). https://doi.org/10.1007/978-3-031-61057-8_8
5. Aversano, L., Bernardi, M.L., Cimitile, M., Iammarino, M., Verdone, C.: A data-aware explainable deep learning approach for next activity prediction. Eng. Appl. Artif. Intell. **126**, 106758 (2023)
6. Bukhsh, Z.A., Saeed, A., Dijkman, R.M.: ProcessTransformer: predictive business process monitoring with transformer network. arXiv:2104.00721 (2021)
7. Camargo, M., Dumas, M., González-Rojas, O.: Learning accurate lstm models of business processes. In: Hildebrandt, T., van Dongen, B., Röglinger, M., Mendling, J. (eds.) Hildebrandt, T., Van Dongen, B.F., Röglinger, M., Mendling, J. (eds.) Business Process Management, LNCS, vol. 11675. Springer (2019). https://doi.org/10.1007/978-3-030-26619-6_19
8. Ceravolo, P., Comuzzi, M., De Weerdt, J., Di Francescomarino, C., Maggi, F.M.: Predictive process monitoring: concepts, challenges, and future research directions. Process Sci. **1**(1), 2 (2024)
9. Chen, Y., Zou, L., Qin, Z.: Gated relational graph neural network for semi-supervised learning on knowledge graphs. In: Cheng, R., Mamoulis, N., Sun, Y., Huang, X. (eds.) Web Information Systems Engineering – WISE 2019. LNCS, vol. 11881, pp. 617–629. Springer, Cham (2019). https://doi.org/10.1007/978-3-030-34223-4_39
10. Chiorrini, A., Diamantini, C., Genga, L., Potena, D.: Multi-perspective enriched instance graphs for next activity prediction through graph neural network. J. Intell. Inf. Syst. **61**(1), 5–25 (2023)
11. Costa, A., Mäkelburg, J.: ana-luisa-costa/GNN4PPM: NewRelease (NewRelease) (2026). https://doi.org/10.5281/zenodo.20490368
12. Mauro, D., Appice, N., Basile, A., T.M.A.: Activity prediction of business process instances with inception CNN models. In: Alviano, M., Greco, G., Scarcello, F. (eds.) Alviano, M., Greco, G., Scarcello, F. (eds.) AI 2019 - Advances in Artificial Intelligence. LNCS, vol. 11946, pp. 348–361. Springer (2019). https://doi.org/10.1007/978-3-030-35166-3_25
13. Dimou, A., Vander Sande, M., Colpaert, P., Verborgh, R., Mannens, E., Van de Walle, R.: RML: a generic language for integrated RDF mappings of heterogeneous data. LDOW **1184** (2014)
14. Dissegna, S., Di Francescomarino, C., Ronzani, M.: Multi-perspective next event prediction in ppm via heterogeneous graph neural networks. In: Grabis, J., Vos, T.E.J., Escalona, M.J., Pastor, O. (eds.) Research Challenges in Information Science, vol. 547. Springer. LNBIP. (2025) https://doi.org/10.1007/978-3-031-92474-3_22
15. van Dongen, B.: BPI challenge 2017 (2017). DOIurlhttps://doi.org/10.4121/uuid:5f3067df-f10b-45da-b98b-86ae4c7a310b

16. Duong, L.T., Travé-Massuyès, L., Subias, A., Merle, C.: Remaining cycle time prediction with graph neural networks for predictive process monitoring. In: Proceedings of the 2023 8th International Conference on Machine Learning Technologies, ACM (2023)
17. Evermann, J., Rehse, J.R., Fettke, P.: Predicting process behaviour using deep learning. Decis. Support Syst. **100**, 129–140 (2017)
18. Francescomarino, C.D., Ghidini, C.: Predictive process monitoring. In: van der Aalst, W.M.P., Carmona, J. (eds.) Process Mining Handbook. LNBIP, vol. 448, pp. 320–346. Springer, Cham (2022). https://doi.org/10.1007/978-3-031-08848-3_10
19. Grover, A., Leskovec, J.: node2vec: scalable feature learning for networks. In: Proceedings of the 22nd ACM SIGKDD International Conference on Knowledge Discovery and Data Mining, pp. 855–864 (2016)
20. Gunther, C.W., Verbeek, H.: XES standard definition version 2.0, Technische Universiteit Eindhoven / BPMcenter.org (2014). BPM-14-09 Technical Report
21. Harl, M., Weinzierl, S., Stierle, M., Matzner, M.: Explainable predictive business process monitoring using gated graph neural networks. J. Decis. Syst. **29**(sup1), 312–327 (2020)
22. He, H., Garcia, E.A.: Learning from imbalanced data. IEEE Trans. Knowl. Data Eng. **21**(9), 1263–1284 (2009)
23. Hogan, A., et al.: Knowledge graphs. ACM Comput. Surv. (CSUR) **54**(4), 1–37 (2021)
24. Huber, P.J.: Robust estimation of a location parameter. In: Kotz, S., Johnson, N.L. (eds.) Breakthroughs in statistics: methodology and distribution, pp. 492–518. Springer (1992). https://doi.org/10.1007/978-1-4612-4380-9_35
25. Lin, L., Wen, L., Wang, J.: MM-PRED: a deep predictive model for multi-attribute event sequence. In: Proceedings of the 2019 SIAM International Conference on Data Mining, pp. 118–126., SIAM (2019)
26. Maggi, F.M., Di Francescomarino, C., Dumas, M., Ghidini, C.: Predictive monitoring of business processes. In: Jarke, M., et al. (eds.) Advanced Information Systems Engineering (CAiSE 2014). LNCS, vol. 8484, pp. 457–472. Springer (2014). https://doi.org/10.1007/978-3-319-07881-6_31
27. Mehdiyev, N., Majlatow, M., Fettke, P.: Interpretable and explainable machine learning methods for predictive process monitoring: a systematic literature review. Artif. Intell. Rev. **58**(12), 378 (2025)
28. Mikolov, T., Chen, K., Corrado, G., Dean, J.: Efficient estimation of word representations in vector space. arXiv:1301.3781 (2013)
29. Mäkelburg, J., Peng, Y., Alam, M., Weller, T., Acosta, M.: Markov process–based graph convolutional networks for entity classification in knowledge graphs. In: 17th International Conference on Agents and Artificial Intelligence (ICAART 2025). Proceedings of ICAART, vol. 2, pp. 429–440 (2025)
30. Osborne, J.: Notes on the use of data transformations. Pract. Assess. Res. Eval. **8**(1) (2002)
31. Oyamada, R.S., Tavares, G.M., Junior, S.B., Ceravolo, P.: Enhancing predictive process monitoring with time-related feature engineering. In: Guizzardi, G., Santoro, F., Mouratidis, H., Soffer, P. (eds.) Advanced Information Systems Engineering, vol. 14663, pp. 71–86. Springer Nature Switzerland, Cham (2024). https://doi.org/10.1007/978-3-031-61057-8_5
32. Pasquadibisceglie, V., Appice, A., Castellano, G., Malerba, D.: Using convolutional neural networks for predictive process analytics. In: 2019 International Conference on Process Mining (ICPM), pp. 129–136. IEEE (2019)

33. Pasquadibisceglie, V., Appice, A., Castellano, G., Malerba, D.: Leveraging multi-view deep learning for next activity prediction. In: Proceedings of the 3rd International Workshop on Artificial Intelligence for Business Process Management. CEUR Workshop Proceedings, vol. 2952. CEUR-WS.org (2021)
34. Pasquadibisceglie, V., Appice, A., Castellano, G., Malerba, D.: A multi-view deep learning approach for predictive business process monitoring. IEEE Trans. Serv. Comput. **15**(4), 2382–2395 (2022). https://doi.org/10.1109/TSC.2021.3051771
35. Pasquadibisceglie, V., Appice, A., Malerba, D.: Lupin: a LLM approach for activity suffix prediction in business process event logs. In: Proceedings of the 6th International Conference on Process Mining (ICPM), pp. 1–8. IEEE (2024)
36. Pasquadibisceglie, V., Scaringi, R., Appice, A., Castellano, G., Malerba, D.: Prophet: explainable predictive process monitoring with heterogeneous graph neural networks. IEEE Trans. Serv. Comput. **17**(6), 4111–4124 (2024)
37. Rama-Maneiro, E., Vidal, J., Lama, M.: Deep learning for predictive business process monitoring: review and benchmark. IEEE Trans. Serv. Comput. (2022)
38. Rama-Maneiro, E., Vidal, J.C., Lama, M.: Embedding graph convolutional networks in recurrent neural networks for predictive monitoring (2021)
39. Rauch, S., Frey, C.M.M., Maldonado, A., Seidl, T.: Best: bilaterally expanding subtrace tree for event sequence prediction. In: Senderovich, A., Cabanillas, C., Vanderfeesten, I., Reijers, H.A. (eds.) Business Process Management, LNCS, vol. 16044, pp. 415–432. Springer (2026). https://doi.org/10.1007/978-3-032-02867-9_25
40. RDF working group: RDF standard model: W3C recommendations (2024). https://www.w3.org/RDF/
41. Ristoski, P., Paulheim, H.: RDF2VEC: RDF graph embeddings for data mining. In: Proceedings of the 10th International Conference on Semantic Systems, pp. 498–507. (2016)
42. Schlichtkrull, M., Kipf, T.N., Bloem, P., van den Berg, R., Titov, I., Welling, M.: Modeling relational data with graph convolutional networks. In: The Semantic Web – ESWC 2018. LNCS, vol. 10843, pp. 593–607. Springer (2018). https://doi.org/10.1007/978-3-319-93417-4_38
43. Tax, N., Verenich, I., La Rosa, M., Dumas, M.: Predictive business process monitoring with LSTM neural networks. In: Advanced Information Systems Engineering (CAiSE 2017). LNCS, vol. 10253, pp. 477–492. Springer (2017). https://doi.org/10.1007/978-3-319-59536-8_30
44. Theis, J., Darabi, H.: Decay replay mining to predict next process events. IEEE Access **7**, 119787–119803 (2019)
45. Van Assche, D., Delva, T., Heyvaert, P., De Meester, B., Dimou, A.: Towards a more human-friendly knowledge graph generation & publication. In: ISWC2021, The International Semantic Web Conference, vol. 2980. CEUR (2021)
46. Venugopal, I., Töllich, J., Fairbank, M., Scherp, A.: A comparison of deep-learning methods for analysing and predicting business processes (2021)
47. Weinzierl, S.: Exploring gated graph sequence neural networks for predicting next process activities. In: Marrella, A., Weber, B. (eds.) Business Process Management Workshops, Lecture Notes in Business Information Processing, vol. 436, Springer (2022). https://doi.org/10.1007/978-3-030-94343-1_3
48. Wuyts, B., Vanden Broucke, S., De Weerdt, J.: SUTRAN: an encoder-decoder transformer for full-context-aware suffix prediction of business processes. In: Proceedings of the 6th International Conference on Process Mining (ICPM), pp. 17–24. IEEE (2024)

Making Sense of Continuous Process Dynamics

Sana Dodangeh[1,2](✉), Martin Kabierski[1], and Han van der Aa[1]

[1] Faculty of Computer Science, University of Vienna, Vienna, Austria
{sana.dodangeh,martin.kabierski,han.van.der.aa}@univie.ac.at
[2] UniVie Doctoral School Computer Science DoCS,
University of Vienna, Vienna, Austria

Abstract. Operating in dynamic and often unpredictable environments, business processes are exposed to changes that cause their execution to fluctuate over time. Since changes are inseparable parts of process execution, capturing them is crucial for gaining reliable insights into how they behave during their execution. Although prior research has studied continuous fluctuations of processes, most studies have focused primarily on visual representations of such changes. However, drawing more actionable insights requires a structured analysis approach that goes beyond visualization. To this end, this paper explores the potential of leveraging established time series analysis techniques—such as decomposition and correlation analysis—to derive meaningful insights into such changes. We evaluate our approach through backward-looking and forward-looking analysis: the former demonstrates the type of insights it can provide into continuous process dynamics by applying it to multiple event logs, while the latter examines whether the approach is beneficial for a downstream process mining task, the prediction of future process behavior.

Keywords: process dynamics · time series analysis · continuous fluctuations

1 Introduction

Business processes are typically analyzed using data extracted from information systems that record their execution. These event logs form the foundation of process mining techniques, which aims to understand how processes behave in reality. By analyzing such data, process mining provides valuable insights into process performance and behavioral patterns, supporting various applications [1].

The analysis of such data needs to consider the inherently dynamic nature of business processes, which continuously change over time [2,3] due to intentional and unintentional adaptations, and are influenced by environmental changes, technological progress, seasonal trends, and other factors [4–6]. A concrete example can be found in the healthcare domain: A hospital check-in process is adjusted

F. M. Maggi et al. (Eds.): BPM 2026, LNBIP 599, pp. 131–149, 2027.
https://doi.org/10.1007/978-3-032-37877-4_8

annually at the beginning of each flu season by modifying procedures to adapt to this season [7]. Failing to take into account such systematic, non-random process changes during analysis will distort analysis outcomes, leading to inaccurate and unreliable insights, and, ultimately, suboptimal decision-making [8].

Given the dynamic nature of processes, prior work has recognized the need to capture such continuous changes over time. However, this has so far only been addressed through visualization-based approaches [5,9,10]. These approaches typically depict the progression of process characteristics over time to reveal phenomena such as gradual changes and recurring patterns. While such approaches provide initial insights, they fall short in two key analytical aspects:[1]

1. They fail to separate systematic, structural changes from random fluctuations, rendering interpretation ambiguous, incomplete, or outright false.
2. They are unable to capture interrelations between process characteristics, making it difficult to understand how changes in one relate to others.

To tackle these limitations, this paper explores the potential of leveraging established time series analysis techniques to gain deeper insights into the continuous dynamics of processes. Recognizing that understanding such dynamics is key to making informed adjustments and decisions in process management, we operationalize these time series analysis techniques for process mining through a structured, three-step algorithmic approach that focuses on (i) the derivation of time series for different process perspectives, (ii) the in-depth analysis of individual time series through decomposition and quantitative characterization (tackling Limitation 1 above), and (iii) the analysis of interrelations between time series through correlation analysis (addressing Limitations 2). This approach provides business analysts with structured information for interpreting continuous process dynamics, supporting downstream tasks in process mining such as analyzing past process dynamics and predicting future process behavior [6]. We evaluate the proposed approach through two complementary analyses. A backward-looking analysis evaluates how the approach supports understanding systematic process dynamics and interrelations between process characteristics. A forward-looking analysis then evaluates how separating these dynamics can support the prediction of future process behavior.

The remainder of this paper is structured as follows: Sect. 2 presents the motivation behind this work, while Sect. 3 introduces the proposed algorithmic approach. Section 4 reports on our evaluation. Section 5 provides an overview of related work, and finally, Sect. 6 concludes the paper.

2 Motivation

To motivate our work, we consider current practices for gaining insights into continuous process dynamics. This is so far achieved in a visual manner, especially through time series visualization [10] and drift maps [5]. Figure 1 shows

[1] We illustrate and elaborate on these limitations in Sect. 2.

an example of the former idea, highlighting the progression of the case load for the *Road Traffic Fine Management* event log.[2] It shows that the number of new and active cases exhibit temporal fluctuations, with an apparent seasonal pattern. While such visualizations are useful for revealing that process characteristics change over time, deriving more concrete insights from them requires analysts to manually inspect and interpret the observed patterns. This makes the analysis highly dependent on visual judgment, which may lead to ambiguous or inaccurate interpretations. Specifically, we recognize that purely visual approaches lack two key aspects that are necessary to properly understand the dynamics of processes:

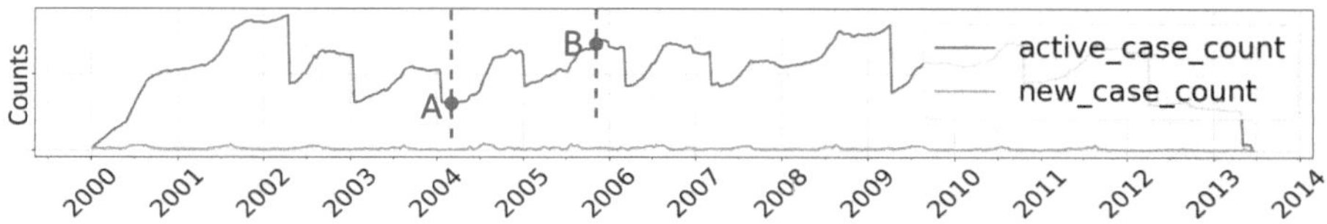

Fig. 1. Time series of new and active case counts in the *Road Traffic Fine Management* process.

C1: Separating Systematic from Random Fluctuations. Process dynamics can be influenced by various factors, including seasonal patterns, long-term trends, and irregular fluctuations. The interplay of such factors makes interpretations of visualizations particularly problematic, and ignoring these may yield misleading results. For instance, points A and B in Fig. 1 mark time points with different active case counts. A direct comparison shows an increase in active cases from point A to B. However, from the visualization alone, it remains unclear whether this increase mainly reflects a persistent upward movement in the underlying trend or the position of the two points within a recurring seasonal pattern. Furthermore, while it is also apparent that not all fluctuations are due to seasonality, it is unclear to what extent factors such as seasonality, long-term trends, or inherent randomness affect the case count, and whether such patterns are consistent over time. All of these insights are critical for properly understanding how a process behaves (and can be expected to behave) over time. Therefore, to support more accurate analysis, it is necessary to systematically separate fluctuations (e.g., seasonality and trends) from changes that occur in a more ad-hoc manner. Moreover, understanding such dynamics is not only important for interpreting past process behavior, but also for some downstream tasks such as predicting how processes may evolve in the future.

C2: Identifying Interrelated Patterns. The comprehensive understanding of process dynamics requires insight into the interrelations of fluctuations across

[2] Available on https://data.4tu.nl/categories/13500?categories=13503.

different aspects. For example, Fig. 1 suggests that increases in case arrivals tend to coincide with higher workloads (i.e., active cases). However, the strength and consistency of this correlation—and thus the extent to which workload aligns with variations in arrival rates—are difficult to assess through visual inspection alone. Similarly, analysts may want to investigate whether periods of increased workload tend to coincide with variations in process performance, for instance, changes in cycle times. This indicates a clear need for analytical support in identifying and quantifying the interrelations between different process dynamics.

3 Algorithmic Approach

This section investigates how established time series analysis techniques can be integrated into the analysis of process dynamics, thereby addressing the challenges of purely visual approaches, laid out in Sect. 2. In particular, we propose a three-step algorithmic approach for sense-making of continuous process dynamics, which is illustrated in Fig. 2. Beginning with an event log, the approach consists of the following steps:

1. *Derive time series* that capture the temporal behavior of a range of process characteristics of interest.
2. *Analyze individual time series* by extracting statistical summaries and decomposing the series to separate structured from irregular fluctuations.
3. *Analyze the interrelations among time series* to uncover dependencies and related changes that reveal broader behavioral structures.

These steps result in visualizations of decomposed time series, and statistical summaries and dependency measures of process characteristics. Together, they provide analysts with insights into behavioral patterns, irregular fluctuations, and interrelations between different process characteristics.

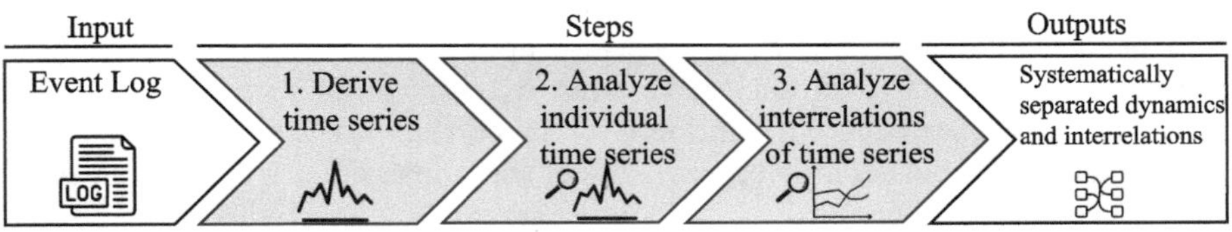

Fig. 2. Overview of the approach for obtaining insights into process dynamics

3.1 Derive Time Series

First, the approach transforms the event log into a set of time series, each capturing how a specific process characteristic of interest varies over time. This preserves their temporal evolution and enables analysis of individual series and their interrelations. We operationalize this step based on existing works [10, 11].

Input. The approach takes an *event log* L as input, defined as a collection of recorded events. Each event $e \in L$ is a tuple $e = (id, act, ts, D)$, where id is the case ID, act the executed activity, ts the recorded timestamp, and D a set of attribute-value pairs capturing data attributes. A trace $\pi = \langle e_1, ..., e_n \rangle$ is a sequence of events that belong to the same case, ordered by their timestamps.

Process Characteristics. The approach constructs time series for a user-defined set C of process characteristics. A characteristic $c \in C$ can be defined as any aspect of a process that can be measured for a given time window (see also DyloPro for an overview [10]), such as new or active case counts, frequencies of activities, the number of active resources, or the average cycle time of cases. The exact selection for C depends on the analysis goals of a user.

Windowing. We segment the timeline of the event log L into equal-length, non-overlapping time windows using time-based tumbling windowing [12, p.354], which results in a sequence of disjoint windows $W_l = \langle w_1, \ldots, w_n \rangle$, where each window w_t covers a fixed duration l (e.g., a day). Each event $e \in L$ is assigned to a window based on its timestamp.

Time Series Construction. For each characteristic $c \in C$, we construct a time series $y_c = \{y_{c,t} \mid t \in \{1, \ldots, |W_l|\}$, where $y_{c,t} \in \mathbb{R}$ reflects the behavior of characteristic c within window w_t. Examples are shown in Fig. 1. We refer to the collection of time series of all considered characteristics as Y_C.

Warm-Up and Cool-Down Elimination. Event logs may contain incomplete case executions at the beginning or end of the observation period, which leads to distorted observations in the resulting time series of a process characteristic $c \in C$ (i.e., warm-up and cool-down, respectively). To detect such phases, we follow an existing warm-up and cool-down elimination procedure proposed in prior work [11]. In particular, we use the time series of active case counts, as it reflects the overall workload of the process over time. From the distribution of its values, we compute a percentile threshold δ. Subsequently, windows in W_l at the beginning and end of the active case count time series whose values fall below this threshold are identified as warm-up or cool-down periods. These windows are then removed from all time series in Y_C before further analysis.

3.2 Analyze Individual Time Series

The goal of this step is to support understanding the dynamics of each process characteristic. To this end, each of the time series in Y_C is analyzed in isolation. The step first uses descriptive statistics to summarize the overall variability of the series and then applies decomposition to separate systematic from random fluctuations through decomposition, thereby addressing C1 in Sect. 2.

General Statistics. We first obtain a high-level overview of the magnitude and variability of each process characteristic. For this, we compute descriptive statistics such as the *mean*, *standard deviation*, and *coefficient of variation* (CV), defined as the ratio of the standard deviation to the mean.

Time Series Decomposition. To better understand the temporal dynamics of each process characteristic and distinguish the sources of variation discussed in C1, the approach breaks down the time series into components that capture distinct types of variation [13]. For a process characteristic $c \in C$:

- Trend (T_c): systematic, long-term changes in the characteristic over time.
- Seasonality (S_c): systematic, recurring short-term patterns periodically.
- Residual (R_c): random fluctuations not explained by trend or seasonality.

Since time series decomposition is a widely studied task, a range of existing methods can be used. Currently, the approach considers three decomposition methods: classical additive, classical multiplicative, and Seasonal-Trend decomposition using Loess (STL), where Loess stands for locally estimated scatterplot smoothing. In the additive case, the value $y_{c,t}$ of a characteristic $c \in C$ in time window w_t is expressed as the sum of its components, $y_{c,t} = T_{c,t} + S_{c,t} + R_{c,t}$, whereas in the multiplicative case it is expressed as their product, $y_{c,t} = T_{c,t} \times S_{c,t} \times R_{c,t}$[14, Ch. 3]. While the classical methods estimate the trend and seasonal components using moving averages, STL performs locally weighted regression to capture trend and seasonal patterns [15] In all cases, a seasonal period must be specified to capture recurring patterns in the time series.

Figure 3 shows the multiplicative decomposition of the new case count time series, revealing yearly seasonality, a stable trend, and residual fluctuations that are not explained by other components.

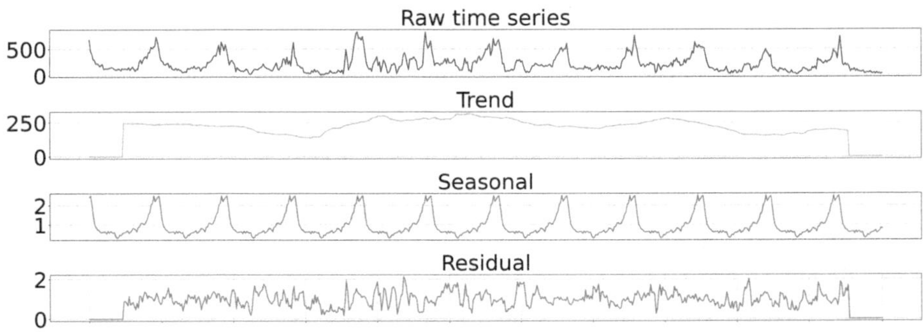

Fig. 3. Decomposed time series for new case count in the *Road Traffic Fine Management* process.

Decomposition Analysis. Beyond visualizing the decomposed time series, it is crucial to determine to what extent the trends and seasonal patterns explain the observed temporal dynamics, and how much of the variations correspond to random fluctuations. To this end, we use the following analyses:

Component Variance Fraction. To understand how much each component contributes to the overall dynamics of a characteristic, the fraction of total variation associated with the trend, seasonal, and residual components can be quantified. We refer to this measure as the component variance fraction. We use the component variance fraction instead of trend- or seasonality-strength measures because our aim is not to assess whether trend or seasonality is strong relative to random fluctuations. Rather, we aim to determine to what extent each component accounts for the observed dynamics of a process characteristic.

For this, the relative contribution of a given component to the summed variance of all components can be computed. For example, for the trend this yields:

$$\mathrm{vf}(T_c) = \frac{\mathrm{Var}(T_c)}{\mathrm{Var}(T_c) + \mathrm{Var}(S_c) + \mathrm{Var}(R_c)} \tag{1}$$

The formulas for the seasonal and residual components follow the same structure, replacing T_c in the numerator with S_c and R_c, respectively. Note that the denominator does not necessarily equal $\mathrm{Var}(y_c)$, as covariances between the components may be non-zero.

Taking the new case counts in Fig. 3 as an example, suppose that the trend, seasonality, and residual components explain 60%, 30%, and 10% of the variance, respectively. This would indicate that both long-term shifts and recurring cycles contribute significantly to the dynamics of new cases, whereas irregular fluctuations would play a comparatively minor role.

Ljung-Box Test. To assess whether any structured patterns remain in the residuals after decomposition, the Ljung-Box test can be applied to each residual time series [14, Ch. 5] This test evaluates whether the residuals can be considered white noise at a predefined significance level α, i.e., random fluctuations with no autocorrelation. It takes a predefined maximum lag h for which residual autocorrelation is jointly assessed. In this test, the null hypothesis is that the residual autocorrelations up to lag h are jointly zero. It thus complements the component variance fractions by indicating whether the residuals still contain autocorrelated structure after decomposition.

Decomposition Method Selection. As different time series may be better represented by different decomposition methods, the approach selects a decomposition method for each time series y_c. To this end, the approach applies the classical additive, classical multiplicative, and STL decomposition methods and computes the variance fractions of the resulting trend, seasonal, and residual components. The method whose decomposition yields the smallest residual variance fraction ($\mathrm{vf}(R_c)$) is then selected. This criterion is aligned with our goal of explaining as much of the observed dynamics as possible through systematic components (trend and seasonality), leaving only a limited share of unexplained fluctuations in the residual component.

3.3 Analyze Time Series Interrelations

Next, we explore how the investigation of multiple time series can be supported beyond visualization. This responds to C2 in Sect. 2: the need to identify interrelated patterns across different process characteristics. After analyzing each time series individually, this step examines whether pairs of process characteristics move together over the same time windows. To this end, we applyzero-lag Pearson correlation analysis, which quantifies the strength and direction of the linear relationship between two time series $y_a, y_b \in Y_C$ within the same time window. The correlation coefficient $r_{a,b} \in [-1, 1]$ indicates whether the series move in the same direction (positive correlation), in opposite directions (negative correlation), or show no linear relationship (values close to 0). It is computed as follows [14, ch.2]:

$$r_{a,b} = \frac{\sum_t (y_{a,t} - \overline{y_a})(y_{b,t} - \overline{y_b})}{\sqrt{\sum_t (y_{a,t} - \overline{y_a})^2 \sum_t (y_{b,t} - \overline{y_b})^2}} \tag{2}$$

Here $\overline{y_a}$ and $\overline{y_b}$ denote the means of the respective time series. Note that delayed interrelations could be studied with lag-aware methods such as cross-correlation. However, this would require comparing multiple lags and carefully interpreting whether a high correlation at a given lag reflects a meaningful delay or is influenced by temporal patterns such as trend and seasonality. We therefore leave such analysis as future work.

3.4 Output

Overall, the approach produces a detailed profile of continuous process dynamics. For each characteristic in C, it provides time series, decomposed components, statistical summaries, component variance fractions, and residual diagnostics. Together with correlation measures between characteristics, these outputs help analysts interpret both individual dynamics and how they evolve in relation to each other over time.

4 Evaluation

This section presents evaluation experiments to systematically assess the proposed approach in two complementary experiments. In the first experiment, we conduct a backward-looking analysis to examine how the approach supports understanding systematic process dynamics and interrelations between process characteristics. We then perform a forward-looking analysis to evaluate whether separating systematic dynamics improves the accuracy of predicting future process behavior, compared to predicting the raw time series. For reproducibility, all code and supplementary results are available in our project repository.[3]

[3] https://git01lab.cs.univie.ac.at/sanad95/cbf.

4.1 Experiment 1: Backward-Looking Analysis

In the first experiment, we assess the ability of our approach to make sense of continuous process dynamics. Specifically, we determine the extent to which the systematically separated dynamics explain observed variations, and to what degree changes in one process characteristic relate to changes in others.

Data. For this experiment, we use four real-life event logs publicly available through the 4TU Research Data repository[4], namely BPIC18, BPIC15-1, Hospital Billing, and International Declarations (IntDec). They originate from different domains and differ substantially in their characteristics. Specifically, the mean workload across the logs ranges from about 78 to over 14,500 cases per time window. The mean number of arriving cases varies from less than one to more than 40 per time window, while the mean number of active resources ranges from about two to over 60 per time window. This diversity allows us to evaluate the approach across processes with different scales and operational dynamics.

Setup. To conduct this experiment, we use the following setup.

Approach Implementation. We implemented the proposed approach in Python, leveraging the *PM4Py* and *statsmodels* libraries for event log handling and time series analysis, respectively.

Time Series Construction. To conduct the evaluation across different aspects of processes, we construct time series for multiple process characteristics using Step 1 of our approach.

- *New case counts:* A case-level process characteristic indicating the number of new cases in time window w_t.
- *Active case counts:* A case-level process characteristic representing the number of cases that have started but have not yet completed during time window w_t. This characteristic quantifies the workload of the process.
- *Average throughput time:* A performance-related process characteristic capturing the mean completion time of cases in window w_t.
- *Active resource counts:* A resource-related process characteristic representing the number of distinct resources involved in the process during window w_t.

Parameter Settings. For the evaluation, following Step 1, we use daily windowing with a seasonal period of 7 days, and remove warm-up and cool-down phases using a percentile-based threshold δ, computed from the active case counts as the value below which 25% of observations fall. Moreover, in Step 2 of our approach, we use $\alpha = 0.05$ and $h = 10$ for the Ljung–Box test.

Results. We first examine the overall variability of individual process characteristics to understand how strongly they fluctuate over time. We then assess

[4] https://data.4tu.nl/categories/13500?categories=13503.

how much of their observed variation is explained by the systematically separated dynamics identified by our approach. Finally, we examine interrelations between characteristics to determine whether our approach reveals systematic relationships between their changes.

High-level Variability of Process Characteristics. Table 1 summarizes the CVs (coefficients of variation) for all characteristics across the considered event logs. After examining the statistics of all event logs, we observe substantial differences in the degree of variability across logs and process characteristics. Overall, the CVs range from 0.07 to 3.45, indicating that some processes exhibit relatively stable behavior while others exhibit larger dynamics. Among all event logs, the Hospital Billing log consistently shows the most stable behavior across all characteristics, with low CVs (CV = 0.07 for active cases, 0.14 for case arrivals, 0.26 for average throughput time, and 0.12 for active resources). In contrast, several logs exhibit substantially stronger fluctuations in specific characteristics. For example, BPIC18 shows the strongest variability in case arrivals and active resources (CV = 3.45 and CV = 0.94, respectively). Similarly, International Declarations and BPIC15-1 exhibit the highest variability in active cases and average throughput time (CV = 0.63 and CV = 2.41, respectively). As shown in Fig. 4, the raw (non-decomposed) time series of active resource counts shows a stable pattern in the Hospital Billing log, while it fluctuates strongly in BPIC18.

Explained Variability Through Decomposition. To gain a nuanced understanding of these fluctuations, we analyze the variance fractions of the decomposed components, as reported in Table 1. The table also includes the results of the white noise test of the residuals and the automatically selected decomposition methods.

The results reveal a clear pattern: almost all characteristics with low variability (CV) are largely explained by the trend component. We can observe this pattern in Hospital Billing log, where all characteristics exhibit very small CVs. For these series, the approach selects the classical multiplicative decomposition and attributes nearly all variance to the trend component (rounded to 1.00 for all four characteristics), while seasonal and residual contributions remain negligible. As illustrated in Fig. 4a, although the approach separates seasonal patterns and random fluctuations of the active resource counts in this event log, their magnitudes remain much smaller than the trend. A similar pattern can also be observed in the International Declarations log, where active case counts are almost entirely explained by the trend component (1.00), suggesting that fluctuations are primarily driven by long-term developments. This can encourage analysts to focus on long-term trends rather than short-term variations.

In contrast, BPIC18 exhibits more diverse dynamics across most characteristics, where in active case counts the trend dominates, while in others multiple components contribute substantially to the observed variability. For instance, new case counts are explained by a combination of trend (0.76) and seasonal (0.24) components. Similarly, as shown in Fig. 4b, for active resource counts the seasonal component (0.59) dominates the trend (0.31), with both seasonal and residual components visibly contributing to the observed fluctuations. This sug-

Table 1. Variance fractions of trend (T_c), seasonal (S_c), and residual (R_c) components for the considered process characteristics. WN indicates if the residuals are white noise, and DM the selected decomposition method (STL, Additive, Multi. for multiplicative).

Log	Characteristic	CV	$\mathbf{vf(T_c)}$	$\mathbf{vf(S_c)}$	$\mathbf{vf(R_c)}$	WN	DM
BPIC18	Active case count	0.35	1.00	0.00	0.00	No	Multi.
	New case count	3.45	0.76	0.24	0.00	No	STL
	Avg throughput time	1.18	0.36	0.46	0.18	No	STL
	Active resource count	0.94	0.31	0.59	0.10	No	STL
BPIC15-1	Active case count	0.14	1.00	0.00	0.00	No	Multi.
	New case count	1.52	0.18	0.13	0.69	No	Additive
	Avg throughput time	2.41	0.04	0.25	0.71	Yes	STL
	Active resource count	0.83	0.10	0.67	0.23	No	STL
Hospital Billing	Active case count	0.07	1.00	0.00	0.00	No	Multi.
	New case count	0.14	1.00	0.00	0.00	No	Multi.
	Avg throughput time	0.26	1.00	0.00	0.00	No	Multi.
	Active resource count	0.12	1.00	0.00	0.00	No	Multi.
IntDec	Active case count	0.63	1.00	0.00	0.00	No	Multi.
	New case count	1.13	0.46	0.34	0.20	No	STL
	Avg throughput time	1.42	0.10	0.42	0.48	Yes	STL
	Active resource count	0.30	0.17	0.48	0.35	Yes	STL

gests that fluctuations are largely driven by systematic seasonal effects, which may encourage analysts to explicitly consider such patterns in resource planning.

Finally, a different pattern emerges when the residual component dominates the variability. Specifically, we observe this in BPIC15-1 for average throughput time, where the residual explains most of the variance (0.71), followed by smaller contributions from the seasonal (0.25) and trend (0.04) components. In this case, the residual behaves as white noise, indicating that most systematic dynamics are effectively captured by the decomposition, suggesting that the remaining fluctuations are largely irregular and therefore difficult for analysts to anticipate.

Interrelations Between Process Characteristics. To complement the analysis of individual dynamics, we next examine how process characteristics relate to one another through correlation analysis, as summarized in Table 2.

Across all event logs except for the International Declarations log, the correlation between new case counts and active case counts is relatively weak, despite the intuitive expectation that increases in arrivals should immediately lead to more active cases. For example, in the BPIC18 log, where new case counts are highly dynamic, this correlation is slightly negative (-0.17), suggesting that case arrivals and workload do not necessarily increase or decrease simultaneously. This likely reflects temporal delays between incoming cases and their accumulation. In contrast, in the International Declarations log, where active case counts

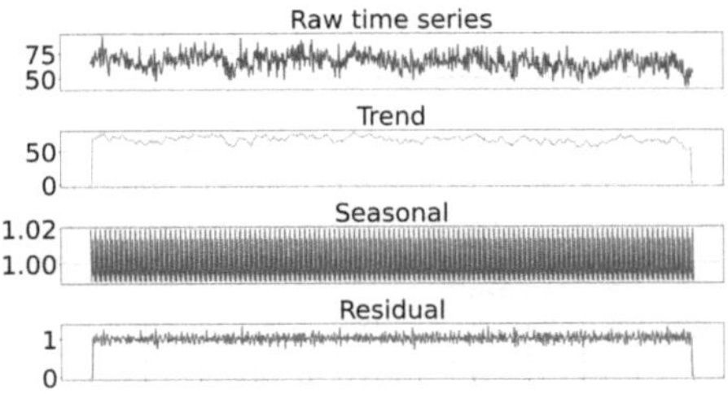

(a) Hospital Billing

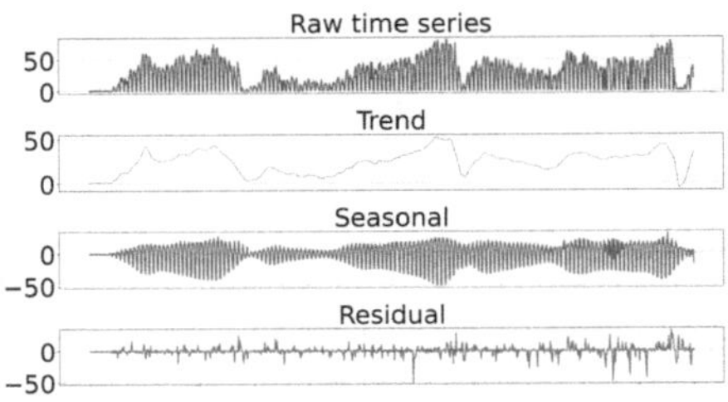

(b) BPIC18

Fig. 4. Extracted components of active resource counts in two event logs.

fluctuate strongly, the correlation becomes much stronger (0.57), indicating a more immediate accumulation of active cases. This can enable analysts to identify whether increases in case arrivals are instantly reflected in the workload or whether delays occur before cases accumulate in the system, helping them better understand and address such delays.

Table 2. Pairwise correlations between process characteristics across the analyzed event logs. NC = new case count, AC = active case count, AR = active resource count, and TT = average throughput time.

Log	$r_{NC,AC}$	$r_{NC,AR}$	$r_{AC,AR}$	$r_{TT,AC}$	$r_{TT,NC}$	$r_{TT,AR}$
BPI Challenge 2018	−0.17	−0.15	0.17	0.03	0.04	0.50
BPIC15-1	0.15	0.51	0.23	−0.02	0.07	0.27
Hospital Billing	0.12	0.67	0.18	0.31	−0.35	−0.18
International Declarations	0.57	0.41	0.34	0.18	0.16	0.29

At the same time, active resource counts often show stronger associations with incoming cases than with the current number of active cases. In the Hospital Billing log, for instance, the correlation between new case counts and active resource counts reaches 0.67, while the correlation with active case counts remains much weaker (0.18). This suggests that resource availability fluctuates more in line with case arrivals than with the current workload. Specifically, such patterns may help analysts understand whether resource allocation is primarily driven by incoming demand rather than by the number of ongoing cases.

Moreover, average throughput time is often more strongly associated with active resource counts than with new or active case counts. In the BPIC15-1 log, where throughput time is highly dynamic, the correlation with active resource counts is 0.27, while correlations with new case counts (0.07) and active case counts (-0.02) remain negligible. This allows analysts to identify whether fluctuations in average throughput time co-occur with changes in resource availability rather than with variations in case arrivals or workload, providing insights into operational factors associated with performance dynamics.

Overall, the backward-looking analysis demonstrates that the proposed approach can support analysts in obtaining insights into both the individual dynamics of process characteristics and their interrelations. Such insights can in turn guide further investigations into potential process improvements.

4.2 Experiment 2: Forward-Looking Analysis

This experiment further evaluates the usefulness of our approach by assessing whether systematically separating process dynamics through decomposition improves predictions of future process behavior. Specifically, we compare the accuracy of predictive models trained on decomposed time series to those trained on the original series, determining which models provide better insights into the future values of characteristics such as case arrivals, workload, resource availability, and process performance.

Setup. We use the same data and setup as for Experiment 1, unless stated otherwise below. We further consider the following aspects in Experiment 2:

Prediction Models. We consider two widely-used time series models: AutoRegressive Integrated Moving Average (ARIMA) and Seasonal Autoregressive Integrated Moving Average (SARIMA). ARIMA captures non-seasonal patterns such as trends, whereas SARIMA additionally models seasonality [13, Chs. 3,7,8]. In our experiment, both models are configured with one seasonal cycle, while all other parameters follow their default settings.

Train-Test Split. After removing the warm-up and cool-down phases in Step 1 of our approach, we split each time series chronologically into training and test segments. We train the models on the first 80% of observations and evaluate them on the remaining 20%, i.e., the test segment.

Baseline Prediction Models. We establish baselines by applying ARIMA and SARIMA directly to the original time series using the train-test split described above. This allows us to assess how well commonly used time series models capture process dynamics without separating systematic and irregular fluctuations.

Decomposition-Based Prediction Models. For decomposition-based prediction models, we decompose the training segment into trend, seasonal, and residual components. We employ STL decomposition, as the other decomposition methods produce undefined trend values at the boundaries of the series, which prevents generating predictions over the full time horizon.

After decomposition, we represent each time series by combinations of its three component series. We can model these components either separately or in a grouped form, which leads to multiple possible configurations. In this study, we focus on three configurations that reflect different assumptions about the role of the residual component: we either model it independently, combine it with the trend, or exclude it as an irregular fluctuation. For each configuration, we train one model per component group on the training segment and predict the test segment. We then combine these predictions to reconstruct the original series.

- *Independent residual* (T_c, S_c, R_c): We predict the components independently: trend and residual using ARIMA and the seasonal component using SARIMA.
- *Residual merged with trend* (T_c+R_c, S_c): We aggregate the residual with the trend and predict them jointly using ARIMA, while the seasonal component is predicted separately using SARIMA. This configuration aims to reduce error accumulation from predicting the residual independently.
- *Residual excluded* (T_c, S_c): We predict only the systematic components: the trend using an ARIMA model and the seasonal component using a SARIMA model, while the residual is not modeled.

Evaluation Metric. To compare the prediction accuracy of the baselines and decomposition-based predictions, we use Root Mean Squared Error (RMSE), which captures deviations at individual time points and emphasizes larger errors [13, Ch. 5]. This metric is defined as follows:

$$RMSE = \sqrt{\frac{1}{n}\sum_{t=1}^{n}(y_{c,t} - \hat{y}_{c,t})^2} \tag{3}$$

where $y_{c,t}$ denotes the observed value of time series c in time window w_t, $\hat{y}_{c,t}$ the corresponding prediction, and n the number of test observations.

Results. Table 3 reports the RMSE obtained by predicting the process characteristics across all event logs using the raw baselines as well as the decomposition-based predictions. Out of the 16 time series, decomposition outperforms both baselines in six and improves over at least one baseline in nine. This indicates that decomposition can improve prediction accuracy in some settings, but does not consistently outperform the raw-series baselines. Overall, the results show that when decomposition improves over at least one baseline, the reduction in RMSE is often substantial: for the (T_c+R_c, S_c) configuration, the average improvement is 24.03%. In contrast, when it does not outperform the baselines, the accuracy decrease remains limited, with an average degradation of only 3.48% for this configuration. This can be illustrated by individual examples. For instance, in the BPIC18 log, the (T_c+R_c, S_c) configuration reduces the RMSE for the average throughput time from 7707.44 (SARIMA) to 5789.62. Conversely, for the active case count in the International Declarations log, its RMSE (764.50) is only slightly higher than that of ARIMA (749.46).

To understand when decomposition improves prediction accuracy, we analyze the variance fractions of the decomposed components. The corresponding values are available in the accompanying repository. Among time series with substantial seasonal fluctuations ($vf(S_c) > 0.30$), decomposition outperforms at least one baseline in six out of seven time series. This indicates that decomposition

Table 3. RMSE values for baselines and decomposition-based predictions across all event logs and process characteristics.

Log	Characteristic	Baselines		Decomposition-based		
		ARIMA	SARIMA	(T,S,R)	(T+R,S)	(T,S)
BPIC18	new case counts	1.07	48.12	0.47	**0.44**	0.47
	active case counts	**4573.30**	4583.98	4596.81	4643.27	4593.68
	avg throughput time	7948.89	7707.44	5866.70	**5789.62**	5944.21
	active resource counts	34.95	**19.95**	27.73	20.31	27.75
BPIC15-1	new case counts	0.84	**0.83**	1.01	0.85	0.91
	active case counts	10.31	10.36	7.97	9.66	**7.95**
	avg throughput time	**2466.50**	2471.61	2529.38	2470.77	2599.60
	active resource counts	1.54	1.52	1.66	**1.21**	1.68
Hospital Billing	new case counts	**12.66**	12.67	24.48	13.36	24.55
	active case counts	1884.04	1903.66	1765.75	1917.35	**1758.21**
	avg throughput time	**631.71**	632.19	647.93	645.30	649.90
	active resource counts	8.06	**8.05**	9.26	8.92	9.34
IntDec	new case counts	**10.13**	10.19	15.91	10.83	15.82
	active case counts	**749.46**	756.96	795.31	764.50	796.25
	avg throughput time	1411.54	1400.61	1442.48	**1369.33**	1473.47
	active resource counts	0.66	**0.65**	0.67	0.66	0.71

tends to improve prediction when seasonal patterns contribute substantially to the dynamics. For instance, Fig. 5 shows the average throughput time in the BPIC18 log, where strong seasonal dynamics are present ($vf(S_c)$=0.41). Here, neither ARIMA (RMSE = 7948.89) nor SARIMA (RMSE = 7707.44) captures the recurring fluctuations effectively. In contrast, the (T_c+R_c, S_c) configuration (RMSE= 5789.62), shown as a representative decomposition-based prediction, captures the seasonal structure more accurately and therefore produces substantially better predictions.

The grouping configurations also exhibit different behaviors. Overall, the (T_c+R_c, S_c) configuration performs best among the evaluated decompositions in four time series and outperforming at least one baseline in eight. This suggests that combining the residual with the trend while modeling the seasonal component separately often improves prediction accuracy. In contrast, the (T_c, S_c, R_c) configuration does not achieve the best overall result for any time series, although it still improves prediction accuracy over at least one baseline in five. Similarly, the (T_c, S_c) configuration exhibits the best performance in two time series and also outperforms at least one baseline in five. A closer inspection shows that both (T_c, S_c, R_c) and (T_c, S_c) mainly improve prediction when the residual variance fraction is very small and the trend variance fraction is large. For example, for the active case count in the Hospital Billing log (trend 0.99, residual 0.01),

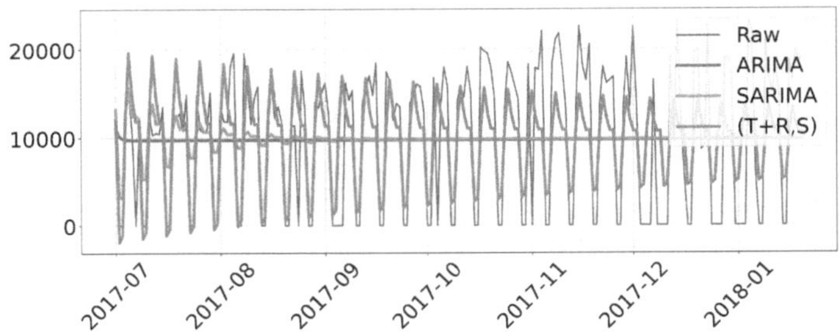

Fig. 5. Predictions for average throughput time in BPIC18 log for baselines and (T+R,S) decomposition-based prediction in the test segment.

(T_c, S_c, R_c) yields an RMSE of 1765.75, while (T_c, S_c) achieves a slightly lower RMSE of 1758.21, indicating that explicitly modeling the systematic components helps capture the main structure of the series.

Overall, decomposition is most beneficial for predicting future behavior of a process when seasonality explains a large share of the dynamics. The $(T_c{+}R_c, S_c)$ grouping often performs best, highlighting the importance of appropriate component groupings. Moreover, decomposition rarely harms prediction accuracy and can yield substantial improvements when strong seasonal dynamics are present.

5 Related Work

Our work relates to research on the analysis of process dynamics and the use of time series in process mining.

Analysis of Process Dynamics. As the dynamic nature of business processes has gained increasing attention, several approaches have emerged to support the analysis and visualization of behavioral changes over time. Process Evolution Analysis (PEA) [5] was among those that track how clusters of process executions emerge and shift across the timeline. Building on this, Visual Drift Detection (VDD) [9] facilitates the visual identification of behavioral drifts by highlighting changes in constraints over time. More recently, DyLoPro [10] was introduced as a visual analytics tool that allows for the exploration of changes across multiple process aspects. While these approaches offer support for analyzing process dynamics, they primarily rely on visual inspection and lack systematic, data-driven analysis of continuous fluctuations. However, current data-driven analyses remain largely descriptive and focus on aggregated behavior, offering limited support for understanding fine-grained, temporal dynamics.

To address these limitations, our work provides a systematic, time-series-based perspective for capturing continuous behavioral change in business processes, moving beyond visual exploration toward deeper analytical insight.

The Use of Time Series in Process Mining. Time series analysis has become a valuable technique in process mining, as it enables the investigation of temporal patterns across various process characteristics. Many studies have transformed event logs into time series that track the evolution of metrics such as activity frequencies, case volumes, process variants, and event counts [2,10,16]. Initially, early efforts explored the use of such temporal representations to monitor behavioral changes, treating event logs as time-ordered sequences of traces to detect significant shifts over time [16].

Beyond descriptive monitoring, time series have served as the basis for more advanced forms of analysis, including steady-state detection [2] and visual exploration of evolving process aspects [10]. Furthermore, they have enabled predictive tasks–for instance, by forecasting structural changes in control-flow by modeling temporal patterns of DFG relations [17]. In addition, multivariate time series analysis has supported the examination of interdependencies across perspectives such as workload, performance, and resource usage, shedding light on how process dimensions influence one another during short-term disruptions [11].

While prior work applies time series techniques in process mining to raw time series, our approach is the first to leverage time series decomposition to detect and analyze structured changes in process behavior.

6 Conclusion

This paper explores how time series analysis can provide deeper insights into the continuous dynamics of processes. Compared to prior work, our approach focuses on separating systematic from random fluctuations and identifying interrelated dynamics. Our evaluation shows that this can reveal the sources of temporal fluctuations and dependencies between process characteristics. In particular, a forward-looking analysis indicates that separating systematic dynamics can expose informative predictive signals: decomposition-based predictions are often comparable to or better than predictions on the raw time series, especially when seasonality explains a substantial share of the observed variance.

These findings should be interpreted in light of several threats to validity. First, the evaluation is based on four event logs, and the findings may therefore not generalize to all types of processes and process dynamics. Second, the results depend on several design choices, including the selected process characteristics, the window size, and other parameter settings for both backward-looking and forward-looking analyses. Third, in the forward-looking analysis, the choice of prediction models itself may influence the comparison. Other prediction approaches, such as SARIMAX, VAR, VARIMA, etc., may yield different results Finally, although the approach can provide meaningful insights into dynamic patterns, it has not yet been tested by users. We aim to address this in future work to evaluate its practical usefulness. Moreover, future work could extend the predictive analysis by incorporating uncertainty estimates, for example by using prediction intervals instead of point predictions to capture uncertainty in future process dynamics. Overall, our results demonstrate that integrating time series decomposition and correlation analysis into process mining provides a structured perspective for understanding and anticipating continuous process dynamics.

References

1. Van der Aalst, W.: Process Mining: Data Science in Action. Springer (2016). https://doi.org/10.1007/978-3-662-49851-4
2. Kraus, A., Elyasi, K.A., van der Aa, H.: On the use of steady-state detection for process mining: Achieving more accurate insights. In: Krogstie, J., Rinderle-Ma, S., Kappel, G., Proper, H.A. (eds.) CAiSE, pp. 204–220. Springer (2025). https://doi.org/10.1007/978-3-031-94569-4_12
3. Grisold, T., Janiesch, C., Röglinger, M., Wynn, M.T.: Managing dynamics in and around business processes. BISE **66**(5), 533–540 (2024)
4. Bose, R.J.C., Van Der Aalst, W.M., Žliobaitė, I., Pechenizkiy, M.: Dealing with concept drifts in process mining. IEEE Trans. Neural Netw. Learn. Syst. **25**(1), 154–171 (2013)
5. Yeshchenko, A., Bayomie, D., Gross, S., Mendling, J.: Visualizing business process evolution. In: Nurcan, S., Reinhartz-Berger, I., Soffer, P., Zdravkovic, J. (eds.) BPMDS, pp. 185–192. Springer (2020). https://doi.org/10.1007/978-3-030-49418-6_12
6. Franzoi, S., Hartl, S., Grisold, T., van der Aa, H., Mendling, J., vom Brocke, J.: Explaining process dynamics: a process mining context taxonomy for sense-making. Process Sci. **2**(1), 2 (2025)
7. Pentland, B., Vaast, E., Wolf, J.R.: Theorizing process dynamics with directed graphs: a diachronic analysis of digital trace data. MIS Quart. **45**(2) (2021)
8. Browne, C.: A change dynamic model for the online detection of gradual change. arXiv preprint arXiv:2205.01054 (2022)
9. Yeshchenko, A., Di Ciccio, C., Mendling, J., Polyvyanyy, A.: Visual drift detection for event sequence data of business processes. IEEE Trans. Visual Comput. Graph. **28**(8), 3050–3068 (2021)
10. Wuyts, B., Weytjens, H., vanden Broucke, S., De Weerdt, J.: Dylopro: profiling the dynamics of event logs. In: Di Francescomarino, C., Burattin, A., Janiesch, C., Sadiq, S. (eds.) BPM, pp. 146–162. Springer (2023). https://doi.org/10.1007/978-3-031-41620-0_9
11. Kraus, A., Rehse, J.-R., van der Aa, H.: Data-driven assessment of business process resilience. Process Sci. **1**(1), 4 (2024)
12. Van Der Aalst, W.M., Carmona, J.: Process Mining Handbook, Springer (2022). https://doi.org/10.1007/978-3-031-08848-3
13. Hyndman, R.J., Athanasopoulos, G.: Forecasting: Principles and Practice,. 2nd edn. OTexts (2018)
14. Hyndman, R.J., Athanasopoulos, G.: Forecasting: Principles and Practice, 3rd edn. OTexts (2021)
15. Cleveland, R.B., Cleveland, W.S., McRae, J.E., Terpenning, I.: STL: a seasonal-trend decomposition procedure based on loess. JOS **6**, 3–73 (1990)

16. Bose, R.J.C., van der Aalst, W.M., Žliobaitė, I., Pechenizkiy, M.: Handling Concept Drift in Process Mining In: Mouratidis, H., Rolland, C. (eds.) CAiSE. Springer, 2011, pp. 391–405. https://doi.org/10.1007/978-3-642-21640-4_30
17. De Smedt, J., Yeshchenko, A., Polyvyanyy, A., De Weerdt, J., Mendling, J.: Process model forecasting and change exploration using time series analysis of event sequence data. Data Knowl. Eng. **145**, 102145 (2023)

Multimodal Process Monitoring with On-demand Disambiguation

Marco Franceschetti[1(✉)], Sejma Sijaric[1], Raúl Jiménez Cruz[2], Olaf Zimmermann[1], César Torres-Huitzil[2], and Barbara Weber[1]

[1] University of St.Gallen, St.Gallen, Switzerland
marco.franceschetti@unisg.ch
[2] Tecnológico de Monterrey, Puebla, Mexico

Abstract. Process monitoring is increasingly supported by Internet-of-Things (IoT) environments. However, the abstraction of sensor data into process events often leads to ambiguity. Existing approaches that leverage IoT data do not address how to resolve such ambiguity dynamically. To address this challenge, we present MMPMOD, a novel architecture for multimodal process monitoring that detects ambiguity in IoT-based events and orchestrates contextual sensing modalities on-demand to resolve it. Upon ambiguity detection, it activates cameras to capture the visual context, and applies scene understanding to disambiguate events. We implement and evaluate the architecture in a phlebotomy scenario, demonstrating the feasibility of orchestration of complementary sensing modalities for disambiguation in online monitoring in this scenario.

Keywords: Ambiguity · Multimodal process mining · Computer vision

1 Introduction

Internet of Things (IoT) technologies are increasingly integrated into business processes executed in the physical world. This integration promises unprecedented insights into process executions based on sensor data in near-real time [13]. While this allows monitoring process executions without relying on a process engine or information system producing execution data, it also introduces challenges in abstracting sensor data into meaningful process events [19]. Specifically, IoT data quality issues and imprecise abstraction procedures can lead to ambiguity: multiple process events relating to the same IoT data pattern. This, in turn, reduces the reliability of derived event logs and downstream analysis tasks [9].

The underlying reason for such ambiguity is that IoT data typically provides only local information [13]. For example, in a blood-extraction setting, IoT data may indicate that *something* is moving within the sensing range of a motion

F. M. Maggi et al. (Eds.): BPM 2026, LNBIP 599, pp. 150–167, 2027.
https://doi.org/10.1007/978-3-032-37877-4_9

sensor, not that a syringe is being inserted into an arm. Such local information alone does not allow distinguishing between process events described by the same IoT pattern, e.g., syringe insertion and arm disinfection. In contrast, contextual information can provide cues that help disambiguate process events. Knowing that a nurse is manipulating a syringe, for instance, helps rule out arm disinfection as the interpretation of the motion sensor's data. The disambiguation potential of contextual information motivates integrating IoT data with complementary modalities able to contextualize IoT-based process events [11]. Video data provides such a modality by capturing a visual representation of the process environment. Unlike IoT sensors that measure localized signals, cameras can capture a broader view of the context in which process events occur. Prior work has explored the use of video data for process mining and activity recognition [16,18,25]. However, existing approaches typically assume continuous acquisition and analysis of video streams, which yields substantial computational overhead and may raise privacy concerns in many operational settings. Moreover, multimodal monitoring approaches generally treat sensing modalities as parallel inputs to activity recognition pipelines rather than exploiting them selectively.

These limitations point to the opportunity to leverage complementary sensing modalities in a more adaptive manner, investigating *how on-demand multimodal sensing orchestration supports efficient disambiguation of IoT-based process events.* Instead of continuously processing video data, contextual sensing modalities can be activated only when ambiguity arises. We therefore treat ambiguity detection as a trigger for orchestrating contextual sensing modalities, using video data as contextual evidence to disambiguate events in near-real time, as demonstrated in [22]. Because disambiguation often depends on recognizing the objects involved in an activity and how they are manipulated [26], we rely on scene understanding to identify relevant objects and human-object interactions. Following Design Science Research principles [21], we design, instantiate, and evaluate *MMPMOD*, a novel multimodal process monitoring architecture that combines IoT sensing with on-demand contextualization to disambiguate events. By improving the reliability of IoT-based event abstraction, MMPMOD supports the generation of more accurate event logs for downstream tasks. We instantiate MMPMOD in a realistic lab simulation environment for an IoT-monitored phlebotomy (blood-extraction) process, where contextualization is realized through computer vision. This setup provides a representative setting where people and physical objects interact, causing ambiguity. We evaluate MMPMOD's disambiguation accuracy and run-time performance, showing the effectiveness of on-demand disambiguation, with latencies appropriate for the specific domain.

While prior work [24] focuses on abstracting IoT events into process events, we address the complementary problem of resolving ambiguities that arise during event abstraction. We make three contributions: (1) a multimodal process monitoring architecture that detects ambiguity in IoT-based events and orchestrates contextual sensing modalities on-demand to resolve it; (2) a vision model trained on a phlebotomy dataset for recognizing objects and interactions characteristic of phlebotomy activities; (3) five generalized, actionable design principles for building adaptive, ambiguity-aware multimodal process monitoring systems.

In the remainder of the paper, Sect. 2 introduces a motivating scenario; Sect. 3 presents the architecture and vision model; Sect. 4 discusses implementation and evaluation; Sect. 5 reviews related work; Sect. 6 concludes the paper.

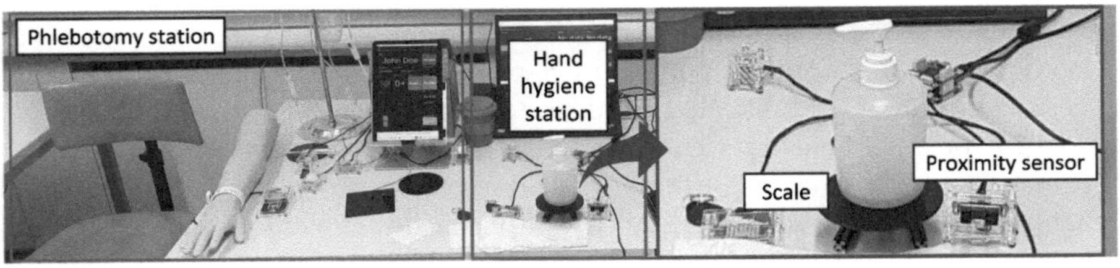

Fig. 1. IoT-augmented environment monitoring a simulated phlebotomy process

2 Motivating Scenario: Ambiguity in Event Abstraction

As a motivating scenario, we consider an IoT-augmented environment for monitoring healthcare worker (HCW) trainees practicing a simplified phlebotomy process using a training arm. The phlebotomy process and the environment implementation in our lab (cf. Fig. 1) have been validated by healthcare domain experts (cf. [8] for further details). Thus, it represents a realistic example of the challenge of ambiguity hindering monitoring in IoT-augmented environments.

Process monitoring in this environment is realized via IoT event abstraction, implemented through *activity detection services* [24]. In short, for each activity in the process specification (e.g., *HCW check-in*, *Perform hand hygiene*), a dedicated Web service runs in the background that, at run time, listens to IoT events generated by relevant sensors in the process environment. Upon a match of the run-time IoT events with a pattern of IoT events pre-specified at design time by a domain expert, each service emits a process event signifying the occurrence of the respective activity. For instance, a domain expert knows that hand hygiene events occur when a HCW is in proximity to the hand hygiene station and applies pressure to the sanitizer dispenser. This results in a detection service with an associated pattern `proximity_sensor <= 30` cm && `scale >= 1000` g, i.e., a pattern of IoT events emitted by a proximity sensor and a scale (cf. Fig. 1). Once deployed, whenever the IoT data produced by these two sensors satisfies the pattern conditions, the service emits a process event *Perform hand hygiene*.

Due to the lack of means (e.g., additional sensors) able to distinguish IoT event patterns at a finer-granular level, multiple services might be associated with the same pattern, each emitting the respective process event, thereby leading to ambiguity [8]. The setup in Fig. 1 leads to ambiguity in event abstraction. Here, we focus on three activities occurring multiple times in a process instance–namely, *Disinfection*, *Tourniquet application*, *Injection*–that are all matched by the same IoT event pattern involving proximity, motion, and illuminance sensors. This is because they are all performed at the same location (the phlebotomy station) and involve similar movements and changes in light conditions. As a result, their IoT patterns are indistinguishable and cannot be uniquely mapped to a single process event. Whenever one of these activities is performed, three services–one per activity–each emit their associated process event. This results in an incorrect trace in the event log, misrepresenting the actual enactment.

We propose to contextualize ambiguous process events by leveraging an additional sensing modality to disambiguate events based on the execution context. Assuming that the physical execution environment is visually observable, we propose to use cameras and scene understanding to identify objects and human-object interactions in the visual scene depicting an ambiguous event. This is motivated by two facts. First, adequately positioned cameras can capture a broad view over the area where a process unfolds, including the human actors and the objects they interact with. Second, often disambiguation depends on the identification of the objects involved in an activity and how they are manipulated [26]–a task that modern computer vision techniques can carry out in near-real time. In this setting, IoT sensing provides candidate process events but cannot distinguish between activities that share indistinguishable IoT patterns, while visual sensing provides complementary contextual evidence for process event disambiguation.

3 Architecture with On-demand Disambiguation

In line with Design Science Research principles [21], we specify design objectives addressing the identified problem and let them guide our architecture design.

3.1 Design Objectives

Drawing on the ambiguity challenges observed in the phlebotomy environment and on insights from IoT-BPM literature (cf. [13,19]), we establish the following non-exhaustive functional (FDO) and non-functional (NDO) design objectives.

- **FDO1: Detect Ambiguity.** In IoT-augmented environments, event abstraction mechanisms such as the activity detection services in [24] often operate independently and may emit multiple process events for the same IoT event pattern. As these mechanisms are typically unaware of ambiguities they introduce by operating this way, the architecture should include functionality for detecting ambiguities introduced during event abstraction.
- **FDO2: Disambiguate On-demand.** Introducing an additional sensing modality for disambiguation yields computational overhead. To limit the impact on efficiency and scalability, additional sensing modalities should be used only upon ambiguity detection, instead of continuously acquiring and analyzing data.
- **FDO3: Disambiguate Based on Context.** Even with additional sensing modalities, data patterns may still appear similar across activities that are distinct (e.g., *tourniquet application* and *injection* both involve contact with the

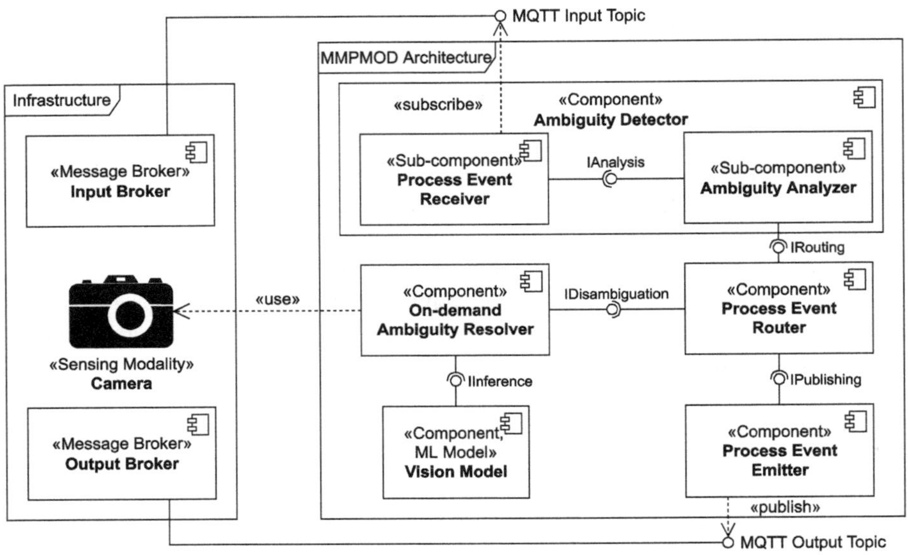

Fig. 2. UML Component diagram for the MMPMOD architecture

training arm and tool use). To distinguish between such cases and avoid misinterpretations, the architecture should rely on distinguishing contextual features such as relevant objects and human-object interactions.

- **NDO1: Performance.** The architecture should support online process monitoring and feedback provision for ongoing executions. Ambiguity detection and resolution should be performed online rather than offline, with acceptable tradeoffs between accuracy and latencies appropriate for the domain.

- **NDO2: Integrability.** As existing IoT setups are often complex and costly to modify, the architecture should integrate existing IoT-augmented environments without requiring extensive reconfiguration.

- **NDO3: Robustness, resilience, and recoverability.** The architecture should remain operational even when disambiguation cannot determine a single event interpretation. Moreover, it should support recoverability so that process event flow can continue or resume despite temporary failures.

- **NDO4: Security and privacy.** Since additional sensing modalities such as cameras may capture sensitive information (e.g., images of individuals' faces), the architecture should prevent such information from being exposed.

- **NDO5: Modularity.** As process environments evolve over time, the architecture should allow introducing new sensing modalities and ambiguity detection and resolution methods. Thus, it should comprise loosely coupled, interchangeable components that can be updated or replaced independently.

3.2 MMPMOD Architecture

The proposed architecture addresses the challenge of integrating heterogeneous sensing modalities to disambiguate events in an adaptive manner, without continuously processing all data sources. It comprises two external interfaces and five components, illustrated in Fig. 2 and described in detail in the following.

Ambiguity Detector. Ambiguity detection is a critical design objective (**FDO1**) of the architecture. The Ambiguity Detector is the component that fulfills this design objective: without it, the system would be unable to distinguish between ambiguous and unambiguous events. This component ingests a stream of IoT-based process events and identifies cases where two or more process event candidates exist for the same IoT event pattern. It consists of two sub-components: (i) a *Process Event Receiver*, which provides access to the stream of IoT-based process events through an input channel (e.g., MQTT-based), and (ii) an *Ambiguity Analyzer*, which analyzes the incoming event stream to detect ambiguity.

The Ambiguity Analyzer can employ different strategies to detect ambiguity. A general, implementation-agnostic approach is based on windowing the event stream. Depending on the process characteristics, this may involve multiple concurrent windows–one per process instance or execution thread–identified and correlated through attributes such as case IDs. This strategy supports processes with concurrency or parallelism. Under more restrictive assumptions, for example when the process guarantees no concurrent instances or parallel executions, the windowing approach can be simplified to a single window. The window type and size depend on the process characteristics; ambiguity manifests when a window contains multiple events–one per interpretation of an IoT event pattern. If the implementation of the activity detection procedures can be inspected easily, a static analysis of their code is an alternative strategy. By identifying overlapping conditions in the patterns matched by the procedures, this analysis can reveal ambiguities among the associated process events. The Ambiguity Analyzer can use these results to determine when disambiguation is required. The Analyzer hands control to the Process Event Router to initiate disambiguation if ambiguity is detected, and to initiate event republishing otherwise.

Process Event Router. A core requirement of the architecture is modularity: the system should be structured so that components can be developed, replaced, or extended without affecting others. While the decomposition into an Ambiguity Detector, an On-demand Ambiguity Resolver, and a Process Event Emitter already contributes to this goal, modularity also requires that these components do not embed assumptions about each other's control flow. To this end, the Process Event Router centralizes the coordination logic. It defines when ambiguity resolution is triggered, how results are processed, and when events are forwarded, thereby preventing the functional components from having to call or depend on one another. This design localizes control-flow changes to a single component: modifications to the handling of ambiguity such as alternative resolution strategies or different policies for unresolved cases affect only the Process Event Router rather than requiring changes in multiple modules (cf. **NDO5**). Concretely, this component handles three cases: (i) it receives requests from the Ambiguity Analyzer to resolve ambiguous events and invokes the On-demand Ambiguity Resolver; (ii) it receives disambiguation results from the On-demand Ambiguity Resolver and forwards the resulting unambiguous or remaining ambiguous events to the Process Event Emitter; and (iii) it receives requests from the Ambiguity Analyzer to publish unambiguous events and forwards these to the Process Event Emitter. By encapsulating coordination decisions, the architecture preserves modularity across detection, resolution, and emission components.

On-demand Ambiguity Resolver. On-demand disambiguation (cf. **FDO2**) and context-based disambiguation (cf. **FDO3**) are two other critical functional design objectives. The On-demand Ambiguity Resolver is the component that satisfies them. This component is invoked by the Process Event Router after the Ambiguity Analyzer identifies ambiguity in the input event stream. Once invoked, it activates the additional modality to acquire data about the current context. Next, it uses this data to attempt disambiguation. In case of successful disambiguation, it communicates the result, i.e., a unique event label, to the Process Event Router for publishing a corresponding disambiguated event. In case of unsuccessful disambiguation, it passes all ambiguous events–to be flagged as such–to the Process Event Router to let downstream tasks handle them. How such unresolved ambiguities are handled by these tasks is out of scope.

This component can adopt various modalities to support contextualization and disambiguation. In our lab setup, we use cameras that capture static images to acquire a visual representation of the current context. Specifically, we use only one downward-facing camera positioned above the training arm to avoid capturing sensitive information such as individuals' faces, but in principle multiple cameras could be used to capture images from different positions and angles.

Vision Model. A fine-tuned Vision Model (described in detail in Sect. 4.2) is invoked by the On-demand Ambiguity Resolver to disambiguate ambiguous process events. The Resolver provides the Vision Model with the relevant image data–captured around the time the ambiguity was detected–and the model performs a classification task that distinguishes between the possible interpretations of the IoT event pattern. The set of labels used in this task corresponds to the activity types defined in the monitored process model; these labels are either specified as part of a process ontology or derived from annotated training data collected during system setup. Although our implementation uses computer vision for this task, alternative modalities such as audio signals or normative process specifications (when run-time deviations are impossible or can be controlled for) can be integrated analogously into the Ambiguity Resolver.

Process Event Emitter. Analyzed process events–whether unambiguous or the result of a disambiguation attempt–must be made available to external systems consuming them (cf. **NDO2**). The Process Event Emitter fulfills this role. It receives process events forwarded by the Process Event Router, originating either from the Ambiguity Analyzer (for events deemed unambiguous) or from the On-demand Ambiguity Resolver (for events for which disambiguation was attempted). It then publishes each event as an individual message to an output channel. To ensure clarity for downstream consumers, the Emitter marks disambiguated events as unambiguous, while events for which disambiguation was inconclusive are explicitly flagged as ambiguous. This information facilitates subsequent analysis such as manual disambiguation [22] or algorithms that must account for uncertainty. In this way, the Process Event Emitter provides external systems, including conformance checking algorithms, with an event stream that reflects the current system-level interpretation of the IoT data.

4 Implementation and Evaluation

Here, in line with the demonstration and evaluation phases of [21], we discuss the architecture implementation, the phlebotomy vision model, and their evaluation.

4.1 Architecture Implementation

To demonstrate the feasibility of the proposed architecture and evaluate it, we implemented a prototype[1] for the phlebotomy scenario described in Sect. 2.

The prototype implements the Ambiguity Detector, On-demand Ambiguity Resolver, Process Event Router, and Process Event Emitter in Java (Spring Boot, Modular Monolith). The On-demand Ambiguity Resolver uses a camera as an additional data source and computer vision to achieve disambiguation. Its sub-components responsible for image capture and scene understanding are implemented in Python to leverage libraries dedicated to these tasks (cf. Sect. 4.2).

To integrate our preexisting phlebotomy environment (cf. **NDO2**), the Process Event Receiver in the Ambiguity Detector subscribes to an existing MQTT topic where activity detection services publish process events [24]. Likewise, the Process Event Emitter publishes messages encapsulating process events on another MQTT topic. This allows the new components to consume the preexisting input and produce the output without modifying the original infrastructure.

The Ambiguity Detector finds ambiguities via a *windowing strategy* based on three characteristics of the phlebotomy process in our lab setup. First, the process does not allow for parallel activity executions (cf. *parallel split* in [5]); likewise, blood extractions are carried out on one training arm at a time, i.e., there are no concurrent instances. Therefore, only one process event can occur at any given point in time. Second, based on our experience with the process enactment by HCW trainees [8], each activity lasts at least five seconds. Third, the activity detection services typically emit process events without significant latencies. Consequently, we treat multiple events observed within a one-second window as candidate interpretations of the same IoT event pattern originating from different activity detection services, i.e., to be ambiguous. Thus, the Ambiguity Detector leverages a one-second session window to detect ambiguity.

[1] The source code is publicly available at https://github.com/ics-unisg/mmpmod.

4.2 Vision Model for the Phlebotomy Process

The vision model focuses on three activities of the phlebotomy process: *Disinfection*, *Tourniquet application*, and *Injection*. To enable scene-based disambiguation, the model follows a three-stage pipeline comprising: (1) object segmentation and detection, (2) spatial interaction and temporal aggregation and reasoning, and (3) rule-based activity reasoning to infer clinically meaningful activities from camera images. This design provides an explainable and efficient mechanism for disambiguating events. The model is based on YOLOv11, selected for its trade-off between real-time inference and instance segmentation accuracy [23].

Object Segmentation and Detection. The model was trained on a dataset from [14], which contains 11,884 annotated images of realistic blood-drawing scenes with five annotated object classes: syringe, tourniquet, disinfectant wipe, training arm, and gloves. Each object instance is labeled with bounding boxes and pixel-level masks in YOLO format. During inference, the model outputs both bounding boxes and segmentation masks for each object, enabling fine-grained spatial reasoning, such as measuring relative distances and contact areas. This allows the model to capture manipulation cues that IoT data alone cannot capture [18].

Interaction and Temporal Reasoning. Detected gloves act as proxies for the HCW's hands. When both gloves are present, their segmentation masks are merged into a single manipulation region. For each detected object, the model computes a fused overlap score, balancing spatial precision and geometric coverage by weighting the intersection over union between masks and bounding boxes. The score gives 70% weight to mask overlap and 30% to bounding-box overlap, increasing robustness under partial occlusions. This score is complemented by the normalized Euclidean distance between glove and object centroids, capturing proximity cues. Within a fixed aggregation window of ten images (one every 100 ms for sufficient temporal separation between interaction states, yielding a 1 s window for stable interaction capture), detected activities are combined via majority voting.

Activity Reasoning. Once glove-object interactions are established, a symbolic reasoning layer interprets them into phlebotomy activities. The system identifies *injection* when gloves interact with both the *training arm* and the *syringe*, *tourniquet application* when gloves interact with the *training arm* and the *tourniquet* but not the syringe, and *disinfection* when gloves interact with the *training arm* and optionally the *disinfectant wipe* while the syringe is absent. Each activity's confidence is calculated as the mean detection confidence of its contributing objects, further adjusted by temporal consistency across the aggregation window. A robust confidence value is then derived by subtracting half of the standard deviation from the mean, penalizing unstable detections. This rule-based reasoning ensures explainability aligned with the semantics of the phlebotomy process.

4.3 Evaluation

The evaluation assesses the extent to which the proposed architecture fulfills the design objectives, using MMPMOD's implementation and its integration in an end-to-end monitoring scenario. In the spirit of transparency and reproducibility, the complete datasets and evaluation results are available in an online appendix[2].

Vision Model Evaluation. We first evaluated the vision model's ability to detect phlebotomy-relevant objects reliably in isolation using the test set with ground-truth annotations from [14]. Since YOLO inference is deterministic for fixed model weights and inputs, a single evaluation run was sufficient. The model achieved high performance across all objects. Gloves and the training arm were detected with near-perfect precision and recall; the syringe and tourniquet also showed strong performance, each exceeding an F1-score of 0.88. The disinfectant wipe proved most challenging due to its small size and motion blur, yet its recall of 0.69 was sufficient for robust recognition when combined with temporal reasoning. Qualitative inspection of full video sequences further confirmed that the model correctly identifies the temporal order of actions and transitions between activities. Most false detections occurred during brief ambiguous movements, for instance when gloves passed near the syringe without manipulation. Temporal smoothing and contextual reasoning effectively filtered out these inconsistencies.

End-to-End Evaluation. To evaluate the run-time performance of MMPMOD, we instantiated it in our lab setup. We deployed the prototype on a MacBook Pro M3 Max (36 GB RAM), connected with a Logitech StreamCam camera (1920×1080 resolution). We performed 50 complete phlebotomy enactments, thereby generating, through existing activity detection services, 1766 process events that the prototype ingested on the fly. We recorded the time required for each disambiguation, which includes camera activation overhead (a fixed average time of 0.85 s intrinsic to our technical setup), capture of ten images (one every 100 ms, cf. Sect. 4.2), and label inference. Moreover, we saved the events in output to compare them with a manually-defined ground truth to compute disambiguation accuracy. The ground truth was derived from predefined enactment scripts specifying the exact sequence of activities to be performed during each enactment. During enactment, two authors monitored adherence to the script in real time to detect deviations. Since the activity sequence was predetermined and verified during enactment, no post-hoc interpretive annotation process was required.

During process enactment, we witnessed 259 occurrences of multiple events appearing in the same session window, which were flagged as ambiguous by the Ambiguity Detector. 18 out of these 259 flagged ambiguities were false ambiguities caused by erroneous sensor readings rather than by an ambiguous activity execution. For these 18 false ambiguity occurrences, the Ambiguity Resolver was triggered and correctly returned no unique label in 16 occurrences; it incorrectly returned an activity label in 2 occurrences. The remaining 241 occurrences of

[2] https://doi.org/10.5281/zenodo.18892612.

flagged ambiguous events corresponded to genuine ambiguities that required resolution: of the respective disambiguation attempts, 22 did not result in a correct label. Specifically, in 16 attempts the system could not determine a label, and in 6 it produced an incorrect one. For all the other 219 attempts, the system returned the correct label, resulting in a disambiguation accuracy of $219/241 = 0.91$.

Unambiguous events were typically emitted within 10 ms. The time required for disambiguation attempts was on average 3.69 s (median: 3.63, σ: 0.17)–a value lower than the typical duration of the ambiguous activities. As disambiguation always completed before the start of the next activity, we conclude that on-demand vision-based disambiguation meets latency requirements in this scenario.

Ablation Study. To assess the impact of temporal visual aggregation on disambiguation, we conducted a post-hoc ablation study. In the implementation, disambiguation is performed by aggregating predictions from ten images captured over one second. To isolate the contribution of temporal aggregation, we re-executed disambiguation while varying the number of images used to derive a prediction. Specifically, we evaluated settings using zero images (using a common *null-model* baseline that randomly selects an interpretation, akin to [20]), one image (vision model without temporal aggregation), two, three, four, and five images (reduced temporal aggregation). In each case, all other components remained unchanged. This design allows examining how the amount of temporal visual context affects disambiguation accuracy and time. Because the images captured during process enactment were stored, we were able to perform this analysis ex post by re-running the vision model on them. For the one-image condition, we used the first captured image to predict the activity label. For larger image counts we aggregated predictions from the first $N \in \{2..5\}$ stored images.

Fig. 3 summarizes the accuracy results of the ablation study; the full results, including execution times, are in the online appendix. The null-model baseline achieved an accuracy of 0.31 (95% Wilson CI [0.26, 0.37]), while using visual contextualization with a single image reached 0.79 (95% Wilson CI [0.73, 0.84]). Aggregating multiple images further improved accuracy, with most of the gain obtained within five images, achieving 0.91 (95% Wilson CI [0.87, 0.94]), like with ten images. Since the five-image and ten-image conditions were evaluated on the same ambiguity instances, we compared them using McNemar's test. The test results indicated no statistically significant difference between the two conditions ($p = 1.0$). The two conditions differed in only three cases, suggesting that aggregating more than five images yielded no additional benefit in our setup.

The time for label inference decreased moderately and approximately linearly when reducing the number of images processed. For example, inference alone using a single image required 81% of the ten-image inference time; using a five-image aggregation required 89% of the ten-image inference time but already achieved the same accuracy. Thus, an accuracy of 0.91 can already be attained with five-image aggregates, each taking 2.99 s (0.85 s fixed camera activation overhead per disambiguation attempt, 0.5 s image acquisition, and 1.64 s inference) compared to 3.69 s for ten images. These results indicate that IoT data alone does not provide sufficient information for reliable disambiguation in our

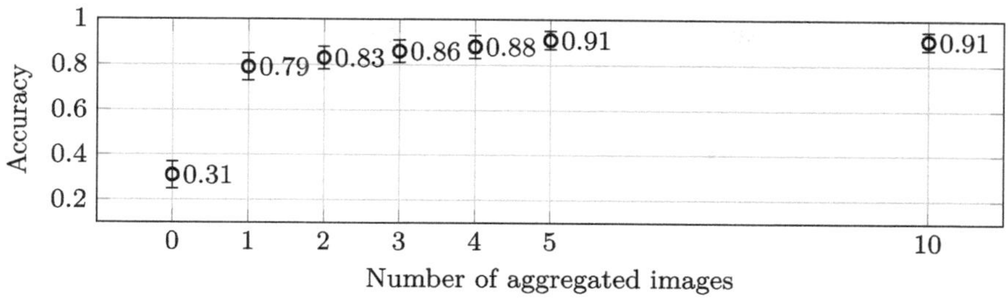

Fig. 3. Disambiguation accuracy for different temporal aggregation settings with 95% Wilson confidence intervals

setup, while visual context provides substantial benefit. In our setup most of this benefit is already achieved with a five-image aggregation.

4.4 Discussion

Fulfillment of design objectives. The evaluation indicates that MMPMOD fulfills the functional design objectives (**FDO1, FDO2, FDO3**) set out in Sect. 3.1. Moreover, it indicates that the non-functional design objectives can be satisfied by implementations of the architecture. The one presented in Sect. 4 was able to operate with latencies appropriate for the domain, fulfilling **NDO1**. We could deploy MMPMOD in our preexisting lab setup without reconfiguration; the Process Event Receiver consumes the same inputs and the Process Event Emitter produces outputs in the same format as the original setup. Therefore, **NDO2** is met. **NDO3** calls for robustness and recoverability, achieved with MQTT topics serving as the only external interfaces and messages as reliable processing snapshots; forwarding events that cannot be disambiguated–due to either sensor-induced false ambiguities or the inability to determine an activity–promotes resilience. Using a downward-facing camera, we can achieve confidentiality without compromising video data integrity and availability, thus fulfilling **NDO4**. Through its modular design, the architecture also meets **NDO5**.

Transferability and generality of the architecture. Except for the Vision Model, the components of MMPMOD are domain-agnostic. The architecture is therefore designed to be extensible to domains beyond phlebotomy. Its Vision Model, however, must be domain-specific, as it requires training on domain-relevant objects. From an external interface perspective, only two channels are needed: one for ingesting process events and another for publishing disambiguated events.

Our phlebotomy scenario does not involve concurrency or parallel execution; accordingly, in MMPMOD's implementation we rely on a single detection and resolution window. In settings that allow concurrent process instances or parallelism, the architecture can employ multiple windows, provided that events can be correlated with their respective instance or execution thread, for example via a *case ID* attribute (see Sect. 3.2). Shifting to a multi-window strategy requires changing the implementation of the *Ambiguity Analyzer* to ensure the isolated analysis of each individual window for instance- or thread-specific ambiguity detection. The mechanisms used to analyze the windows are domain-specific, since some processes may exhibit true parallelism, where multiple events can occur simultaneously in concurrent execution threads. For some processes, a minimum temporal separation between events may be guaranteed, which can be leveraged when defining the window size. Therefore, domain-specific strategies are required to isolate, correlate, and analyze events within their respective windows. Because no changes to the interface of the *Ambiguity Analyzer* are required, the remaining architectural components and their interactions remain unaffected. Overall, the architecture is equally flexible with respect to process changes and to the strategies used for ambiguity detection and resolution. Alternative detection mechanisms (e.g., heuristic, rule-based) and resolution strategies (e.g., audio-based analysis) can be integrated without changing the overall control flow. Additional modalities such as audio streams can be added, with integration effort arising primarily from incorporating the respective devices and adapters, adjusting detection and resolution strategies, and introducing or training new siblings of the Vision Model. This requires the ambiguous events to be known to train the models accordingly; no changes to the interfaces between components are required. Importantly, the Process Event Router and the Process Event Emitter remain unaffected, as MMPMOD operates on process events rather than raw IoT data.

Limitations. Our evaluation focuses on a single process with three recurring ambiguities in a realistic IoT-augmented environment. This process does not exhibit concurrency, therefore we adopted a single-window ambiguity detection strategy and did not empirically evaluate multi-window strategies (cf. Sect. 3.2). While this setting allows controlled assessment of ambiguity detection and on-demand disambiguation, further studies across additional domains and more complex processes including multi-window strategies would help assess the architecture's generality and scalability. Moreover, the evaluation focuses on how contextual sensing complements IoT-based event abstraction rather than comparing with monitoring approaches that continuously acquire video data. This design choice was deliberate and resulted in only 241 out of 1766 (13.6%) process events requiring video data acquisition, limiting the associated processing. Nevertheless, such comparisons represent an interesting direction for future work to assess the architecture's strengths and limitations in relation to state-of-the-art approaches.

From Design Objectives and Decisions to Design Principles. Reflecting on our problem and requirements analysis, our architecture design, and the evaluation of the architecture implementation, we derived five guiding principles for designing multimodal process monitoring systems with disambiguation support:

- **DP1: Selective Modality Activation.** *Activate additional sensing modalities when and only when needed, taking the confidence in the current modality and the detection of ambiguity into account.* This principle generalizes the on-demand activation of the Ambiguity Resolver to reduce computational overhead and latency. In MMPMOD and its implementation, a reactive, rule-based mechanism activates the vision modality upon ambiguity detection. On a more general level of knowledge, we can see this reusable design as a combination of two Enterprise Integration Patterns [12]: Selective Consumer and Content-Based Router.

- DP2: Explicit Modeling and Management of Ambiguity. *Treat ambiguity as a first-class concern detected, communicated, and handled holistically.* This principle generalizes the ambiguity detection and disambiguation functionalities of the Ambiguity Detector and the On-demand Ambiguity Resolver. It emphasizes that ambiguity should not be treated merely as a technical side effect of IoT data abstraction, but as an explicit phenomenon addressed across the BPM life cycle. Current practice rarely models ambiguity explicitly; doing so provides a shared conceptual foundation for structuring detection procedures and resolution strategies. Domain-Driven Design [7] offers one way to support this principle: the domain model can include elements for ambiguity types, detection rules, and resolution strategies, forming a Ubiquitous Language that supports consistent communication and interdisciplinary stakeholder interactions.

- DP3: Contextualized Disambiguation. *Disambiguate by incorporating contextual information such as detected objects and human-object interactions obtained from complementary sensing modalities.* This principle generalizes the use of contextual cues, such as objects and human-object interactions, for event interpretation. It emphasizes the role of context as an enabler of higher disambiguation accuracy. Like ambiguity, contextualization should be a first-class design concern. The context dimensions should be defined as part of the requirements elicitation and domain modeling [7]. Domain-specific event patterns also require attention: recurring combinations of event sequences as well as provenance data (e.g., sensor type, configuration, and placement) typically constitute these patterns. Time series databases, event sourcing and streaming, and other event processing concepts [6] are eligible design options to detect event pattern occurrences. The Strategy pattern [10] can be leveraged to account for contextual differences, flexible event pattern detection, and resolution of ambiguities.

- DP4: High Cohesion in, Loose Coupling of Process Monitoring Components. *Structure multimodal monitoring systems with independently deployable components.* The deployability of multimodal monitoring architectures benefits from structuring the system as a set of independently deployable components, each exposing a single responsibility and interacting through loosely coupled, open interfaces. This domain-specific refinement of general software engineering principles supports maintainability and manageability by allowing components to scale, change, or be replaced independently, without impacting the rest of the system. In multimodal monitoring systems, it facilitates integration with existing infrastructures and the addition of sensing modalities. Contemporary architectural patterns such as microservices and modular monoliths, ports-and-adapters [3], and asynchronous messaging [12] make this principle actionable.

- DP5: Value- and Cost-Driven Prioritization of Disambiguation Requirements. *Identify modality-specific conflicts between quality goals and find case-specific tradeoffs resolving them.* Tradeoffs between latency and accuracy are inherent to handling ambiguities in context, as acquiring and processing contextual information improves disambiguation accuracy but introduces additional delays. For low latency, our On-demand Ambiguity Resolver can be configured to capture fewer camera images that are processed faster but with limited accuracy; if accuracy is critical, more images can be used to disambiguate an event at the price of higher latency. Our ablation study illustrates this principle empirically: while aggregating more images improved disambiguation accuracy, most gains were achieved with five images, highlighting the need to balance latency and accuracy. Such quality goal conflicts should be identified early in any process monitoring system so that relative priorities can be negotiated, tradeoffs be found and informed decisions be made. Attribute-Driven Design and Architecture Tradeoff Analysis Method [1], customized (i.e., extended or trimmed down) for the engineering of continuously changing systems, may support this principle.

Each principle abstracts a design rationale that is essential for fulfilling the functional and non-functional design objectives. Collectively, the principles constitute tangible design knowledge for crafting adaptive multimodal process monitoring systems with on-demand disambiguation, which is general enough to be applicable across domains and scenarios. The principles are detailed enough to be actionable and testable: explicit ambiguity management (**DP2**) enables selective activation (**DP1**), which employs contextualized disambiguation (**DP3**), realized within a modular architecture (**DP4**) with adequate tradeoffs (**DP5**).

5 Related Work

Multimodal process mining (MMPM) is an emerging sub-discipline that investigates the integration of data of different types–e.g., sensor, textual, visual, auditory–for a holistic process mining [11]. The proposed MMPMOD architecture can be seen as an instance of MMPM specifically focused on process monitoring, which integrates IoT sensor data and video data. A related instance is found in [22], which proposes a multimodal activity recognition approach to process discovery. Similar to MMPMOD, it explicitly addresses ambiguity introduced with event abstraction. Unlike our on-demand architecture, it continuously acquires video data, requiring dedicated resources for video analysis. Moreover, it performs full-frame rather than semantics-informed classification. A similar approach to detect activities from pixel values in specific regions of camera images without extracting semantic information is found in [15]. Another instance of MMPM is the pipeline for activity recognition in robotic systems proposed in [4]. It fuses visual and auditory robotic data with foundational models by leveraging the models' zero-shot reasoning capabilities. While zero-shot reasoning alleviates the need to train a model–which is a prerequisite in our prototype–, it is currently challenging to leverage foundational models beyond offline settings: further advancements of these models, however, could enable online applications.

Beyond MMPM, process mining has successfully leveraged video data analysis before, for instance for conformance checking [2] and behavioral pattern discovery [18]. Unlike approaches that derive process events directly from video data [11,18], MMPMOD assumes IoT-based event abstraction and uses computer vision only as contextual evidence for disambiguating candidate events. This makes direct quantitative comparisons difficult, as these systems address different monitoring assumptions and data acquisition strategies. In [17], a reference architecture (ViProMiRA) has been proposed to extract process information from video data, thereby enabling post-mortem analysis tasks such as conformance checking and process discovery. While ViProMiRA leverages supervised learning, the more recent RAVEE [25] advances the former by enabling extracting process information from video data in an unsupervised setting. These advancements offer an opportunity to alleviate the effort to train a fine-tuned model in our prototype. More generally, along with other work, e.g., [4,16], they point to the opportunity to integrate our MMPMOD architecture with models able to reliably analyze data from additional modalities without specific training.

6 Conclusion

We proposed MMPMOD as an objectives-driven component architecture for multimodal process monitoring with on-demand disambiguation. Its instantiation in a phlebotomy scenario demonstrated the feasibility of orchestrating complementary sensing modalities for disambiguation in online monitoring. Further investigation of the latency-disambiguation accuracy tradeoff remains a future research opportunity. While our vision model proved reliable in detecting objects and interactions, we plan to further enhance its detection capabilities. Finally, although we instantiated MMPMOD in a single scenario, its modular design suggests potential applicability beyond phlebotomy and may serve as a design blueprint for similar systems, which should be validated in future work. Our five generalized, actionable design principles support such reuse.

Acknowledgments. This work has received funding from the Swiss National Science Foundation under Grant No. IZSTZ0_208497 and under Grant No. 10002384.

Disclosure of Interests. The authors have no competing interests to declare.

References

1. Bass, L., Clements, P., Kazman, R.: Software Architecture in Practice, 4th edn., Addison-Wesley Professional (2021)
2. Chen, S., Zou, M., Cao, R., Zhao, Z., Zeng, Q.: Video process mining and model matching for intelligent development: conformance checking. Sensors **23**(8), 3812 (2023)

3. Cockburn, A., de Paz, J.M.G.: Hexagonal architecture explained: how the ports and adapters architecture simplifies your life, and how to implement it, Updated 1st Edn. Humans and Technology Inc. (2025)
4. Corradini, F., Pettinari, S., Re, B., Rossi, L., Sampaolo, M.: Multimodal zero-shot activity recognition for process mining of robotic systems. In: ER, M., et al. (eds.) International Conference on Business Process Management, pp. 263–275. Springer (2025). https://doi.org/10.1007/978-3-032-02936-2_19
5. van Der Aalst, W.M., Ter Hofstede, A.H., Kiepuszewski, B., Barros, A.P.: Workflow patterns. Distrib. Parallel Databases **14**(1), 5–51 (2003)
6. Etzion, O., Niblett, P.: Event Processing in Action, 1st edn. Manning Publications Co., USA (2010)
7. Evans, E.: Domain-Driven Design: Tackling Complexity in the Heart of Software. Addison-Wesley Longman Publishing Co., Inc, USA (2003)
8. Franceschetti, M., Seiger, R., García-Bañuelos, L., Weber, B.: Proambition, reloaded: a two-year retrospection. Research Projects Exhibition@ CAiSE, 2025 (2025)
9. Franceschetti, M., Seiger, R., López, H.A., Burattin, A., García-Bañuelos, L., Weber, B.: A characterisation of ambiguity in BPM. In: Almeida, J.P.A., Borbinha, J., Guizzardi, G., Link, S., Zdravkovic, J. (eds.) International Conference on Conceptual Modeling, pp. 277–295. Springer (2023). https://doi.org/10.1007/978-3-031-47262-6_15
10. Gamma, E., Helm, R., Johnson, R., Vlissides, J.: Design Patterns: Elements of Reusable Object-Oriented Software. Addison-Wesley Longman, USA (1995)
11. Gavric, A., Bork, D., Proper, H.: Multimodal process mining. In: 2024 26th International Conference on Business Informatics (CBI), pp. 99–108. IEEE (2024)
12. Hohpe, G., Woolf, B.: Enterprise Integration Patterns: Designing, Building, and Deploying Messaging Solutions. Addison-Wesley Longman Publishing (2003)
13. Janiesch, C., Koschmider, A., Mecella, M., Weber, B., Burattin, A., Di Ciccio, C., et al.: The internet of things meets business process management: a manifesto. IEEE Syst. Man Cybern. Mag. **6**(4), 34–44 (2020)
14. Jimenez Cruz, R., et al.: Labeled images of a simulated phlebotomy procedure on a training arm (2025). https://doi.org/10.5281/zenodo.16924786
15. Knoch, S., Ponpathirkoottam, S., Fettke, P., Loos, P.: Technology-enhanced process elicitation of worker activities in manufacturing. In: Teniente, E., Weidlich, M. (eds.) International Conference on Business Process Management, pp. 273–284. Springer (2017). https://doi.org/10.1007/978-3-319-74030-0_20
16. Knoch, S., Ponpathirkoottam, S., Schwartz, T.: Video-to-model: unsupervised trace extraction from videos for process discovery and conformance checking in manual assembly. In: Fahland, D., Ghidini, C., Becker, J., Dumas, M. (eds.) International Conference on Business Process Management, pp. 291–308. Springer (2020). https://doi.org/10.1007/978-3-030-58666-9_17
17. Kratsch, W., König, F., Röglinger, M.: Shedding light on blind spots-developing a reference architecture to leverage video data for process mining. Decis. Support Syst. **158**, 113794 (2022)
18. Lepsien, A., Koschmider, A., Kratsch, W.: Analytics pipeline for process mining on video data. In: Di Francescomarino, C., Burattin, A., Janiesch, C., Sadiq, S. (eds.) International Conference on Business Process Management, pp. 196–213. Springer (2023). https://doi.org/10.1007/978-3-031-41623-1_12

19. Mangler, J., Seiger, R., Benzin, J.V., Grüger, J., Kirikkayis, Y., Gallik, F.: From internet of things data to business processes: challenges and a framework. In: Fortino, G., Mecella, M. (eds.) Internet of Things Meets Business Process Management: A Synergistic Integration, pp. 25–61. Springer (2026). https://doi.org/10.1007/978-3-031-90746-3_2
20. Metzner, C., et al.: Classification at the accuracy limit: facing the problem of data ambiguity. Sci. Rep. **12**(1), 22121 (2022)
21. Peffers, K., Tuunanen, T., Rothenberger, M.A., Chatterjee, S.: A design science research methodology for information systems research. J. Manag. Inf. Syst. **24**(3), 45–77 (2007)
22. Rebmann, A., Emrich, A., Fettke, P.: Enabling the discovery of manual processes using a multi-modal activity recognition approach. In: Di Francescomarino, C., Dijkman, R., Zdun, U. (eds.) B International Conference on Business Process Management, pp. 130–141. Springer (2019). https://doi.org/10.1007/978-3-030-37453-2_12
23. Redmon, J., Divvala, S., Girshick, R., Farhadi, A.: You only look once: unified, real-time object detection. In: Proceedings of the IEEE Conference on Computer Vision and Pattern Recognition (CVPR), pp. 779–788 (2016)
24. Seiger, R., Kurz, A.F., Franceschetti, M.: Online detection of process activity executions from IoT sensors using generated event processing services. Futur. Gener. Comput. Syst. **174**, 107987 (2026)
25. Wördehoff, N., Egger, A., Kratsch, W., König, F., Röglinger, M.: Beyond assumptions: a reference architecture to enable unsupervised process discovery from video data. Decis. Support Syst., 114544 (2025)
26. Wu, J., Osuntogun, A., Choudhury, T., Philipose, M., Rehg, J.M.: A scalable approach to activity recognition based on object use. In: International Conference on Computer Vision, pp. 1–8. IEEE (2007)

A Hybrid LLM-Based Framework for Automated Security Annotation Generation in Business Process Models

Md Kamrul Islam[1(✉)], Tiphaine Henry[2], Mattia Salnitri[3], Julius Köpke[4], and Sami Souihi[2]

[1] CentraleSupélec, Gif-sur-Yvette, France
mdkamrul.islam@student-cs.fr
[2] Université Paris-Saclay, Palaiseau, France
{tiphaine.henry,sami.souihi}@cea.fr
[3] University of Bergamo, Bergamo, Italy
mattia.salnitri@unibg.it
[4] University of Klagenfurt, Klagenfurt, Austria
julius.koepke@aau.at

Abstract. The modelling and analysis of secure business processes require the incorporation of security annotations into process models. Although BPMN extensions, including SecBPMN2, exist for this purpose, the derivation of accurate and complete security annotations from natural-language specifications remains a manual, expert-intensive, and error-prone task. This paper presents a hybrid framework that takes a BPMN process model and a security requirements document as input and automatically generates security annotations adhering to the SecBPMN2 specification. The approach combines Large Language Model (LLM)–based semantic extraction with schema-constrained mapping, rule-based normalization, and deterministic validation. The framework is evaluated comprehensively on a curated dataset of 27 process models from various domains. The results indicate that it consistently produces structurally valid SecBPMN2 annotations with high schema completeness. Compared to human security analysts, the system achieves substantially higher precision (0.58 vs. 0.29) while maintaining comparable recall (0.52 vs. 0.50) and reduces erroneous or misplaced annotations by nearly 50%. In addition, annotation generation is significantly faster than manual annotation. These findings demonstrate that hybrid LLM- and rule-based automation can reduce modeling effort while improving consistency and reliability, thereby providing a scalable foundation for security-by-design BPM.

Keywords: SecBPMN2 · Large Language Models · Information Extraction

F. M. Maggi et al. (Eds.): BPM 2026, LNBIP 599, pp. 168–184, 2027.
https://doi.org/10.1007/978-3-032-37877-4_10

1 Introduction

BPMN 2.0[1] is the established industry standard notation for modeling business processes across the BPM lifecycle. However, BPMN lacks native support for expressing security requirements, which is increasingly problematic in regulated and data-sensitive environments [3]. To address this limitation, several extensions-including SecureBPMN [1], SecBPMN [15], and PE-BPMN [13] have been proposed to enable security-by-design by introducing explicit security annotations at the process-modeling level. Despite their expressive power, these extensions have seen limited adoption in practice. This limited adoption is largely due to the fact that security annotations are typically derived manually from natural-language descriptions, a task that requires specialized expertise, is prone to inconsistency, and scales poorly in complex and evolving regulatory contexts [14].

Recent advances in LLMs have demonstrated strong capabilities in interpreting unstructured text and generating structured process representations [7,11]. However, security-annotated process modeling poses additional challenges, as security requirements are often implicit, context-dependent, and linguistically heterogeneous. In this work, we focus on SecBPMN2, which extends BPMN with semantically grounded vocabulary of security annotations and precisely defined attachment rules, enabling explicit and verifiable specification of properties such as confidentiality, integrity, accountability, and separation of duties [15].

While these constraints are essential for systematic validation, their rich semantics imposes extensive effort for human modelers and substantial challenges for LLM-based text-to-model generation. These challenges motivate the following research question.

> **RQ.** *How can an LLM augmented system automatically annotate BPMN models with SecBPMN2-compliant security annotations based on textual process descriptions?*

To answer this question, this paper makes three contributions. First, we propose a hybrid automation pipeline that derives SecBPMN2 annotations from natural-language descriptions by coupling LLM-based extraction and mapping with rule-based normalization and validation. Second, we introduce a benchmark dataset aligning textual process descriptions with expert-designed SecBPMN2 ground-truth models. Finally, we present an empirical evaluation across multiple LLMs and prompting strategies, including Retrieval-Augmented Generation (RAG), using quantitative metrics and human baselines. Together, these contributions advance LLM-based support for security-aware process modeling by reducing manual effort, improving consistency, and enabling the systematic operationalization of SecBPMN2 annotations.

The remainder of the paper is organized as follows. Section 2 reviews related work, Sect. 3 presents a motivating example, and Sect. 4 describes the proposed automation pipeline. Section 5 outlines the prototype implementation, Sect. 6 reports the experimental evaluation, and Sect. 7 concludes the paper.

[1] https://www.omg.org/spec/BPMN.

2 Related Work

Table 1. Comparison of related work (NL = natural language).

Approach	NL Input	Auto-mation	Security Semantics	Process Integration	Constraint Enforcement
Security extensions [1,2,13,15]	No	Manual	Explicit	BPMN/SecBPMN2	Yes
Rule-based generation [4]	Yes	Partial	None	BPMN	No
LLM-based generation [7–9,11]	Yes	Yes	None	BPMN	No
Security req. extraction [16,19]	Yes	Yes	Explicit (text)	None	Partial
This work	Yes	Yes	Explicit (SecBPMN2)	BPMN/SecBPMN2	Yes

Research on security-aware BPM spans three main directions: BPMN security extensions, automated BPMN generation from text, and extraction of structured security requirements. These directions differ along five dimensions- n natural-language input, automation, security semantics, integration with BPMN artifacts, and constraint enforcement-as summarized in Table 1.

The first line of research extends BPMN with explicit security constructs. BPMN primarily captures functional behavior and provides limited support for security and privacy concerns [3]. Extensions address this gap: SecureBPMN introduces authorization and duty constraints [1], and PE-BPMN focuses on privacy and data protection [13]. SecBPMN and SecBPMN2 provide a comprehensive framework for security modeling. SecBPMN2-ml defines a structured vocabulary of security goals grounded in RMIAS [2,15], while SecBPMN-Q enables formal policy verification. These approaches provide explicit semantics, tight BPMN integration, and constraint enforcement, but rely on manual annotation and do not support natural-language input.

The second line focuses on automating BPMN generation from textual descriptions. Early approaches rely on syntactic parsing and rule-based extraction [4], while recent work leverages large language models, including translation pipelines [8], iterative prompting [9], conversational modeling [7], and direct BPMN XML generation [11]. These methods support natural-language input and automated model construction, but focus on functional structure and do not incorporate explicit security semantics or constraint enforcement.

The third line concentrates on extracting structured security requirements from natural-language specifications. Empirical studies show that LLMs can identify entities and relations but struggle to construct complete and semantically consistent formal models under controlled benchmarks [19]. Domain-specific approaches derive structured requirements from standards such as ISO/IEC 27001 [16]. While these methods produce explicit security representations at the textual level, they operate largely independently of BPMN artifacts and provide limited process-level constraint enforcement.

In summary, existing work addresses either security semantics, automation from text, or structured requirement extraction. No existing approach integrates natural-language input, automated generation, SecBPMN2 semantics, BPMN-level integration, and systematic constraint enforcement within a framework.

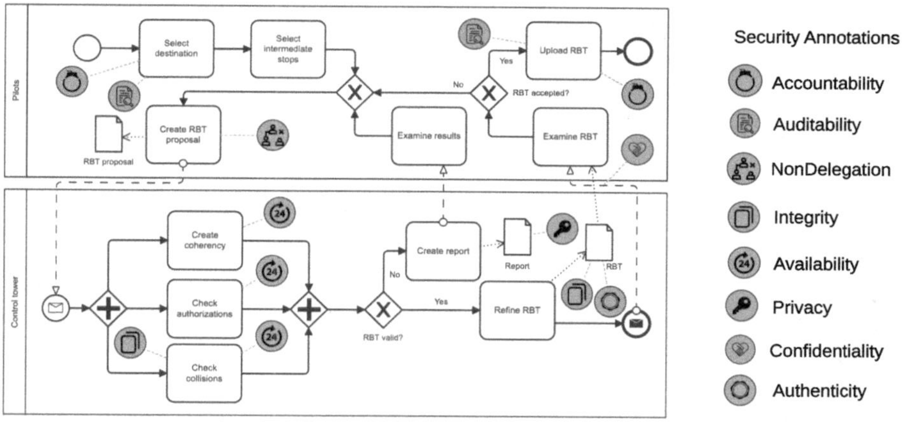

Fig. 1. Simplified RBT negotiation process annotated with SecBPMN2 [15].

3 Motivating Example

Automated security annotation is particularly important in settings that involve sensitive or safety-critical information. A representative example is the Route-Based Trajectory (RBT) negotiation process introduced by Salnitri et al. [15], which models the iterative exchange between pilots and air-traffic controllers during the proposal, revision, and approval of flight trajectories. Figure 1 depicts a simplified version of this process and serves as the motivating example used throughout this paper. The RBT process exchanges highly sensitive trajectory data and therefore requires confidentiality of flight messages, authenticity of proposed and approved trajectories, accountability of approving actors, auditability of critical decisions, and continuous availability of safety checks. In practice, these requirements are expressed informally (e.g., restricting message access, logging approvals, or ensuring continuous collision checks) and are not directly translatable into formal security annotations. Mapping such requirements to SecBPMN2 requires identifying relevant BPMN elements, selecting appropriate security goals, and instantiating schema-valid parameters in accordance with SecBPMN2 attachment rules. Even for compact models, this manual translation requires substantial expertise and effort and is prone to inconsistency and omissions.

The RBT example highlights a fundamental challenge in security-aware BPM. Security requirements are often expressed in flexible, implicit, and context-dependent natural language, whereas SecBPMN2 requires explicit, schema-constrained annotations attached to specific BPMN elements.

4 Approach

Building on the challenge identified in the motivating example, this section details the proposed approach for automatically generating SecBPMN2 secu-

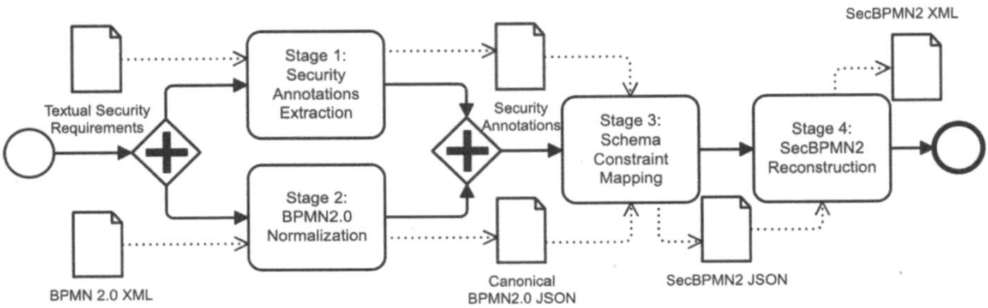

Fig. 2. Four-stage hybrid pipeline for deriving SecBPMN2 annotations from textual security requirements.

rity annotations from natural-language process descriptions. The approach takes as input (i) a syntactically valid BPMN 2.0 model and (ii) a natural-language description of security-relevant requirements. It produces a SecBPMN2-compliant process model in which security annotations are instantiated, validated, and attached to the appropriate BPMN elements. To bridge this gap, the approach decouples semantic interpretation from structural validation. LLM components perform security-goal extraction from text, while rule-based components enforce SecBPMN2 attachment rules, schema constraints, and structural consistency.

In the following, (i) *security requirements* denote textual security requirements expressed in the input documentation; (ii) *Security goals* denote the abstract security properties (e.g., confidentiality, integrity, or availability) derived from these requirements; (iii) *Security annotations* denote the SecBPMN2 constructs that instantiate these goals and are attached to BPMN elements.

4.1 Overview of the Hybrid Approach

The proposed methodology is a four-stage pipeline, illustrated in Fig. 2. Stages 1 and 2 operate in parallel. Stage 1 analyzes the natural-language process description to extract security goals without committing to concrete BPMN elements. Stage 2 transforms the BPMN 2.0 XML model into a normalized, graph-oriented JSON representation that exposes the structural primitives required for downstream reasoning. In Stage 3, the outputs of these two stages are combined through an LLM-assisted, schema-constrained mapping procedure that enforces SecBPMN2 attachment rules. Stage 4 then reconstructs the enriched model as SecBPMN2-compliant XML using rule-based templates.

The pipeline's design is guided by two principles. The semantic interpretation and structural validation are decoupled to prevent linguistic ambiguity from propagating into violations of SecBPMN2 annotations. The intermediate representations are reversible. This ensures the traceability from textual requirements to process models. The usage of JSON allows current LLMs to produce outputs that practically fully comply with the schemas [7]. However, the particular serialization format is an implementation issue that can be adopted based on the capabilities of particular LLMs.

Before detailing the individual stages, we formalize the modeling assumptions and annotation schema that define the scope and constraints of the approach.

4.2 Modeling Assumptions and Security Annotation Schema

Assumptions. The method operates on BPMN 2.0 within the subset supported by SecBPMN2. The input model is assumed to be syntactically valid and to expose the BPMN elements relevant for security annotation (tasks, events, gateways, message flows, data objects, and organizational elements).

The accompanying textual description is expected to reference these elements either directly or indirectly through actors, activities, communications, or information artifacts. The approach does not require references to match BPMN element names or types verbatim; for example, actor-centric descriptions such as "pilots are accountable for selecting a destination" can be aligned with the corresponding BPMN activity through semantic mapping. Semantic alignment is established during Stage 3 using textual anchors and process context. Security requirements are assumed to be expressible using SecBPMN2 annotation types [6]. Descriptions that lack any identifiable process-level anchor may lead to ambiguous mappings and reduced annotation accuracy.

Schema and Notation. Let J_{bpmn} denote the normalized BPMN-JSON representation of the input model (produced in Stage 2), $\mathcal{L}$ the set of candidate security annotations extracted from text (Stage 1), and J_{sec} the resulting SecBPMN-JSON model after mapping and validation (Stage 3). The annotation schema enforced by our approach follows SecBPMN2-ML and represents a security annotation as the tuple $\alpha = (e, g, p)$, where $e \in \mathcal{E}$ denotes a BPMN element, g a security goal, and p a goal-specific parameter set. BPMN elements are partitioned as $\mathcal{E} = \mathcal{E}_{\text{Act}} \cup \mathcal{E}_{\text{DO}} \cup \mathcal{E}_{\text{MF}} \cup \mathcal{E}_{\text{GW}} \cup \mathcal{E}_{\text{ORG}}$. Schema-compliant security associations satisfy a compatibility predicate $\mathit{compat}(g, \mathit{type}(e))$, which enforces type-level consistency between security goals and BPMN element categories (e.g., confidentiality applies only to data objects and message flows). These definitions lay the ground for the algorithmic enforcement of the schema during mapping stage.

Each SecBPMN2 annotation type encodes its admissible BPMN element category through a suffix $s \in \{\text{Act}, \text{DO}, \text{MF}, \text{GW}, \text{ORG}\}$, which corresponds to the respective partition of $\mathcal{E}$. This suffix determines the permissible target category type(e) for a constraint attached to a BPMN element.

4.3 Stage 1: Security Annotations Extraction

Stage 1 derives a set of candidate security annotations $\mathcal{L}$ from the natural-language process description provided by the security expert. The objective is to identify security goals g (e.g., accountability, auditability, integrity) together with an inferred target type(e) $\in \mathcal{E}$ (e.g., activity, data object, message flow, gateway, participant), while avoiding the concrete BPMN 2.0 element identifiers. At this stage, extracted annotations are semantic and model-agnostic: each annotation encodes a security goal g, goal-specific parameter placeholders p, and textual anchors (e.g., referenced actions, data items, or actors), but is not bound to a specific BPMN element e. This separation is a deliberate design choice motivated by the distinction between semantic interpretation and structural validation. Security requirements are often expressed independently of concrete BPMN element identifiers, whereas SecBPMN2 annotations must satisfy strict attachment constraints. By first extracting model-agnostic security annotations and postponing element-level attachment, the approach decouples semantic reasoning from structural mapping, enabling deterministic validation and reducing the risk of schema-incompatible annotations.

A role-based structured prompt combined with chain-of-thought reasoning and few-shot examples extracts a set of candidate security annotations $\mathcal{L}$ under type and parameter constraints [18]. The prompt guides the LLM through a structured socio-technical analysis aligned with STS-ML [12]. It extracts three perspectives: (i) a *social view* describing actors, roles, and their relationships, (ii) an *information view* identifying relevant data objects and exchanges, and (iii) an *authorization view* capturing access rights and delegation assumptions relevant for extracting security annotations. It incorporates a security annotation catalog that maps each SecBPMN2 annotation type-identified by a suffix $s \in \{\mathrm{Act}, \mathrm{DO}, \mathrm{MF}, \mathrm{GW}, \mathrm{ORG}\}$ - to its admissible BPMN element category in $\mathcal{E}$ and required parameter set p.

The output of Stage 1 is a schema-conformant JSON representation of $\mathcal{L}$. Each label specifies a candidate security goal g with goal-specific parameters p, together with an inferred target category $\mathrm{type}(e) \in \mathcal{E}$ and associated textual anchors. For the RBT negotiation process, Stage 1 yields annotations corresponding to goals such as accountability, auditability, and integrity, instantiated as SecBPMN2 annotation types (e.g., `accountabilityAct`, `auditabilityAct`, `integrityAct`).

Multiple extraction configurations are supported, including alternative prompting strategies and retrieval-augmented extraction. When enabled, concise definitions from SecBPMN2 documentation are retrieved and incorporated into the prompt to improve alignment with the intended security semantics.

4.4 Stage 2: BPMN 2.0 Normalization

Stage 2 constructs the normalized BPMN-JSON representation J_{bpmn} from the BPMN 2.0 XML input. The normalization yields a canonical, token-efficient representation that serves as the structural basis for downstream processing.

It retains all BPMN constructs, including tasks and events, gateways, message flows, data objects and data object references, as well as participants (pools), lanes, message definitions, and data associations. Each element is represented using a stable identifier and explicit connectivity (e.g., incoming and outgoing relations), while XML-specific syntactic overhead, such as namespaces, deeply nested tags, and presentation-oriented BPMN-DI layout metadata, is omitted. The original BPMN XML, including BPMN-DI, is preserved for visualization and export in later stages.

Beyond flattening element encodings, normalization reconstructs an explicit control-flow structure through a linear-time traversal of the BPMN control-flow graph encoded in the model. Gateways are represented as structured branching nodes with explicit paths and retained conditions. The resulting J_{bpmn} provides a compact, machine-oriented representation that reduces token overhead while preserving the identifiers and relations required for deterministic reconstruction and unambiguous label-to-element mapping in Stage 3, aligning with evidence that lightweight, text-centric encodings outperform verbose formats such as XML for LLM-based structured reasoning [20]. Although alternative graph-based representations could also be employed, the chosen format offers a practical balance between structural fidelity, reversibility, and integration with subsequent validation and reconstruction steps.

4.5 Stage 3: Mapping Security Annotations to BPMN Elements

Stage 3 maps the security annotations $\mathcal{L}$ from Stage 1 to BPMN2.0 elements in the normalized BPMN-JSON representation J_{bpmn} (Stage2). Given $(J_{\mathrm{bpmn}}, \mathcal{L})$, the objective is to construct a SecBPMN2-JSON model J_{sec} that preserves the BPMN structure while augmenting it with security annotations (represented as `secConstraints`) and corresponding `secAssociations`. Each candidate annotation is instantiated as a concrete SecBPMN2 annotation $\alpha = (e, g, p)$ by associating it with a BPMN element identifier that satisfies the schema constraints defined in Sect. 4.1.

Mapping is performed using a single low-temperature LLM invocation ($T = 0.1$) followed by deterministic validation. The prompt consists of (i) the normalized BPMN-JSON representation, (ii) a compact index containing BPMN element identifiers, labels, types, and participant information, and (iii) the candidate security annotations extracted in Stage 1. The LLM is not tasked with generating a BPMN model; instead, it preserves the BPMN representation unchanged and generates only SecBPMN2 security constraints and their associations (Listing 10.1).

Listing 10.1. Representative Stage 3 prompt.

```
Input: BPMN Base, Compact BPMN Index, Security Labels

Task: Copy the BPMN Base unchanged. Map security labels to
schema-compatible BPMN element identifiers using participant
and type constraints. Generate only securityConstraints and
securityAssociations.
```

Since Stage 1 provides only textual anchors (e.g., references to actors, activities, data items, or communications), the LLM resolves paraphrases and semantic matches under BPMN type constraints and participant scoping. The resulting mappings are subsequently validated by checking target existence, attachment

Algorithm 1. Stage 3: Security annotation mapping and validation.

Require: BPMN model J_{bpmn}, annotations $\mathcal{L}$, LLM M, validator V
Ensure: SecBPMN-JSON model J_{sec}

$P \leftarrow \text{RenderPrompt}(J_{\text{bpmn}}, \mathcal{L})$
$r \leftarrow \text{CallLLM}(\mathsf{M}, P)$
$J_{\text{sec}} \leftarrow \text{ParseJSON}(r)$
if J_{sec} invalid **then**
    $J_{\text{sec}} \leftarrow \text{FallbackMapping}(J_{\text{bpmn}}, \mathcal{L})$
end if
$J_{\text{sec}} \leftarrow \mathsf{V}.\text{ValidateAndFilter}(J_{\text{sec}}, J_{\text{bpmn}})$
return J_{sec}

cardinality, and type compatibility through the predicate $compat(g, type(e))$. Any invalid mappings are removed, yielding a validated SecBPMN2-JSON model J_{sec}.

Example. In the RBT negotiation process (Fig. 1), a requirement such as the revision communication employs advanced encryption methods" is extracted as a confidentiality annotation with an intended message-flow target. During Stage 3, the LLM maps the textual anchor "revision communication" to the corresponding BPMN message flow, while validation ensures that the resulting confidentiality annotation is attached only to a schema-compatible message-flow element. This combination of semantic mapping and deterministic validation yields structurally valid SecBPMN2 annotations.

4.6 Stage 4: SecBPMN2 XML Reconstruction

Stage 4 reconstructs the validated SecBPMN-JSON model J_{sec} as SecBPMN2-compliant BPMN 2.0 XML. This stage restores the BPMN 2.0 and SecBPMN2 namespaces, materializes SecBPMN2 extension elements for each entry in `secConstraints` and `secAssociations`, and serializes the collaboration and process structures into a well-formed XML document. Because J_{sec} preserves the original BPMN element (e.g., tasks, gateways, message flows, participants), the resulting XML retains the same BPMN 2.0 control-flow and interaction structure as the input model, augmented with security annotations.

5 Prototype Implementation

To support reproducibility and demonstrate the practical realization of the proposed approach, we implemented a prototype that operationalizes the SecBPMN2 automation pipeline within a BPMN modeling environment. The prototype, extending an open-source BPMN assistant[2], follows a three-layer architecture presented in Fig. 3. The user interaction layer comprises (1) a BPMN

[2] https://github.com/jtlicardo/bpmn-assistant.

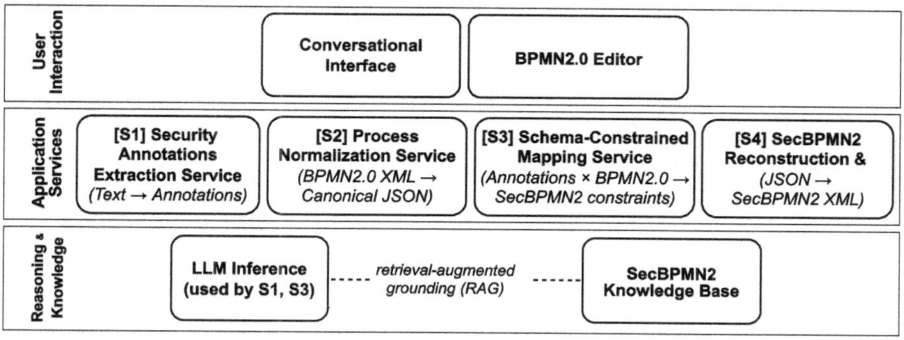

Fig. 3. Architecture of the SecBPMN2 annotation prototype.

modeling interface that enables users to supply process models and (2) a conversational interface in which the user can provide the textual security requirements and inspect the SecBPMN2 annotations identified during Stage 1. To preserve compatibility with standard BPMN tools, the exported models retain the original BPMN structure and are augmented solely with SecBPMN2 extensions. The application services layer executes the four stages of the pipeline, including security annotation extraction, BPMN normalization, schema-constrained mapping, and SecBPMN2-compliant reconstruction. The reasoning and knowledge layer, which handles LLM inference requests, is invoked during security annotation extraction (Stage 1) and schema constraint mapping (Stage 3), while all structural validation and reconstruction steps are performed deterministically. The prototype implementation is publicly available to support reproducibility[3].

6 Evaluation

The evaluation assesses the effectiveness and practicality of the SecBPMN2 automation pipeline based on the generated SecBPMN2 artifacts. As the implementation is designed for experimentation and reproducibility, the evaluation focuses on artifact quality rather than interface usability. Specifically, it considers five objectives: (E1) assessing whether the pipeline generates SecBPMN2 annotations that conform to the meta-model and attachment constraints across process complexity levels; (E2) measuring how accurately the pipeline extracts and maps security annotations from natural-language specifications relative to expert-designed ground truth; (E3) evaluating the quality and consistency of SecBPMN2 Assistantgenerated annotations compared to security experts; (E4) analyzing how extraction performance varies across different categories of security goals; and (E5) assessing the computational cost of the pipeline and how runtime and token usage scale with process complexity.

[3] Software artifact: https://doi.org/10.5281/zenodo.19063772.

6.1 Experimental Setup

The evaluation is conducted on a curated dataset[4] designed to align natural-language security requirements with SecBPMN2 process models. Because no public benchmark provides such paired artifacts, we constructed a dataset of 27 process models spanning aviation (9), healthcare (7), finance (6), public services (3), and hospitality (2). The models were collected from published SecBPMN2 case studies (20) and industrial examples (7). While several source models provided SecBPMN2 annotations, none included aligned natural-language security descriptions. We therefore manually authored one description per model by systematically translating BPMN constructs (e.g., participants, activities, gateways, message flows, and data objects) into domain-language narratives. The generated descriptions follow the assumptions introduced in 4.2 and were reviewed against the corresponding SecBPMN2 models to ensure coverage of all security-relevant process elements and annotated security requirements. Descriptions range from 97 to 670 words. Each dataset instance therefore consists of a BPMN process model, a natural-language security description, and a SecBPMN2 ground-truth annotation model, enabling reproducible evaluation of security annotation pipelines. Table 2 summarizes the dataset characteristics by complexity tier (simple, medium, and complex), defined by the number of BPMN elements and security annotations.

We evaluate two extraction strategies: prompt-based and RAG, where prompts are enriched with passages retrieved from a knowledge base constructed from SecBPMN2, SecBPMN2BC, and STS-ML documentation [6,12,17]. Experiments are conducted with GPT-4.1-mini and Mistral Small 3.2 – using identical decoding settings (temperature = 0.1, max_tokens = 30000). The temperature = 0.1 was used to balance reproducibility with robust annotation extraction and mapping by reducing stochastic variation while retaining limited generation flexibility. Retrieval follows a fixed hybrid BM25–FAISS configuration [5], with top-$k = 16$ and fusion weight $\alpha = 0.35$ for BM25 and $1 - \alpha$ for FAISS. Unless restricted to a single document type, we allocate retrieved context across sources (70% SecBPMN2, 20% SecBPMN2BC, 10% STS-ML). FAISS embeddings are computed using `text-embedding-3-small`. All generated models are validated against the SecBPMN2 schema and compared with annotations produced by three participants: two experienced cybersecurity professionals and one Master's student in cybersecurity, none of whom had prior SecBPMN2 experience.

The evaluation operationalizes the objectives defined in Sect. 6 using six metrics. Structural validity evaluates E1 by measuring conformance to the SecBPMN2 meta-model and correctness of annotation attachment. Precision, recall, and F1 score measure E2 and E3 by quantifying the correctness and completeness of extracted and mapped security annotations relative to expert annotations [7,9]. For evaluation purposes, each security annotation is represented by three components: (i) the security goal, (ii) the SecBPMN2 annotation type, and (iii) the BPMN element to which the annotation is attached.

[4] Dataset: https://doi.org/10.5281/zenodo.19064046.

Table 2. Dataset summary aggregated by complexity tier. Percentages report the distribution of security targets.

Tier	#	BPMN elems.	Sec. ann.	ACT	DO	MF	GW/ORG
Simple	10	9.7 (4-17)	3.1 (1-6)	87.1%	3.2%	9.7%	0.0%
Medium	7	24.4 (11-38)	4.0 (1-8)	35.7%	46.4%	17.9%	0.0%
Complex	9	54.0 (29-116)	23.7 (7-64)	46.9%	31.5%	18.8%	2.8%

Precision, recall, and F1 are computed using strict exact matching: a generated annotation is counted as a true positive only if all three components match a ground-truth annotation exactly. Generated annotations without a matching ground-truth annotation are counted as false positives, whereas ground-truth annotations that are not generated are counted as false negatives. Consequently, these metrics assess not only structural validity but also semantic correctness with respect to expert-designed SecBPMN2 annotations, as incorrectly inferred security goals, misplaced annotations, and hallucinated annotations are reflected as false positives or false negatives. Category-level recall supports E4 by analyzing extraction performance across different security goal classes. Finally, computational efficiency and reproducibility address E5 by measuring runtime cost, annotation effort, and stability of results under LLM nondeterminism [11].

To address E1E5, we conduct three experiments: (i) structural validity and extraction performance, (ii) human baseline and category-level analysis, and (iii) computational cost and scalability. All experiments use identical decoding parameters for comparability.

6.2 Results

This subsection reports the results with respect to the evaluation objectives defined in Sect. 6. Key results are summarized in Table 3.

Structural Validity and Extraction Performance (E1–E2). Table 3a reports schema validity of generated SecBPMN2 annotations after mapping and validation. Schema validity is evaluated first, as it indicates whether the pipeline produces well-formed and attachable security annotations. Across all complexity tiers, both GPT-4.1-mini and Mistral-small achieve consistently high validity, confirming that the hybrid extractionmapping pipeline reliably enforces SecBPMN2 structural constraints. RAG yields the strongest results overall, with GPT-4.1-mini attaining the highest validity on simple (0.987), medium (0.911), and complex (0.924) models, demonstrating that structural correctness is preserved as process complexity increases.

Table 3b reports F1, precision, and recall across models, extraction strategies, and complexity tiers. Extraction performance declines with increasing model complexity, reflecting the richer and more heterogeneous security annotations of multi-actor processes. For GPT-4.1-mini, RAG achieves the highest F1 on simple workflows (0.73), while prompt-based extraction performs better on complex

Table 3. Structural validity, extraction performance, and human baseline.

(a) Schema validity after mapping, aggregated by model and complexity tier.

Complexity	Prompt-based		RAG	
	GPT	Mistral	GPT	Mistral
Simple (avg.)	0.926	0.954	**0.987**	0.905
Medium (avg.)	0.801	0.857	**0.911**	0.808
Complex (avg.)	0.900	0.897	**0.924**	0.865

(b) Core extraction performance (F1, Precision, Recall) across methods and tiers.

Model	Method	Tier	F1	Precision	Recall
GPT-4.1-mini	Prompt-based	Simple	0.60	0.59	0.74
	RAG	Simple	**0.73**	0.76	0.72
GPT-4.1-mini	Prompt-based	Medium	0.36	0.30	0.46
	RAG	Medium	**0.36**	0.36	0.36
GPT-4.1-mini	Prompt-based	Complex	**0.34**	0.44	0.30
	RAG	Complex	0.23	0.46	0.18
Mistral-small	Prompt-based	All tiers (avg.)	0.34	0.35	0.41
	RAG	All tiers (avg.)	**0.40**	0.47	0.39

(c) Inter-annotator agreement among human experts (10 pairwise comparisons).

Statistic	Jaccard	Cohen's κ
Mean	0.416	0.409
Standard deviation	± 0.189	± 0.258
Range	[0.083, 0.714]	[−0.056, 0.746]

(d) Annotation performance: human experts vs. SecBPMN2 Assistant.

Annotator	F1	Precision	Recall
Human expert average	0.33	0.29	0.50
SecBPMN2 Assistant (GPT-4.1-mini, RAG)	**0.52**	**0.58**	**0.52**

models (0.34 vs. 0.23). On medium-tier processes, both strategies perform comparably. For Mistral-small, results are reported as averages across tiers; although overall F1 remains lower than GPT-4.1-mini, RAG improves performance (0.40 vs. 0.34), indicating that lightweight models benefit from external context.

Interestingly, retrieval augmentation is not uniformly beneficial across process complexity tiers. While RAG substantially improves performance on simple models, GPT-4.1-mini achieves lower F1 on complex processes than the prompt-based configuration. A possible explanation is that, as process complexity increases, the retrieved SecBPMN2 and STS-ML documentation adds contextual information that competes with process-specific cues required for accurate annotation recovery. Prior work has shown that language models often struggle to effectively utilize relevant information distributed within long contexts and may underuse information embedded among large amounts of surrounding content [10]. Consequently, retrieval may encourage more conservative annotation decisions in complex models, reducing the number of recovered annotations despite improving schema grounding. More generally, the observed performance trend suggests that extraction difficulty is influenced not only by process size but also by the composition of the annotation task. In particular, medium and complex models contain a more diverse mix of annotation targets, including activities, data objects, and message flows, which increases the complexity of mapping security requirements to schema-compatible BPMN elements.

Precisionrecall trade-offs reveal complementary failure modes. On complex processes, GPT-4.1-mini favors precision over recall, omitting a non-trivial fraction of intended annotations, which in security-by-design settings corresponds to unmitigated risks. Mistral-small exhibits a more balanced precisionrecall profile, with RAG favoring precision over recall and producing more conservative but structurally cleaner annotations. While this reduces over-specification, it may still result in missed annotations, increasing the need for expert review. These results indicate that GPT-based configurations are preferable when minimizing omissions is paramount, whereas Mistral-small is better suited to scenarios prioritizing precision and controlled annotation generation.

Human Baseline Comparison (E3). Table 3c reports inter-annotator agreement and humansystem performance. The human baseline was constructed from annotations independently produced by three experts with prior SecBPMN modeling experience. Each annotator was randomly assigned ten process models spanning simple, medium, and complex tiers to ensure balanced coverage across different process complexity. Human annotations exhibit moderate agreement (mean Jaccard 0.416, mean Cohen's κ 0.409), indicating that the task is challenging but yields a reasonable level of consistency among practitioners. The remaining variation reflects the interpretive nature of mapping security requirements to BPMN elements, particularly in complex models. Human annotators required on average nine minutes per process (ranging from approximately two minutes for simple workflows to over twenty minutes for complex models). Against this baseline, the SecBPMN2 Assistant achieves higher F1 and precision with comparable recall while requiring substantially less annotation time. The higher precision further suggests that the system generates fewer semantically incorrect or unsupported annotations than the average human annotator.

Category Analysis (E4). Figure 4 shows category-level recall across security goals. Recall is the highest for CIA-triad (confidentiality, integrity, and availability) objectives up to 0.79, which are frequent and typically expressed explicitly in text. By contrast, goals such as authenticity, accountability, and auditability occur less often and are commonly expressed implicitly, resulting in lower recall. This pattern indicates that the pipeline reliably captures protection-oriented requirements, while assurance- and compliance-oriented annotations benefit from expert oversight.

Token Cost, Runtime, and Scalability (E5). GPT-4.1-mini consistently uses more tokens ($\approx$18.6k20.6k per run) and incurs higher latency than Mistral-

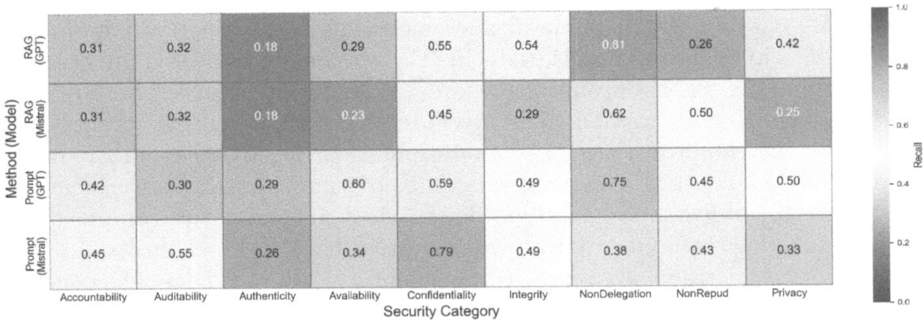

Fig. 4. GPT and Mistral category-level recall per security goal (Prompt-based, RAG).

small (≈15.7k19.4k tokens), reflecting its larger capacity and broader reasoning. RAG further increases token usage and latency due to retrieved context and longer outputs. For both models, latency scales predictably with process complexity and extraction strategy. In deployment, Mistral-small enables faster, lightweight inference for interactive or high-throughput scenarios, whereas GPT-4.1-mini trades higher computational cost for greater accuracy and closer alignment with human annotations.

Overall, the results indicate that the proposed pipeline satisfies the evaluation objectives. It consistently produces structurally valid SecBPMN2 annotations (E1), achieves competitive extraction accuracy relative to expert annotations (E2E3), captures the majority of explicit security requirements across goal categories (E4), and scales predictably with process complexity while maintaining practical computational cost (E5).

7 Discussion and Conclusion

This paper introduced a hybrid LLM- and rule-based pipeline for automatically deriving SecBPMN2 security annotations from natural-language process descriptions. The approach combines semantic interpretation of textual requirements with rule-based enforcement of SecBPMN2 schema constraints, enabling the generation of structurally valid security-aware BPMN models. By decoupling semantic extraction from schema validation, the pipeline reconciles the flexibility of natural-language specifications with the strict structural requirements of security-oriented process modeling.

An empirical evaluation on 27 curated textmodel pairs shows that schema-aware prompting and compatibility-constrained mapping improve annotation quality across models and extraction strategies. The pipeline consistently produces SecBPMN2-compliant annotations, achieving higher precision than human annotators while maintaining comparable recall with substantially less manual effort. The low inter-annotator agreement among human experts highlights the inherent ambiguity and cognitive burden of manual security modeling, underscoring the practical value of automated, schema-constrained assistance. Category-level analysis further indicates that the pipeline performs best for CIA-triad goals, such as confidentiality, integrity, and availability, whereas assurance- and compliance-oriented annotations, including accountability and auditability, remain more challenging and benefit from expert review.

Several limitations should be considered. Support for the full range of BPMN constructs remains partial, as the normalized representation abstracts away certain modeling details. While structurally inconsistent or non-attachable annotations can be detected and filtered, the system does not yet support automatic repair or conflict resolution. In addition, the evaluation is based on a curated dataset of 27 expert-annotated process models aggregated from the literature. Although the dataset spans multiple domains and was systematically aligned with textual security descriptions, its size reflects the current scarcity of publicly available SecBPMN2 benchmarks and may limit the generalizability of the findings.

Future work will focus on improving the accuracy and robustness of the proposed pipeline through larger benchmark datasets and the evaluation of the latest foundation models. We also plan to investigate retrieval strategies for complex process models, including methods for improving the relevance and utilization of retrieved security knowledge in long-context settings, as well as alternative intermediate representations such as graph-based encodings and their impact on token efficiency, annotation quality, and scalability. Furthermore, a more detailed analysis of the relationship between process complexity, runtime, token consumption, and computational cost would provide deeper insights into the scalability characteristics of the approach. Another promising direction concerns collaborative security modeling, where multiple stakeholders contribute requirements that must be consolidated into a shared process model, including mechanisms for detecting and resolving conflicts between security annotations.

References

1. Brucker, A.D., Hang, I., Lückemeyer, G., Ruparel, R.: SecureBPMN: modeling and enforcing access control requirements in business processes. In: ACM SACMAT, pp. 123–126 (2012)
2. Cherdantseva, Y., Hilton, J.: A reference model of information assurance & security. In: 2013 International Conference on Availability, Reliability and Security, pp. 546–555. IEEE (2013)
3. Chinosi, M., Trombetta, A.: BPMN: an introduction to the standard. Comput. Stand. Interfaces **34**(1), 124–134 (2012)
4. Friedrich, F., Mendling, J., Puhlmann, F.: Process model generation from natural language text. In: Mouratidis, H., Rolland, C. (eds.) International Conference on Advanced Information Systems Engineering, pp. 482–496. Springer (2011). https://doi.org/10.1007/978-3-642-21640-4_36
5. Johnson, J., Douze, M., Jégou, H.: Billion-scale similarity search with GPUS. IEEE Trans. Big Data **7**(3), 535–547 (2019)
6. Köpke, J., Meroni, G., Salnitri, M.: Designing secure business processes for blockchains with SECBPMN2BC. Futur. Gener. Comput. Syst. (2023)

7. Köpke, J., Safan, A.: Efficient LLM-based conversational process modeling. In: Gdowska, K., Gómez-López, M.T., Rehse, JR. (eds.) BPM, pp. 259–270. Springer (2024). https://doi.org/10.1007/978-3-031-78666-2_20
8. Kourani, H., Berti, A., Schuster, D., van der Aalst, W.M.: Process modeling with large language models. In: van der Aa, H., Bork, D., Schmidt, R., Sturm, A. (eds.) International Conference on Business Process Modeling, Development and Support, pp. 229–244. Springer (2024). https://doi.org/10.1007/978-3-031-61007-3_18
9. Kourani, H., Berti, A., Schuster, D., van der Aalst, W.M.: Evaluating large language models on business process modeling: framework, benchmark, and self-improvement analysis. Softw. Syst. Model., 1–36 (2025)
10. Liu, N.F., et al.: Lost in the middle: how language models use long contexts. Trans. Assoc. Comput. Linguist. **12**, 157–173 (2024)
11. Nour Eldin, A., Assy, N., Anesini, O., Dalmas, B., Gaaloul, W.: NALA2BPMN: automating bpmn model generation with large language models. In: Comuzzi, M., Grigori, D., Sellami, M., Zhou, Z. (eds.) International Conference on Cooperative Information Systems, pp. 398–404. Springer (2024). https://doi.org/10.1007/978-3-031-81375-7_27
12. Paja, E., Dalpiaz, F., Giorgini, P.: Modelling and reasoning about security requirements in socio-technical systems. Data Knowl. Eng. (2015)
13. Pullonen, P., Tom, J., Matulevičius, R., Toots, A.: Privacy-enhanced BPMN: enabling data privacy analysis in business processes models. Softw. Syst. Model. **18**(6), 3235–3264 (2019)
14. Ramadan, Q., Strüber, D., Salnitri, M., Jürjens, J.: A semi-automated BPMN-based framework for detecting conflicts between security, data-minimization, and fairness requirements. Softw. Syst. Model. (2020)
15. Salnitri, M., Dalpiaz, F., Giorgini, P.: Designing secure business processes with SECBPMN. Softw. Syst. Model. **16**(3), 737–757 (2017)
16. Siavvas, M., Xanthopoulou, G., Kalouptsoglou, I., Kehagias, D., Tzovaras, D.: Digital transformation of security standards: requirements extraction using large language models. In: DSA, pp. 430–431. IEEE (2024)
17. STS-tool development team: Security requirements modeling tool: SecBPMN2 elements reference guide (rev. 1.0) for STS-Tool Version, vol. 2, p. 1 (2015)
18. Sultan, M.A., Ganhotra, J., Astudillo, R.F.: Structured chain-of-thought prompting for few-shot generation of content-grounded QA conversations. In: Findings of ACL: EMNLP, pp. 16172–16187 (2024)
19. Van Woensel, W., Motie, S.: NLP4PBM: a systematic review on process extraction using natural language processing with rule-based, machine and deep learning methods. Enterprise Inf. Syst. **18**(11), 2417404 (2024)
20. Yang, J., Jiang, D., He, T.: StructEval: benchmarking LLMs' capabilities to generate structural outputs. Trans. Mach. Learn. Res. (2026)

CV-Driven Window Selection for Multi-perspective Concept Drift Detection in Event Logs

Samira Khraiwesh(✉), Dogukan Dagtekin, and Luise Pufahl

CIT, Technical University of Munich,
Bildungscampus 2, 74076 Heilbronn, Germany
{samira.khraiwesh,luise.pufahl}@tum.de

Abstract. Process mining leverages event logs recorded in information systems to discover and analyze business processes. However, rapid technological and market changes often lead to evolving process behavior, which manifests as concept drift in event logs. While recent research has focused on detecting and explaining such drifts, many existing approaches focus on individual perspectives of process behavior, often emphasizing control-flow, while other perspectives receive less attention. In this paper, we propose a multi-perspective concept drift detection approach. The approach models process behavior as a time series and introduces a coefficient-of-variation-based strategy to automatically determine suitable window sizes for drift detection. This adaptive window selection enables more stable change-point detection across different process perspectives. We evaluate the approach against established baselines using synthetic and benchmark datasets. The results demonstrate competitive detection performance and show that the proposed method effectively identifies drifts across multiple perspectives of process behavior.

Keywords: Process Mining · Concept Drift · Multi-Perspective Analysis · Window selection

1 Introduction

Information systems have become essential infrastructures supporting modern organizations [2]. They record detailed information about the execution of business processes in the form of event logs [28], which constitute the primary input for process mining techniques. Process mining (PM) is a family of techniques to analyze, monitor, and improve business processes across various domains [28]. Core PM tasks include process discovery, conformance checking, and performance analysis [27]. Most of these techniques implicitly assume that process behavior remains relatively stable over time [4]. In reality, however, business processes operate in dynamic environments. Regulatory updates, technological changes, and organizational restructuring continuously reshape process behavior [4]. These changes are reflected in event logs and may alter control-flow,

F. M. Maggi et al. (Eds.): BPM 2026, LNBIP 599, pp. 185–202, 2027.
https://doi.org/10.1007/978-3-032-37877-4_11

temporal, or resource-related patterns and is referred to as *concept drift* [23]. Concept drift may be planned or unexpected and is recognized as a fundamental challenge in process mining [29]. Numerous approaches have been proposed to detect concept drift in event logs [4,6,30]. However, most focus on the control-flow perspective [23], while methods incorporating additional process perspectives remain limited [2,23]. Yet, drift can also occur in temporal and resource perspectives, affecting the correctness of downstream PM applications, such as business process simulation and predictive monitoring.

Since drift detection commonly relies on comparing behavioral statistics over time, *window-based segmentation* plays a central role in many approaches [23]. Existing methods typically rely on either fixed or adaptive windowing strategies to segment the event log before comparing behavioral statistics. While fixed windows are simple and computationally efficient, their effectiveness strongly depends on the manually selected window size: small windows may be sensitive to noise, whereas large windows may smooth behavioral variations and delay or even miss drift points [13]. As a result, the effectiveness of many drift detection techniques strongly depends on the selected window size. Despite its critical influence on detection performance, the problem of automatically determining suitable window sizes from event logs has received limited attention in existing PM research [23]. Adaptive approaches mitigate this limitation by dynamically adjusting the window length [19], but are typically applied globally and do not account for the heterogeneous evolution of different process perspectives.

To address these limitations, we introduce a novel concept drift detection approach with two main contributions. First, we propose a multi-perspective drift detection framework **(CVDrift)** that analyzes control-flow behavior, activity durations, inter-arrival time, and resourceactivity relationships, enabling a more comprehensive and behavior-aware detection of process evolution. Second, we introduce a data-driven, component-specific window selection mechanism that determines suitable window sizes independently for different behavioral time series, enabling finer-grained and more robust drift detection. The remainder of the paper is organized as follows. Section 2 introduces the preliminaries, Sect. 3 reviews related work, Sect. 4 presents the proposed approach, Sect. 5 reports the experimental evaluation, and Sect. 6 concludes the paper.

2 Preliminaries

This section introduces the fundamental concepts for this paper, including event logs as the primary data source for concept drift detection, followed by the notion of concept drift and the process perspectives considered in this work.

Event Logs. In PM, an *event log* represents the recorded execution history. An event log L consists of a set of cases $L = \{\sigma_1, \sigma_2, \ldots, \sigma_n\}$ where each case $\sigma = \langle e_1, e_2, \ldots, e_k \rangle$ is an ordered sequence of events. An event can be represented as a tuple $e = (c, a, t_s, t_e, r)$ where c is the case identifier, a the activity, t_s the start timestamp, t_e the end timestamp, and r the resource performing the activity.

We define a projection function $\pi_x(e)$ that returns the value of attribute $x \in \{c, a, t_s, t_e, r\}$, where, e.g., $\pi_a(e)$ returns the activity of e. Additionally, we define the first-event projection as $\mathit{first}(\sigma) = e_1$. Accordingly, the start timestamp of a case is defined as $s(\sigma_i) = \pi_{t_s}(\mathit{first}(\sigma_i))$.

Concept Drift. Concept drift refers to changes in process behavior over time. Let B_t denote the process behavior at time t. A drift occurs when $B_{t_1} \neq B_{t_2}$ for $t_1 < t_2$. Drift may appear in different forms: *sudden drift*, where behavior changes abruptly at a point t_d; *gradual drift*, where the transition occurs over an interval; and *recurring drift*, where a previously observed behavior reappears.

2.1 Process Perspectives

Control-Flow Perspective. The control-flow projection of a case is defined as $CF(\sigma) = \langle \pi_a(e_1), \ldots, \pi_a(e_n) \rangle$. *Routing behavior* captures the probability that activity b directly follows a is

$$P(a \to b) = \frac{|\{\sigma \in L,\ e_i, e_{i+1} \in \sigma \mid \pi_a(e_i) = a,\ \pi_a(e_{i+1}) = b\}|}{|\{\sigma \in L,\ e_i, e_{i+1} \in \sigma \mid \pi_a(e_i) = a\}|}$$

Temporal Perspective. The duration of an event e is defined as $dur(e) = \pi_{t_e}(e) - \pi_{t_s}(e)$. The *inter-arrival* time between two consecutive cases is defined as $t_{\text{arr}}(\sigma_i) = s(\sigma_i) - s(\sigma_{i-1})$.

Resource Perspective. The relationship between activities and resources is defined as the probability that activity a is performed by resource r, denoted by $P(a \to r)$, where

$$P(a \to r) = \frac{|\{\sigma \in L,\ e \in \sigma \mid \pi_a(e) = a,\ \pi_r(e) = r\}|}{|\{\sigma \in L,\ e \in \sigma \mid \pi_a(e) = a\}|}.$$

The extracted time series are organized as case-indexed time series ordered by case start times, enabling the detection of sudden changes in different process perspectives.

3 Related Work

Concept drift is a well-known challenge in process mining, and detecting such changes in event logs has attracted significant research attention [23]. Prior work mainly follows two directions: (i) detecting drifts and supporting their analysis, and (ii) developing online PM techniques for continuously evolving environments [23]. In this section, we focus on the first direction and review approaches that identify change points in event logs. Table 1 summarizes representative methods.

Existing approaches for change point detection in process mining can be broadly categorized into three groups: (i) approaches that propose new drift

Table 1. Overview of concept drift detection approaches in PM

Method	Perspective	Window Strategy	Window Scope
Bose et al. [4,5]	CF	Fixed	Global
Lin et al. [17]	CF	Fixed / Adaptive	Global
Zheng et al. [31]	CF	Fixed	Global
Maaradji et al. [18]	CF	Adaptive	Global
Martjushev et al. [20]	CF	Adaptive	Global
Neto et al. [21]	CF	Fixed	Global
Gallego-Fontenla et al. [12]	CF	Fixed	Global
Kraus and van der Aa [16]	CF	Fixed	Global
Brockhoff et al. [6]	CF, T	Fixed	Global
Adams et al. [2]	CF, T, R	Fixed	Global
Cao et al. [7]	CF, R	Fixed	Global
Klijn et al. [15]	CF, R	Fixed	Global

detection techniques, (ii) approaches that focus on adaptive window size strategies, and (iii) approaches that investigate multi-perspective drift detection.

The first category introduces techniques for detecting concept drift in event logs. Bose et al. [4,5] propose statistical features based on activity relations, namely Window Count and J-Measure, to detect changes across consecutive windows. Similarly, Lin et al. [17] analyze variations in directly-follows relations, while Zheng et al. [31] represent activity relations using Boolean matrices to identify behavioral changes over time. Other approaches rely on *alternative representations of process behavior*: Maaradji et al. [18] propose ProDrift, which applies statistical hypothesis testing on run distributions. Neto et al. [21] use transition matrices to capture control-flow behavior, Gallego-Fontenla et al. [12] combine process discovery and conformance checking, and Kraus and van der Aa [16] apply deep learning on image representations of event logs to detect drift locations and types. Despite this diversity, most approaches focus primarily on control-flow changes and rely on window-based log segmentation. Their performance therefore strongly depends on the chosen window size: small windows may be sensitive to noise, whereas large windows may smooth variations and delay or miss drift points.

To reduce the dependency on predefined window sizes, a second line of research investigates adaptive windowing strategies. Martjushev et al. [20] introduce an adaptive mechanism that dynamically adjusts the window size based on statistical evidence of change. Their method applies hypothesis testing to identify potential drift regions and progressively refines the analyzed window boundaries in order to better localize drift points. Although such strategies alleviate the need for a single fixed window size, they rely on reactive drift-detection mechanisms that dynamically update a single window over a data stream, which may overlook the heterogeneous evolution of different process perspectives.

A third line of research investigates concept drift detection from multiple process perspectives, recognizing that process evolution may occur not only in control-flow but also in other behavioral dimensions. For example, Brockhoff et al. [6] incorporate temporal characteristics using an Earth Mover's Distance (EMD)-based approach, while Adams et al. [2] propose an explainable framework analyzing multiple process measures. Cao et al. [7] introduce the MDD approach, integrating perspectives such as activities, resources, and transitions using graph-based representations, and [15] extends drift detection to the organizational perspective by incorporating actor-related information. Although these approaches demonstrate the benefits of multi-perspective analysis, they typically rely on a single global window configuration to segment the event log. However, control-flow relations, temporal patterns, and resource behavior may evolve at different temporal scales, making a uniform window configuration insufficient to capture the heterogeneous dynamics of process evolution.

We propose a data-driven mechanism that determines window sizes per perspective based on the stability of the coefficient of variation (CV), enabling robust multi-perspective drift detection.

4 Approach

To detect drifts across the control flow, temporal and resource perspectives introduced in Sect. 2, the proposed method employs a detection pipeline consisting of three main stages: (i) event log preparation and initial time-series construction, (ii) CV-based window selection, and (iii) change-point detection.

4.1 Event Log Preparation and Initial Time-Series Construction

The proposed pipeline transforms an event log L into a set of case-indexed time series capturing different process perspectives. Let $L = \{\sigma_1, \ldots, \sigma_n\}$ denote the set of cases, ordered by their start timestamps $s(\sigma_i)$. Within each case, events are ordered also according to their start timestamps. This ordering induces a temporal index over cases, enabling their sequential analysis.

For each case σ_i, a single numerical observation is derived per perspective. This results in a collection of aligned, case-indexed time series, where each observation is associated with the start timestamp $s(\sigma_i)$ of its corresponding case. While the previous definitions capture global log-level characteristics, the following constructs derive case-indexed time series that approximate these measures locally over time and serve as input for the subsequent drift detection procedure.

Temporal Perspective. For each case index $i \in \{1, \ldots, n\}$, an activity-specific time series for the *duration* of activity a is defined as
$t_a(i) = \text{median}\{\, dur(e) \mid e \in \sigma_i,\ \pi_a(e) = a \,\}$. If activity $\pi_a(e) = a$ appears multiple times within a case, the median duration is used to reduce the influence of extreme values. Cases in which the activity does not occur are assigned missing values. An example event log and derived initial time series is shown in Fig. 1.

The *inter-arrival* time series is defined as $t_{\text{arr}}(i) = s(\sigma_i) - s(\sigma_{i-1})$. To avoid distortions caused by overnight gaps or system downtime, arrivals spanning different calendar days or exceeding a predefined maximum gap are excluded.

Case ID	Activity	Start	End	Resource
i	a	t_s	t_e	r
1	A	10:00:00	10:05:00	R1
1		...		
2	A	10:10:00	10:17:00	R1
2		...		
n	A	12:00:00	12:10:00	R1

$$t_A = \{t_A(1),\ t_A(2),\ \ldots,\ t_A(n)\}$$

$$t_A = \{5,\ 7,\ \ldots,\ 10\}$$

Fig. 1. Example event log and derived initial time series for activities A

Control-Flow Perspective. Let $\mathcal{P} \subseteq \mathcal{A} \times \mathcal{A}$ denote the set of activity pairs observed in the directly-follows relation. For each pair $(A, B) \in \mathcal{P}$ and each case index $i \in \{1, \ldots, n\}$, the routing time series is defined as

$$t_{A,B}(i) = \frac{|\{e_j, e_{j+1} \in \sigma_i \mid \pi_a(e_j) = A,\ \pi_a(e_{j+1}) = B\}|}{|\{e_j, e_{j+1} \in \sigma_i \mid \pi_a(e_j) = A\}|}.$$

Cases in which activity a does not occur are assigned missing values.

Resource Perspective. For each activity–resource pair (A,R), the corresponding time series captures the proportion of executions in which resource R performs activity A:

$$t_{A,R}(i) = \frac{|\{e \in \sigma_i \mid \pi_a(e) = A,\ \pi_r(e) = R\}|}{|\{e \in \sigma_i \mid \pi_a(e) = A\}|},$$

4.2 CV-Based Window Selection

The case-indexed time series constructed in the previous step serve as the input for the window-selection stage. Let $t = \{t(1), \ldots, t(n)\}$ denote one of the derived time series. Given a set of candidate window sizes $W = \{w_1, \ldots, w_m\}$, a rolling transformation is applied to t. For a window size $w \in W$, a rolling window of w consecutive cases is moved across the series to produce a smoothed sequence

$$r_w(t) = \{s_w(1), \ldots, s_w(n - w + 1)\},$$

where $s_w(j)$ denotes a summary statistic computed over the window $\{t(j), \ldots, t(j+w-1)\}$. The choice of the statistic depends on the process perspective. For temporal series (activity duration and inter-arrival time), the median is used to provide robustness against extreme values, whereas for control-flow and resource activity–resource relationships perspective, the mean is used to estimate

Candidate window size w	$\mathrm{CV}\left(\ell_A^{(w)}\right)$
2	CV_2
3	CV_3
$\vdots$	$\vdots$
w_i	CV_{w_i}

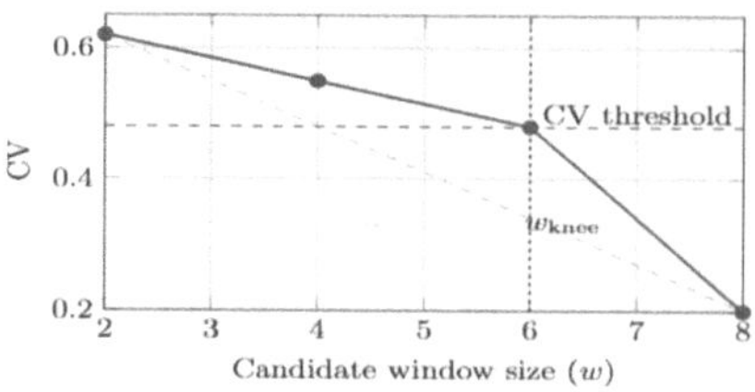

Fig. 2. Knee-point detection based on CV values

the relative frequency of the observed relation. For each candidate window size w, the stability of the resulting rolling series is evaluated using the coefficient of variation (CV) [10], defined as

$$CV_w = \frac{\sigma(r_w(t))}{|\mu(r_w(t))|},$$

where $\sigma(\cdot)$ and $\mu(\cdot)$ denote the standard deviation and mean of the rolling values, respectively. The CV measures the relative variability of the smoothed signal: smaller values indicate a more stable series, while larger values reflect stronger fluctuations. The candidate window sizes and their corresponding CV values form a curve $\{(w_i, CV_{w_i})\}_{i=1}^{m}$ (see Fig. 2). Inspired by geometric knee-point detection methods [24], the knee of the normalized CV curve is identified as the point with the maximum perpendicular distance from the line connecting the first and last points as shown in Fig. 2. Let the CV curve consist of points $P_i = (w_i, CV_{w_i})$ for $i = 1, \ldots, m$. After normalizing both axes, the perpendicular distance of each point to the line connecting the first and last points is computed as

$$d_i = \frac{|(B_x - A_x)(P_{i,y} - A_y) - (B_y - A_y)(P_{i,x} - A_x)|}{\sqrt{(B_x - A_x)^2 + (B_y - A_y)^2}} \tag{1}$$

where $A = (w'_1, CV'_{w_1})$ and $B = (w'_m, CV'_{w_m})$ denote the first and last points of the normalized curve. The knee point is then selected as

$$k = \arg\max_i d_i.$$

The CV value at this point defines a threshold, and the smallest candidate window whose CV does not exceed this value is selected. Each perspective, and more specifically each component, is treated independently; for example, in the activity duration perspective, each activity is assigned its own window size, which is used to construct the final smoothed series for subsequent change-point detection. The procedure is summarized in Algorithm 1.

The window size can be determined using two strategies. The **CVPer-Pair** strategy assigns a separate window size to each activity, routing relation, or resource relationship based on its behavior. In contrast, the **ModeWindow**

Algorithm 1: CV-Based Window Selection

Input: Time-series set T, candidate window sizes $W = \{w_1, \ldots, w_m\}$
Output: Selected window size w_t^* for each series $t \in T$
foreach $t \in T$ **do**
    Initialize $CV\_scores \leftarrow \emptyset$;
    foreach $w \in W$ **do**
        Compute rolling series $r_w(t)$ using window size w;
        Compute coefficient of variation $CV_w = \sigma(r_w(t))/|\mu(r_w(t))|$;
        Store (w, CV_w) in CV_scores;
    Construct CV curve $\{(w_i, CV_{w_i})\}$;
    Identify knee point k using the maximum perpendicular distance criterion (Eq. 1);
    Select w_t^* as the smallest w such that $CV_w \leq CV_k$;

strategy selects a single window size per perspective, defined as the most frequently occurring window size across all pairs within that perspective. This window is specific for a process perspective rather than global to the entire event log.

4.3 Change-Point Detection

After selecting the window size, a rolling statistic is computed over consecutive cases for each series. Change-point detection is then performed on the resulting rolling series using the Pruned Exact Linear Time (PELT) algorithm [14]. To improve robustness, each series is analyzed at multiple window resolutions, nearby detections within a series are merged, and very small shifts are removed using a minimum effect-size threshold. Nearby detections across related series are clustered to identify consistent changes. Although the window size is selected automatically, the PELT algorithm still requires a penalty parameter that influences the number of detected change points. The proposed pipeline reduces manual effort in window-size selection, but it is not fully parameter-free. The influence of the PELT penalty and the effect-size threshold is examined in the sensitivity analyses in Sect. 5.

5 Evaluation

This section presents two experiments evaluating the proposed approach: Experiment 1 focuses on the temporal perspective, while Experiment 2 targets the control-flow perspective. Similar to prior work [7], experiments are conducted separately, as datasets typically contain drifts affecting specific process perspectives. The datasets, implementation, and results are publicly available in our repository The proposed framework supports several process perspectives, including control-flow, temporal, and resource-related behavior. In the evaluation, we focus on the temporal and control-flow perspectives because suitable benchmark settings and ground-truth drift points are available for these perspectives. The resource perspective is not included in the evaluation due to the lack of publicly available datasets with labeled resource-perspective drifts. We therefore interpret the evaluation as evidence for selected perspectives rather than as a complete validation of all possible process perspectives.

5.1 Experiment 1: Temporal Perspective

In this experiment, we evaluate drift in activity durations[1]. As no publicly available datasets contain controlled temporal drifts with ground truth, we generated synthetic datasets using a custom script[2]

Datasets. We constructed a synthetic event log corpus using PM4Py [1] based on the process model. The corpus consists of 72 event logs, including 18 clean logs and 54 noisy logs, covering six drift magnitudes, three log sizes, and four noise levels (including 0%). The clean logs were generated using activity durations following a normal distribution with a 20% standard deviation, while inter-activity waiting times range from one to three minutes within working hours (08:0015:00). This configuration reflects typical real-world process behavior. To introduce duration drift, we selected activities A, B, and D – spanning different branches of the process model to ensure that the drift affects multiple execution paths rather than a single bottleneck. At two predefined change points, located at 37% and 75% of the total log length, the durations of these activities are increased by a fixed amount. The asymmetric placement allows evaluation of both early and late drift detection. Six drift magnitudes were applied (10, 20, 30, 40, 50, and 60 min), ranging from subtle shifts comparable to the natural variance to large disruptions. Each configuration was generated at three log sizes (100, 1,000, and 3,000 cases), yielding 18 clean logs. To evaluate robustness to noise, we applied the FLAWD noise injection framework [11] to each clean log at three noise ratios (5%, 10%, and 20%). The noise injection targeted two types of anomalies: (i) random character-level typographical errors in activity names, simulating manual data-entry mistakes, and (ii) stochastic shifts in event timestamps, mimicking real-world temporal anomalies such as system delays or late data entries. These noise levels represent low, moderate, and high data quality degradation commonly encountered in practice, resulting in 54 noisy logs.

Baseline. We compare our approach with the **OCPA** technique [2], which constructs time series from process measures and applies the PELT change-point detection algorithm. The method relies on a fixed global time window. We used the authors' original implementation and developed a wrapper to evaluate it on our benchmark datasets. Since the code requires Object-Centric Event Logs (OCEL), standard XES logs were converted into a pseudo-OCEL format by mapping each case identifier to a single object type, following the authors' description. We used the parameter configuration provided in the authors' drift detection example, where a fixed window size of 7 d is applied, as it was reported

[1] Drift detection for arrival times is also implemented in our repository.

[2] Available at https://zenodo.org/records/20497480.

to yield the best results. Although [7] aims to support multiple perspectives, the available implementation covers only control flow and therefore cannot serve as a baseline for the temporal perspective in this study. For **CVDrift**, 15 candidate window sizes ranging from 15 to 5000 cases were considered for the window selection step. Change-point detection was performed using the PELT algorithm with a penalty parameter $\lambda = 3$ and a minimum mean difference threshold of 0.00. The impact of these parameters is examined in the sensitivity analysis presented later.

Evaluation Metrics. Detection performance is evaluated using precision, recall, and F1-score by matching detected drift points with ground-truth drifts. To account for detection delays, we apply a tolerance window. A detected drift within this tolerance is counted as a true positive. OCPA and CVDrift operate in different indexing spaces: CVDrift outputs case-indexed changepoints, whereas OCPA detects changes on fixed-duration time windows and maps them to median case IDs, which introduces localization uncertainty. Accordingly, evaluation is performed using tolerance-based matching. We use 10% of the dataset size as the primary tolerance window, while CVDrift also shows strong performance at stricter tolerances.[3]

Table 2. Results of P (precision), R (recall), and F1-score by Drift Magnitude

Technique	10 min			20 min			30 min		
	P	R	F1	P	R	F1	P	R	F1
CVDrift	**1.000**	**0.944**	**0.971**	0.810	**0.773**	**0.791**	**0.857**	**0.750**	**0.800**
OCPA	0.833	0.555	0.667	**0.875**	0.636	0.737	0.842	0.667	0.744
Technique	40 min			50 min			60 min		
	P	R	F1	P	R	F1	P	R	F1
CVDrift	**0.850**	**0.708**	**0.773**	**0.875**	**0.875**	**0.875**	**0.870**	**0.833**	**0.851**
OCPA	0.842	0.667	0.744	0.254	0.625	0.361	0.068	0.750	0.125

Tables 2 and 3 present the precision, recall, and F1-score comparison between our approach and OCPA by drift magnitude and noise level. The results in Table 2 show that CVDrift achieves consistently high performance across all drift magnitudes, with F1 scores ranging from 0.800 to 1.000. This stability is due to its window selection mechanism, which adapts to the statistical properties of each time series. In contrast, OCPA achieves competitive precision for

[3] Our algorithm achieves an F1 score of 0.88 (5%, 3%, and 2% tolerances). Full results are available in the repository.

small-to-moderate drifts; its precision drops significantly at larger magnitudes, indicating over-segmentation. This aligns with prior findings [3], where smaller drifts require larger observation windows, while larger drifts can be detected with fewer observations. CVDrift implicitly captures this relationship through its data-driven window selection. As shown in Table 3, CVDrift demonstrates robust performance across noise levels, with only a gradual decline as noise increases, and consistently outperforms the baseline (OCPA). We have tested our approach used for detecting the concept drift using *CVPerPair* and *ModeWindow* stratiegies and both provides the same results

Table 3. Results of P (precision), R (recall), and F1-score by Noise Level

Technique	Clean			5%			10%			20%		
	P	R	F1	P	R	F1	P	R	F1	P	R	F1
CVDrift	**1.000**	**0.917**	**0.957**	**0.839**	**0.813**	**0.825**	**0.839**	**0.765**	**0.800**	**0.807**	**0.735**	**0.769**
OCPA	0.452	0.666	0.539	0.166	0.625	0.263	0.221	0.677	0.333	0.198	0.647	0.303

Sensitivity Analysis. To examine the impact of the PELT penalty scale and the effect-size filter, we conducted a sensitivity analysis over a grid of penalty scales ($\beta \in \{2, 3, 5, 7\}$) and minimum effect-size thresholds ($d_{\min} \in \{0.00, 0.10, 0.15, 0.30\}$), resulting in 16 configurations. These parameters were selected because they directly control the sensitivity of change-point detection, determining how many change points are detected and which of them are retained after filtering. The aggregated results are shown in Fig. 3. Overall, the results indicate that detection performance is primarily influenced by the penalty scale, while the effect-size filter provides limited additional benefit. The best configuration ($\beta = 3$, $d_{\min} = 0.00$) achieves an F1 score of 0.925 with a high recall (0.993). This indicates that no effect-size filtering is required, as the algorithm already reliably detects relevant change points. Lower penalties increase false positives, whereas higher penalties suppress true drift points. The tested penalty values follow BIC-type recommendations [26], and the effect-size thresholds follow standard conventions [9]. The full factorial grid assesses both parameters and their interaction.

5.2 Experiment 2: Control-Flow Perspective

In this section, we evaluate our approach's ability to detect change points in changes in routing behavior from a control-flow perspective. The performance of our approach is compared with existing baselines using synthetic datasets.

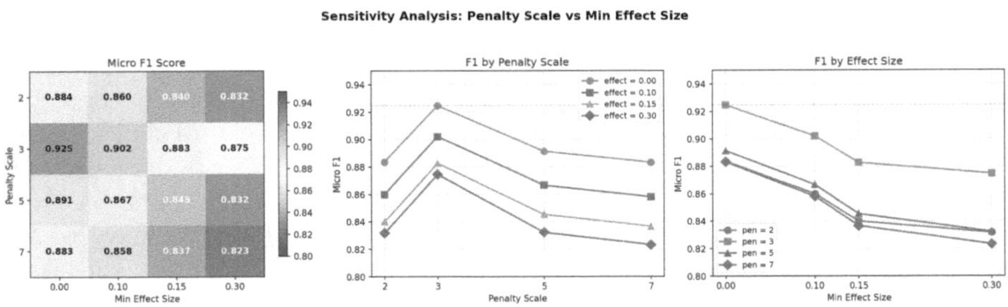

Fig. 3. Sensitivity results across penalty scales and effect-size thresholds.

Datasets. We use a publicly available benchmark dataset[4] previously employed for evaluating concept drift detection methods [1]. The dataset consists of synthetic event logs from three studies [5,8,22], with about 1700 traces on average and one to four annotated change points. To enable direct comparison, we applied our approach to the same logs and compared the results with those reported in the benchmark.

Baselines. We compare our method against seven concept drift detection techniques that were evaluated in the benchmark study of Adams et al. [1]: **Bose/J** [5], **Adwin/J** [20], **ProGraphs** [25], **ProDrift** [19], **RINV** [31], **EMD**, [6], and **LCDD** [17]. For **CVDrift**, 15 candidate window sizes ranging from 15 to 5000 cases were considered for the window selection step. Change-point detection was performed using the PELT algorithm with a penalty parameter $\lambda = 7$ and a minimum mean difference threshold of 0.15 and using *ModeWindow* strategy. The impact of these parameters, including the strategy, is further examined in the sensitivity analysis presented later.

Evaluation Metrics. To evaluate the performance of the proposed approach, we use the evaluation metrics from the benchmark study by Adams et al. [1], focusing on detection accuracy based on a lag window corresponding to the equivalent time of 200 process executions. In addition, we consider versatility across different drift types since this dimension is influenced by the chosen window size, as different change patterns manifest at different temporal scales.

Results. Figure 4 shows the comparison of the evaluated approaches in terms of F1 score on the noiseless datasets. The results indicate that CVDrift achieves competitive performance compared to the other approaches. In particular, it performs comparably to methods relying on fixed window sizes, while outperforming several approaches that employ adaptive windowing strategies.

Figure 5 presents performance across varying noise levels using non-fitting events (datasets from [22]). CVDrift maintains stable performance across all

[4] https://github.com/cpitsch/cdrift-evaluation.

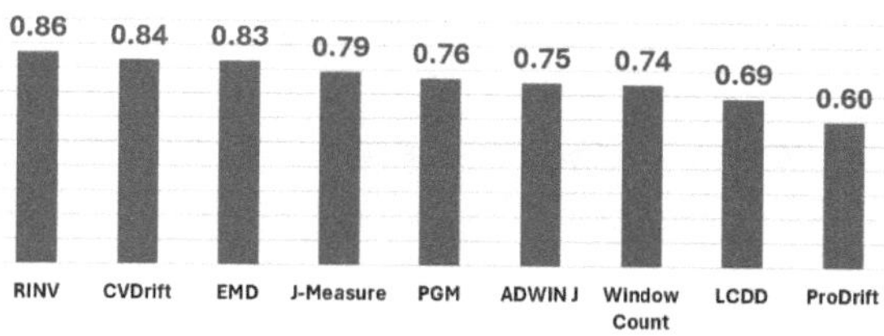

Fig. 4. Average F1-score comparison of drift detection algorithms across the benchmark datasets.

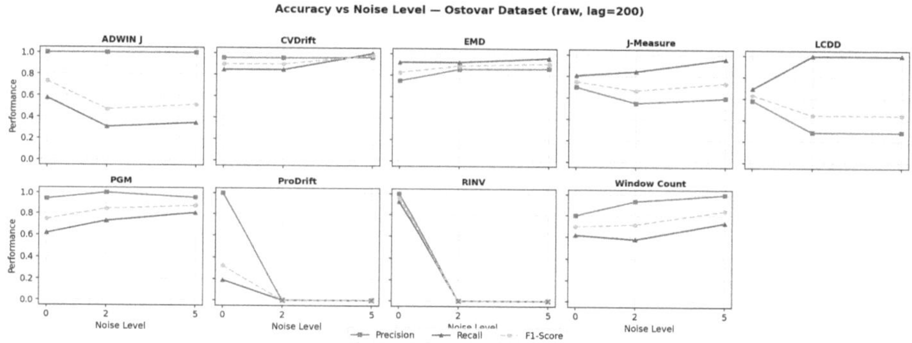

Fig. 5. Performance comparison of drift detection algorithms across noise levels.

noise levels and shows more consistent behavior than adaptive window-based approaches such as ADWIN, PGM, and LCDD, which exhibit noticeable fluctuations. Its performance is also comparable to fixed-window methods such as Window Count and EMD, while maintaining a balanced trade-off between precision and recall. Figure 6 illustrates the versatility of the evaluated approaches across different change patterns. Most window-based methods exhibit uneven performance, reflecting the biasvariance trade-off inherent to fixed observation windows, where a window size suitable for one change pattern may be suboptimal for another. Overall, CVDrift achieves competitive versatility (V = 0.83), with performance comparable to EMD and RINV and superior to methods that rely on adaptive window strategies. However, the approach shows weaker performance for frequency (fr) and conditional removal (cre) patterns. For frequency changes, this may be explained by the fact that CVDrift models routing behavior using probabilities rather than absolute frequencies, making subtle frequency variations less visible in the derived signal. In the case of CRE, the change affects only a subset of traces where an activity is conditionally skipped, while the overall process structure remains largely unchanged, resulting in a relatively weak signal that is harder to detect.

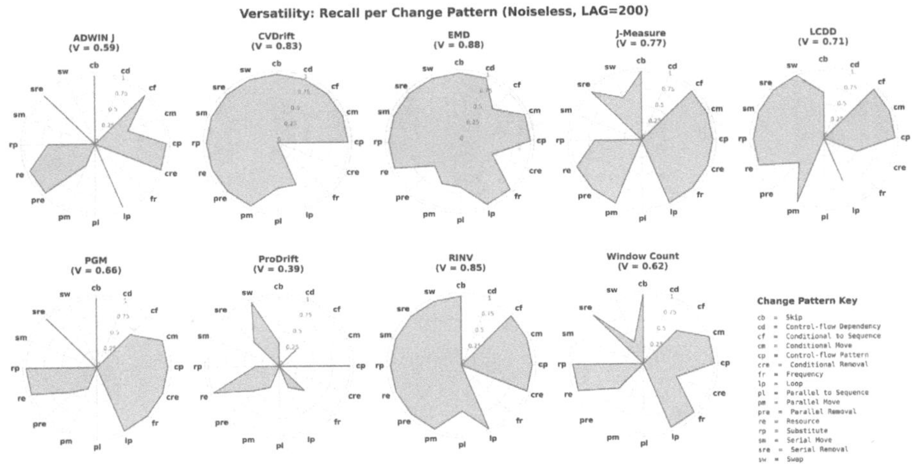

Fig. 6. Versatility: Recall per Change Pattern

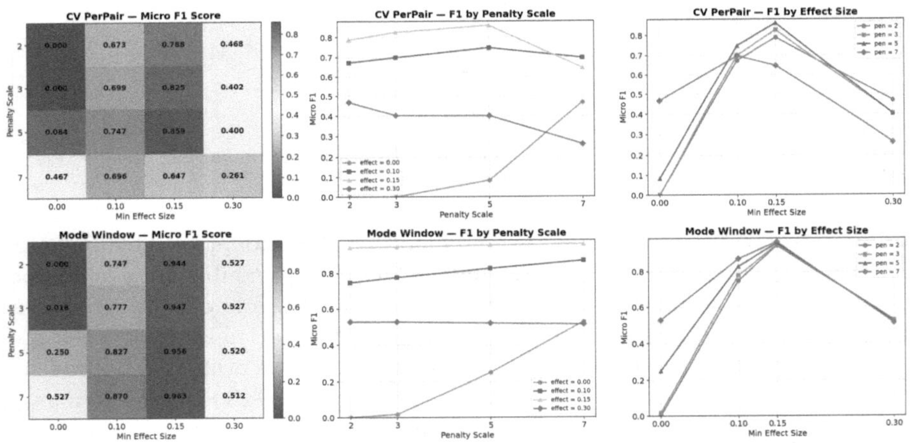

Fig. 7. Sensitivity results across penalty scales, effect-size thresholds, and strategies.

Sensitivity Analysis To examine the impact of the PELT penalty scale, the effect-size filter, and the window strategies, we conducted a sensitivity analysis. Specifically, we evaluated penalty scales of 2, 3, 5, and 7, and effect-size thresholds of 0.00, 0.10, 0.15, and 0.30, resulting in 16 configurations. These configurations were tested using two window strategies (*CVPerPair* and *ModeWindow*), yielding a total of 32 configurations. This sensitivity analysis is done on Ostovar dataset [22]. The results shown in Fig. 7 indicate that the minimum effect size has the strongest influence on the detection performance. Very small thresholds (e.g., 0.00) lead to poor results due to the detection of minor fluctuations, while large thresholds (e.g., 0.30) may suppress relevant changes. In contrast, moderate values between 0.10 and 0.15 consistently yield the best performance. The penalty parameter shows a smaller impact on the results, with comparable performance observed across several values. Furthermore, the *ModeWindow* strategy produces more consistent results compared to the per-pair approach, supporting its use for structural perspectives.

6 Discussion and Conclusion

The experimental results show that the proposed CVDrift approach achieves competitive and often superior performance in detecting concept drift across multiple process perspectives. Although it relies on the widely used PELT change-point detection algorithm, its effectiveness stems from generating different time series for multiple process perspectives and applying a data-driven window selection strategy based on the characteristics of each time series. These results highlight that drift detection performance depends not only on the underlying algorithm but also strongly on how the observation window is defined. Another observation is that the proposed approach performs very well for the temporal perspective (activity durations) without requiring additional effect-size filtering, indicating that the detected changes are sufficiently strong and reliable. In contrast, applying the approach to the control-flow perspective requires a minimum effect-size threshold to achieve optimal results. This difference can be attributed to the nature of the underlying time series. Temporal time series are typically unbounded and exhibit higher variability, allowing drift to manifest as pronounced absolute changes. In contrast, control-flow time series are bounded and often represented as probabilities, which compress their magnitude and make them less distinguishable from noise. The same reasoning also explains why the *ModeWindow* strategy performs better than the *CVPerPair* strategy for the control-flow perspective. A key contribution of this study is the introduction of a data-driven window selection strategy that eliminates the need for manual window size specification. This represents an important step toward more automated drift detection in process mining settings, where event logs vary significantly in size, granularity, and the nature of underlying process changes. While the window-selection component is entirely data-driven, the PELT algorithm still requires a penalty parameter. The sensitivity analyses indicate that the penalty parameter has a smaller impact on detection performance than other configuration choices, with comparable performance observed across several penalty values. In contrast, inappropriate window sizes may directly lead to missed or delayed drift detections, as discussed in Sect. 1. Furthermore, the knee-point strategy selects the *smallest* window achieving sufficient stability rather than the globally smoothest window, thereby preserving sensitivity to genuine drift signals and reducing the risk of smoothing away meaningful behavioral changes. The penalty parameter relates solely to the change-point detection algorithm and could be replaced by alternative algorithms or adaptive parameter selection strategies in future work. This separation of concerns—automated window selection combined with configurable change-point detection—provides a foundation for further automation. Future research could integrate the proposed strategy with recent approaches that require little or no manual configuration beyond window definition, such as the approach proposed by Kraus et al. [16]. The current evaluation focuses on sudden drifts, where ground truth is well-defined and results can be compared objectively across methods. Extending CVDrift to gradual, recurring, and overlapping drifts remains an important direction for future work. Similarly, while the temporal evaluation relies on synthetic datasets, the use of logs with known drift locations and magnitudes enables rigorous and objective assessment. Additional validation on real-world event logs is required to assess the approach under more complex and heterogeneous process environments. A significant research opportunity concerns evaluation infrastructure. Currently, publicly available datasets and generators provide limited coverage of multi-perspective concept drifts with labeled ground truth. Developing comprehensive multi-perspective drift generators and standardized benchmarking frameworks, comparable to the benchmark by Adams et al. [1] but covering multiple perspectives, would substantially improve the reproducibility and comparability of future research. The current behavioral representations also point toward concrete and actionable extensions. For the control-flow perspective, CVDrift relies on probability-based directly-follows representations. This representation is intentionally simple and interpretable, and the analysis in Sect. 5 characterizes where it reaches its limits—particularly for frequency-related changes and conditional removals that affect only a subset of traces. Incorporating richer representations, such as eventually-follows relations or explicit decision points, could directly address these patterns. Similarly, the current resource perspective models activity–resource relationships under the assumption that a single resource is associated with each activity execution. Extending this representation to incorporate role-based behavior, workload distribution, or collaborative resource patterns would improve the framework's ability to capture more complex organizational changes. Finally, the perspective-specific window selection opens interesting optimization opportunities. CVDrift evaluates multiple candidate window sizes for each generated time series before applying change-point detection, which enables component-specific and perspective-specific adaptation. However, this also introduces additional computational overhead compared to approaches that rely on a single predefined window size. Investigating efficient pruning strategies, scalability characteristics, and parallelization approaches therefore represents an important direction for future work. Overall, the results demonstrate that perspective-specific, data-driven window selection is a promising direction for improving concept drift detection in event logs. By reducing dependence on manually selected window sizes and supporting multiple process perspectives, CVDrift contributes toward more robust and practical drift detection in dynamic business process environments.

References

1. Adams, J.N., Pitsch, C., Brockhoff, T., Van Der Aalst, W.M.: An experimental evaluation of process concept drift detection. Proc. VLDB Endowment **16**(8), 1856–1869 (2023)
2. Adams, J.N., van Zelst, S.J., Rose, T., van der Aalst, W.M.: Explainable concept drift in process mining. Inf. Syst. **114**, 102177 (2023)
3. Bifet, A., Gavalda, R.: Learning from time-changing data with adaptive windowing. In: Proceedings of the 2007 SIAM International Conference on Data Mining, pp. 443–448. SIAM (2007)
4. Bose, R.J.C., van der Aalst, W.M., Žliobaitė, I., Pechenizkiy, M.: Handling concept drift in process mining. In: International Conference on Advanced Information Systems Engineering, pp. 391–405. Springer (2011)
5. Bose, R.J.C., Van Der Aalst, W.M., Žliobaitė, I., Pechenizkiy, M.: Dealing with concept drifts in process mining. IEEE Trans. Neural Netw. Learn. Syst. **25**(1), 154–171 (2013)
6. Brockhoff, T., Uysal, M.S., van der Aalst, W.M.: Time-aware concept drift detection using the earth mover's distance. In: 2020 2nd International Conference on Process Mining (ICPM), pp. 33–40. IEEE (2020)
7. Cao, Y., Lin, L., Di, Y., Li, X., Chen, W.: MDD: process drift detection in event logs integrating multiple perspectives. In: 2024 IEEE International Conference on Web Services (ICWS), pp. 1125–1134. IEEE (2024)
8. Ceravolo, P., Tavares, G.M., Junior, S.B., Damiani, E.: Evaluation goals for online process mining: a concept drift perspective. IEEE Trans. Serv. Comput. **15**(4), 2473–2489 (2020)
9. Cohen, J.: Statistical power analysis for the behavioral sciences. Routledge (2013)
10. Colin Koeniguer, E., Nicolas, J.M.: Change detection based on the coefficient of variation in SAR time-series of urban areas. Remote Sens. **12**(13), 2089 (2020)
11. Comuzzi, M., Ko, J., Maggi, F.: A language to model and simulate data quality issues in process mining. ACM J. Data Inf. Qual. **17**(2), 1–36 (2025)
12. Gallego-Fontenla, V., Vidal, J.C., Lama, M.: A conformance checking-based approach for sudden drift detection in business processes. IEEE Trans. Serv. Comput. **16**(1), 13–26 (2021)
13. Gama, J., Žliobaitė, I., Bifet, A., Pechenizkiy, M., Bouchachia, A.: A survey on concept drift adaptation. ACM Comput. Surv. (CSUR) **46**(4), 1–37 (2014)
14. Killick, R., Fearnhead, P., Eckley, I.A.: Optimal detection of changepoints with a linear computational cost. J. Am. Stat. Assoc. **107**(500), 1590–1598 (2012)
15. Klijn, E.L., Mannhardt, F., Fahland, D.: Multi-perspective concept drift detection: including the actor perspective. In: International Conference on Advanced Information Systems Engineering, pp. 141–157. Springer (2024)
16. Kraus, A., Van der Aa, H.: Machine learning-based detection of concept drift in business processes. Process Sci. **2**(1), 5 (2025)
17. Lin, L., Wen, L., Lin, L., Pei, J., Yang, H.: LCDD: detecting business process drifts based on local completeness. IEEE Trans. Serv. Comput. **15**(4), 2086–2099 (2020)
18. Maaradji, A., Dumas, M., La Rosa, M., Ostovar, A.: Fast and accurate business process drift detection. In: International Conference on Business Process Management, pp. 406–422. Springer (2015)
19. Maaradji, A., Dumas, M., La Rosa, M., Ostovar, A.: Detecting sudden and gradual drifts in business processes from execution traces. IEEE Trans. Knowl. Data Eng. **29**(10), 2140–2154 (2017)

20. Martjushev, J., Bose, R.J.C., Van Der Aalst, W.M.: Change point detection and dealing with gradual and multi-order dynamics in process mining. In: International Conference on Business Informatics Research, pp. 161–178. Springer (2015)
21. Meira Neto, A.C., de Sousa, R.G., Fantinato, M., Peres, S.M.: Revisiting the transition matrix-based concept drift approach: Improving the detection task reliability through additional experimentation. SN Comput. Sci. **5**(1), 188 (2024)
22. Ostovar, A., Maaradji, A., La Rosa, M., ter Hofstede, A.H., van Dongen, B.F.: Detecting drift from event streams of unpredictable business processes. In: International Conference on Conceptual Modeling, pp. 330–346. Springer (2016)
23. Sato, D.M.V., De Freitas, S.C., Barddal, J.P., Scalabrin, E.E.: A survey on concept drift in process mining. ACM Comput. Surv. (CSUR) **54**(9), 1–38 (2021)
24. Satopaa, V., Albrecht, J., Irwin, D., Raghavan, B.: Finding a" kneedle" in a haystack: Detecting knee points in system behavior. In: 2011 31st International Conference on Distributed Computing Systems Workshops, pp. 166–171. IEEE (2011)
25. Seeliger, A., Nolle, T., Mühlhäuser, M.: Detecting concept drift in processes using graph metrics on process graphs. In: Proceedings of the 9th Conference on Subject-Oriented Business Process Management, pp. 1–10. (2017)
26. Truong, C., Oudre, L., Vayatis, N.: Selective review of offline change point detection methods. Signal Process **167**, 107299 (2020)
27. Van Der Aalst, W.: Process mining: overview and opportunities. ACM Trans. Manag. Inf. Syst. (TMIS) **3**(2), 1–17 (2012)
28. Van Der Aalst, W.: Data science in action. In: Process mining: data science in action, pp. 3–23. Springer (2016)
29. Van Der Aalst, W., et al.: Process mining manifesto. In: International Conference on Business Process Management, pp. 169–194. Springer (2011)
30. Yeshchenko, A., Di Ciccio, C., Mendling, J., Polyvyanyy, A.: Visual drift detection for event sequence data of business processes. IEEE Trans. Visual Comput. Graph. **28**(8), 3050–3068 (2021)
31. Zheng, C., Wen, L., Wang, J.: Detecting process concept drifts from event logs. In: OTM Confederated International Conferences" On the Move to Meaningful Internet Systems", pp. 524–542. Springer (2017)

Connecting the Dots: Path-Based Entity Linking in Object-Centric Event Data

Aaron Küsters(✉) and Wil M. P. van der Aalst

Chair of Process and Data Science (PADS), RWTH Aachen University, Aachen, Germany
{kuesters,wvdaalst}@pads.rwth-aachen.de

Abstract. How long does it take from placing an order to its full delivery, and does the number of packages it was shipped in affect this duration? In object-centric event data, answering such questions requires connecting objects and events that have no direct relationship but are linked through chains of intermediate entities. Existing techniques analyze control flow within individual object types yet provide no systematic means to discover or quantify these cross-type connections. We introduce *path schemas*: type-level descriptions of how any two entities (events or objects) can be connected through the data graph. We automatically enumerate path schemas from a compact type graph and rank them by metrics such as selectivity and coverage to separate meaningful connections from noise. Path schemas enable end-to-end performance analysis across object types (e.g., order-to-delivery throughput) and structural connection discovery (e.g., tracing how orders relate to packages through items). We evaluate our open-source implementation on five datasets, including two real-life ones, and show that path schemas uncover process structure often overlooked by existing techniques and that targeted analysis scales to logs with over a million events.

Keywords: Object-Centric Process Mining · Event Knowledge Graph · Path Schemas · Entity Linking · Performance Analysis

1 Introduction

Real-world business processes often involve multiple, interacting objects. For instance, an order might comprise several items, where each item is first picked and packed into a package, which is then shipped to the customer. While for decades, techniques have focused on analyzing processes from a single fixed perspective (e.g., looking at each order in isolation), more recently approaches analyzing multiple perspectives have gained popularity [1,6]. Many of these techniques are based on the notion of *Object-Centric Event Data* (OCED) [1], where each event can relate to multiple objects of different types, and objects can relate to each other, forming a rich graph of entities and relationships. With that flexibility, a natural question arises: *How can we systematically connect*

F. M. Maggi et al. (Eds.): BPM 2026, LNBIP 599, pp. 203–220, 2027.
https://doi.org/10.1007/978-3-032-37877-4_12

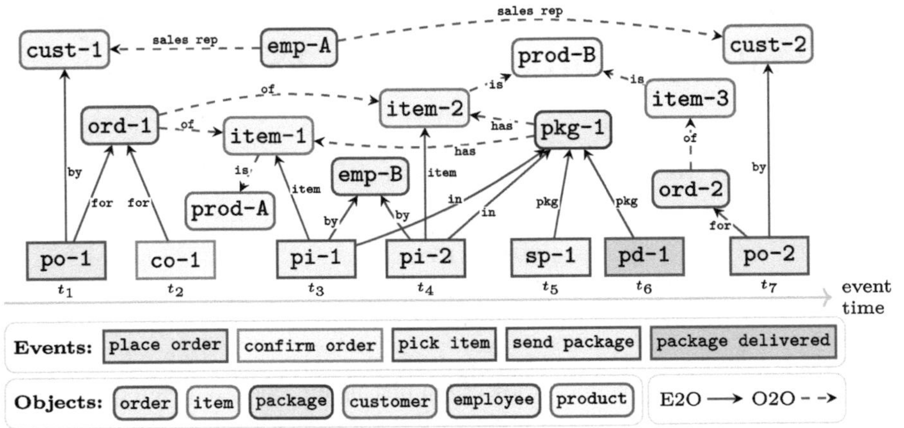

Fig. 1. Instance-level graph of object-centric event data, showing an order management process. Multiple (undirected) paths connect `po-1` and `pd-1`, e.g., over `item-1`, `pi-2`, or even `cust-2`.

entities (events or objects) that are not directly related but are linked through chains of intermediate hops? For example, measuring order-to-delivery throughput requires connecting a `place order` event to a `package delivered` event through intermediate steps, e.g., `item` and `package` objects. More generally, the analytical task is to systematically identify, quantify, and compare these indirect connections between a chosen source and target entity type, instead of manually guessing which intermediate objects link the two. Techniques like Object-Centric Petri Nets [1] or OC-DFGs [18] often focus on control flow *within* an object type and do not discover or quantify such indirect, cross-type connections. Moreover, they require discovering a complete process model before providing meaningful insights, which can be challenging for complex real-life processes.

Motivating Example. Consider Fig. 1: Event `place order` (`po-1`) relates to an `order`, which relates to two `items`; event `package delivered` (`pd-1`) relates to a `package`. Although `po-1` and `pd-1` share no object, they are connected through the order's items and the delivery's package. We call a type-level description of such a connection a *path schema*: a sequence of entity types linked by qualified, directed steps ($>$ forward, $<$ reverse). Schema P connects each `place order` event to related `package delivered` events via three intermediate types:

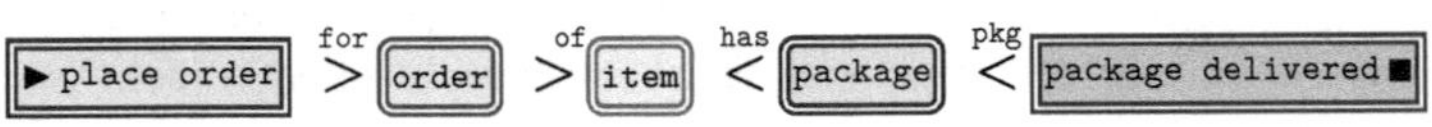

Multiple schemas can connect the same entity types, differing vastly in quality: some yield tight, meaningful connections while others produce noise. In Sect. 3, we show that path schemas automatically identify order fragmentation across packages as strongly associated with delivery delay, entirely unsupervised.

Path schemas are versatile analytical primitives that enable various applications for object-centric event data:

- **Cross-Type Performance Analysis:** Connecting events across types enables end-to-end throughput measurement (e.g., order placement to delivery), where different schemas reveal distinct performance characteristics.
- **Structural Connection Discovery:** Connection patterns reveal the relational structure of a process (e.g., how orders map to packages); equivalence classes can correspond to structural invariants.
- **Correlation Analysis:** Connection patterns can be correlated with performance measures to identify structural factors associated with delays or undesired process behavior.

Path schemas can describe links between any pair of entity types (E2E, O2O, E2O, O2E). We primarily evaluate E2E connections to demonstrate their potential for performance analysis.

Analysis Workflow. Given a source and target entity type of interest, the analyst (1) enumerates candidate path schemas from the type graph, (2) ranks and prunes them using structural metrics such as selectivity and coverage, (3) applies temporal and event-selection filters to fix causal direction and performance semantics, (4) inspects the top-ranked schemas, and (5) uses them for end-to-end performance measurement or unsupervised correlation analysis.

Contributions. Our contributions are as follows:

1. We formalize *Instance* and *Type Graphs* and introduce *Path Schemas* as type-level descriptions of connection patterns between any pair of entity types.
2. We define properties (dead schemas, equivalence), metrics (selectivity, coverage, reach, exclusivity), and filters (temporal constraints, event selection) to assess and refine schemas.
3. We provide an open-source implementation of path schemas in Rust4PM, and integrate it into the OCPQ tool. Moreover, we evaluate our approach on five datasets, showing that path schemas surface and rank cross-type process structure that existing techniques cannot discover automatically, and that analysis scales to logs with millions of events.

The remainder is organized as follows. Section 2 introduces path schemas and their metrics, Sect. 3 presents the evaluation, Sect. 4 the related work, and Sect. 5 concludes.

2 Type and Instance Paths

Let $\mathcal{ET}$, $\mathcal{OT}$, $\mathcal{Q}$ be universes of event types, object types, and qualifiers; $\mathcal{E}$, $\mathcal{O}$ universes of events and objects; $\mathbb{T}$ timestamps; and $\mathcal{P}(X)$ the power set of X.

Definition 1. (OCED) *Object-Centric Event Data* (OCED) is a tuple $L = (T_E, T_O, Q, E, O, type, time, R)$, where $T_E \subseteq \mathcal{ET}$ and $T_O \subseteq \mathcal{OT}$ are event and object types, $E \subseteq \mathcal{E}$ is a set of events, $O \subseteq \mathcal{O}$ a set of objects, $Q \subseteq \mathcal{Q}$ is a set of qualifiers (relationship types) used, $type\colon E \cup O \to T_E \cup T_O$ maps events and objects to their type, such that $type(e) \in T_E$ for $e \in E$ and $type(o) \in T_O$ for $o \in O$, $time\colon E \to \mathbb{T}$ assigns timestamps to events, and $R \subseteq (E \times Q \times O) \cup (O \times Q \times O)$ specifies qualified event-to-object (E2O) and object-to-object (O2O) relationships.

2.1 OCED Instance and Type Graph

Definition 2. (OCED Instance Graph) Given an OCED $L = (T_E, T_O, Q, E, O, type, time, R)$, its *Instance Graph* is a directed labeled graph $G = (V, R, \lambda, t)$:

- $V = E \cup O$ is the set of nodes, representing both events and objects.
- R is the set of directed, labeled edges (inherited from L).
- $\lambda\colon V \to T_E \cup T_O$ maps each node to its type, i.e., $\lambda(v) = type(v)$.
- $t\colon V \to \mathbb{T} \cup \{\perp\}$, where $t(e) = time(e)$ for $e \in E$ and $t(o) = \perp$ for $o \in O$.

To reason about connections at a structural level, we lift this to a Type Graph.

Definition 3. (Type Graph) A *Type Graph* is a directed, labeled graph $G_T = (V_T, R_T)$ of object or event types as nodes $V_T \subseteq \mathcal{ET} \cup \mathcal{OT}$ and edge relations $R_T \subseteq (\mathcal{ET} \times \mathcal{Q} \times \mathcal{OT}) \cup (\mathcal{OT} \times \mathcal{Q} \times \mathcal{OT})$ between event and object types (E2O) or two object types (O2O) labeled with a relationship qualifier.

The Type Graph of an Instance Graph can be constructed by projecting nodes and edges to their types:

Definition 4. (Type of Instance Graph) The *Type Graph* of an OCED Instance Graph $G = (V, R, \lambda, t)$ is the graph $G_T = (V_T, R_T)$ where $V_T = \{\lambda(v) \mid v \in V\}$ is the set of type nodes, and $R_T = \{(\lambda(a), q, \lambda(b)) \mid (a, q, b) \in R\}$ is the set of edges between types, labeled with qualifiers.

Figure 2 shows the Type Graph for our running example.

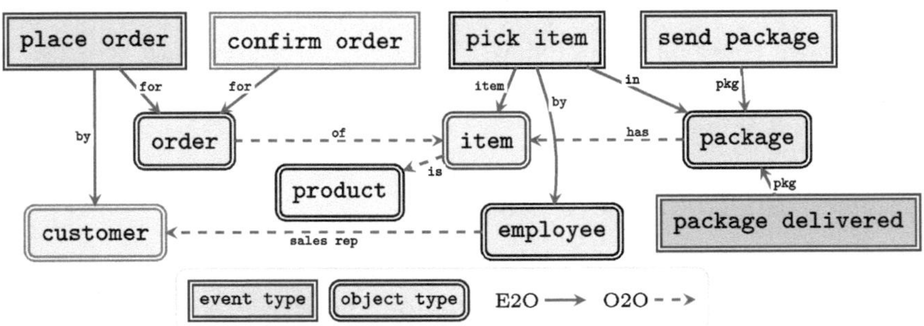

Fig. 2. Type Graph for Fig. 1. Double borders denote type-level nodes. The `employee` type exhibits role specialization, causing dead schemas (Definition 8).

2.2 Path Schemas and Their Instances

Intuitively, a path schema is a sequence of entity types where each consecutive pair is connected by a qualifier and a traversal direction (forward $>$ or reverse $<$), as illustrated by schema P in Sect. 1. The two instance paths of P through concrete entities from Fig. 1 are: p_1: ▶ `po-1` $>$ `ord-1` $>$ `item-1` $<$ `pkg-1` $<$ `pd-1` ■; p_2: ▶ `po-1` $>$ `ord-1` $>$ `item-2` $<$ `pkg-1` $<$ `pd-1` ■.

Definition 5. (Path Schema) A *Path Schema* of length n over a Type Graph $G_T = (V_T, R_T)$ is a sequence $P = (T_1, s_1, T_2, s_2, \dots, s_{n-1}, T_n)$, where:

- $T_i \in V_T$ are entity types (nodes in G_T) for $1 \leq i \leq n$.
- Each step $s_i = (q_i, d_i)$ for $1 \leq i < n$ consists of a qualifier $q_i \in \mathcal{Q}$ and direction $d_i \in \{>, <\}$ and must correspond to an edge in R_T:
  - If $d_i = >$, then $(T_i, q_i, T_{i+1}) \in R_T$.
  - If $d_i = <$, then $(T_{i+1}, q_i, T_i) \in R_T$.

We write $P_{\blacktriangleright} = T_1$ and $P_{\blacksquare} = T_n$ for the source and target type of P, respectively. For $n = 1$, $P = (T_1)$ is a trivial path schema with no steps.

We write $k = n-1$ for the number of *steps* (edges) in a path schema and use k throughout the evaluation; a schema with k steps connects $k+1$ entity types. Reverse traversal is crucial, since E2O relationships are directed from events to objects, and connecting back requires following edges in reverse.

Definition 6. (Instance Path) Let $G = (V, R, \lambda, t)$ be an OCED Instance Graph and $P = (T_1, s_1, T_2, s_2, \dots, s_{n-1}, T_n)$ a path schema over its Type Graph. An *Instance Path* p of P is a sequence of entities $p = (v_1, \dots, v_n) \in V^n$ with:

- ***Type Consistency:*** $\lambda(v_i) = T_i$ for all $1 \leq i \leq n$.
- ***Edge Consistency:*** For each step $s_i = (q_i, d_i)$ with $1 \leq i < n$, an edge with qualifier q_i exists between v_i and v_{i+1} in the direction specified by d_i:
  - If $d_i = >$, then $(v_i, q_i, v_{i+1}) \in R$.
  - If $d_i = <$, then $(v_{i+1}, q_i, v_i) \in R$.
- ***Acyclicity:*** All entities $v_1, \dots, v_n$ are distinct.

All instance paths of a path schema can be enumerated for a given dataset. The *connection set* of the resulting paths is the set of endpoints linked through them.

Definition 7. (Instance and Connection Set) Let G be an OCED Instance Graph and P a path schema of length n on its Type Graph. We write $\mathcal{I}_P^G$ for the set of all instance paths of P in G. Moreover, we define the *connection set* of P on G as $C_P^G = \{(v_1, v_n) \mid (v_1, \dots, v_n) \in \mathcal{I}_P^G\}$, the set of all endpoint pairs.

Often, we are interested in instance paths or end entities per source entity. In Fig. 1, instance paths p_1 and p_2 for path schema P yield the connection set $\{(\texttt{po-1}, \texttt{pd-1})\}$ and target set $\{\texttt{pd-1}\}$ for source `po-1`.

Attributes. Entity attributes (i.e., object and event attributes) integrate in two complementary ways. As *filters*, attribute predicates can restrict which connections are retained (e.g., only orders above a value threshold), requiring no discretization. As *labels*, a discretized attribute (e.g., region or product category) can replace the entity type as the categorical dimension over which schemas are enumerated and compared, although we focus on the type as the default choice.

2.3 Schema Properties

Not all type-level schemas correspond to actual connections. We identify two structural properties.

Definition 8. (Realized and Dead Schemas) Let G be an OCED Instance Graph and P a path schema on its Type Graph. We say that P is *realized* if $\mathcal{I}_P^G \neq \emptyset$. Otherwise, i.e., if $\mathcal{I}_P^G = \emptyset$, we say P is *dead*. The *realizability* of a set of schemas $\mathcal{S}$ is $\frac{|\{P \in \mathcal{S} | \mathcal{I}_P^G \neq \emptyset\}|}{|\mathcal{S}|} \in [0, 1]$.

Dead schemas typically arise from *role specialization*. In Fig. 2, the `employee` type connects to both `pick item` (as picker) and `customer` (as sales rep), but these are disjoint populations, so schemas traversing both roles are dead. Low realizability signals that the type graph overapproximates true connectivity.

Definition 9. (Schema Equivalence) Two schemas P_1, P_2 are *equivalent* ($P_1 \equiv P_2$) if their connection set is the same (i.e., $C_{P_1} = C_{P_2}$).

Equivalence classes help avoid redundant analysis. Moreover, they can correspond to structural constraints: In Fig. 1, schemas connecting `place order` to `send package` either via O2O or via event co-occurrence are equivalent, implying that items and their pick events always involve the same package. Violations would indicate a problem (e.g., an item packed in the wrong package).

2.4 Metrics

Realized schemas can still differ vastly: some connect each source to few targets while others produce noise. The *support* $|C_P|$ counts distinct (source, target) pairs; we define four normalized metrics to quantify these differences.

Definition 10. (Schema Metrics) Given a path schema P, let C_P be the connection set, $\mathcal{S}_P = \{v_1 \mid (v_1, v_n) \in C_P\}$ the active sources, $\mathcal{T}_P = \{v_n \mid (v_1, v_n) \in C_P\}$ the active targets, $N_1 = \{v \in V \mid \lambda(v) = P_{\blacktriangleright}\}$ all source-type entities, and $N_n = \{v \in V \mid \lambda(v) = P_{\blacksquare}\}$ all target-type entities. We define:

- *Coverage*: $cov(P) = |\mathcal{S}_P|/|N_1|$, the fraction of source-type entities that participate as a source in at least one connection.
- *Selectivity*: $sel(P) = |\mathcal{S}_P|/|C_P|$, the ratio of active sources to connections. Equivalently, $1/sel(P)$ is the average number of targets per active source.

- *Reach* (i.e., target-side coverage): $reach(P) = |\mathcal{T}_P|/|N_n|$, the fraction of target-type entities reached by at least one connection.
- *Exclusivity* (i.e., target-side selectivity): $excl(P) = |\mathcal{T}_P|/|C_P|$, the ratio of active targets to connections. Equivalently, $1/excl(P)$ is the average number of sources per active target.

Figure 3 illustrates these four metrics with a concrete example.

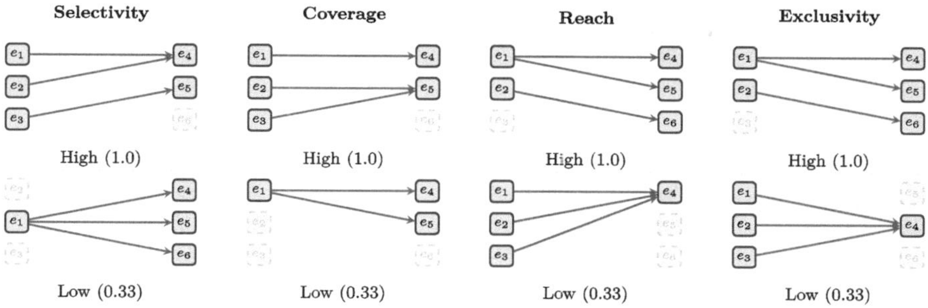

Fig. 3. Illustrations of schema metrics, contrasting low and high values. High *selectivity* connects each source to few targets; low selectivity manifests in a fan-out pattern. High *coverage* means all sources participate; low coverage leaves sources unconnected (dashed). *Reach* mirrors coverage for the target side; high values indicate that most target entities are reached. High *exclusivity* means each target is reached by few sources; low values indicate a fan-in pattern.

2.5 Temporal Constraints and Event Selection

So far, path schemas treat all edges equally regardless of when events occurred. However, connections where the target event precedes the source rarely represent meaningful process flow. We therefore introduce a filtering mechanism for instance path sets with two concrete families of filters. In the following, let $G = (V, R, \lambda, t)$ be a fixed OCED Instance Graph and P a path schema on its Type Graph. As with path schemas, we also write $p_{\blacktriangleright} = v_1$ and $p_{\blacksquare} = v_n$ for an instance path $p = (v_1, \ldots, v_n)$. For a set of instance paths $\mathcal{J}$, we write $\mathcal{J}_v = \{p \in \mathcal{J} \mid p_{\blacktriangleright} = v\}$ for the paths in $\mathcal{J}$ originating from source v.

Definition 11. (Instance Path Filter) An *instance path filter* on schema P and graph G is a function $f\colon \mathcal{P}(\mathcal{I}_P^G) \to \mathcal{P}(\mathcal{I}_P^G)$ satisfying $f(\mathcal{J}) \subseteq \mathcal{J}$ for all $\mathcal{J} \subseteq \mathcal{I}_P^G$. Filters compose by function composition: $(f_1 \circ f_2)(\mathcal{J}) = f_1(f_2(\mathcal{J}))$. We write $C_{P,f}^G = \{(p_{\blacktriangleright}, p_{\blacksquare}) \mid p \in f(\mathcal{I}_P^G)\}$ for the *filtered connection set.*

Definition 12. (Temporal Filter) A *temporal filter* restricts instance paths to those satisfying a temporal predicate. For a path $(v_1, \ldots, v_n) \in \mathcal{J}$, let $E_p = \{v_i \mid v_i \in \mathcal{E}\}$ denote its event nodes. We define $\phi_{\rightarrow}$ and ϕ_{Δ} (only if $v_1 \in \mathcal{E}$):

- $\phi_{\rightarrow}(\mathcal{J}) = \{(v_1, \ldots, v_n) \in \mathcal{J} \mid \forall\, v_i, v_j \in E_p,\ i < j\colon t(v_i) \leq t(v_j)\}$ (*forward*)
- $\phi_{\Delta}(\mathcal{J}) = \{(v_1, \ldots, v_n) \in \mathcal{J} \mid \forall\, v_i \in E_p\colon |t(v_i) - t(v_1)| \leq \Delta\}$ (*bounded*)

Note that the bounded filter assumes $p_\blacktriangleright \in \mathcal{E}$, i.e., that the source is an event with a defined timestamp. The forward filter prunes non-causal connections; the bounded filter limits the time horizon. The second family selects representative targets when a source reaches multiple events.

Definition 13. (Selection Filter) Let $\mathcal{J}$ be a set of instance paths ending in events, i.e., $p_\blacksquare \in \mathcal{E}$ for all $p \in \mathcal{J}$. A *selection filter* selects, for each source, the instance paths whose target satisfies an extremal criterion:

- $\sigma_{\mathit{first}}(\mathcal{J}) = \{p \in \mathcal{J} \mid t(p_\blacksquare) = \min\{t(p'_\blacksquare) \mid p' \in \mathcal{J}_{p_\blacktriangleright}\}\}$ (earliest)
- $\sigma_{\mathit{last}}(\mathcal{J}) = \{p \in \mathcal{J} \mid t(p_\blacksquare) = \max\{t(p'_\blacksquare) \mid p' \in \mathcal{J}_{p_\blacktriangleright}\}\}$ (latest)
- If, additionally, $p_\blacktriangleright \in \mathcal{E}$ for all $p \in \mathcal{J}$ (E2E), we define: (soonest)
- $\sigma_{\mathit{soonest}}(\mathcal{J}) = \{p \in \mathcal{J} \mid |t(p_\blacksquare) - t(p_\blacktriangleright)| = \min\{|t(p'_\blacksquare) - t(p_\blacktriangleright)| \mid p' \in \mathcal{J}_{p_\blacktriangleright}\}\}$

In practice, a temporal filter ensures causal plausibility, followed by a selection filter. For example, $C^G_{P,\sigma_{\mathit{last}} \circ \phi_\rightarrow}$ yields the last causally related target per source, useful for measuring when an order is *fully* delivered. For E2E schemas (both source and target are events), the filtered connection set induces a natural performance measure: the *throughput time* of a connection $(e_s, e_t) \in C^G_{P,f}$ is $t(e_t) - t(e_s)$. The schema determines *which* entities are connected; the filter determines *which* connections are retained per source.

Anchored Paths. Selection so far picks an extremal target per source. Instead, an intermediate hop can be used as an *anchor* (e.g., `order` objects), which splits the schema into a source arm and a target arm evaluated independently. Applying σ_{first} or σ_{last} to each arm and pairing the selected events yields one span per anchor entity. This expresses the four first/last combinations, e.g., from the first `place order` to the last `package delivered` of an order, or from the last confirmation to the first shipment. Each arm carries its own metrics.

2.6 Counting Schemas

Beyond connecting pairs of entities, path schemas can also serve as a counting mechanism. For a given source, how many distinct target entities are reachable along a path schema?

Definition 14. (Counting Schema) The *count* of v_1 under path schema P is $rc_P(v_1) = |\{v_n \mid (v_1, v_n) \in C_P\}|$, the number of target entities reached from v_1.

For single-hop schemas this reduces to a direct count (e.g., items per order). Multi-hop counting schemas reveal structural information not available from any single relationship. For example, there is no direct `order`→`package` edge in Fig. 1, but the two-hop schema ▶ `order` $\overset{\text{of}}{>}$ `item` $\overset{\text{has}}{<}$ `package` ■ yields a per-order package count that characterizes the fulfillment structure. The distribution of reachability counts can be correlated with performance measures to identify structural factors associated with delay, as demonstrated in Sect. 3.

2.7 Algorithms

Schema Enumeration. Given a source type and maximum length k, schemas are enumerated by Breadth-First Search (BFS) on the type graph, treating edges as bidirectional and requiring simple paths, i.e., no repeated types (so a type-revisiting connection, such as an `item` linked to the other `items` in its `package`, is excluded).

Instance Path Finding. For a given schema and source entity, instance paths are found via schema-guided step-wise expansion, extending partial paths one step at a time while pruning acyclicity or temporal violations early. The per-source worst-case complexity is $O(d_{\max}^k)$, where $d_{\max}$ is the maximum node degree in the instance graph; in practice, type-guided expansion and selectivity-based pruning keep runtimes well below this bound.

3 Evaluation

We evaluate path schemas along three research questions: **(RQ1)** What structural properties do path schemas exhibit across datasets, and how effectively do metrics separate meaningful schemas from noise? **(RQ2)** How does the approach scale? **(RQ3)** Do path schemas enable cross-type performance analysis and unsupervised discovery of performance correlations?

3.1 Implementation

We implemented path schema discovery in the open-source Rust4PM library [14] with data parallelization. Two optimizations keep discovery tractable: *selectivity-*

Fig. 4. The `place order` → `package delivered` scenario in the tool. The discriminating `orders→items` schema (blue) and the noisy `products` schema (red) in the type graph. Bottom right: throughput distribution under the *last* filter.

based early termination maintains an upper bound and stops once it falls below the pruning threshold, and only connection endpoints and timestamps are materialized, not full intermediate paths. We integrate this into the object-centric query tool OCPQ [15] (Fig. 4) for interactive analysis: The analyst imports an OCEL 2.0 log [6], selects source and target types on the type graph, and the tool ranks schemas in a sortable table with metric heatmaps. Selecting a schema highlights its path and shows details and throughput; temporal constraints and event-selection strategies can be configured interactively. A schema can also be edited as an OCPQ query, bridging automatic discovery and manual querying. Figure 4 revisits the running example of Sect. 1 (as one of five RQ3 scenarios, see Table 1). Selectivity ranks the `orders`→`items` path above the noisy `products` path. The initial implementation, evaluation code and results are publicly available at https://github.com/aarkue/oced-path-con.

3.2 Datasets and Setup

Beyond this example, we evaluate path schemas systematically: we use five OCEL 2.0 [6] datasets spanning different domains and scales (Table 1). **Order Management (OM)**, **Container Logistics (CL)**, and **Procure-to-Pay (P2P)** are standard simulated object-centric logs covering retail, logistics, and finance, respectively[1]. **BPIC2017** [9] and **BPIC2019** [10] are real-life datasets converted to OCED [13]; at more than 1M events and 100K objects they serve as scalability stress tests. Together, the five datasets form a broad spectrum: Simulated logs provide controlled structural properties (role specialization, hub types, O2O edges), while BPIC2019 lacks O2O relationships entirely, requiring cross-type connections to be recovered through event co-occurrence at k=4. Table 1 also shows one specific E2E scenario (i.e., pair of event types) per dataset that will be further analyzed to address RQ3.

Table 1. An overview of the evaluation datasets and evaluation scenarios.

	Log Characteristics				Evaluation Scenarios		
Dataset	$\|\mathcal{E}\|$	$\|\mathcal{O}\|$	$\|\mathcal{ET}\|$	$\|\mathcal{OT}\|$	Scenario	Source	Target
P2P	14,671	9,543	10	7	cr → ep	`Create Purch. Req.`	`Execute Payment`
OM	21,008	10,840	11	6	po → pd	`place order`	`package delivered`
CL	35,413	13,910	14	7	ro → dp	`Register Cu. Order`	`Depart`
BPIC2017	1,202,267	106,162	26	4	ca → oa	`A_Create Appl.`	`O_Accepted`
BPIC2019	1,595,923	330,685	41	4	cpr → rgr	`Create Pur.Req.It.`	`Record Goods Rec.`

3.3 Results

RQ1: Structural Properties and Metric Distributions. We enumerated all path schemas across the five datasets for $k \in \{2, 3, 4\}$, pruning schemas with a selectivity below 0.01 early. For BPIC2019 at k=4, exhaustive enumeration of all

[1] Sourced from https://www.ocel-standard.org/event-logs/overview/.

>2,000 type pairs exceeds practical time limits; we report results based on a 10% stratified sample of type pairs (sampled proportionally across E2E, E2O, O2E, O2O categories). In total, 201,246 schemas were identified, growing from 10,136 at k=2 to 134,356 at k=4 as longer paths traverse more type-graph combinations. Of these, 81,750 (40.6%) are dead, yielding an overall realizability of 0.594. Realizability varies across datasets: P2P (0.988) and CL (0.990) show near-complete realization, OM (0.732) and BPIC2017 (0.906) exhibit moderate rates, while BPIC2019 (0.491) has the lowest, reflecting its large type graph (41 event types) combined with heavy role specialization of the `Resource` object type. A further 37,083 alive schemas (18.4% of all) have selectivity below 0.01 and are pruned as uninformative, leaving 82,413 alive, non-pruned schemas for analysis. Among these, E2E schemas account for 14–83% depending on dataset and k; the remainder are distributed among E2O, O2E, and O2O schemas, confirming that cross-type connections are widespread across all pair kinds.

Dead Schemas serve as a diagnostic: They signal that the type graph over-approximates true connectivity. In OM, a schema from `package delivered` to `place order` via `employees` and `customers` is dead because the `employees` type connects to deliveries as *shipper* and to customers as *sales rep*, but these are disjoint populations. In BPIC2017, `A_Accepted` → `Case_R` → `A_Submitted` is dead while the same pair via `Application` is alive (sel. 1.0). The `Case_R` objects associated with `A_Submitted` events are disjoint from those linked to later lifecycle events. In P2P, a schema linking approval to delegation through shared objects is dead: Requisitions are either approved directly or delegated, but never both. Figure 5 shows metric distributions across the alive, non-pruned schemas. **Selectivity** and **exclusivity** are bimodal, with mass at both extremes, providing the most differentiation between schemas: Schemas traversing a discriminating type (1:1 or 1:few mapping) cluster near 1, while those traversing a resource hub type

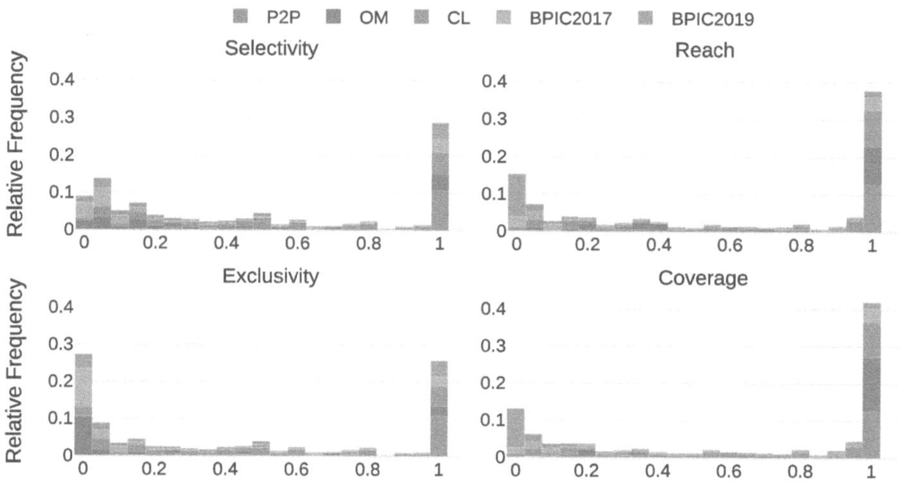

Fig. 5. Distribution of selectivity, reach, exclusivity, and coverage across all alive, non-pruned schemas, each dataset contributing equally.

(1:many) cluster near 0. **Reach** and **coverage** concentrate near 1.0, indicating that most non-pruned schemas cover nearly all source and target entities. BPIC2019 departs from this pattern: Coverage and reach frequently take values near 0, reflecting its structural properties (many infrequent types).

Equivalence. All datasets exhibit non-trivial equivalence classes. In OM and P2P, more than 75% of type pairs have a compression factor ≥ 2 (i.e., $2r$ path schemas collapse to $\leq r$ equivalence classes). Beyond compression, equivalence classes reveal structural invariants: In OM, three schemas connecting `place order` to `package delivered` (via `items` directly, via `items`→`packages`, and via `orders`→`items`) are equivalent (all sel. 0.47), implying that every item reachable from an order via the O2O relationship also co-occurs with that order in events. If this equivalence did not hold, it would signal items linked to orders through O2O edges but absent from order events, exposing an inconsistency in the data.

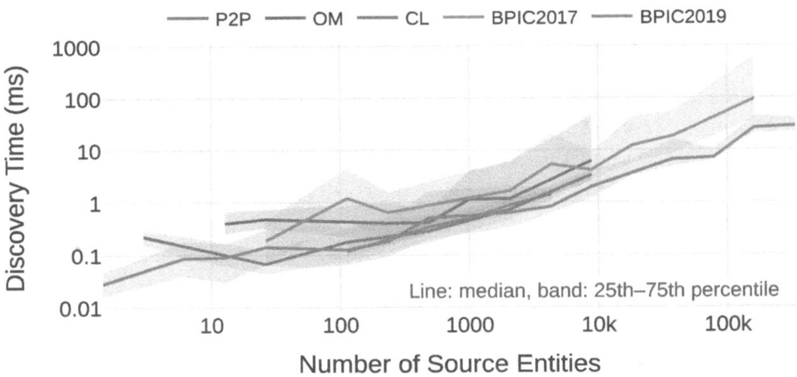

Fig. 6. Discovery time vs. number of source entities (log-log) across all entity type pairs and datasets for $k \in \{2, 3, 4\}$. The median discovery duration and 25th–75th IQR are shown, indicating a linear relationship.

RQ2: Runtime and Scalability. Schema enumeration completes in under 1 ms even at $k = 4$. The cost lies in instance-level connection discovery: Fig. 6 plots discovery time against source count for all 82,413 non-pruned schemas, showing a linear trend in log-log space. **Runtimes.** For targeted scenario analysis (including data loading), P2P, OM, and CL complete in under 3 s at k=3; BPIC2017 takes 3.3 s and BPIC2019 takes 60 s. For BPIC2019 at k=4 (466 enumerated schemas), the analysis completes in 5.5 min with a selectivity threshold of 0.05 and 16.3 min with 0.01. Selectivity-based pruning is also effective: At threshold 0.05, 201 of 466 schemas are dead; selectivity pruning cuts the remaining 265 by more than half, leaving only 125 for the analyst to inspect.

RQ3: Cross-Type Performance and Correlation Analysis. We now demonstrate the analysis workflow of Sect. 1 end to end: metrics rank schemas, temporal and selection filters set causal direction and performance semantics, and counting schemas surface structural correlates of throughput.

Schema Selection and Metrics. Table 2 reports distinct path schemas for the E2E scenarios from Table 1 at $k = 3$, grouped by equivalence classes. Schemas are pruned during discovery when forward-temporal selectivity falls below 0.001, deliberately lower than the 0.01 cutoff of RQ1 so that the table retains low-selectivity schemas for contrast.

Table 2. Path schema metrics (forward temporal, k=3). Schemas with identical metrics are grouped. Fwd./Bnd. red. show path count reduction under forward and bounded (14 d) temporal constraints. Throughput columns show mean throughput time across sources for first/last event selection

Scenario	Via Types	Supp.	Fwd. red.	Bnd. red.	Metrics (fwd.) Cov.	Rea.	Sel.	Excl.	Time (ms)	Throughput First	Last
cr→ep	Mat., Pur. Ord.	931	0%	92%	1.00	0.80	0.78	1.00	0.3	22.3 d	23.3 d
	Mat., Goods Rec.	537	0%	90%	0.57	0.46	0.77	1.00	0.3	21.6 d	22.6 d
po→pd	Orders, Items	4,264	0%	21%	1.00	1.00	0.47	0.26	0.9	5.3 d	14.6 d
	Products	744,823	48%	92%	1.00	1.00	0.00	0.00	54	0.7 d	228.4 d
ro→dp	Cu. Ord., Tr. Doc.	754	0%	62%	0.95	1.00	0.76	0.17	0.2	16.0 d	19.6 d
ca→oa	Appl., Offer	17,228	0%	54%	0.55	1.00	1.00	1.00	7	18.1 d	18.1 d
	Case_R	155,240	33%	93%	0.02	0.68	0.00	0.08	11	92.5 d	208.8 d
cpr→rgr	POItem	46,752	0%	43%	0.92	0.15	0.92	1.00	15	15.3 d	15.5 d
	Po	748,683	0%	26%	0.95	0.18	0.06	0.08	84	13.6 d	20.7 d
	Resource	615,260	71%	90%	0.11	0.05	0.01	0.02	54	6.4 d	62.6 d

Selectivity is the most discriminating metric. In OM, the schema via `orders→items` (sel. 0.47) connects each order to roughly two deliveries, reflecting physical item-to-package routing, while the alternative via `products→items` (sel. ≈ 0) links *every* order of the same product to *every* delivery (744K connections). BPIC2017 shows the same contrast: sel. 1.0 via `Application→Offer` versus sel. ≈ 0 through shared `Case_R` objects. Similar patterns hold for BPIC2019 (`POItem` sel. 0.92 vs. `Resource` sel. 0.01 with 615K connections). CL has only a single non-dead schema (sel. 0.76), which already provides a meaningful connection. These results highlight where path schemas go beyond prior work (cf. Sect. 4): OPerA [17] requires a shared object between measured events, and OC-DFG analysis [5] stays within one object type. Where a shared object exists (OM via `items`, BPIC2019 via `POItem`), our single-hop schema recovers the same connection these approaches would use. However, the P2P and CL scenarios have no shared object at all and require multi-hop schemas ($k = 3$); the discriminating BPIC2017 schema likewise needs two hops (Application→Offer, $k = 3$), as the only shared object (`Case_R`) yields a near-zero-selectivity connection. Manual EKG queries [11] could express any of these traversals but require the analyst to know the path structure *a priori*; path schemas instead discover and rank all of them automatically, including multi-hop connections that no existing approach surfaces without such prior knowledge. When the relevant path is already known, a manual query may suffice; path schemas help most when many candidate connections must be compared without knowing which is relevant a priori.

Coverage complements selectivity: OM reaches full coverage, while the discriminating BPIC2017 schema covers only 55%, reflecting that not all applications lead to an accepted offer. **Event Selection** captures different performance semantics: In OM, σ_{first} yields 5.3 d mean throughput vs. 14.6 d for σ_{last}. This 9.3-day gap quantifies the additional time until all packages of a multi-item order are delivered. **Temporal Constraints** serve two roles. The forward filter prunes acausal connections: Broad schemas see up to 71% reduction, while discriminating ones see 0%, suggesting that these specific high-selectivity schemas are already temporally coherent. The bounded 14 d filter limits the time horizon and reaches reduction values of 21–93%.

O2O Recovery Through Co-occurrence. BPIC2019 lacks O2O relationships, so cross-type E2E connections require k=4 with event co-occurrence as the bridge. At k=3, schemas traverse a single object type, where `POItem` is the most discriminating (sel. 0.92). At k=4 (threshold 0.01), 466 schemas are enumerated (201 dead, 102 pruned, 163 alive); all high-selectivity schemas connect through `POItem` or `PO`, confirming that meaningful cross-type connections can be recovered from event co-occurrence alone when O2O relationships are absent.

Counting Schema Correlation. For each E2E scenario, we enumerate all counting schemas (Definition 14) from the source event type. Each source event yields one reachability count and one throughput time (σ_{last}); we rank counting schemas by Spearman correlation across source events, collapsing equivalence classes (Fig. 7, left). In OM, equivalence classes reduce 37 counting schemas (k=2) to 20 distinct classes. The top-ranked class counts *packages per order* (4 equivalent schemas, $\rho = 0.714$, $p < 0.001$): single-package orders have a median throughput of 5.9 d vs. 26.0 d for 5-package orders (Fig. 7, right). No direct `order–package` relationship exists; the 2-hop schema through `item` discovers this connection automatically, outperforming the best single-hop baseline (*items per order*, $\rho = 0.386$). The second-strongest predictor is *item out of stock* ($\rho = 0.709$, count $\geq$1 for 54% of sources): orders with at least one out-of-stock item have a mean throughput of 20.3 d vs. 7.9 d otherwise. For BPIC2017, the number of `A_Incomplete` events per application ($\rho = 0.406$, count $\geq$1 for 73% of sources) is the strongest predictor (Fig. 7, right), suggesting rework-driven delays where repeated resubmission of incomplete information prolongs processing. For P2P, delegated vs. direct approval schemas show opposing correlations ($\rho = \pm 0.185$), linking the approval route to throughput; for CL, the 1:1 structure yields zero variance, so no correlations emerge. For BPIC2019, the strongest predictor counts `POItems` reachable via `Clear Invoice` on the shared `PO` ($\rho = -0.410$, count $\geq$1 for 54% of sources), likely reflecting that items on larger POs complete faster (11.9 d vs. 19.8 d). All these patterns were discovered automatically without prior domain knowledge of the process structure.

	Counting Schema	ρ	Non-zero %
cr → ep	Cre. PR $\xrightarrow{\text{mat}}$ Mat. $\xleftarrow{\text{mat}}$ Del. PR Appr.	-0.185	17%
	Cre. PR $\xrightarrow{\text{mat}}$ Mat. $\xleftarrow{\text{mat}}$ Appr. PR	0.185	83%
	Cre. PR $\xrightarrow{\text{mat}}$ Mat.	0.142	100%
po → pd	Pla. Ord. $\xrightarrow{\text{it}}$ It. $\xleftarrow{\text{con}}$ Pkg.	0.714	100%
	Pla. Ord. $\xrightarrow{\text{it}}$ It. $\xleftarrow{\text{it}}$ It. OOS	0.709	54%
	Pla. Ord. $\xrightarrow{\text{it}}$ It.	0.386	100%
ca → oa	Cr. App. → App. ← App. Incomp.	0.406	73%
	Cr. App. → App. ← App. Valid.	0.290	100%
	Cr. App. → App. → Offer	0.270	100%
cpr → rgr	Cre. PRI → PO ← Clear Inv. → POItem	-0.410	54%
	Cre. PRI → PO ← Clear Inv.	-0.408	54%
	Cre. PRI → PO ← Rec. Inv. Rec. → POItem	-0.406	88%

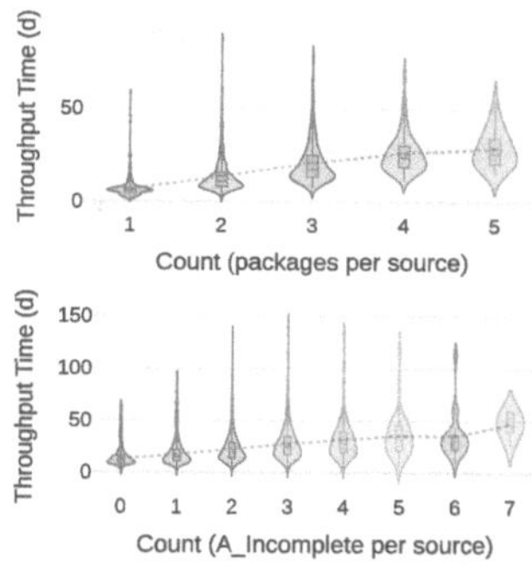

Fig. 7. Automatic correlation discovery. All enumerated counting schemas ranked by Spearman correlation with E2E throughput ($p < 0.01$ for all); equivalent schemas are collapsed. Right: Visualization for the top-ranked OM (top) and BPIC2017 (bottom) counting schema against throughput time.

3.4 Threats to Validity

Selectivity and coverage are proxies for connection meaningfulness; no ground truth exists for schema relevance. Still, high-selectivity schemas consistently yield domain-interpretable connections (e.g., item-to-package routing in OM), while low-selectivity ones produce combinatorial noise. Domain-expert validation remains future work. The RQ3 counting-schema correlations are associative, not causal; we guard against spurious associations with significance thresholds ($p < 0.01$, Bonferroni-corrected), but causal interpretation requires domain knowledge. Three datasets are simulated; the two real-life logs (BPIC2017, BPIC2019) are BPI challenge data converted to OCED rather than originally object-centric. Both still offer realistic scale (> 1M events), and BPIC2019's lack of O2O relationships stress-tests the approach. Validation on object-centric industrial data would strengthen generalizability. We restrict schemas to acyclic paths; bounded revisitation is a natural extension. At $k \geq 5$, combinatorial growth can make exhaustive enumeration impractical.

4 Related Work

Performance analysis is well-established in case-centric process mining, with mature frameworks for defining and computing KPIs over single-case traces [7]. In Object-Centric Process Mining (OCPM) [1], performance analysis is less developed. Most object-centric discovery techniques (OC-DFGs [5], OC Petri nets [1]) model control flow *within* individual object types. OPerA [17] measures sojourn times and synchronization delays within a single object's lifecycle, requiring a perfectly fitting object-centric Petri net as input. Park et al. [18] extend OC-DFGs with waiting and service times at activities. Both approaches compute performance along an object's lifecycle trajectory, so they require that the measured events share a common object; they cannot express durations where no single object participates in both endpoints (e.g., order placement to package delivery, linked only through items). De Leoni and Volpato [16] show that such implicit interferences between object-centric executions significantly affect prediction accuracy, motivating the need to make hidden connections explicit.

More expressive frameworks can capture these cross-type relationships: Esser and Fahland [11] proposed Event Knowledge Graphs (EKGs), storing events and objects as labeled property graphs and supporting hand-crafted Cypher queries for cross-object performance measures. OCPQ [15] provides a visual query language for selecting and querying combinations of objects and events. However, both approaches require manually specifying the exact traversal path, which assumes prior knowledge of the data's structure. A generic Cypher query could enumerate instance paths up to a fixed length without such prior knowledge, but returns raw paths rather than ranked typed schemas, without dead-schema or equivalence detection, temporal filters, or counting schemas. Path schemas automate this by enumerating all type-level schemas from the type graph, which is derived automatically from the log rather than supplied, and ranking them by quality metrics, as demonstrated in Sect. 3, where metrics rank meaningful routes ahead of spurious ones. Discovered schemas could also serve as a starting point for EKG or OCPQ analysis by generating candidate queries. Table 3 summarizes the positioning of path schemas against these approaches.

Table 3. Positioning against existing OCED approaches for performance analysis. EKG/OCPQ require manual specification of cross-type connections. OC-DFG is based on DFGs built from the data, but requires no further model as input.

	OPerA [17]	OC-DFG [5,18]	EKG [11]/OCPQ [15]	**Path Schemas**
Metric scope	activity	activity pair	flexible	flexible
Cross-type	×	×	✓	✓
Model-free	×	$\sim^{\text{DFG}}$	✓	✓
Auto. discovery	N/A	N/A	×	✓

Prescriptive formalisms such as Proclets [12] and Object-Centric Behavioral Constraints (OCBC) [3] define inter-object interactions at design time. Path schemas take a *descriptive* perspective, discovering how entities connect *across* types and enabling cross-type performance measurement directly from data.

Adams et al. [2] define process executions as connected subgraphs where the analyst selects participating object types. Van Detten et al. [8] generalize this into a framework for case construction. Basmer et al. [4] take a more exploratory approach, materializing views from different context definitions to help analysts discover relevant perspectives on the data. All three define connections at the level of case notions or contexts; path schemas instead operate at the entity-to-entity level, discovering and quantifying specific connection patterns.

The concept of type-level path templates also appears in heterogeneous information networks as meta-paths [19], where selecting meaningful paths for a given task is an active research challenge addressed through learning-based methods. Instead, we apply exhaustive enumeration combined with interpretable quality metrics, providing a ranking without requiring predefined training data. Temporal filters and structural properties like dead schemas and equivalence classes arise specifically from the structure of *object-centric event data.*

5 Conclusion

We introduced path schemas as a new analytical primitive for object-centric process mining that discovers and quantifies cross-type entity connections without requiring a process model or domain knowledge. Our evaluation on five datasets shows that structural properties expose type-graph overapproximation, that the introduced metrics rank discriminating schemas ahead of spurious ones, that reachability counts can reveal structural correlates of performance without supervision, and that targeted analysis scales to logs with millions of events.

Future work. Dead schemas and equivalence classes are promising diagnostics deserving deeper investigation. Further extensions include attribute-awareness, bounded cyclic paths that allow recurring types such as items packaged together, and validation with domain experts on object-centric industrial data.

Acknowledgments. The authors gratefully acknowledge the German Federal Ministry of Research, Technology and Space (BMFTR) and the state government of North Rhine-Westphalia for supporting this work as part of the NHR funding.

References

1. van der Aalst, W.M.P., Berti, A.: Discovering object-centric Petri nets. Fundam. Informaticae **175**(1–4), 1–40 (2020)
2. Adams, J.N., Schuster, D., Schmitz, S., Schuh, G., van der Aalst, W.M.P.: Defining cases and variants for object-centric event data. In: ICPM, pp. 128–135. IEEE (2022)
3. Artale, A., Kovtunova, A., Montali, M., van der Aalst, W.M.P.: Modeling and reasoning over declarative data-aware processes with object-centric behavioral constraints. In: BPM, LNCS, vol. 11675, pp. 139–156. Springer (2019)
4. Basmer, M., Ueck, H., Fahland, D., Weidlich, M.: MANTA: materializing views on event data for context exploration in process analysis. In: BPM, LNCS, vol. 16044, pp. 51–68. Springer (2025)
5. Berti, A., van der Aalst, W.M.P.: OC-PM: analyzing object-centric event logs and process models. Int. J. Softw. Tools Technol. Transf. **25**(1), 1–17 (2023)
6. Berti, A., et al.: OCEL (Object-Centric Event Log) 2.0 Specification. CoRR abs/2403.01975
7. del-Río-Ortega, A., Resinas, M., Cabanillas, C., Cortés, A.R.: On the definition and design-time analysis of process performance indicators. Inf. Syst. **38**(4), 470–490 (2013)
8. van Detten, J.N., Schumacher, P., Leemans, S.J.J.: A framework for advanced case notions in object-centric process mining. In: ICPM Workshops, LNBIP, vol. 533, pp. 402–414. Springer (2024)
9. van Dongen, B.: BPI Challenge 2017 (2017). https://data.4tu.nl/articles/dataset/BPI_Challenge_2017/12696884/1

10. van Dongen, B.: BPI Challenge 2019 (2019). https://data.4tu.nl/articles/dataset/BPI_Challenge_2019/12715853/1
11. Esser, S., Fahland, D.: Multi-dimensional event data in graph databases. J. Data Semant. **10**(1–2), 109–141 (2021)
12. Fahland, D.: Describing behavior of processes with many-to-many interactions. In: Petri Nets, LNCS, vol. 11522, pp. 3–24. Springer (2019)
13. Khayatbashi, S., Hartig, O., Jalali, A.: Transforming event knowledge graph to object-centric event logs: a comparative study for multi-dimensional process analysis. In: ER, LNCS, vol. 14320, pp. 220–238. Springer (2023)
14. Küsters, A., van der Aalst, W.M.P.: Rust4PM: a versatile process mining library for when performance matters. In: BPM (Demos / Resources Forum). CEUR Workshop Proceedings, vol. 3758, pp. 91–95. CEUR-WS.org (2024)
15. Küsters, A., van der Aalst, W.M.P.: OCPQ: object-centric process querying and constraints. In: RCIS (1), LNBIP, vol. 547, pp. 383–400. Springer (2025)
16. de Leoni, M., Volpato, P.: Global predictive monitoring of object-centric processes. In: BPM, LNCS, vol. 16044, pp. 255–272. Springer (2025)
17. Park, G., Adams, J.N., van der Aalst, W.M.P.: OPerA: object-centric performance analysis. In: ER, LNCS, vol. 13607, pp. 281–292. Springer (2022)
18. Park, G., Adams, J.N., van der Aalst, W.M.P.: Conformance checking and performance analysis using object-centric directly-follows graphs. In: BPM (Forum), LNBIP, vol. 526, pp. 179–196. Springer (2024)
19. Sun, Y., Han, J., Yan, X., Yu, P.S., Wu, T.: Heterogeneous information networks: the past, the present, and the future. Proc. VLDB Endow. **15**(12), 3807–3811 (2022)

Process Binoculars: Resilient Log Generation from Multi-video Sources

Julian Neuberger(✉), Siyuan Deng, and Hugo A. López

Technical University of Denmark, Kongens Lyngby, Denmark
{jtone,hulo}@dtu.dk

Abstract. Process mining links event data to process models, typically assuming structured logs. However, many actual processes do not leave digital traces and need to be discovered in the field. Existing approaches mitigate this by integrating logs with multiple sources, including observations of actual processes. Recent works have pioneered process discovery methods from video sources, but these methods often require manual labeling and produce brittle logs dependent on a single viewpoint, risking incomplete capture of activities. We introduce the Process BINOCULARS, a technique for generating event logs from multiple video sources that provides 360-degree coverage of activities. The approach automates activity identification using representation learning and semantic abstraction. Micro-actions are first extracted from video streams, enabling generalization across processes. Learned representations over event windows improve robustness to camera errors, while clustering and abstraction yield human-readable activities. Evaluation on a large-scale assembly dataset shows that Process BINOCULARS produces accurate process models and remains resilient under controlled noise. Our code and benchmarks are publicly released (See https://github.com/JulianNeuberger/pm-binoculars).

Keywords: Video Process Mining · Event-log generation · Semantic Abstraction · Representation Learning

1 Introduction

Knowledge of internal business processes is vital for organizations to stay competitive. To this end, formal models of business processes are needed, which allow, among others, to reason about process outcomes, detect possible improvements, or simply document the process. In most cases, a BPM initiative does not start with the implementation of entirely new processes, but with the identification and discovery of existing processes in the organization [5]. *Process Mining* [30] is an enabling technology to do so, and, provided structured information (in the shape on an *event log*), it can facilitate the definition of formal representations that can later be improved and automated. However, getting to an event log is already a complex task. First, because more than 80% of the information in an organization is unstructured [14], and second, because in

F. M. Maggi et al. (Eds.): BPM 2026, LNBIP 599, pp. 221–237, 2027.
https://doi.org/10.1007/978-3-032-37877-4_13

some cases, the nature of the decisions may be so complex and nuanced that operators would prefer not to document them [24]. Therefore, process identification normally relies on additional practices such as interviews, ethnographies, and brown-paper exercises, which, on top of process mining, help to identify candidate models.

This paper explores one of these alternative approaches: the incorporation of video recordings into event logs to be used for process discovery. Compared to interviews and ethnographies, the introduction of video data will provide logs of what actually happened in the process, rather than what the subject recalls. Compared to both interviews and ethnographies, video-recordings can incorporate multiple viewpoints, revealing process information that may be ignored by the interviewer or the subject. For instance, an interviewee may recall when they *"screw the wheel into the chassis"* and forget that to perform this action, they needed to fetch specialized tools. While activity detection has been explored in the process mining community in the past (e.g., [16,17,22]), no prior work has considered a holistic approach where multiple viewpoints can be integrated. This combination is not a simple task for existing frameworks, which normally require manual labeling of areas of interest representing process activities [16], training deep learning systems [13,31] to detect specific activities that cannot generalize to other domains, or that result in activities at such a fine level of abstraction that generated models are difficult to understand. Returning to our example, the activity *"fetch screwdriver"* may run completely ignored if the tool is either outside the area of interest or the camera is obstructed (c.f., Fig. 1).

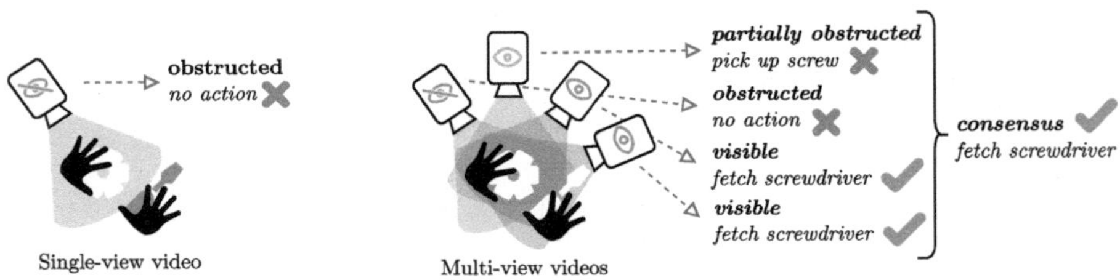

Fig. 1. Multiple camera angles of the same process provide resilience to obstructions, while redundant cameras enable compensating for computer vision errors.

We propose the Process BINOCULARS, a novel technique that combines multiple viewpoints to generate event logs, bridging the gap between micro-actions extracted from raw video and high-level activities. Our technique works in five distinct steps. First, micro-actions are extracted from raw video. Sequences of micro-actions are then embedded into a high-dimensional space via a contrastively trained transformer. High-dimensional representations are then clustered based on similarity. Clusters are assigned a high-level label, which is generated by rephrasing a representative sequence through an Large Language Model (LLM). Finally, an event log is constructed that can be mined using standard algorithms. Compared with prior work, Process Binoculars employs an unsupervised micro-task framework and a semantic abstraction layer, enabling reuse without retraining. Moreover, our technique is parameterizable by the user, who can define the level of abstraction of the generated event log.

Our experiments show that the Process Binoculars are robust in several key characteristics. First, we show that they are resilient to camera malfunctions across camera angle, color, or obstructions, for a given number of cameras. In our experiments, the Process Binoculars are resilient to up to 67% of camera failures before performance degrades. Second, we show that computer vision errors do not propagate in the downstream prediction tasks. Moreover, our experiments suggest that errors coming from computer vision can be partially compensated. While the micro-tasks extracted by the computer vision model are only 51% accurate, the activities generated from them by the Process Binoculars are 65% accurate compared to the log generated from ground-truth micro-tasks. Third, we showcase the applicability of our technique with a case study on the mining of manufacturing processes recorded from multiple viewpoints.

The rest of this paper is structured as follows. In Sect. 2, we define the task of generating an event log from raw video, fit for business process tasks. In Sect. 3 we detail our proposed technique. We validate the Process Binoculars through a case study in Sect. 4 and discuss threats to validity in Sect. 5. Section 6 concludes the paper.

2 Problem Description

Process Mining (PM) relies on structured data in the form of an *event log*, i.e., a set of previous business process executions [29]. A specific instance of such an execution is called *case* and consists of *events*, which are instants with significance to the process. It is a process analyst's task to generate an event log by finding a suitable level of abstraction [12]. This event log is usually generated from execution logs in an information system, e.g., an ERP system, but recently, generating them from video sources garnered increased attention.

Knoch et al. [13] study a pipeline for extracting a process model from video recorded with a single, stationary overhead camera. They use a Convolutional Neural Network (CNN) to extract the trajectories of a worker's hands while they assemble a product. These trajectories are clustered and then assigned to predefined process activities. Melfsen et al. [22] present an approach to mine models describing the behavior patterns of pigs, based on previous work by Lepsien et al. [18]. They train a CNN-based machine learning model to predict the activities of pigs from a single, stationary overhead camera. Patterns in scores are mapped to activities, which are then used to mine a process model. Chen et al. [2] study how a CNN can directly detect the activities of an employee in a process. Activities are extracted from a single, stationary overhead camera, which is positioned behind the worker. They then check the conformance of extracted traces with a predefined process model. Kratsch et al. [17] propose a pipeline for event log generation from video, which extracts micro-actions from a single video source, abstracts them to activities through a rule-based system, and finally constructs the event log. To the best of our knowledge, Zhang et al. [31] present the first approach to video-to-process, which uses video from a non-stationary camera as input. They use a CNN to extract activities directly from a single recording of a camera mounted to a construction crane. An event log is constructed from these predictions and mined to obtain process models. Gavric et al. [6] propose a multi-model pipeline for mining process models from heterogeneous sources. They study how one part of this pipeline, video-to-log, can be used to obtain an event log from a single, stationary camera feed. To this end, they cluster frames in the video and pass representatives of these clusters to a LLM to obtain an activity that is likely to be depicted by it.

However, none of these approaches consider multiple videos when generating event logs from video. This approach comes with issues pertaining to the limited field of view of a single camera, as our example in Sect. 1 explained. Using multiple videos instead, as we propose, grants resilience against these issues.

We formally define the components of multi-video log generation as follows. Let $\mathcal{V} = \{V_1, \ldots, V_m\}$ be a set of videos. Each video $V_i = (f_t^i)_{t \in T_i}$ is a time-indexed sequence of frames, where each video in $\mathcal{V}$ shares a synchronized timeline T. An event log is a multiset of traces $\mathcal{L} = \{\sigma_1, \ldots, \sigma_n\}$ where each trace is a sequence of events $\sigma = \langle e_1, \ldots, e_k \rangle$, and an event is defined as $e = (act, time, case)$ where $act \in \mathcal{A}$ is an activity label, $time \in \mathbb{R}_{\geq 0}$ a timestamp, and $case \in \mathcal{C}$ a case identifier. Then, the event-log generation problem consists of learning a mapping $\Phi : \mathcal{V} \rightarrow \mathcal{L}$.

Enriching event logs with multimodal evidence [7], as well as with IoT sensors and Images [25], will be instrumental for the identification and discovery of processes with Manual Assembly Tasks, with potential for the robotization of activities risky for human operators [9].

3 Event Log Generation from Multi-view Video

The Process BINOCULARS learn the mapping Φ in five distinct stages. (*1*) We first use a computer vision model to extract micro-action sequences from multiple synchronized video recordings $\mathcal{V}$ (Sect. 3.1). Then, (*2*) we slide a window over these micro-action sequences and embed them into a high-dimensional vector space to obtain numeric representations (Sect. 3.2). (*3*) Vector representations are then clustered based on their semantic similarity (Sect. 3.3). (*4*) Clustered micro-action windows are assigned a human-readable label (Sect. 3.4). Finally, (*5*) an event log is constructed from the extracted information (Sect. 3.5). The output of the Process BINOCULARS is an event log which can be used in downstream tasks, such as discovery or conformance checking. An overview of the Process BINOCULARS is shown in Fig. 2. We will now describe each of the five stages in more detail.

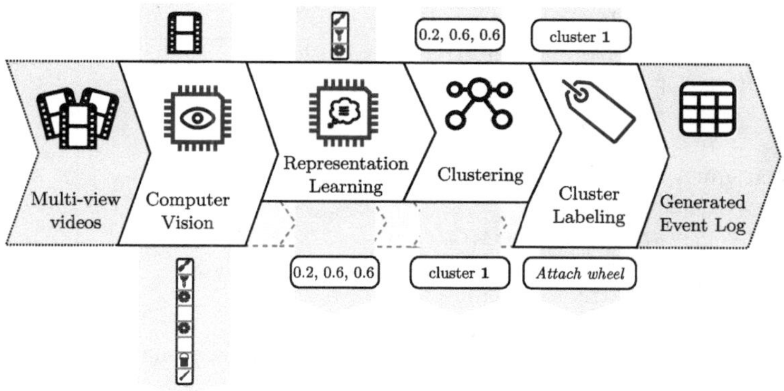

Fig. 2. High-level overview of the Process BINOCULARS.

3.1 Computer Vision

The first step in the Process BINOCULARS is turning a set of synchronized videos $\mathcal{V}$ into a sequence of micro-actions. We assume that the video is shot with a known, constant frame rate, to be able to assign time stamps to micro-actions based on the frame they were detected in. The video views can be in color, grayscale, or a mix thereof. Views can be stationary, fixed angle, or moving, e.g., mounted to a tripod or a headset camera. Using multiple camera views as an input to training the computer vision model lets it learn a generalizable abstraction of micro-action recognition. For instance, using both stationary and head-mounted camera viewpoints allows one to use the same model in settings where only one of these two setups is used. Similarly, using video streams with different resolutions, chroma, and luma during training allows flexible deployment agnostic of the concrete camera sensor used.

Algorithm 1 illustrates how sequences of micro-actions are extracted from a multi-view video $\mathcal{V}$. Each view $v \in \mathcal{V}$ is first sampled to obtain still frames, which are then passed to the action and object extraction models. The top predictions of these two models are concatenated to obtain a single micro-action. Algorithm 1 then returns the most frequently predicted micro-action for each frame.

We assume the computer vision model is trained to recognize objects and actions separately, similar to previous work in this field [28]. This allows us to generalize to a large number of possible micro-action classes. For instance, combining the action set $A = \{\text{pick up}, \text{attach}\}$ with the object set $O = \{\text{screw}, \text{wheel}, \text{screwdriver}\}$ would already result in six possible classes of micro-actions, e.g., *pick up screw*. This approach allows us to differentiate between detecting common motions (e.g., for picking something up) and recognizing manipulated objects (e.g., the screw). Previous work in video process mining shows that detecting the trajectories of a worker's hands can be done with modern machine learning [13]. Similarly, detecting objects in video is a well-studied task in computer vision [26]. We argue that this approach is more flexible than

Algorithm 1. Micro-action Detection

Input: $\mathcal{V}$, actionExtractor, objectExtractor, sampleRate
Output: $\mathcal{M}$: list of micro-action sequences for each video $v \in \mathcal{V}$

```
votes ← { }                                   ▷ Hash map
videoId ← 0
for each v in 𝒱 do
    votes[videoId] = [ ]
    frames ← sample(v, sample_rate)
    for each f in frames do
        a ← actionExtractor.predict(f)
        b ← objectExtractor.predict(f)
        append a b to votes[videoId]          ▷ Concatenate action and object
    end for
    videoId ← videoId + 1
end for
ℳ ← [ ]
for each v in values of votes do              ▷ Majority vote
    append the most frequent item of v to ℳ
end for
return ℳ
```

training a model to recognize a fixed set of classes of micro-actions, making it open for extension with novel techniques.

During inference, the computer vision model predicts micro-actions from each video separately and finds the consensus for each by averaging predictions. In a single-view setup, the failure or obstruction of a camera would be non-recoverable. Through finding consensus between multiple views, the Process BINOCULARS can recover, albeit with potentially reduced accuracy (see Sect. 4.4). The computer vision model itself may be a source of errors if its predictions are inaccurate. Current state-of-the-art computer vision models are up to 33% accurate in recognition of micro-actions on relevant datasets [28]. While multiple views help reduce prediction errors, i.e., through averaging predictions across several views, they cannot be eliminated at the computer vision stage. We use representation learning and clustering to compensate for and prevent the error from propagating in the generated event logs.

3.2 Representation Learning

The second step in the Process BINOCULARS involves mapping a sequence of micro-actions to a high-dimensional latent space, which then serves as an abstract numerical representation. It assumes that conceptually similar sequences are also represented by similar high-dimensional vectors (representations).

The representation learning step assumes sequences of micro-actions as input. These micro-actions are represented by their human-readable labels, e.g., "*pick up screw*". Each label consists of several tokens, which can, for this paper, be thought of as the words making up the label, e.g., "*pick*", "*up*", and "*screw*".

During training, we separate micro-actions in the input sequence with a special separation token, **<ACT>**. A window with fixed width is then slid over this sequence to produce sub-sequences, which are fed into a transformer-based architecture. The training task is inspired by that of bi-directional encoders, such as BERT [4]. Here, randomly selected micro-actions in the input are masked with a special masking token, **[MASK]**. The representation model then has to unmask the micro-actions, i.e., replace the masking token with the original, masked tokens of the micro-actions, by taking the surrounding micro-actions into account. For example, the masked micro-actions in "*pick up screw* <ACT> *pick up screw driver* <ACT> [MASK] <ACT> *attach wheel*" should be unmasked to produce "*pick up wheel*". Similar to the original masked language modeling training task of BERT, the representation model learns an intermediate, latent representation of the input, which is then decoded to the correct unmasked sequence.

To promote separation in latent space between representations of semantically different sequences, we additionally employ a contrastive learning objective. It is realized by adding a regularization term to the training loss, penalizing similarity (measured in cosine distance) between semantically different sequences. The regularization penalizes putting two windows containing different high-level activities close to one another, while rewarding putting windows of the same activity close to one another in the vector space. To read each window's latent vector representation, we prepend the special **[CLS]** token. Through contrastive learning, we train the model to learn a useful vector representation of the [CLS] token for distinguishing between windows. For inference on the trained representation model, we follow Algorithm 2. Each window of micro-actions in the sequence of micro-actions $\mathcal{M}$ is first preprocessed, by appending a special class action, and interleaving separating tokens between micro-actions. They are then embedded using the previously trained model and appended to the return list. While the latent representation of micro-action sequences is not directly interpretable for a human, it has properties that make it useful for the following stages in the Process BINOCULARS. For instance, the sequences "*pick up screw* <ACT> *pick up screwdriver*" and "*pick up screwdriver* <ACT> *pick up screw*" should have very similar latent representations, making the model invariant to the order of certain micro-actions, given that the training data supports this.

Sequences of micro-actions are inherently noisy, even when the computer vision model produces perfect predictions. Mainly, the order of micro-actions is not fixed. For a human observer, it is clear that picking up a screw and a screwdriver has the same result, regardless of order. Yet, a digital system cannot directly quantify the similarity of sequences where the order is swapped. Instead, only through projecting them into a latent vector space are we able to produce a numerical similarity score, e.g., by calculating the cosine distance between the vectors representing the sequences. Similarly, representation learning helps mitigate noisy and imperfect predictions by the computer vision model. If it

Algorithm 2. Micro-Action Window Representations

Input: $\mathcal{M}$, model, windowSize
Output: $\mathcal{R}$: vector representations of micro-action windows

$\mathcal{R} \leftarrow [\,]$ ▷ List of vector representations
for each *seq* in $\mathcal{M}$ **do**
  for each *w* in slidingWindow(*seq*, *windowSize*) **do**
    input ← '[CLS]' ▷ Special class token
    for each *ma* in *w* **do**
      append *ma* to *input*
      append <ACT> to *input* ▷ Special action separator token
    end for
    repr ← model(*input*) ▷ Embedded vector representation
    append *repr* to $\mathcal{R}$
  end for
end for
return $\mathcal{R}$

misses or mis-predicts certain micro-actions, the resulting vector representation can still be close to the correct sequence. Consider, for example, the (imperfect) sequence "*pick up screwdriver* <ACT> *pick up* **nail** <ACT> *screw in wheel*", which resulted from misclassification of "*screw*" as a "*nail*". The training data is unlikely to have this sequence, as it usually would not appear in a normal process. Considering the context of picking up a *screwdriver* and *screwing* in a wheel makes picking up a screw much more likely. The latent vector representing that sequence reflects that: The cosine similarity between these two vectors is 0.99, when using the representation model we train (Sect. 4.2). Finally, micro-actions that appear often in the training data, regardless of context, will have a lower effect on the embedded representation. The model, therefore, naturally learns to ignore them during representation learning, as they do not contribute contextual information. Latent representations of micro-action windows are a numeric representation of their semantic content, and thus are well-suited for being clustered into groups of windows with similar meaning in the following step.

3.3 Clustering

The clustering step in the Process BINOCULARS groups micro-action sequences based on their representation's similarity. This results in clusters of semantically similar micro-action sequences, which are likely candidates for process-relevant activities. We assume the input to the clustering step of the Process BINOCULARS to be a set of vectors of uniform length. These vectors are high-dimensional, typically more than 500 dimensions. This informs our choice of clustering algorithm and distance function, as calculating distance for high-dimensional vectors becomes infeasible (cf. curse of dimensionality [15]). Further, we assume the number of unique activities in the process, and therefore the number of clusters, to

Algorithm 3. Automated Labeling

Input: $\mathcal{C}, \mathcal{M}$, languageModel, windowSize
Output: $\mathcal{A}$: clusters mapped to activity labels

```
A ← empty hash map
for each c in C do
    input ← [ ]                                          ▷ List of strings
    for each m in M do
        for each w in slidingWindow(m, windowSize) do
            if w ∈ c then
                append w to input                        ▷ String representation of w
            end if
        end for
    end for
    label ← languageModel(input)
    A[id of c] = label
end for
return A
```

be unknown. We cluster the outputs from Alg. 2 to obtain a list $\mathcal{C}$ of cluster assignments for each vector $r \in \mathcal{R}$.

The Process BINOCULARS use hierarchical clustering, which allows treating the number of clusters as a parameter tunable by a process analyst post-hoc. This way, the desired level of abstraction can be controlled, as discussed in the case study in Sect. 4. The root cluster of the hierarchy represents the whole process (too coarse), and leaf nodes represent steps in an activity (too fine). By picking the correct number of clusters, activities of the correct level of abstraction are produced, e.g., *attach wheel*, instead of *assemble toy car* or *pick up screwdriver*.

3.4 Semantic Labeling

The fourth step in generating an event log with the Process BINOCULARS is assigning a human-readable label to each cluster of micro-actions. This step is saves the process analyst the effort of going through all clusters and trying to recognize the common task in clustered windows. Algorithm 3 illustrates our approach to this step. As input we assume clusters $\mathcal{C}$ over windows of the micro-action sequences in $\mathcal{M}$, which is the output of the previous step, described in Sect. 3.3. First, we concatenate each window w over the sequences of micro-actions m in a given cluster $c \in \mathcal{C}$ line by line. Each window is represented as a string, similar to the input for representation learning (see Sect. 3.2). A LLM is then prompted to find the common, process-relevant activity that is described by the windows of micro-actions, using the following prompt.

> **Prompt for LLM-based semantic abstraction**
>
> Below is a list of sequences of low-level actions that are performed during a {{domain}} process. [...] Can you find a succinct label for the activity, they most likely relate to? The granularity of this label should be an activity, that is one semantically significant step in the process, not a description of the entire process.
> Output only the label, nothing else.
> {{event_list}}

Algorithm 4. Trace Generation

Input: $\mathcal{C}, \mathcal{M}, \mathcal{A}, windowSize, videoId$
Output: σ: trace of process execution corresponding to $\mathcal{M}$

$\sigma \leftarrow \{\ \}$
for each m in $\mathcal{M}$ **do**
  for each w in slidingWindow($m, windowSize$) **do**
    $c \leftarrow$ cluster $c \in \mathcal{C}$ corresponding to w
    $activity \leftarrow a \in \mathcal{A}$ corresponding to c ▷ From Alg. 3
    if $activity =$ last activity in σ **then continue** ▷ Aggregate windows
    end if
    $timestamp \leftarrow$ time of first frame $f \in w$
    append ($videoId, activity, timeStamp$) to σ
  end for
end for
return σ

The output is automatically parsed and assigned to the mapping $\mathcal{A}$. Since the LLM gets semantic information about the process only via the list of micro-actions, the prompt generalizes to new scenarios.

3.5 Event Log Generation

The last step of the Process Binoculars generates an event log in XES[1] format. To this end we transform each sequence of micro-actions $m \in \mathcal{M}$ into its corresponding trace σ, as shown in Algorithm 4. Every window w over the sequence of micro-actions m was already assigned a cluster, and thus a human-readable label in the previous Algorithm 3. Since windows overlap, we aggregate repeated windows with the same label into a single event, and assign it the timestamp corresponding to the first frame in the window.

[1] See https://www.xes-standard.org/, last accessed March 20, 2026.

4 Case Study

To evaluate the efficacy of the Process BINOCULARS, we apply it to extract event logs from video recordings of toy vehicle assembly processes. We first discuss the dataset used in this case study and present implementation details. We then apply the Process BINOCULARS to extract event logs in different scenarios, compare them, and finally qualitatively study the process models mined from these logs.

4.1 Dataset

As a basis for our case study, we select Assembly101 [28], a large-scale dataset for action recognition in synchronized multi-view video recordings. This dataset consists of recordings of standardized assembly procedures of toy cars. The videos are annotated with both fine-grained micro-actions and coarse-grained activities. Figure 3 shows example frames that can be found in the Assembly101 dataset. Here, a mobile crane is assembled by a human operator. The dataset annotates the videos with micro-actions and activities on a frame level, for example *"pick up screwdriver"*, *"unscrew first wheel"*, and *"remove bolt"*.

Fig. 3. Example scene from Assembly101, while an operator is attaching wheels to an excavator. The left image shows a monochromatic head-mounted camera, while the center and right images are fixed angular views. The task is obstructed in the third view, but can be compensated for by the other two views.

4.2 Implementation

We implement all stages of the Process BINOCULARS in Python and make the code publicly available[2].

Computer Vision. We use a version of *Temporal Shift Module* [19] pretrained by the authors of Assembly101 [28]. This model employs shifting video features by a certain number of frames during training. This facilitates learning temporal features in a standard two-dimensional Convolutional Neural Network, whereas three dimensions would otherwise be needed to model time. In our experiments,

[2] See https://github.com/JulianNeuberger/pm-binoculars.

we do not focus on the accuracy of this model, as we consider it a feature extractor, rather than a fixed component of the Process BINOCULARS. This makes them extensible, if better and more accurate models for action recognition from video are proposed in the future.

Representation Learning. We employ a pretrained BERT model [4] for learning representations, specifically *bert-base-uncased* from HuggingFace[3]. We implement batch-wise contrastive loss [11] in PyTorch, and set its temperature to $\tau = 0.01$. Further, we implement masked language modeling loss, but always mask entire micro-actions, not random tokens, as shown below.

Whole-event masking

Original	<ACT> pick up screw <ACT> screw in wheel
Token masking	<ACT> pick up [MSK] <ACT> screw in wheel
Event masking	<ACT> [MSK] <ACT> screw in wheel

The final loss function for fine-tuning is $\mathcal{L} = 0.1 * \mathcal{L}_{contr} + 0.9 * \mathcal{L}_{mlm}$, where $\mathcal{L}_{contr}$ is the contrastive loss and $\mathcal{L}_{mlm}$ is the masked language modeling loss.

Clustering. We use HDBSCAN [21], a hierarchical version of the density-based clustering algorithm DBSCAN. Unlike DBSCAN, which needs the number of clusters a priori, HDBSCAN results in a hierarchy of clusters. The number of clusters can then be picked post hoc to optimize given clustering metrics. This parametrizes the level of abstraction in the event logs, and lets a process analyst adjust the Process BINOCULARS to best fit their application scenario.

Metrics. To describe the makeup of generated event logs, we use common descriptive measures, such as the average length of traces and average events per trace. Additionally, we count the number of unique traces and calculate the Trace Similarity Rate (TSR) [10]. TSR can be understood as the average number of edits needed to transform one variant into another [10]. As such, it is a useful metric for gauging the complexity of an event log. We use Kullback-Leibler (KL) to measure the difference in event frequency distribution of two event logs. KL is a metric from information theory, which estimates the entropy between two distributions, in our case event frequencies. It is defined for two (discrete) probability distributions P and Q over a random variable x as $D_{KL}(P||Q) = \sum_x P(x) log(\frac{P(x)}{Q(x)})$ [20]. Finally, we use two metrics to quantify the accuracy with which the Process BINOCULARS can generate activities from micro-tasks. First, we define the *event accuracy* as $ACC(P, G) = \#ok/\#gold$, where P and G are the sets of predicted and expected events respectively, $\#gold = |G|$, and $\#ok \subseteq P$ is the set of predicted events, which could be matched to an equivalent expected event. We match predicted events to expected events by calculating

[3] See https://huggingface.co/google-bert/bert-base-uncased, last access Mar 16, 2026.

their semantic similarity using Sentence-BERT [27] and linear sum assignment. Similar, the event F_1 score for events is defined as $F_1 = 2 * P * R/(P + R)$, where $P = \#ok/\#pred$ and $R = \#ok/\#gold$ are precision and recall, respectively, with $\#pred = |P|$.

4.3 Generated Event Logs

Table 1 shows descriptive statistics for the event logs generated by the Process BINOCULARS. To test the effects of errors during predicting micro-actions in the computer vision stage, we additionally run the Process BINOCULARS with the ground truth data for action recognition. We find that while there are changes in the makeup of generated event logs, they are minor. This is a promising result, considering that the computer vision model is 56% accurate in action recognition[4].

Table 1. Descriptive statistics for the generated event logs. * Results from running the pipeline end-to-end, i.e., with micro-actions predicted by the computer vision model. † Compared to log generated from ground-truth micro-actions.

	toy	mean trace length	events/trace	traces	variants	TSR	D_{KL} †	Event accuracy†	Event F_1 †
assembly	c08a	17.33	14.00	3	3	1.17	—	—	—
	c08a*	15.67	12.00	3	3	0.98	2.50	0.65	0.69
	c08b	12.00	7.00	3	3	1.50	—	—	—
	c08b*	11.00	7.00	3	3	0.41	1.12	0.61	0.64
	c08c	9.00	8.20	5	5	5.43	—	—	—
	c08c*	8.60	7.60	5	5	4.56	4.81	0.60	0.61
disassembly	c08a	6.00	4.67	3	3	1.43	—	—	—
	c08a*	8.00	6.67	3	3	1.25	0.62	0.72	0.62
	c08b	7.67	6.33	3	3	1.88	—	—	—
	c08b*	10.33	8.00	3	3	1.56	2.06	0.74	0.63
	c08c	6.67	5.33	6	6	6.18	—	—	—
	c08c*	8.17	6.67	6	6	3.56	1.56	0.57	0.52

All traces generated by the Process BINOCULARS are unique, as shown by *variants*, but the TSR suggests that most of them differ only slightly from one another. While they are not the same, they are similar enough so, for example, the traces for assembling *c08a* could be made equivalent, by changing $\tilde{1}$ event[5]. D_{KL} quantifies the difference in activity frequencies. It has a lower bound of 0 if both event logs have exactly the same activity frequencies. It has no upper bound, but instead gives a measure of how "surprising" changes in probability distributions are [20]. One notable exception in the overall results is toy variation *c08c*. While trace length, the average number of events per trace, and F_1 are not exceptional, D_{KL} and TSR are. This suggests differences in the order of events. Preliminary analysis of clusters and traces did not reveal possible root causes, and future investigation is needed.

4.4 Resilience to Failed Cameras

[4] Note that this relates to exact action match, *pick up screwdriver* and *attempt to pick up screwdriver* are considered different here.

[5] TSR normalizes edits to the longer trace, which is why values < 1 are possible.

We study how resilient the Process Binoculars are to failing cameras by removing an increasing number of randomly selected cameras, up to eleven out of twelve cameras. Results for F_1 scores are shown in Fig. 4. We find that the Process Binoculars are failing gracefully, with event F_1 being stable around 50% up to 8 (67%) failed cameras, and then dropping to $F_1 = 25\%$, after 10 (83%) cameras failed. F_1 recovers slightly, after all but one camera failed. This could happen if the last two cameras contain both an unobstructed view and an obstructed view. Due to the majority vote being tied in this case, there is no way for the unobstructed view to compensate for the obstructed one. This notion is confirmed when we run the Process Binoculars with camera 12 only: The F_1 score drops to 0.26, instead of 0.42 (camera 11). Considering confidence scores of predictions in voting could alleviate this issue.

4.5 Mined Process Models

We now study the models obtained from mining event logs generated with the Process Binoculars. As each of the generated traces was unique and slightly different from the others, and human operators tend to deviate slightly from fixed plans, we decided to focus our analysis on the discovery of behavioral constraints obeyed by the operators. We selected DCR graphs [8] as a suitable process modelling technique for this matter. We use DCR-js [3] to extract the process models from the generated event logs. Figure 5 shows a fragment of the model mined from the event log generated for the videos showing the assembly of toy *c08a*. The model mined from gold standard data (*left*) shows an artifact of clustering, where two activities (*install rear base* and *install rear door*) are merged. Fine-tuning the clustering stage could fix this issue. The model shows some obvious relations (e.g., you cannot install a crane arm twice) as well as identifying some not-so-obvious ones (e.g., the installation of a bumper is mandatory after the installation of the interior, but it requires the installation of a cabin first). At the same time, the model evidences some ambiguities in the generated labels, where the modeller would need to discern the difference between *install crane arm* and *install crane arm and fasten with screws*. The model mined from predicted micro-actions (*right*) is very similar, and some behavioral constraints

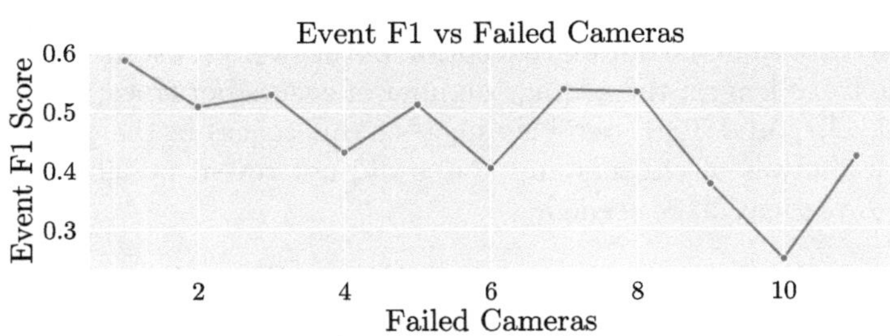

Fig. 4. Event F_1 score by number of failed cameras. We observe a gradual degradation of the event F_1 score, with a sharp drop after 8 cameras failed.

coincide, for instance, the relations applied to *attach front bumper*. However, the predicted micro-actions generate potentially very similar events (e.g. *mount crane arm and fasten with screws* and *attach crane arm to body*). The use of different labels for the same event may have generated duplicated activities, and exclusive relations (in red) denote that these cannot happen in the same trace.

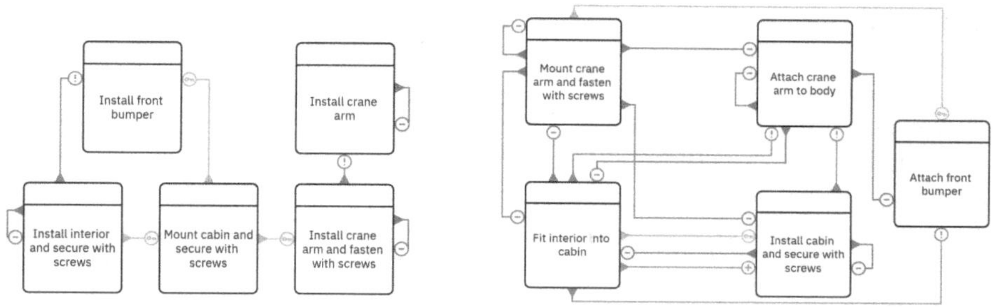

Fig. 5. DCR graphs discovered from ground-truth micro-actions (left) and predicted micro-actions (right).

5 Threats to Validity

While our results show that the Process BINOCULARS are an effective technique for unlocking unstructured data in videos, one must consider a few points that threaten the validity of our results. First, our validation was based on videos showing the assembly of toys. While the setting is realistic, the process is no real-world scenario. However, we argue it still serves as a good proxy in the absence of a high-quality, large-scale dataset from a real assembly line. It is important to note that while the complexity of the process may have detrimental effects on the performance of process mining algorithms, it affects the detection and abstraction of micro-actions less. Further, the Process BINOCULARS are currently limited to one resource per process. In future work, they could be extended to use video recordings showing multiple resources at once. Similarly, we do not track data objects from video data. It would be interesting to develop extensions to the Process BINOCULARS, that would allow an object-centric view on processes. When considering the results of evaluating the Process BINOCULARS' resilience towards camera failure, note that the cameras selected are not randomized, but made to fail in order of their id, which could limit the validity of results.

6 Concluding Remarks

We present the Process BINOCULARS, a novel technique for generating event logs from multiple video sources. It entails five distinct steps: *(1)* extracting sequences of micro-tasks from video, *(2)* learning representations of windows over these sequences, *(3)* clustering windows based on their representation, *(4)* labeling clustered windows, and finally *(5)* generating an event log. We find that our technique prevents propagation of errors in computer vision, is resilient towards camera failure, and enables the process mining of generated event logs.

Currently repeated activities are aggregated directly. In future work, we plan to filter noise, like single mislabeled micro-action sequence windows. While we currently focus on videos showing one process execution, future work should consider case correlation in the video-to-log task. Additionally, we would like to explore how giving the LLM more context could improve the accuracy of assigned labels. Hallucinations are a major concern when using LLMs for any task. While the process analyst is always able to reject automatically generated labels, future work could investigate the applicability of topic modeling techniques, such as Latent Dirichlet Allocation [1], which are not prone to hallucinations. This also reduces the reliance on external providers or expensive hardware, which is critical for privacy and sustainability [23]. The Process BINOCULARS are targeted at the elicitation of process models from videos, however the technology behind them could be used for monitoring and tracking employees. We are aware of these privacy and ethical concerns and advocate the responsible use of our work.

Acknowledgments. This work was supported by VILLUM FONDEN (grant VIL57420) through the Center for Digital Compliance (DICE), the Innovation Fund Denmark projects "Explainable Hybrid-AI for Computational Law and Accurate Legal Chatbots" (XHAILe, grant 4355-00018B) and Predictive and Prescriptive Process Analytics for Industry 4.0 (P3AI4, grant 4105-00045B).

References

1. Blei, D.M., Ng, A.Y., Jordan, M.I.: Latent dirichlet allocation. JMLR (2003)
2. Chen, S., Zou, M., Cao, R., Zhao, Z., Zeng, Q.: Video process mining and model matching for intelligent development: conformance checking. Sensors (2023)
3. Christfort, A.K., López, H.A.: DCR-JS: an online environment for declarative process mining. In: BPM (2025)
4. Devlin, J., Chang, M.W., Lee, K., Toutanova, K.: BERT: pre-training of deep bidirectional transformers for language understanding. arXiv:1810.04805 (2018)
5. Dumas, M., Rosa, L.M., Mendling, J., Reijers, A.H.: Fundamentals of business process management, Springer (2018)
6. Gavric, A., Bork, D., Proper, H.: Multimodal process mining. In: 2024 26th International Conference on Business Informatics (CBI), pp. 99–108. IEEE (2024)
7. Gavric, A., Bork, D., Proper, H.A.: Enriching business process event logs with multimodal evidence. Presented at the (2024)
8. Hildebrandt, T.T., Mukkamala, R.R.: Declarative event-based workflow as distributed dynamic condition response graphs. arXiv (2011)
9. Jeong, Y., Wiktorsson, M., Park, D., Gans, J., Svensson, L.: Data preparation for ai-assisted video analysis in manual assembly task: a step towards industry 5.0. In: IFIP (2023)
10. Kherbouche, M.O., Laga, N., Masse, P.A.: Towards a better assessment of event logs quality. In: SSCI (2016)

11. Khosla, P., et al.: Supervised contrastive learning. In: NeurIPS (2020)
12. Klessascheck, F., et al.: Domain-specific event abstraction. In: BIS (2021)
13. Knoch, S., Ponpathirkoottam, S., Schwartz, T.: Video-to-Model: unsupervised trace extraction from videos for process discovery and conformance checking in manual assembly. In: BPM (2020)
14. Kobayashi, V.B., Mol, S.T., Berkers, H.A., Kismihók, G., Den Hartog, D.N.: Text mining in organizational research. Organ. Res. Methods (2018)
15. Köppen, M.: The curse of dimensionality. In: WSC5 (2000)
16. Koschmider, A., et al.: Process mining for unstructured data: challenges and research directions. arXiv (2023)
17. Kratsch, W., König, F., Röglinger, M.: Shedding light on blind spots-developing a reference architecture to leverage video data for process mining. In: DSS (2022)
18. Lepsien, A., Bosselmann, J., Melfsen, A., Koschmider, A.: Process mining on video data. ZEUS **3113**, 56–62 (2022)
19. Lin, J., Gan, C., Han, S.: TSM: temporal shift module for efficient video understanding. In: IEEE/CVF (2019)
20. MacKay, D.J.: Information theory, inference and learning algorithms. Cambridge university press (2003)
21. McInnes, L., Healy, J., Astels, S.: hdbscan: Hierarchical density based clustering. J. Open Source Softw. (2017)
22. Melfsen, A., Lepsien, A., Bosselmann, J., Koschmider, A., Hartung, E.: Describing behavior sequences of fattening pigs using process mining on video data and automated pig behavior recognition. Agriculture **13**(8), 1639 (2023)
23. Neuberger, J., der Aa, H.v., Khrop, I., López, H.A.: Enabling small language models for text-to-process extraction: balancing accuracy and efficiency through distant supervision. pp. 148–166 (2026)
24. Petersen, A.C., Christensen, L.R., Harper, R., Hildebrandt, T.: "we would never write that down" classifications of unemployed and data challenges for AI. In: PACMHCI (2021)
25. Rebmann, A., Emrich, A., Fettke, P.: Enabling the discovery of manual processes using a multi-modal activity recognition approach. In: BPM Workshops (2019)
26. Redmon, J., Divvala, S., Girshick, R., Farhadi, A.: You only look once: unified, real-time object detection. In: CVPR (2016)
27. Reimers, N., Gurevych, I.: Sentence-BERT: sentence embeddings using siamese BERT-networks. In: EMNLP (2019)
28. Sener, F., et al.: Assembly101: a large-scale multi-view video dataset for understanding procedural activities. In: CVF (2022)
29. Van Der Aalst, W.: Data science in action. In: Process mining: data science in action. pp. 3–23. Springer (2016)
30. van Der Aalst, W., et al.: Process mining manifesto. In: BPM (2011)
31. Zhang, H., Lin, C.: Computer vision and process mining-based simulation model generation for construction planning and control. Autom. Constr. (2025)

Feature Attribution-Based Explainability Analysis of Deep Learning Models in Predictive Process Monitoring

Kseniya Sahatova[1,2(✉)], Rafael Seidi Oyamada[2], Xuefei Lu[1], and Johannes De Smedt[2]

[1] SKEMA Business School, 92156 Suresnes, France
{kseniya.sahatova,xuefei.lu}@skema.edu
[2] KU Leuven, b-3000 Leuven, Belgium
rafael.oyamada@kuleuven.be

Abstract. Predictive process monitoring supports the optimization and control of operational business processes by forecasting the future state or outcome of ongoing cases. While deep neural networks have achieved strong performance for these tasks by modeling sequential dependencies in event logs, their black-box nature limits trust and practical adoption. Feature attribution methods are often used to address this, but applying them directly poses a dilemma: event-level attributions impose high computational complexity for long traces, while explanations based on aggregated trace representations often fail to capture the underlying control-flow dynamics. To address this issue, we propose a local post-hoc explainability method for deep neural networks in outcome prediction. The method relies on a control-flow-aware segmentation algorithm that partitions a trace into meaningful segments and supports the computation of segment-level SHAP explanations. This makes it possible to identify which parts of a trace influence a prediction and which change points steer the case toward the predicted outcome. We assess the proposed segmentation method on a synthetic dataset with known process logic, where meaningful change points can be explicitly verified, and we demonstrate its usefulness on real-world event logs from a loan application process and an administrative process of a Dutch municipality.

Keywords: Predictive Process Monitoring · SHAP · Control-flow-aware Trace Segmentation

1 Introduction

Businesses increasingly use Artificial Intelligence (AI) to anticipate issues such as delays or failures, enabling optimization and prevention. Predictive process monitoring (PPM), a branch of Process Mining (PM), focuses on predicting the outcome of an ongoing business process instance, estimating the remaining time before completion, or forecasting the sequence of activities in the future. To

F. M. Maggi et al. (Eds.): BPM 2026, LNBIP 599, pp. 238–254, 2027.
https://doi.org/10.1007/978-3-032-37877-4_14

address these predictive tasks, researchers have adopted a variety of sequence modeling architectures [27,32], such as recurrent neural networks (RNNs) and transformers, which have emerged as a dominant paradigm due to their ability to capture the sequential and temporal dependencies inherent in event logs.

Despite their strong predictive performance, sequential models in PPM remain largely obscure: their internal representations are difficult to interpret, limiting practical adoption in high-stakes operational settings. This challenge has motivated growing interest in explainability methods that facilitate understanding of how models trained on event logs arrive at their predictions [4,5,11,13,25]. Recent studies have explored the use of feature attribution techniques such as LIME (local interpretable model-agnostic explanations) and SHAP (Shapley additive explanations) to enhance the transparency of deep learning models in PPM [10,11,21,31]. However, current approaches still explain predictions either at the level of individual events or at the level of the full aggregated trace. Both views are problematic for PPM: event-level explanations are often too fragmented to reveal meaningful process behavior, whereas trace-level explanations operate at a broader level of abstraction and introduce information loss through aggregation, making it harder to identify which part of the execution drives the prediction. Consequently, existing methods may assign importance scores without preserving the local control-flow context that gives those scores process meaning.

Event logs exhibit sequential and temporal dependencies, which makes them partly related to time-series data. This connection is relevant because the time-series literature has proposed several feature attribution methods that improve interpretability by grouping adjacent observations, in contrast to interpreting shallow event-level or fully aggregated data, into segments [6,9,16,22,24]. In coalition-based methods such as Shapley values, this strategy also reduces the computational cost, which otherwise grows exponentially with the number of timesteps. However, event logs are not generic time series: they contain process-specific information such as control-flow behavior, activity semantics, and case progression, and are irregular in time, since events are recorded upon activity execution rather than at fixed sampling intervals as in regular time series. As a result, segmentation strategies developed for time series may produce groups of events that do not correspond to meaningful process fragments, for example by splitting a coherent process stage or merging unrelated control-flow behavior.

To address this gap, we propose a control-flow-aware segmentation algorithm that improves the computational efficiency of feature attribution-based explanations while preserving the control-flow constraints of business processes. Our method partitions traces into coherent subsequences of activities by identifying shifts in transition probabilities, thereby yielding more interpretable units. We compute segment-level SHAP values for sequential classifiers in a binary prediction setting. For stakeholders and practitioners, event-level feature attributions provide limited interpretation because the impact of an activity on the outcome often depends on its execution context. As a result, attributing predictive importance to individual activities may provide limited insight in isolation.

We argue that computing SHAP values at the segment level groups activities that share similar transition dynamics, and that the underlying transition probabilities within each segment offer additional interpretive context beyond the attribution scores themselves.

The paper is organized as follows. Section 2 reviews existing work on explainability in PPM and motivates our study. Section 3 introduces the main definitions required to facilitate understanding of the methods used in this paper. Section 4 describes the proposed control-flow-aware segmentation algorithm and the metrics used for evaluating segmentation quality. Sections 5 and 6 introduce the experimental setup and present the interpretation of the obtained results, while Sect. 7 discusses the findings summarized in Sect. 8.

2 Related Works

Various explainability methods have been applied to enhance transparency and improve understanding of predictive models in process monitoring. In the work [21], LIME explanations were used to improve outcome prediction. Event-level feature attributions for false positives and false negatives were analyzed with rule mining to identify regions contributing to errors. The study [11] shows how Shapley values can be adapted to explain an LSTM model in the remaining-time prediction task generating explanations for each feature per timestep. In the study [31] the stability of LIME and the model-specific SHAP value estimation methods are examined in the outcome classification task across different types of trace encoding. Similar research [10] estimates the effect of design choices in predictive models and pre-processing techniques on global explainability methods, examining SHAP values through the dependence plot representation. In the study [4], KernelSHAP is adapted to explain RNNs at the feature, event, and cell levels, where a cell corresponds to a specific feature of a specific event; that is, explanations are generated for each feature within a subset of events. To enable efficient computation for long sequences, the authors introduce a pruning algorithm based on the assumption that only recent events are relevant. Older events are therefore grouped into a single coalition and treated as less important for fine-grained explanations.

The previously described approaches compute explanations for a single event and its associated features, proposing algorithms for efficient computation while disregarding the correlations between adjacent events inherent in event logs. These temporal dependencies make event logs analogous to time series, enabling the adaptation of explainability approaches originally developed for time-series data to the field of PM. Active research on the explainability of sequential models for time series has introduced various techniques that identify semantically meaningful segments and subsequently generate explanations for them. The LIME-Segment framework [24] was developed to explain time-series segments based on shape patterns and statistical properties. TS-MULE [22] utilizes LIME to explain segments of univariate and multivariate time series, employing matrix profile and SAX transformations for segmentation. Unlike in time series, consecutive activities in process traces may exhibit strong correlations that indicate structured process behavior, or they may show little to no correlation, suggesting loops, branching, or unpredictable behavior—i.e., process unstructuredness.

WindowSHAP method [18] was introduced to reduce computational complexity and variance in point-wise explanations. In this approach, adjacent data points are grouped using windowing techniques with fixed- and variable-length time windows. Jiang et al. [13] adapt KernelSHAP for subsequence-level explanations by relying on the definition of sessions in recommendation system, where events can be grouped based on the time windows in which they occur. While such an approach is suitable for data, which temporal patterns are clearly distinguishable, it has limitations in identifying meaningful subsequences when the primary analytical focus is on the process control flow—i.e., when subsequences need to be grouped based on the types of activities performed. While segment-based explanation methods have shown promise in the time-series domain, their direct application to event logs disregards the process-specific dependencies. This motivates our work, which bridges control-flow-aware trace segmentation with feature attribution to produce explanations aligned with the control-flow constraints of business processes.

3 Preliminaries

In this section, we introduce the main concepts from predictive process monitoring and feature attribution used throughout this work.

Definition 1 (Event, Trace, Prefix, Event Log). *Let $\mathcal{A}$ be the universe of activities and $\mathcal{C}$ the universe of case identifiers. An event is defined as a tuple $e = (a, c, t, D)$, where $a \in \mathcal{A}$ is the activity, $c \in \mathcal{C}$ is the case identifier, t is the timestamp, and D is a set of event- or case-level attributes. We denote the universe of all possible events by $\mathcal{E}$.*

A trace $\sigma = \langle e_1, e_2, \ldots, e_n \rangle$ is a non-empty ordered sequence of events such that $e_i \in \mathcal{E}$ for all $1 \leq i \leq n$, and $t_i \leq t_{i+1}$ for all $1 \leq i < n$. Given a trace $\sigma = \langle e_1, \ldots, e_n \rangle$ and an integer $1 \leq m \leq n$, the prefix of length m is defined as $prefix(\sigma, m) = \langle e_1, \ldots, e_m \rangle$. The set of all possible traces over $\mathcal{E}$ is denoted by $\mathcal{E}^*$. Finally, an event log L is a multiset of traces.

In this work, we consider binary outcome prediction only, that is, the task of predicting the final outcome of an ongoing case from a trace prefix.

Definition 2 (Outcome Classifier). *A binary outcome classifier is a function $f_{cl} : \mathcal{E}^* \rightarrow \{0, 1\}$. In practice, we consider its associated score function $f_{cls} : \mathcal{E}^* \rightarrow [0, 1]$, which maps a prefix $\sigma_{\leq m} = prefix(\sigma, m)$ to the estimated probability of a positive outcome.*

Our goal is to explain how a classifier arrives at its final prediction. To this end, we compute SHAP values over the segments produced by our proposed directly-follows-relation-based segmentation algorithm.

Definition 3 (Directly-Follows Relation (DFR)). *Let $\sigma = \langle e_1, \ldots, e_n \rangle$ be a trace and let $x, y \in \mathcal{A}$ be two activities. We say that x is in a directly-follows relation with y, denoted by (x, y), if there exist two consecutive events e_i and e_{i+1} in σ such that the activity of e_i is x and the activity of e_{i+1} is y.*

Definition 4 (Shapley Values). *Let $N = \{1, ..., n\}$ be a set of features. Let $v : 2^N \rightarrow \mathbb{R}$ be a value function that assigns to each subset of features $S \subseteq N$ a real number $v(S)$, representing the model output when only features in S are used. The Shapley value ϕ of the feature i quantifies the weighted average marginal contribution to the model prediction over all possible feature subsets as:*

$$\phi_i(v) = \sum_{S \subseteq N \setminus \{i\}} \frac{|S|!(|N| - |S| - 1)!}{|N|!} (v(S \cup \{i\}) - v(S))$$

We utilize the following characteristic function: $v(S) = f_{cls}(x_S, x'_{N \setminus S})$, where x' is a baseline drawn from the training distribution and f_{cls} is the scoring function. In outcome prediction, $v(N)$ corresponds to the classifier output $f_{cls}(\sigma_{\leq m})$. Shapley values satisfy several desirable properties, namely local accuracy, missingness, and consistency; we refer to [1,17,26] for further details.

4 Segment-Based Feature Attribution for Explainability

We capture the evolution of process executions recorded in an event log through *process phases*, in which consecutive events belong to the same local control-flow context. For example, in a loan process, a trace may contain a segment $\hat{\sigma} = \langle e_i, \ldots, e_j \rangle$ ($1 \leq i < j \leq n$, where n is the trace length) characterized as a segment of activities related to application completion, followed by a segment related to offer creation. The former might contain common transitions with high directly-follows probabilities, such as `Submit application` → `Check documents` and `Check documents` → `Validate application`. Later, when the case moves to the offer creation part, we may see a less common transition like `Validate application` → `Create offer`, characterizing a drop of likelihood that can mark the boundary between the two phases. Thus, a natural way to estimate such probabilities is through the discovery of a *directly-follows graph* (DFG).

More specifically, by obtaining the matrix of DFRs, we obtain empirical transition probabilities between activities. Evaluating these probabilities along the events of a trace yields a sequential transition segment that reflects the local control-flow consistency of the observed process execution. Positions where this segment changes can then be interpreted as boundaries between different process phases; we refer to such positions as *change points*. Change point detection is a well-established task in time series analysis [8,12,15,29], where the goal is to identify positions at which the statistical properties of a sequential signal change. In our case, the signal is given by the transition-probability profile derived from the event log, and a change point corresponds to a shift in the control-flow dynamics of the process instance, which can be interpreted as a phase boundary. Thus, let us introduce our concept of trace segmentation.

Definition 5 (Trace Segmentation). *A trace segmentation algorithm is a function* $f_{seg} : \mathcal{E}^* \rightarrow \mathcal{S}^*$ *that partitions a trace* σ *into a sequence of segments* $s = \langle s_1, s_2, \ldots, s_k \rangle$. *Each segment* $s_l \in \mathcal{S}$ *is a sub-trace* $s_l = \langle e_i, \ldots, e_m \rangle$ *where* $1 \leq i \leq m \leq n$, *such that the concatenation of all segments* $s_1 \cdot s_2 \cdot \cdots \cdot s_k$ *equals the original trace* σ.

Using the introduced definition of trace segments and change points, we propose a control-flow-aware segmentation algorithm that identifies the corresponding change points and partitions a trace into segments. The algorithm optimizes a DFR-based cost function using dynamic programming, which we discuss further. Given these segments and a sequence modeling classifier, we provide explanations.

4.1 Control-Flow-Aware Segmentation Algorithm

Our control-flow aware segmentation algorithm combines (i) a cost function derived from directly-follows relations and (ii) the PELT (Pruned Exact Linear Time) algorithm [14], which detects multiple change points by minimizing a penalized objective.

First, from the event log, we discover a DFG, extract the DFR matrix, and normalize it to obtain the transition probabilities between activities. Subsequently, given a trace σ, the goal is to identify k change points that partition the trace into $k+1$ segments. To assess whether a candidate segment is coherent, we assign it a cost based on the directly-follows probabilities. More specifically, transitions that are common in the event log contribute little to the cost, whereas rare or unseen transitions increase it. Therefore, segments that follow the typical control-flow behavior receive low cost, while segments containing unlikely behaviors receive high cost. Formally, for a segment $s = (a_i, \ldots, a_j)$, we define:

$$C(s) = \sum_{i=1}^{m-1} c(a_i, a_{i+1}), \tag{1}$$

where

$$c(a_i, a_{i+1}) = \begin{cases} -\log T[a_i, a_{i+1}] & \text{if } T[a_i, a_{i+1}] > 0, \\ \gamma & \text{otherwise,} \end{cases} \tag{2}$$

where $T[a_i, a_{i+1}]$ denotes the directly-follows probability of activity a_{i+1} after a_i, and γ is some penalty for the transitions that have not been captured by the DFR matrix.

For example, if the transitions `Submit application` → `Check documents` and `Check documents` → `Validate application` are frequent in the event log, then the segment `Submit application` → `Check documents` → `Validate application` will receive a low cost. In contrast, a segment containing a rare or unseen transition will receive a higher cost, indicating a possible shift in local control-flow behavior. Therefore, PELT identifies the change points by minimizing the total segmentation cost over the trace. This objective combines the cost of the resulting segments with a penalty term for introducing additional change points:

$$\min_{\tau_{1:k}} \sum_{i=1}^{k+1} C\big(a_{(\tau_{i-1}+1):\tau_i}\big) + \beta k, \tag{3}$$

where $C(\cdot)$ denotes the segment cost, $\tau_{1:k}$ are the detected change points, and β controls the trade-off between segmentation quality and model complexity. This hyperparameter is selected via grid search by minimizing the Akaike Information Criterion (AIC), which balances goodness of fit against the number of change points, thereby favoring the less costly segmentation for a given trace. In particular, the penalty term discourages over-segmentation, such that new boundaries are introduced only when they lead to a substantial improvement in local control-flow consistency.

4.2 Calculation of Segment-Level SHAP Values

Trace segmentation allows us to explain predictions at the level of coherent trace fragments. This is more informative in process mining than explaining either single events or the entire trace at once. Event-level explanations often ignore the broader local control-flow context, whereas trace-level explanations collapse the whole process instance into one explanatory unit. By using segments instead, we can quantify how different parts of a trace, each corresponding to a locally consistent control-flow region, contribute to the prediction. In other words, each segment is treated as one interpretable process unit in the SHAP computation. The calculation of segment-level SHAP values can be formalized as follows.

Let $\sigma = \langle e_1, \ldots, e_n \rangle$ be a trace of length n. We represent σ as a matrix X of the size $n \times d$, where the i-th row $x_i \in \mathbb{R}^d$ is the feature vector of event e_i, and d is the number of features. Let B be a baseline matrix of the same dimensions $n \times d$. Following the approach in [4], we create a reference event vector $\bar{e} \in \mathbb{R}^d$ that summarizes typical behavior in the training data. For a categorical attribute j, $\bar{e}_j = \text{mode}(X_{*,j})$, where $X_{*,j}$ denotes the j-th column of X. For a numerical attribute j, $\bar{e}_j = \bar{X}_{*,j}$, representing the mean of the j-th column. Further, the baseline B is constructed as $B = \mathbf{1}_n \cdot \bar{e}$. Suppose the trace is partitioned into k contiguous segments s_l, $l = 1, \ldots, k$. To treat segments as explanatory units in the SHAP framework, we define a coalition vector $z \in \{0,1\}^k$, where $z_l = 1$ indicates that segment s_l retains its original values from X, and $z_l = 0$ indicates that it is replaced by the corresponding baseline values from B . For a given coalition z, we construct the mask matrix $M(z) \in \{0,1\}^{n \times d}$, where $M(z)_{i,j} = z_l, \forall i \in s_l, j = 1, \ldots, d$. The perturbation function h is given as $h_s(z) = M(z) \odot X + (J - M(z)) \odot B$, where $J \in \mathbb{R}^{n \times d}$ is an all-ones matrix. Applying KernelSHAP with the k segments as features, we obtain segment-level SHAP values $\phi_1, \ldots, \phi_k$ satisfying the efficiency property $\sum_{l=1}^{k} \phi_l = f(X) - f(B)$, where $f(B)$ serves as the base value.

5 Experimental Setup

This section introduces the employed event logs, the predictive models for predictions, the segmentation baselines, and the evaluation protocol used to study segment-level explanations. The code for reproducing the results is available on GitHub[1].

5.1 Event Logs and Predictive Model

We use one synthetic and two real-world benchmark logs. The synthetic log takes into consideration the challenge behind the evaluation of segmentation strategies due to the absence of ground-truth change points. Thus, we employ the SimBank simulator [7], a tool for benchmarking prescriptive process monitoring methods. The simulator generates synthetic cases of a fictional loan application process and allows control over outcomes via a predefined process model and interventions. We select two samples with clearly defined change points from the simulator output[2]. The first sample is built around the activity *skip_ contact_ hq*, which leads to the *cancellation* outcome. The second sample uses *contact_ hq* and *calculate_ offer*, which lead to the *acceptance* outcome. These known positions are used as ground truth for evaluating segmentation quality.

Regarding the real-world logs, we employ the BPIC15[3] and the BPIC17[4]. For both, we adopt the outcome labeling strategies from the literature [28]. The former contains administrative processes from Dutch municipalities. We use the first municipality, and cases are labeled positive if they satisfy the LTL rule that whenever *send confirmation receipt* occurs, it is eventually followed by *retrieve missing data*; otherwise, it is labeled negative. On the other hand, the BPIC17 describes a loan application process from a financial institution, and we filter only the accepted and canceled cases. For both logs, in order to focus on predictions made before the outcome is explicitly observed, we extract fixed-length prefixes and remove traces that already contain outcome-revealing activities.

As a sequence modeling classifier, we use LSTM models because it is often used in the literature [27]. Since our focus is not to optimize predictive performance but provide explanations from predictions, we opt only for this model and a simple hyperparameter optimization by sweeping the following values: the number of layers $1, 2, 3$, the embedding sizes $32, 64, 128$, the hidden sizes $64, 128$, and the learning rates $1e-4, 5e-5, 1e-5$. As data preprocessing, we exclude outlier cases falling outside the 5th to 95th percentile of all set cases, and use the activity labels and timestamp-related features for training. To evaluate model predictions before the outcome is determined, we extract prefixes of 30 events and exclude cases in which outcome-revealing activities have already occurred. Since the goal of PPM is to provide accurate predictions as early as possible

[1] https://github.com/kSahatova/segment-level-expl-PPM.git.
[2] https://zenodo.org/records/15574272.
[3] https://data.4tu.nl/collections/BPI_Challenge_2015/5065424/1.
[4] https://data.4tu.nl/articles/dataset/BPI_Challenge_2017/12696884.

during case execution, the selected prefix length balances predictive performance and explanatory value. Shorter prefixes often lack sufficient context for accurate predictions, whereas longer prefixes are more likely to contain outcome-revealing events, causing explanations to be dominated by these events and thus reducing their informativeness. The model trained on BPIC15 obtained 92% AUC, and the model trained on BPIC17 achieved 85.14%.

5.2 Segment-Level Explanations and Quantitative Evaluation

Baseline Explainability Methods. We compare our control-flow-aware segmentation against two baselines: the "per-event" explanation, which is the most common approach from the literature [11], and the "distribution-based segmentation" [13]. The latter also proposed for identification of the most probable segments relying on the sessions definition in recommendation system. It depends on several hyperparameters: we set the minimum and maximum window sizes to 3 and 5, respectively, for the initial segmentation stage, and utilize a measuring window width of 4 for segment comparison when evaluating split candidates. Furthermore, distribution-based segmentation relies on a predefined optimal number of segments. Following the study's suggestion, we use the formula $\min(\max(10, n/2), 30)$, where n represents the case length. For both real-world event logs, this results in 15 segments for a 30-event trace prefix. We utilize the same KernelSHAP computation across all differently defined segments. All methods provide explanations for the same feature space, encompassing both control-flow and time-related features. Finally, we note that per-event attribution operates at a finer granularity than the segment-level methods.

Evaluation of Segmentation Quality. We evaluate segmentation quality differently for synthetic and real-world event logs. For the synthetic log, ground-truth change points are available. Therefore, as evaluation metric we employ the RandIndex, Precision, Recall, and F-1 score [29]. These metrics assess how accurately our segmentation algorithm identifies the true segment boundaries. For real-world logs, no ground-truth is available. Therefore, we report descriptive statistics on the resulting segmentation, namely the number and lengths of segments to reflect granularity. Furthermore, we compute the proportion of traces covered by the top-10 most frequent change point patterns as follows. For each trace with change points $\tau_1, \ldots, \tau_k$, we extract the transition boundary $(a_{\tau_i}, a_{\tau_i+1})$ at each detected change point. After computing the frequency of unique patterns, we report the top-10 patterns coverage—how consistent the segmentation is across traces. We also evaluate segmentation quality using entropic relevance (ER) [20], an information-theoretic metric [23]. ER quantifies the number of bits needed to encode traces based on the directly-follows transition probabilities within each segment. Segments with more deterministic transitions yield lower ER scores, indicating that the segmentation captures coherent patterns that require fewer bits to describe. The implementation was adopted from the work [19].

Evaluation of Feature Attributions. In the absence of ground-truth explanations, feature attribution methods are commonly evaluated through perturbation-based metrics [2,3,30]. Following this idea, we assess the faithfulness of segment-level SHAP values using Relative Prediction Change (RPC). We compute RPC for both important (RPCI) and unimportant (RPCU) segments, where importance is defined by the absolute SHAP value. The top-k segments are selected such that they cover 40% of the trace prefix σ, which provides a large enough perturbation to affect the prediction without fully destroying the trace structure. Intuitively, RPC measures how much the model prediction changes after perturbing selected segments. A faithful explanation should assign high importance to segments whose perturbation strongly affects the prediction, and low importance to segments whose perturbation has little effect. To compute RPCI, we perturb the top-k most important segments by replacing their feature values with mean values from 100 randomly sampled training cases. RPCU is computed by perturbing the segments deemed unimportant while keeping the top-k fixed. We report both metrics per predicted class to account for class-dependent effects [3].

6 Experimental Evaluation

This study investigates how meaningful segment-level feature attributions can be derived for predictive process monitoring. We address this by exploring the following research questions:(**RQ1**) how faithful are segment-level SHAP value explanations to the model's output compared to non-segment-level SHAP values? (**RQ2**) Can SHAP values identify meaningful segments within a case that are most influential for a given prediction?

6.1 Trace Segmentation

The results of the segmentation quality performed on the synthetic data are shown in Table 1. The agreement between predicted and ground truth change points, as quantified by the RandIndex, exceeds 0.9 for both canceled and accepted cases with one and two change points, correspondingly. Segmentation accuracy is slightly higher for the former due to the smaller number of segments, achieving an F1 score of 89%.

Table 1. Segmentation quality for the synthetic dataset with the activities specified as change points.

Sample (change points)	RandIndex	Precision	Recall	F1 score
canceled cases (“skipped_hq”)	0.928	0.875	0.925	0.885
accepted cases (“contact_hq”, “calculate_offer”)	0.951	0.8582	0.906	0.874

Regarding the real logs, Table 2 summarizes results for BPIC17 and BPIC15. In the BPIC17, correctly predicted cases contain, on average, three events per segment and approximately eight segments per prefix. The ten most frequent segmentation patterns for True Negative (TN) and False Negative (FN) samples represent the control flow of 81% and 77% of all trace prefixes in the test set. In contrast, canceled cases follow more diverse control-flow paths; the 212 identified patterns result in the ten most frequent patterns covering only 42% of the traces. Finally, the application of the control-flow-aware segmentation algorithm reduces the entropic relevance from 37.81 to 6.68, indicating that the identified segments yield a substantially more compressible, and thus more structured, trace representation.

The BPIC15 event log contains 283 unique activities characterized by low or near-zero transition probabilities between the majority of event pairs. This high log variability, which is also supported by the high ER of 100, causes a more granular segmentation of the trace prefixes, resulting in approximately 13 segments with an average of two events each. These results are distributed uniformly across all samples, which does not allow the identification of any distinct process behaviors. Consequently, the results indicate that each prefix exhibits a unique, case-specific segmentation pattern.

Table 2. Segmentation evaluation on the real-world datasets.

	BPIC17				BPIsC15			
	TP	FP	TN	FN	TP	FP	TN	FN
Sample size	705	4	2004	288	45	7	125	1
Number of segments	8.93 ± 1.26	6.25 ± 2.05	8.21 ± 1.16	7.72 ± 1.56	13.31 ± 0.69	12.86 ± 0.64	13.50 ± 0.74	14 ± 0.0
Segment length	3.44 ± 0.61	5.62 ± 2.58	3.74 ± 0.70	4.14 ± 1.33	2.26 ± 0.125	2.34 ± 0.12	2.23 ± 0.13	2.14 ± 0.0
Top-10 patterns coverage	42.1%	100%	81.4%	77.4%	22.2%	100%	8%	100%
Number of patterns	212	3	49	115	45	7	125	1
ER full log	37.81 ± 6.11				100.1 ± 10.36			
ER after segmentation	6.68 ± 2.50				10.93 ± 2.49			

6.2 Evaluation of Feature Attributions

For the quantitative evaluation of segment-level SHAP explanations, we use perturbation-based metrics described in Sect. 5. Table 3 reports RPCI and RPCU for the three segmentation strategies on BPIC17. A faithful explanation yields high RPCI (important segments substantially affect the prediction) and low RPCU (unimportant segments have minimal effect). The parameter k differs across strategies since each produces segments of varying lengths. To ensure a fair comparison, k is chosen such that the perturbed segments cover 40% of the trace length. The following interpretation of the results addresses our **RQ1**. All three segmentation strategies yield faithful explanations, with higher RPCI and lower RPCU for class 1. RPCI exceeds 1.0 for class 0 cases because their near-zero scores amplify changes when important segments are perturbed, still indicating high faithfulness (e.g., removing completion/validation segments shifts predictions toward cancellation). For class 1, where scores are closer to 1.0, RPCI stays below 1.0 with lower variance, reflecting more consistent explanations across strategies.

The distribution-based and control-flow-aware segmentation strategies sacrifice some faithfulness, which can be attributed to the aggregation of events into semantically coherent segments. In contrast, event-level attributions can precisely target the most individually contributing events. However, it should be noted that RPCU for class 0 decreases as segmentation granularity becomes coarser in the both event logs. For event-level attribution, perturbing individual low-ranked events might disrupt local dependencies between neighboring events, resulting unexpected prediction changes. Segment-level strategies avoid this by perturbing coherent groups of events together, preserving internal structure and producing a cleaner separation between important and unimportant parts of the trace.

Table 3. Evaluation of faithfulness of SHAP value explanations.

	BPIC17			BPIC15		
Sample	k	RPCI	RPCU	k	RPCI	RPCU
Per-event segmentation						
Pred. class 0	12	0.909 ± 0.856	0.525 ± 0.382	12	2.140 ± 2.191	0.311 ± 0.179
Pred. class 1		**0.832 ± 0.097**	**0.005 ± 0.012**		**0.973 ± 0.016**	**0.004 ± 0.005**
Distribution-based segmentation						
Pred. class 0	5	1.352 ± 0.860	0.329 ± 0.290	5	2.139 ± 2.235	0.304 ± 0.166
Pred. class 1		0.773 ± 0.157	0.005 ± 0.013		0.951 ± 0.049	0.018 ± 0.037
Control-flow-aware segmentation						
Pred. class 0	5	1.278 ± 0.777	**0.233 ± 0.230**	6	2.154 ± 2.126	**0.242 ± 0.148**
Pred. class 1		0.769 ± 0.113	0.008 ± 0.018		0.972 ± 0.017	0.007 ± 0.015

6.3 Segment-Level Explanations

In this analysis to address our **RQ2**, we consider the three most frequent segmentation patterns we generated from BPIC17, and the assigned SHAP values for each pattern and each class. In Fig. 1, we illustrate the accepted cases. As previously discussed in the segmentation analysis, these cases follow more regular and recurring control-flow patterns, which indicates a more stable process execution. In these cases, segments containing transitions such as `A_Concept` → `W_Complete_application` consistently push the prediction toward acceptance, as reflected by negative SHAP values. The same holds for later segments related to application completion and validation, such as `A_Complete`. This suggests that the model associates smooth and structured progression through the process with a favorable outcome. On the other hand, in Fig. 2, the canceled cases show a different pattern, mainly driven by repeated offer creation. In particular, segments related to multiple created offers receive high positive SHAP values and increasingly push the prediction toward cancellation. The last created offer often has the strongest effect. For stakeholders, this is meaningful because it points to a potential bottleneck detection: repeated offer generation is strongly associated with unsuccessful case outcomes. A visual comparison of event-level and segment-level feature attributions for these cases is provided in Fig. 3.

We also analyze the wrongly predicted cases, which further support this interpretation: accepted cases wrongly predicted as canceled often already contain multiple created offers in the observed prefix, whereas canceled cases wrongly predicted as accepted lacked any control-flow activities indicating an application completion or validation stage. This is likely influenced by the 30-event prefix, since decisive outcome-related behavior may appear later in the trace.

Overall, these findings show the value of segment-level explanations: instead of attributing predictions to isolated events or to the whole trace, they reveal which *part of the process* drives the prediction and contextualizes particular activities in their segments.

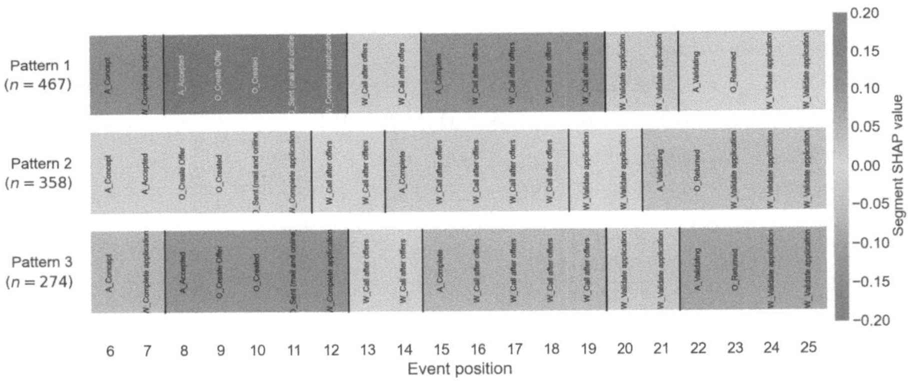

Fig. 1. Visualization of SHAP values for representative traces from the three most frequent segmentation patterns regarding *accepted* cases for the BPIC17.

7 Discussion and Limitations

This study proposes a method to provide explainability of sequential models in the PPM tasks through the estimation of SHAP values at the segment level. The control-flow-aware segmentation algorithm allows trace partitioning into segments preserving structural dependencies. Our experiments show that event logs with moderate control-flow variability, such as BPIC17, exhibit clearly identifiable phases: transition shifts are well-defined due to the log structure and a manageable number of unique activities, allowing the segmentation algorithm to produce meaningful partitions. Aggregating the resulting segmentation patterns

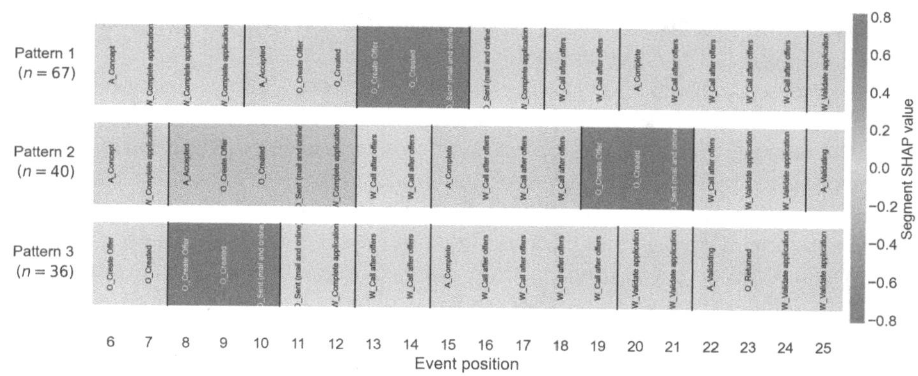

Fig. 2. Visualization of SHAP values for representative traces from the three most frequent segmentation patterns regarding *cancelled* cases for the BPIC17.

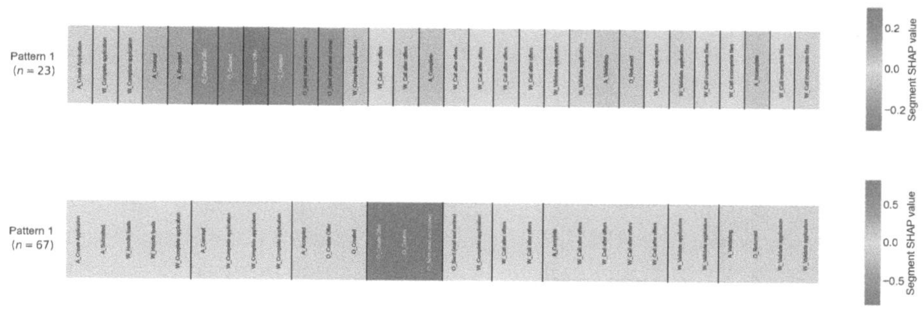

Fig. 3. Comparison of event-level and segment-level SHAP value explanations for the BPIC17 log.

across traces already facilitates a high-level analysis of the event log, revealing common phase structures without requiring inspection of individual traces.

Through the quantitative evaluation of the obtained segment-level feature attributions, we demonstrated that regions meaningful to the model can be identified. Across both datasets and all segmentation strategies, perturbing the top-attributed segments causes substantial prediction changes, while perturbing low-attributed segments has minimal effect. This confirms that the identified regions are influential rather than arbitrary.

Comparing our approach against two alternative segmentation strategies, we demonstrated that grouping activities based on control-flow information leads to more comprehensible explanations that allow the identification of distinct transition patterns. While event-level feature attributions are more faithful to the model's behavior due to their finer granularity, segment-level attributions enable the analysis of model behavior at a coarser level. In particular, they mitigate the interleaving of important and unimportant events, thereby increasing the overall comprehensibility of the explanations.

Since our segmentation method relies on the directly-follows relations among activities, low transition probabilities lead to fine-grained segmentation, which limits the aggregation of segmentation patterns across cases. High variability in the event log control flow further exacerbates this issue, as few cases share similar segment structures, making it difficult to draw generalizable conclusions about model behavior at the segment level. Conversely, event logs with fewer distinct activities and short trace prefixes offer limited benefit from segmentation, as there is insufficient sequential structure to produce meaningful groupings.

8 Conclusion

In this paper, we proposed a feature attribution-based explainability method for PPM that preserves the control-flow perspective of business processes. More specifically, we introduced a control-flow-aware segmentation algorithm that partitions trace prefixes into segments based on shifts in directly-follows transition probabilities, and we used these segments as explanatory units for the computations of segment-level SHAP values. The results on synthetic data showed that the proposed segmentation method is capable of recovering meaningful change points with high agreement to ground truth, while the experiments on real-world event logs showed that it produces interpretable segmentation and faithful explanations. In particular, for BPIC17, the method revealed recurring trace segments associated with accepted and canceled cases, making it possible to relate model predictions to concrete parts of the process rather than to isolated events of fully aggregated traces.

Overall, our findings show that segment-level explanations offer a useful middle ground between fine-grained event-level and trace-level explanations. Although the former remains slightly more faithful due to its finer granularity and simplicity, segment-level explanations improve comprehensibility by preserving local control-flow context and reducing the interleaving of important and unimportant events. At the same time, the study highlights that the benefit of the proposed method depends on the structure of the event log: logs with moderate control-flow variability allow the discovery of meaningful and comparable segments, whereas highly variable logs lead to very fine-grained and case-specific segmentation. In future work, we will extend the method beyond binary outcome prediction enabling analysis of distinct failure modes and supporting more targeted decision-making.

Disclosure of Interests. The authors have no competing interests to declare that are relevant to the content of this article.

References

1. Aas, K., Jullum, M., Løland, A.: Explaining individual predictions when features are dependent: more accurate approximations to shapley values. Artif. Intell. **298** (2021). https://doi.org/10.1016/j.artint.2021.103502
2. Agarwal, C., et al.: OpenXAI: towards a transparent evaluation of post hoc model explanations. In: Proceedings of the 36th International Conference on Neural Information Processing Systems, NIPS '22. Curran Associates Inc., Red Hook, NY, USA (2022)
3. Baer, G., Grau, I., Zhang, C., Van Gorp, P.: Class-dependent perturbation effects in evaluating time series attributions. In: XAI World Conference (2025)
4. Bento, J.A., Saleiro, P., Cruz, A.F., Figueiredo, M.A., Bizarro, P.: Timeshap: explaining recurrent models through sequence perturbations. In: Proceedings of the 27th ACM SIGKDD Conference on Knowledge Discovery & Data Mining, KDD '21, pp. 2565–2573. Association for Computing Machinery, New York, NY, USA (2021). https://doi.org/10.1145/3447548.3467166
5. Buliga, A., Di Francescomarino, C., Ghidini, C., Donadello, I., Maggi, F.M.: Guiding the generation of counterfactual explanations through temporal background knowledge for predictive process monitoring. Data Min. Knowl. Disc. (2025). https://doi.org/10.1007/s10618-025-01117-3
6. Crabbé, J., Van Der Schaar, M.: Explaining time series predictions with dynamic masks. In: ICML (2021). https://doi.org/10.48550/arXiv.2106.05303
7. De Moor, J., Weytjens, H., De Smedt, J., De Weerdt, J.: Simbank: from simulation to solution in prescriptive process monitoring. In: BPM (2025). https://doi.org/10.1007/978-3-032-02929-4_10
8. Deldari, S., Smith, D.V., Sadri, A., Salim, F.: Espresso: entropy and shape aware time-series segmentation for processing heterogeneous sensor data. Proc. ACM Interact. Mobile Wearable Ubiquit. Technol. **4**(3) (2020). https://doi.org/10.1145/3411832
9. Doddaiah, R., Parvatharaju, P., Rundensteiner, E., Hartvigsen, T.: Class-specific explainability for deep time series classifiers. In: ICDM (2022)
10. El-khawaga, G., Abu-Elkheir, M., Reichert, M.: XAI in the context of predictive process monitoring: an empirical analysis framework. Algorithms (2022)
11. Galanti, R., Coma-Puig, B., de Leoni, M., Carmona, J., Navarin, N.: Explainable predictive process monitoring. In: ICPM (2020)
12. Hamed, H., Reis, A.C., Choobar, B.G., Pang, Q., Killick, R., Safari, M.: Change-point detection as a light data-driven approach to battery state-of-health prediction. Cell Rep. Phys. Sci. (2026). https://doi.org/10.1016/j.xcrp.2026.103157
13. Jiang, G., et al.: SeqSHAP: subsequence level Shapley value explanations for sequential predictions. In: DASFAA (2024)
14. Killick, R., Fearnhead, P., Eckley, I.A.: Optimal detection of changepoints with a linear computational cost. J. Am. Stat. Assoc. **107**(500) (2012). https://doi.org/10.1080/01621459.2012.737745
15. Li, J., Fearnhead, P., Fryzlewicz, P., Wang, T.: Automatic change-point detection in time series via deep learning. J. R. Stat. Soc. Ser. B Stat Methodol. **86**(2) (2024)
16. Liu, Z., et al.: Explaining time series via contrastive and locally sparse perturbations. In: ICLR (2024)
17. Lundberg, S.M., Lee, S.I.: A unified approach to interpreting model predictions. In: Proceedings of the 31st International Conference on Neural Information Processing Systems, NIPS'17, pp. 4768–4777. Curran Associates Inc., Red Hook, NY, USA (2017). https://doi.org/10.48550/arXiv.1705.07874

18. Nayebi, A., Tipirneni, S., Reddy, C.K., Foreman, B., Subbian, V.: WindowSHAP: an efficient framework for explaining time-series classifiers based on Shapley values. J. Biomed. Inform. **144**, 104438 (2023)
19. Peeperkorn, J., De Smedt, J., De Weerdt, J.: Model-driven stochastic trace clustering. Inf. Syst. **139** (2026). https://doi.org/10.1016/j.is.2026.102697
20. Polyvyanyy, A., Moffat, A., García-Bañuelos, L.: An entropic relevance measure for stochastic conformance checking in process mining. In: ICPM (2020). https://doi.org/10.1109/ICPM49681.2020.00024
21. Rizzi, W., Di Francescomarino, C., Maggi, F.M.: Explainability in predictive process monitoring: when understanding helps improving. In: BPM (2020). https://doi.org/10.1007/978-3-030-58638-6_9
22. Schlegel, U., Vo, D.L., Keim, D.A., Seebacher, D.: TS-MULE: local interpretable model-agnostic explanations for time series forecast models. In: ECML-PKDD (2021). https://doi.org/10.1007/978-3-030-93736-2_40
23. Shannon, C.E.: A mathematical theory of communication. Bell Syst. Tech. J. **27**(3) (1948)
24. Sivill, T., Flach, P.: LIMESegment: meaningful, realistic time series explanations. In: International Conference on Artificial Intelligence and Statistics (2022)
25. Stevens, A., Ouyang, C., De Smedt, J., Moreira, C.: Generating feasible and plausible counterfactual explanations for outcome prediction of business processes. IEEE TSC (2025). https://doi.org/10.1109/TSC.2025.3609837
26. Sundararajan, M., Najmi, A.: The many Shapley values for model explanation. In: International Conference on Machine Learning (2020). https://doi.org/10.48550/arXiv.1911.06292
27. Tax, N., Verenich, I., La Rosa, M., Dumas, M.: Predictive business process monitoring with LSTM neural networks. In: CAiSE (2017). https://doi.org/10.1007/978-3-319-59536-8_30
28. Teinemaa, I., Dumas, M., Rosa, M.L., Maggi, F.M.: Outcome-oriented predictive process monitoring: review and benchmark. ACM TKDD **13**(2) (2019). https://doi.org/10.1145/3301300
29. Truong, C., Oudre, L., Vayatis, N.: Selective review of offline change point detection methods. Signal Process. **167** (2020). https://doi.org/10.1016/j.sigpro.2019.107299
30. Turbé, H., Bjelogrlic, M., Lovis, C., Mengaldo, G.: Evaluation of post-hoc interpretability methods in time-series classification. Nat. Mach. Intell. **5**(3) (2023). https://doi.org/10.1038/s42256-023-00620-w
31. Velmurugan, M., Ouyang, C., Moreira, C., Sindhgatta, R.: Evaluating stability of post-hoc explanations for business process predictions. In: ICSOC (2021)
32. Weytjens, H., De Weerdt, J.: Process outcome prediction: CNN vs. LSTM (with attention). In: BPM (2020). https://doi.org/10.1007/978-3-030-66498-5_24

Visual Comparative Analysis of Process Variants

María Salas-Urbano[1,2,4(✉)], Cristina Cabanillas[1,2], Paul Zinke[3], Florian Häusler[3], Han van der Aa[4], and Manuel Resinas[1,2]

[1] SCORE Lab, Universidad de Sevilla, Seville, Spain
{msurbano,cristinacabanillas,resinas}@us.es
[2] I3US Institute, Universidad de Sevilla, Seville, Spain
[3] Mehrwerk GmbH, Karlsruhe, Germany
{paul.zinke,florian.haeusler}@mehrwerk.net
[4] Faculty of Computer Science, University of Vienna, Vienna, Austria
han.van.der.aa@univie.ac.at

Abstract. Data visualization plays a central role in process mining, particularly when analysts need to compare multiple process models, often representing different *process variants*, to assess changes, detect deviations, or benchmark performance. However, effectively supporting such comparative analysis remains challenging. To investigate this problem, we combine three perspectives: interviews with nine process mining analysts, established visual analytics principles, and an analysis of existing process mining visualizations. Based on this analysis, we identify key challenges for comparative analysis, including weak support for goal-driven analysis, insufficient scalability, and limited flexibility. Based on these insights, we propose an interactive approach for the flexible, scalable, and goal-driven comparison of multiple process models. The approach enables dynamic, goal-driven highlighting of differences and similarities, allowing analysts to tailor comparison based on specific analysis objectives. We developed a prototype implementing the main concepts and interaction mechanisms, and evaluated its perceived usefulness and usability through two workshops with a total of ten participants (students and experts). The results provide initial design insights for visual analytics techniques supporting process variant comparison and highlight directions for future research.

Keywords: Process mining · Visual process comparison · Process model comparison · Data visualization

1 Introduction

Data visualization is fundamental for analyzing complex datasets and uncovering actionable insights. In process mining, visualization techniques are crucial for understanding process execution and evolution over time [19]. Event logs often capture process executions that can differ significantly along key dimensions,

F. M. Maggi et al. (Eds.): BPM 2026, LNBIP 599, pp. 255–272, 2027.
https://doi.org/10.1007/978-3-032-37877-4_15

such as geographic location, customer type, or time period [1]. These differences give rise to distinct *process variants*. Comparing such process variants is crucial to identify structural differences, inefficiencies, and behavioral deviations [21].

The Directly-Follows Graph (DFG) is one of the most commonly used types of process visualization [11,28]. It provides an intuitive representation of control-flow behavior in an event log, helping users understand the process [30]. However, when several process variants are combined into a single DFG, the resulting model tends to obscure the differences between them, leading analysts to draw misleading conclusions about the true behavior of the process [28]. A common solution is to split the event log, and generate a DFG for each subset [7]. While this technique preserves the distinguishing characteristics of each process variant, it introduces a new challenge: analysts must now interpret a collection of DFGs to uncover meaningful similarities and differences between them.

Despite the value of data visualization in process mining, the ability to view multiple visualizations simultaneously and extract insights remains challenging. Recent work highlights the potential of combining visual analytics with process mining to better support these analytical tasks [19]. However, existing tools are not designed to handle multiple process models or to adequately support visual comparison across multiple process models [23]. Consequently, analysts often rely on manual and error-prone methods to compare models [16]. To better understand these practices, we interviewed nine process analysis experts, which revealed limitations in current tools for flexible, goal-driven, and scalable comparison of multiple models. The aim was to explore how and why they carry out visual comparisons of process models in practice. Through an in-depth analysis of existing solutions and practices, we identified three key challenges that remain insufficiently addressed: (C1) adaptability and goal-driven comparison, tailoring both the comparison and its visualization according to specific goals; (C2) scalability and clarity in multi-model comparison, supporting intuitive exploration and visualization as the number and complexity of models increases; and (C3) flexibility in the analysis of differences and similarities, supporting the identification of both commonalities and deviations without restricting the comparison to a single perspective.

To address these challenges, we propose a set of comparative strategies to support the visualization and exploration of collections of DFGs, together with visual designs that improve cross-model comparison. Our approach is grounded in key visual design principles [12,13] We focus on DFGs as they are the most widely used notation in process mining tools and are well understood by analysts [18,28]. We implemented our approach as a tool prototype that supports analysts in identifying structural similarities and differences across process variants, and in tailoring visualizations to specific analysis goals, such as highlighting control-flow aspects. We evaluated our prototype through two workshops with ten users, including both students and experts, who tested it using a real event log and provided feedback on its usability and usefulness. The evaluation revealed positive user perceptions regarding the usefulness of the proposed interaction mechanisms and visual comparison strategies.

The paper is structured as follows. Section 2 reviews related work. Section 3 describes the research method. Section 4 details the identified challenges. Section 5 presents the conceptual approach to visually compare process models. Section 6 presents the evaluation of the implemented tool functionalities through two workshops. Section 7 discusses implications, opportunities for improvement, and future research. Finally, Sect. 8 concludes the paper.

2 Related Work

Building on representative work in visual process comparison, we examine current approaches in terms of the visualizations they employ, comparison methods, and how these results are reported to users. Presenting results effectively is crucial, as visualization supports interpretation and communication [13]. Then, we distinguish between approaches based on process models and those relying on alternative representations, with limitations discussed in Sect. 4.

Result Representation with Process Models. Several approaches use process models to visualize differences. Ballambettu et al. [3] and Syamsiyah et al. [26] rely on DFGs to compare multiple process models and present the results within a unified DFG, using colors and varying node and edge thicknesses to highlight changes. This can be used to reveal structural differences and changes in DFG annotations, such as frequencies and flow time. While these studies analyze multiple process models simultaneously, the remaining approaches are designed to compare only two models at a time. Jalali et al. [16] incorporate probabilistic information using stochastic DFGs to display behavioral differences with color-coded edges. They compare process models using Markov models and stochastic DFGs, applying a cut-off level to filter insignificant variations. In contrast, other approaches use alternative graph-based representations. For instance, Bolt et al. [5] rely on transition systems to detect statistically significant differences between models in terms of any measurement annotations, and display the similarities and differences of business rules between them. Similarly, Wang et al. [29] use Petri nets to detect structural and behavioral differences between processes, displaying key process structures common between them.

In addition, several commercial process mining tools support visual comparison of business processes using process models[1]. For example, Apromore highlights deviations between two variants with color-coded annotations, while Celonis, LANA (now part of Appian), and Signavio allow users to compare multiple variants. Specifically, Signavio and LANA use color overlays and Celonis shows side-by-side comparisons. In addition, tools like ARIS, Minit, and UiPath support process model overlays and interactive exploration of structural and behavioral differences showing DFGs or BPMN models. Other tools, such as BusinessOptix, EverFlow, mpmX, myInvenio, PAFnow, and QPR Process-Analyzer, provide visual process comparisons considering different scopes and

[1] https://www.processmining-software.com/.

methodologies. For instance, BusinessOptix focuses on scenario-based comparisons, EverFlow and mpmX provide dashboards for variant analysis, and myInvenio and QPR allow structural and behavioral comparisons with some predictive analyses.

Result Representation Without Process Models. Other approaches present process comparison outcomes by means other than process model visualizations. For instance, Nguyen et al. [20] use a matrix-based representation, using perspective graphs for statistical comparison between two process models from multiple attributes. The result is a difference graph visualized as a color-coded matrix, where nodes are columns and edges are cells. Van der Aalst et al. [2] assess how well a model can reproduce event sequences recorded in an event log using Petri nets. The outcome of this analysis is a degree of similarity between them, expressed as a numerical value between 0 and 1. Van Dongen et al. [10] compute the similarity between two process models based on causal footprints, which represent essential behavioral constraints. While such approaches provide numerical outputs, Van Beest et al. [4] capture behavioral differences through statements in natural language based on frequency-enhanced prime event structures, e.g., "In variant 2, b occurs after a, while in variant 1 it does not". Finally, Bin et al. [6] propose a custom notation for describing difference patterns between two process models represented as process structure trees. For example, the notation "∅ VS R" indicates that *task R only exists in the second process model.*

3 Research Method

We follow the principles of Design Study Methodology [24] commonly used in visualization research. Following this methodology, we analyze real-world problems faced by process mining analysts doing comparative analysis of process variants and then design a visualization system to address them. All supplementary materials related to the research method are available in our repository.[2]

3.1 Understanding the Problem

We analyzed the problem from three different perspectives. First, we made a structured review of related work focused on multi-model process comparison considering academic approaches and commercial tools. Second, following the *understanding work practices* approach in visualization research [15], we conducted semi-structured interviews with nine process mining experts to gain insights into how and why visual comparisons of process models are performed in practice. The number of participants is consistent with prior work in this area [15]. The interview protocol included 19 questions in three parts: (i) *demographic information* (3 questions), capturing participants' role and background;

[2] https://doi.org/10.5281/zenodo.20393413.

(ii) *process mining experience* (4 questions), addressing expertise, years of experience, and typical tasks; and (iii) *process mining analysis practices* (12 questions), focusing on exploratory analysis of event logs, use of process models, comparison of multiple process variants, and tool features and limitations. Each interview lasted approximately one hour and was conducted remotely, recorded, and transcribed. Finally, we considered established principles of comparative visualization [12,17] to guide the analysis of existing techniques and to interpret findings from the interviews. The insights obtained in this stage were used to identify challenges and user needs described in Sect. 4

3.2 Conceptualization and Prototype Evaluation

To address the challenges, we introduce a set of comparative strategies and visual designs for multi-model process comparison, grounded in principles of comparative visualization [12,13]. The framework adopted guided the systematic derivation of visualization mechanisms for multi-model comparison through identifying core comparative elements, deriving associated challenges, and linking them to visualization strategies and design choices to manage perceptual and cognitive load. The resulting approach is detailed in Sect. 5.

We implemented a prototype tool to explore and assess the proposed approach. The prototype implements the proposed comparative strategies and designs, and it enables analysts to compare multiple process models. The prototype evolved through several iterative cycles involving internal testing, refinement of interaction mechanisms, and adjustments to the visual comparison strategies based on emerging design considerations. We evaluated the prototype in two workshops: one with five domain experts and another one with five students. The study combined *Visual Data Analysis and Reasoning* and *User Experience* methods [15]. Participants performed analysis tasks using the prototype and provided feedback through open question and questionnaires. Each workshop lasted approximately two hours, and sessions were recorded and transcribed. The goal of the workshops was to obtain exploratory qualitative feedback regarding the usefulness, usability, and applicability of the proposed interaction mechanisms and visual comparison strategies. Further details are provided in Sect. 6.

4 Challenges in Visual Process Comparison

Visual comparison of multiple process models is a central activity in process mining practice. Our interviews show that analysts frequently use process visualizations as part of their exploratory analyses and regularly compare process variants for tasks such as conformance checking, benchmarking and cross-context analysis. DFGs were highlighted as particularly useful for this purpose. As one participant emphasized, *"I find DFGs... one of the most important things in process mining to compare processes together to find where processes differ"*.

Drawing on a critical analysis of existing approaches, established principles of comparative visualization [12,17], and insights from our interview study, we identify three key challenges that reflect both recurring needs highlighted in prior studies [7,16,21] and limitations observed in current tools.

4.1 Adaptability and Goal-Driven Comparison (C1)

Comparison tasks involve two key elements: a set of *targets* (i.e., elements to compare), and an *analysis action* that defines the analytical goal [12]. In our context, the targets are multiple process models, and the actions focus on identifying structural or behavioral similarities and differences between them.

Our interviews reveal that analysts employ both exploratory and hypothesis-driven analysis. Exploratory analyses often begin without a fixed purpose, requiring flexible comparison strategies that support pattern discovery, insight generation, and shifting analytical focus. Even when a clear hypothesis exists, existing tools often fail to align with the analytical intent (cf. Section 2), making it difficult to focus on the most relevant information. This misalignment is evident in many existing approaches that highlight multiple aspects of the comparison between process models without considering the user's analytical intent. For example, Ballambettu et al. [3] simultaneously present structural and behavioral differences without, regardless of whether both are relevant to the task. This lack of focus can introduce visual clutter and increase cognitive load.

The need for adaptive, goal-driven comparison is supported by prior visual analytics research. Von Landesberger et al. [17] highlight algorithms and interactive visualizations for specific tasks within unified systems, and Gleicher et al. [12] argue that interactivity and customization improve the comparison relevance of the visual outputs and support insight generation aligned with the analytical actions. This is consistent with cognitive perspectives on process mining, which emphasize theory-guided analysis driven by the analyst's goals [25].

Challenge C1. Visual process comparison techniques should support goal-driven analysis by allowing users to define and adjust analysis actions, and by adapting the visualization accordingly. This includes enabling focus on specific aspects, such as behavior or performance, while maintaining clarity and supporting effective insight generation.

4.2 Scalability and Clarity in Multi-model Comparison (C2)

Scalability is a well-known challenge in visual analytics, where the number and complexity of targets affect interpretability [12,17]. In process mining, this issue is particularly pronounced: process models are inherently complex, and analysts often need to compare multiple variants derived from event logs (e.g., grouping cases by resource) [23]. This leads to comparison scenarios with many dense models, which are difficult to interpret jointly.

Existing approaches provide limited support for such scenarios. Only a few [3, 26] handle comparing more than two process models at once, restricting the ability to analyze multiple process variants simultaneously. This limitation also appeared in our interviews, where several participants noted that current tools restrict comparisons to two models despite the need to analyze multiple variants.

Scalability and clarity are closely tied: as the number and complexity of process models increases, maintaining clear visualizations becomes more difficult [12]. Approaches for comparing several process models [3,26] typically display all results in a single view, causing visual clutter. One interview participant emphasized the importance of simplifying visualizations and focusing on relevant cases to avoid overwhelming users. Additionally, while seven participants reported that DFGs are useful for comparison, especially to detect activities, loops, or deviations, they also highlighted several limitations related to scalability and clarity, including difficulties comparing many models, managing numerous process variants, and inconsistent node positions across graphs.

Challenge C2. Visual process comparison techniques should scale to large collections of complex process models while maintaining visual clarity. This requires supporting the analysis of multiple variants without overwhelming users.

4.3 Flexibility in Comparing Differences and Similarities (C3)

Prior research emphasizes that comparative visualizations must support balanced exploration of similarities and differences to support reasoning [12,17]. Our interviews reinforce this need. Seven out of nine participants said that highlighting both similarities and differences, for instance through color coding or annotations, would improve the comparison process. However, current visual process comparison approaches mainly focus on identifying differences, limiting the ability of the analysts to recognize similarities between the process models. For instance, Jalali et al. [16] detect behavioral differences based on activity frequency and time between activities. Only a few approaches consider similarities, but they do not rely on process model visualizations [2,10].

Challenge C3. Visual process comparison techniques should allow analysts to switch between exploring similarities and differences across process models. This requires defining actions that explicitly focus on both types of goals and providing explicit encodings that enable analysts to emphasize or adjust the importance of each depending on the task.

5 Conceptual Proposal for Visual Comparisons

Building on established principles of comparative visualization [12,13], particularly Gleicher's framework, our proposal defines comparative elements, their associated analytical challenges, and corresponding visualization strategies and

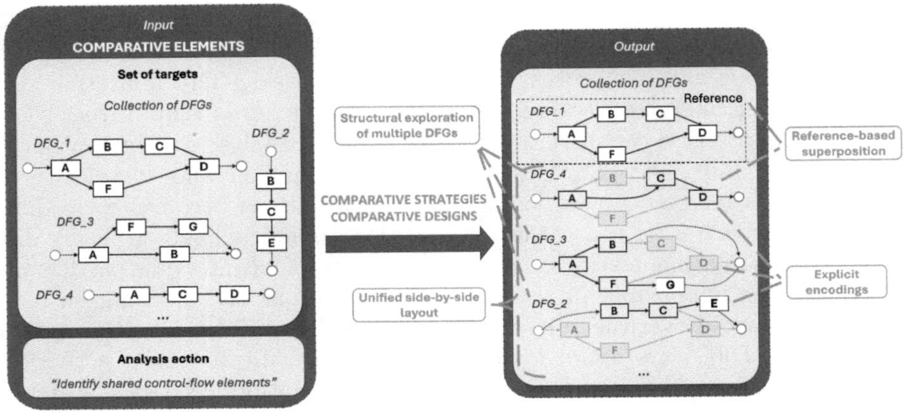

Fig. 1. Overview of the proposed comparative visualization approach.

design choices to support visual multi-model comparison. The supplementary material provides DFGs used in the evaluation to show the application of these strategies and design choices. Figure 1 illustrates these interrelations in our approach, which is described in detail below.

5.1 Comparative Elements

Comparative visualization begins by specifying the inputs in terms of *targets* and *actions* [12]. In our case, the targets are a collection of DFGs generated from several filtering and data manipulation tasks on the event log to create multiple process variants, with each DFG representing one process variant.

Regarding the actions, considering that successful comparison approaches avoid covering too much at once [12], we focus on three control-flow analysis actions: (1) *Identify shared control-flow elements*, which highlights nodes and edges that are shared across the compared models; (2) *Identify exclusive control-flow elements*, which highlights nodes and edges that appear only in a specific model and not in others; and (3) *Identify missing control-flow elements*, which highlights nodes and edges that are absent in a specific model but present in other models. These actions operationalize different comparison goals, enabling analysts to flexibly focus on similarities and differences depending on their analytical intent (C3). They are implemented as selectable options within our approach's foundation, where each choice triggers the corresponding comparative strategies and designs to support the intended analysis (C1).

5.2 Comparative Strategies

To facilitate comparison of large and complex DFG collections (C2), we propose two complementary strategies, grounded in principles of comparative visualization [12]. Comparative strategies define high-level design choices that determine which aspects of the targets are shown and how they are visually encoded.

Structured Exploration of Multiple DFGs. To address the need to handle large collections of process variants (C2), we introduce a *sequential scanning* strategy, where users scroll and examine each DFG individually. To accomplish this, we enable users to sort the collection by a criterion relevant to the analysis task at hand, such as similarity between models or statistical measures like case count or mean duration. Displaying the most informative models first enables users to focus on relevant process variants and stop scanning once sufficient insight is gained, reducing cognitive effort. For instance, when a user is interested in a specific set of activities, models containing the full set can be displayed first, followed by those with most, few, or none of the activities.

Visual Simplification of DFGs. We enable the interactive simplification of DFGs through a set of user-driven mechanisms that reduce complexity (C2) while preserving interpretability. Unlike conventional filters, which are applied before visualizing the results, these mechanisms allow users to iteratively refine DFGs and immediately observe their impact. To this end, our approach integrates standard filters for activities and transitions based on frequency thresholds [28] with three additional visual simplification mechanisms:

Activity hiding reduces DFG size by hiding selected activities. Similar to common filters, it adds edges between preceding and subsequent activities to preserve structure even if transitions did not occur in the original log [28].

Fragment focusing enables users to concentrate on a segment of the DFG between two selected activities for detailed examination. This need has been identified in previous work as a key requirement for effective process analysis [21]. For instance, Fig. 2a displays the full process, while Fig. 2b presents a filtered view that isolates only the portion of the process between activities B and C, hiding all nodes and edges outside this process fragment.

Annotation selection removes irrelevant visual elements, such as colors, activity frequencies, or performance metrics, simplifying the DFG without losing key comparative insights. For example, users focusing on structural differences can hide frequency or performance annotations from nodes and edges.

5.3 Comparative Designs

To visually present similarities and differences between multiple DFGs, we devise two comparative visualization designs for multi-model process comparison grounded on established approaches, namely *juxtaposition*, *superposition*, and *explicit encodings* [13]. These designs are combined and adapted to support different analysis actions and address the challenges identified in Sect. 4.

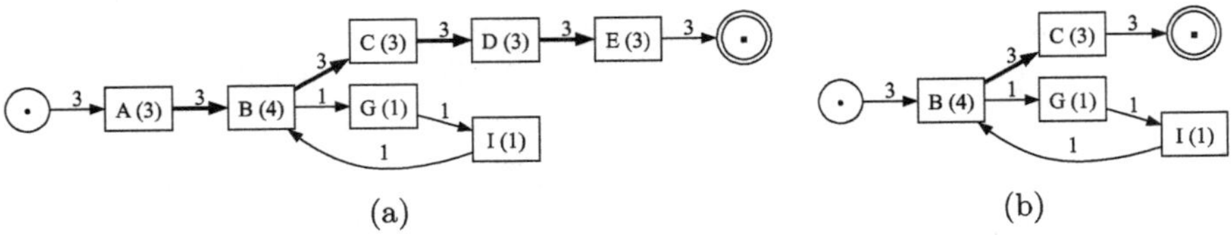

Fig. 2. Example of visual simplification of a DFG. (a) Full process view. (b) Process fragment between activities B and C.

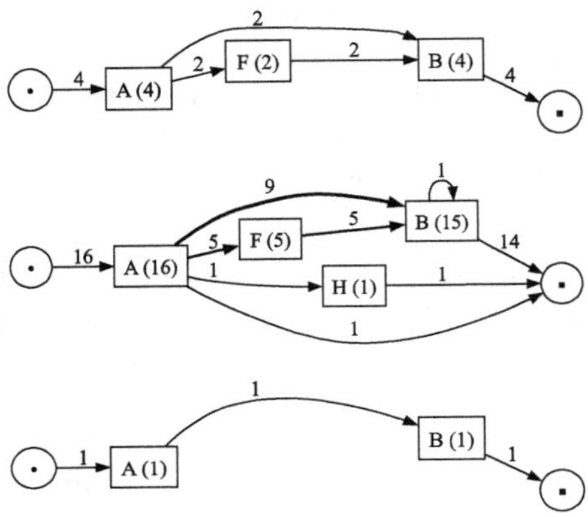

(a) Unified side-by-side layout of three DFGs.

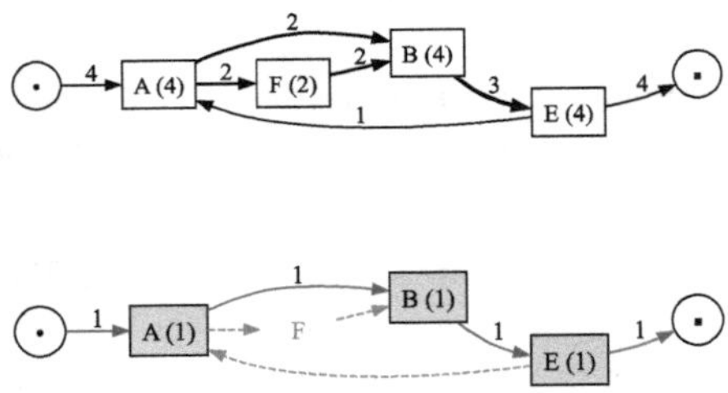

(b) Reference-based superposition of a DFG over a reference model. Elements highlighted in green indicate control-flow elements between them.

Fig. 3. Examples of visual comparative designs.

Unified Side-by-Side Layout. To support sequential scanning of an ordered collection of DFGs (C2), we adopt a *juxtaposition* design by arranging models in a vertical column with an uniform layout [12], i.e., we enforce consistent node and edge positions across all DFGs. To achieve this, we generate a union DFG that aggregates all unique activities and transitions across the collection, and compute a single layout on this unified graph. Each DFG then reuses these positions, displaying only the elements present in that model. This shared layout preserves spatial consistency, enabling efficient scanning and direct structural comparison while reducing cognitive effort [13]. Figure 3a illustrates this approach with a collection of three DFGs representing different process variants.

Reference-Based Superposition with Explicit Encodings. To complement the unified side-by-side layout and support analysis of similarities and differences (C2, C3), we introduce a reference-based *superposition* design with *explicit encodings*. Models are overlaid on a shared layout, preserving spatial coherence and users' mental maps [13].

A key element is the introduction of a *reference model* as a stable visual baseline. Users can select any DFG from the collection, or the union DFG, as the reference. Each model is compared against this reference, reducing visual complexity and avoiding multiple pairwise comparisons (C1, C2). The reference is rendered in the background with unboxed nodes and dashed edges to deemphasize it. A selected model is then superimposed using color encodings to highlight structural similarities and differences while preserving contextual information.

Explicit encodings further support goal-driven analysis and clarity (C1, C3) by allowing users to control which aspects are emphasized. The visualization adapts accordingly, maintaining focus and avoiding clutter. Figure 3b illustrates this design: the top graph shows the reference model, while the bottom graph overlays another DFG. Shared nodes and edges are highlighted in green, while elements only in the reference remain in the background.

Together, these visual designs provide the foundation for comparing multiple process models, combining shared layouts, reference-based comparison, and adaptive visual encodings within a unified framework.

6 Evaluation

To obtain exploratory feedback regarding the usability and applicability of our approach, we implemented it as a prototype integrated into VISCoPro [22], a tool that creates collections of DFGs from event logs. Using Streamlit[3], we extended the tool with the proposed strategies and visual designs, enabling analysts to visually identify structural similarities and differences across DFGs. The source code is publicly accessible in our repository[4].

6.1 Evaluation Method

We evaluated the prototype through two workshops: one with five domain experts with diverse levels of experience in process mining, and another one with five master's students who have prior knowledge in process mining. Four experts also participated in the interview phase, ensuring continuity between problem identification and solution evaluation. In line with common practices in visualization studies, involving three to five domain experts is usually sufficient to gather meaningful qualitative insights [15]. The workshop format was chosen because it enables active interaction between the moderator and participants, allowing misunderstandings to be clarified immediately, and promotes discussions that reflect real-world scenarios they typically face [27]. The goal of the workshops was not to quantitatively measure analytical performance or compare the approach against existing tools, but to gather exploratory qualitative feedback regarding the proposed visual comparison strategies and interaction mechanisms.

[3] https://streamlit.io/.
[4] https://doi.org/10.5281/zenodo.20393413.

Following the guidelines of Thoring et al. [27], the workshop was structured into three parts. Full details are provided as supplemental material in our repository. In the first part, we introduced our conceptual proposal for visually comparing multiple process models based on Gleicher's framework [12]. We then demonstrated how these proposed strategies and designs were implemented in the prototype and how to use them.

In the second part, participants used the tool to perform three tasks that required the comparison of multiple process models. They were encouraged to explore specific functionalities to assess their usability. For these tasks, we employed a simplified version of the Domestic Declarations event log from the 2020 Business Process Intelligence Challenge [9]. These tasks were designed to reflect realistic comparison scenarios and to elicit the use of key features of our approach, such as sequential scanning, subset selection, and reference-based comparison. For example, one task required participants to identify similarities in a specific part of the process across resources. They had to use the appropriate comparison action to highlight commonalities and fragment focusing to inspect that part of the process in detail. As participants worked through the tasks, they shared their screens, allowing the group to observe their visualizations and collaboratively discuss potential improvements, challenges, and suggestions. These exchanges clarified which features were most valuable and which caused difficulties. The DFGs obtained are available in our repository.

In the final part of the workshop, participants completed a structured questionnaire to assess the usefulness and ease of use of the approach. The questionnaire was based on the Technology Acceptance Model (TAM), a well-established framework for explaining and predicting the adoption of new technologies [8]. TAM focused on two key constructs: perceived usefulness and perceived ease of use. Participants rated each item on a 7-point Likert scale ranging from 1 (strongly disagree) to 7 (strongly agree), and responded to four open-ended questions about the strengths and weaknesses of our approach. Details about the evaluation are included in our public repository.

6.2 Evaluation Results

Overall, participants found the comparative features helpful. One participant noted, *"It's a good overview of seeing all the different models side by side to compare more than just two of them"*. The consistent layout was repeatedly highlighted as really beneficial, as it allows users to quickly identify corresponding flows across models. As one participant explained, *"if the position is fixed you can really just look, for example, at Start Trip and check the next model at the same place to see whether it is there or not"*. Most participants emphasized that this feature, combined with color encoding, allows users to make quick visual comparisons, even in complex scenarios.

Participants responded positively to the option of using a reference DFG and displaying it in the background of other process models. They noted that the reduced visual presence of the reference elements, which are rendered in

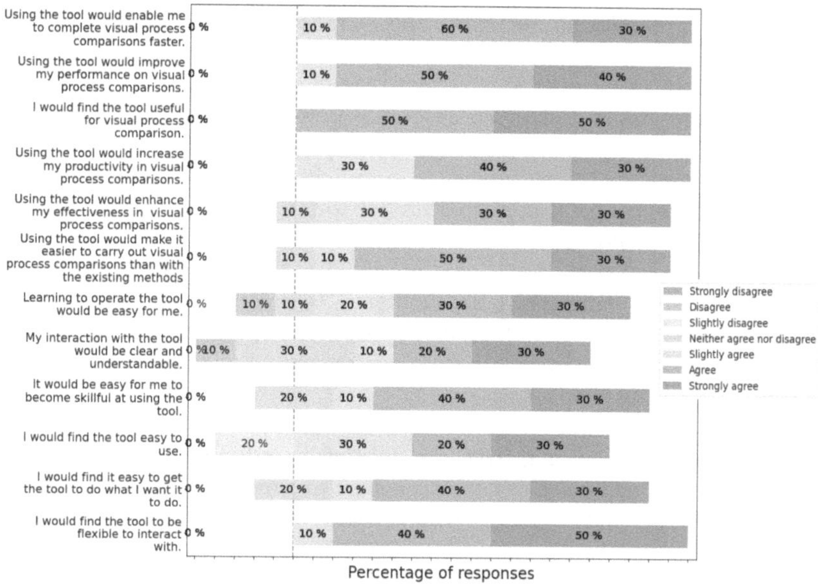

Fig. 4. TAM results for the comparative features. The first six items measure perceived usefulness and the following six perceived ease of use. The vertical dashed line marks the neutral point (4) on the 7-point Likert scale.

the background, helps maintain context without overwhelming the main visualization. However, some participants pointed out that this approach could be harder to interpret with larger or more complex event logs. They suggested that it would be useful to have an option to show or hide the background reference model. Four of the five participants in one workshop suggested this.

The fragment focusing feature was highlighted as particularly valuable when working with larger or more complex processes. Participants noted that comparing entire processes can become impractical when many activities are involved. In these cases, it is beneficial to restrict the analysis to a specific subset of the control flow. As one participant explained, *"usually the processes are really huge with a lot of activities, so just focusing on a small subset is quite good"*. At the same time, participants acknowledged that selecting a fragment may sometimes hide relevant steps that occur just outside of it. However, they generally agreed that this limitation is not specific to visualization.

The activity-hiding feature was considered useful, particularly for exploratory analyses. Participants noted that many activities in real-world event logs may be irrelevant to a specific analysis. In such cases, manually hiding activities allows analysts to simplify the visualization temporarily and focus on the main process while still having the option to restore the hidden elements later.

To assess the perceived usability and ease of use of the comparative features, we analyzed responses to the TAM questionnaire (cf. Figure 4). The results suggest positive perceptions regarding the proposed comparison strategies and designs. The average perceived usefulness was 6.13 (median 6.1), well above the neutral midpoint of the scale, showing that participants perceived the tool as useful for comparative tasks. Perceived ease of use also scored above neutral (mean = 5.68, median = 5.65), suggesting that, while the system is generally easy to use, there remains room for improvement in simplicity and interaction flow. When comparing the two groups of participants, we found that the experts reported a slightly higher perceived usefulness of the system (mean = 6.23, median = 6.3) than the students (mean = 6.03, median = 6.0). However, students rated the system as easier to use (mean = 6.27, median = 6.3) compared to experts (mean = 5.10, median = 4.9). This difference suggests that prior experience with process mining tools may influence expectations regarding usability and interaction complexity.

The responses to the open-ended questions reinforced these findings. Participants praised the tool's comparative capabilities; one described it as the *"fastest way to detect similarities/differences ever seen in process mining"*, while others appreciated the *"fixed positions"* and the use of the reference model. They also suggested enhancements such as adding *"patterns for performance metrics like time between Activity A and B has to be X or lower than..."* to support bottleneck detection and root-cause analysis. Regarding the activity-hiding feature, participants suggested adding *"an overlay of the full process"* to preserve context when activities are hidden, and more control over visual elements, such as hiding or showing reference elements on demand. Finally, participants pointed out limitations related to scalability and visual complexity, noting that *"there could be too much information... if too many activities are involved"* and that comparisons may become *"overcrowded"* for larger datasets. This reinforces the need for future work on abstraction, filtering and scalable visualization techniques.

7 Discussion

In this section, we outline design implications for visual process comparison tools, based on the identified challenges and evaluation feedback, and discuss limitations of our approach.

Supporting Adaptability and Goal-Driven Comparison (C1). Our findings suggest that the process of visual comparison is improved by making analytical goals explicit. Interviews revealed that experts vary between exploratory and hypothesis-driven analyses, often requiring flexible comparison strategies to obtain insights, since current tools do not always align with their analytical goals. Feedback from the workshops further highlighted that, even when using actions to guide comparisons and explicit encodings to emphasise particular aspects, some participants suggested enabling or disabling the display of reference elements in the background. Providing users with greater control can enhance interpretability and align comparisons with their analytical objectives, though it may require additional guidance, especially for less experienced analysts.

Supporting Scalable Exploration of Multiple Models (C2). Our interviews revealed that analysts frequently need to compare more than two process models at a time. This quickly becomes challenging due to increased visual complexity, cognitive load, and limitations in existing tools. Workshop feedback highlighted the importance of mechanisms that reduce visual complexity, such as consistent layouts, activity hiding or fragment focusing. In particular, participants repeatedly emphasized the value of consistent node and edge positions across DFGs. While recent work suggests that preserving the mental map in DFGs may not improve analysis performance in dynamic scenarios [14], our setting differs: we focus on static comparison across multiple variants. In this context, layout consistency plays a different cognitive role, helping alignment attention across models and enhancing the effectiveness of side-by-side comparison. Participants also stressed that effective multi-model comparison requires not only displaying multiple models, but enabling users to progressively narrow their focus while maintaining spatial consistency across views. Activity hiding and fragment focusing proved particularly useful for this purpose. However, our evaluation did not directly compare alternative scalability strategies, leaving their effectiveness in larger, more heterogeneous real-world collections as future work.

Supporting Comparison of Similarities and Differences (C3). Our findings emphasise the importance of flexibly balancing the comparison of similarities and differences in visual process analysis. Workshop participants valued mechanisms that make these aspects explicitly visible. In particular, the use of a reference model combined with color-based encodings was perceived as helpful for highlighting commonalities and deviations while preserving contextual information. Participants also indicated that providing specific comparison actions that directly enable the corresponding visual encodings is beneficial, allowing users to highlight either similarities or differences depending on their analytical goal.

Limitations and Future Directions. The findings should be interpreted within the scope of the present study. First, the evaluation focused on an exploratory assessment of the proposed approach through workshops with a small but diverse group of participants. While this allowed us to gather rich feedback on perceived usefulness, ease of use, and feature relevance, future studies could complement these findings with longitudinal investigations of its use in real-world settings and assessments of analytical performance. Second, the current work focuses on DFGs, which provide an intuitive and widely used abstraction of process behavior, making them a suitable starting point for investigating visual comparison mechanisms. However, they do not capture richer behavioral semantics such as concurrency or non-free-choice behavior. Exploring the applicability of the proposed approach to richer representations, including BPMN models and Petri nets, offers a promising direction for future research. Finally, the evaluation was designed to understand user perceptions and requirements rather than to benchmark the approach against alternative visualization techniques. Future work could therefore incorporate controlled comparisons and additional analytical measures, such as task completion time, and analytical accuracy, to provide a more comprehensive assessment of the approach.

8 Conclusions

This paper addresses the need for effective visual comparison of multiple process models, a central task in process mining practice. Combining expert interviews, analysis of existing approaches, and comparative visualization principles, we identified key challenges related to adaptability, scalability, and the comparison of similarities and differences. To address these challenges, we propose a visual approach that combines structured exploration of multiple DFGs with visual simplification mechanisms (activity hiding, fragment focusing, and annotation selection), enabling users to focus on relevant aspects of the comparison. In addition, we introduce a unified side-by-side layout and a reference-based superposition with explicit encodings to support the identification of similarities and differences across models. We implemented the approach in a prototype and evaluated it in two workshops. The results show that goal-driven comparison actions, consistent layouts, and simplification mechanisms are key for supporting effective multi-model analysis, improving interpretability and user control in practice. These findings provide design insights for interactive systems that support flexible, scalable, and goal-driven visual comparison of process models.

Future work will extend the evaluation to larger and more diverse user groups, include controlled comparative studies, and investigate the applicability of the proposed design implications across different process representations. We also plan to develop automated reference model selection strategies and extend the comparison mechanisms to additional perspectives, such as temporal performance and resources.

Acknowledgment. This paper is part of the projects PID2021-126227NB-C21, PID 2022-140221NB-I00 and PID2024-156482NB-I00 funded by MICIU/AEI/ 10.13039/501100011033 and by ERDF/EU, and by the Junta de Andalucía (Consejería de Universidad, Investigación e Innovación) and co-funded by the EU through the FEDER Andalucía 20212027 Programme under project DGP_PIDI_2024_01609. M. Salas-Urbano is supported by PREP2022-000372 financed by MICIU/AEI/10.13039/501100011033 and ESF+.

References

1. van der Aalst, W.M.P.: Process Mining: Data Science in Action. Springer, Heidelberg (2016)
2. van der Aalst, W.M., de Medeiros, A.K.A., Weijters, A.J.M.M.: Process equivalence: comparing two process models based on observed behavior. In: BPM, pp. 129–144 (2006)

3. Ballambettu, N.P., Suresh, M.A., Bose, R.P.J.C.: Analyzing process variants to understand differences in key performance indices. In: CAiSE, pp. 298–313 (2017)
4. van Beest, N.R.T.P., Dumas, M., García-Bañuelos, L., Rosa, M.L.: Log delta analysis: interpretable differencing of business process event logs. In: BPM, pp. 386–405 (2015)
5. Bolt, A., de Leoni, M., van der Aalst, W.M.: Process variant comparison: using event logs to detect differences in behavior and business rules. Inf. Syst. **74**, 53–66 (2018)
6. Cao, B., et al.: ProDiff: a process difference detection method based on hierarchical decomposition. IEEE Trans. Serv. Comput. **15**(1), 513–526 (2022)
7. Capitán-Agudo, C., Salas-Urbano, M., Cabanillas, C., Resinas, M.: Analyzing how process mining reports answer time performance questions. In: BPM, pp, 234–250 (2022)
8. Davis, F.D.: Perceived usefulness, perceived ease of use, and user acceptance of information technology. MIS Q. **13**(3), 319–340 (1989)
9. van Dongen, B.: BPI Challenge 2020. 4TU.ResearchData (2020). https://doi.org/10.4121/uuid:52fb97d4-4588-43c9-9d04-3604d4613b51
10. van Dongen, B., Dijkman, R., Mendling, J.: Measuring similarity between business process models. In: Seminal Contributions to Information Systems Engineering, pp. 405–419. Springer, Heidelberg (2013). https://doi.org/10.1007/978-3-642-36926-1_33
11. van den Elzen, S., et al.: Towards multi-faceted visual process analytics. Inf. Syst. **133**, 102560 (2025)
12. Gleicher, M.: Considerations for visualizing comparison. IEEE Trans. Vis. Comput. Graph. **24**(1), 413–423 (2017)
13. Gleicher, M., Albers, D., Walker, R., Jusufi, I., Hansen, C.D., Roberts, J.C.: Visual comparison for information visualization. InfoVis **10**(4), 289–309 (2011)
14. Häge, M.C., Rehse, J.R.: Mental maps in process mining: does stabilizing dfgs improve process analysis performance? In: BPM, vol. 16044, pp. 523–540 (2025)
15. Isenberg, T., Isenberg, P., Chen, J., Sedlmair, M., Möller, T.: A systematic review on the practice of evaluating visualization. IEEE Trans. Visual Comput. Graph. **19**(12), 2818–2827 (2013)
16. Jalali, A., et al.: dfgcompare: a library to support process variant analysis through Markov models. BMC Med. Inf. Decis. Mak. **21**(1), 356 (2021)
17. von Landesberger, T.: Insights by visual comparison: the state and challenges. CG&A **38**(3), 140–148 (2018)
18. Leemans, S.J., Poppe, E., Wynn, M.T.: Directly follows-based process mining: exploration & a case study. In: ICPM, pp. 25–32 (2019)
19. Miksch, S., Di Ciccio, C., Soffer, P., Weber, B.: Visual analytics meets process mining: challenges and opportunities. IEEE Comput. Graph. Appl. **44**(6), 132–141 (2025)
20. Nguyen, H., Dumas, M., Rosa, M.L., ter Hofstede, A.H.M.: Multi-perspective comparison of business process variants based on event logs. In: ER, pp. 449–459 (2018)
21. Salas-Urbano, M., Capitán-Agudo, C., Cabanillas, C., Resinas, M.: Designing a user interface to explore collections of directly-follows graphs for process mining analysis. In: BPMDS, pp, 35–47 (2024)
22. Salas-Urbano, M., Capitán-Agudo, C., Cabanillas, C., Resinas, M.: Interactive creation, visualization, and exploration of process model collections. Softw. Syst. Model. 1–26 (2026)

23. Salas-Urbano, M., Capitán-Agudo, C., Cabanillas, C., Resinas, M.: LoVizQL: a query language for visualizing and analyzing business processes from event logs. In: ICSOC, pp. 13–28 (2023)
24. Sedlmair, M., Meyer, M., Munzner, T.: Design study methodology: reflections from the trenches and the stacks. IEEE Trans. Visual Comput. Graph. **18**(12), 2431–2440 (2012)
25. Sorokina, E., Soffer, P., Hadar, I., Leron, U., Zerbato, F., Weber, B.: PEM4PPM: a cognitive perspective on the process of process mining. In: International Conference on Business Process Management, pp. 465–481. Springer, Heidelberg (2023). https://doi.org/10.1007/978-3-031-41620-0_27
26. Syamsiyah, A., et al.: Business process comparison: a methodology and case study. In: BIS, pp. 253–267 (2017)
27. Thoring, K., Mueller, R., Badke-Schaub, P.: Workshops as a research method: guidelines for designing and evaluating artifacts through workshops. In: Proceedings of the 53rd Annual Hawaii International Conference on System Sciences (HICSS 2020) (2020)
28. van der Aalst, W.M.: A practitioner's guide to process mining: limitations of the directly-follows graph. Procedia Comput. Sci. **164**, 321–328 (2019). https://doi.org/10.1016/j.procs.2019.12.189
29. Wang, J., Cao, B., Zheng, X., Tan, D., Fan, J.: Detecting difference between process models using edge network. IEEE Access **7**, 142916–142925 (2019)
30. Zerbato, F., Soffer, P., Weber, B.: Process mining practices: evidence from interviews. In: BPM, pp, 268–285 (2022)

Semantically-Guided LLM Pipeline for Business Architecture Modelling with Incremental Knowledge Acquisition

Iromie Samarasekara[1](✉), Madhushi Bandara[2], Fethi Rabhi[1], and Boualem Benatallah[3]

[1] University of New South Wales, Sydney, Australia
{i.samarasekara,f.rabhi}@unsw.edu.au
[2] University of Technology Sydney, Sydney, Australia
Madhushi.Bandara@uts.edu.au
[3] Dublin City University, Dublin, Ireland
boualem.benatallah@dcu.ie

Abstract. Business Architecture (BA) modelling is a resource-intensive exercise, particularly for abstract concepts such as business capabilities that typically requires iterative workshops that are time-consuming and cognitively demanding for experts. Large Language Models (LLMs) have emerged as a promising solution for automating BA modelling. However, current methods often fail to capture task context, producing semantically invalid models. Although LLMs can be pre-trained or fine-tuned to encode such context, it is costly and difficult to sustain. This paper proposes an LLM pipeline for business capability modeling that integrates rich contextual knowledge by combining a BA semantic model with organisational insights extracted from unstructured business documents and incrementally acquired expert knowledge. We evaluate the approach on three state-of-the-art LLMs by processing 495 statements extracted from business documents, and compare its performance against the baselines combine standard definitions and in-context learning. Our method reports an average performance gain of 65.2% on a commercial LLM and 58.9%–63.8% on two open-source LLMs over the baselines. This work advances the automation of LLM based BA modelling by providing an effective and efficient mechanism for incorporating existing documents and expert feedback into the automation process with minimal expert involvement.

Keywords: LLM · Business Architecture Modelling · Business Capability Maps · Semantics · Knowledge Acquisition

1 Introduction

Business Architecture (BA) modelling helps businesses in establishing a common vocabulary and serves as a framework for defining the scope of the process modelling during the process identification stage [11]. Business capabilities are a core

F. M. Maggi et al. (Eds.): BPM 2026, LNBIP 599, pp. 273–289, 2027.
https://doi.org/10.1007/978-3-032-37877-4_16

element of BA and defined as the ability or capacity that a business may possess or exchange to achieve a specific purpose or outcome [29]. For example, in a manufacturing company, Procurement Management is a high-level (Level 1) business capability that can be further decomposed into a hierarchy of sub-capabilities such as Vendor Management (Level 2) and Vendor Information Management (Level 3).

Capability modelling is a knowledge-intensive exercise that requires a substantial amount of contextual knowledge associated with modelling. This context mainly includes; 1) conceptual knowledge on capabilities such as modelling conventions, semantics and best practices, and 2)organisation-specific knowledge (e.g. local vocabulary, products, processes, and operating constraints) [15,28]. Business architects capture this knowledge from a range of organisational documents such as strategy booklets and corporate plans. It is also followed by multiple iterations of interviews and reviews with organisational stakeholders [29]. Moreover, the abstract nature of business capabilities in contrast to other BA constructs makes the modelling process more challenging [28]. This makes automated tool support a timely and critical need [3].

Large Language Models (LLMs) have been widely adopted for conceptual modelling tasks [26]. However, the current body of literature has not explored how knowledge from unstructured organisational documents and experts can be effectively integrated with an LLM to model abstract concepts such as business capabilities. Previous studies highlight that LLMs find it difficult to grasp the full context of a modelling task due to the lack of this knowledge [13]. While the internalised knowledge of an LLM can be refined through pre-training or fine-tuning, such approaches are often costly and require large volumes of high-quality data [15,32]. Therefore, existing methods mainly use prompts to provide basic context with the help of standard definitions and summaries coupled with few-shot examples [2,14,17,18,25]. Although these methods provide an LLM with a basic conceptual understanding, they do not capture the fine-grained cues and semantics required by the LLM to perform the modelling task [32]. Such knowledge is internalised in the minds of experts [26]. Existing studies leverage techniques such as Retrieval-Augmented Generation (RAG) to extract business process-related entities from textual process descriptions and event logs [2,17]. In contrast, the knowledge required to model abstract concepts such as business capabilities is seldom available in such explicit, structured sources and is instead embedded within heterogeneous strategic and operational documentation. While a few studies investigate modelling capabilities with LLMs, they do not incorporate knowledge from organisational documents and domain experts during the process [22,23].

To address these gaps we pose the following research question: **"How can we incorporate rich contextual knowledge sourced from unstructured text documents and domain experts for LLM-driven capability modelling?"**. Based on existing literature, we have identified four key challenges we should overcome in answering this question:

C1. Organisational documents are not structured to reflect business capability related concepts: Some modelling tasks, such as process modelling, are more amenable to automation because business documents often contain explicit cues (e.g., roles, tasks, and activities) [14]. In contrast, while organisations commonly maintain documents that describe their operational practices, such narratives are not organised around business capability-related concepts [10]. For example, a company's strategy booklet typically outlines high-level goals, objectives, and initiatives that may relate to capabilities, but it does not explicitly articulate business capabilities. Previous studies also point out that LLMs struggle with extracting concepts with such relational complexity despite excelling at understanding specific concepts such as tasks and events [30].

C2. Mismatching level of granularity in the documents: The text documents commonly available in businesses contain varying levels of granularity that often do not align with the needs of capability modelling tasks [10]. For instance, a strategy booklet draws an abstract picture of a company, while task manuals or work profiles are often too detailed for modelling BA constructs at a conceptual level [10]. This is challenging for LLMs, as LLMs often tend to extract information at the abstraction level provided in the input text and do not have the ability reason at a different level of abstraction [33]. For this reason literature often points out the need to interpret and refine the information in these documents before feeding into LLMs for BA modelling [15].

C3. Abstracting the expert knowledge expressed at case level: Modelling business capabilities requires thinking at an abstract level which is challenging even for many expert practitioners [28]. These experts typically think on a case level and often describe the tasks conducted for specific instances [10] . When asked to generalise their expertise they provide answers like "You cannot really generalise, every case is different" [10]. While previous studies point out the need to facilitate effective expert involvement in LLM-driven BA modelling pipelines, how this expert knowledge at case level can be incorporated is an open challenge [15].

C4. Limited semantic validity in automatically produced models: Previous work on LLM-driven BA modelling highlight that the resulting LLM outputs often do not align with the true definitions, relationships and rules associated with BA concepts [13]. Studies also claim that LLMs cannot effectively manage semantics as they are built to make statistical predictions [26], highlighting the need for ensuring semantic validity of produced models.

To address these challenges, we propose a novel LLM pipeline for capability modelling, that combines semantic grounding with incrementally acquired expert knowledge to effectively incorporate context from organisational documents. The pipeline enables LLMs to utilise a semantic model of BA concepts to acquire the most relevant contextual knowledge from unstructured organisational documents addressing C1. Natural language triples extracted from the semantic model are used to map the information extracted from documents to the level of abstraction required for the modelling task, thereby addressing C2. The pipeline also acquires expert knowledge incrementally at case level and preserve as an expert knowledge base, where each rule is a simple IF-THEN statement aimed at rectifying any undesired output given by the LLM, addressing C3. The generated capabilities are further validated with a set of explicit evaluation criteria derived by the LLM from the same semantic model, addressing C4 and ensuring semantic validity of results. We evaluate the effectiveness of the proposed pipeline using real-world organisational artefacts on three LLMs; Llama 3.3-70B and Mistral-Large (open-source) and GPT-4 (commercial). Our key findings are:

- The proposed pipeline is capable of providing the context required for modelling capabilities by effectively leveraging organisational documents and expert knowledge.
- The approach outperforms baseline methods that represent the current practice by reporting average performance gains of around 63% for information extraction and 62% for capability generation tasks respectively across three LLMs.
- The pipeline effectively navigates through the conceptual complexity of capability modelling with minimum expert intervention, reducing the end-to-end modelling latency to seconds compared to an estimated 8–10 person-days for traditional BA modelling,.
- The proposed pipeline matches the performance of state-of-the-art commercial LLMs (GPT-4) while operating on substantially smaller open-source models, namely Llama-3.3-70B and Mistral-Large.

The paper is structured as follows. Section 2 provides an overview of the background and literature. Section 3 elaborates the design of the pipeline followed by the demonstration and evaluation in Sect. 4. Section 5 presents discussion on design implications and study limitations and our paper concludes in Sect. 6.

2 Related Work

This section reviews prior studies in three key areas related to our work, highlighting existing gaps and contrasting them with our approach.

2.1 LLM Applications in BA Modelling

The growth of LLMs has resulted in a steady advancement in the area of conceptual modelling [12,26]. However, literature has predominantly focused on process modelling, with comparatively less attention given to capability modelling [22,23]. Several studies employ techniques such as RAG has been employed in many studies to extract information from external sources [2,17,18], such as textual process descriptions and event logs. Yet there is no guidance on how to leverage unstructured documents with implicitly embedded knowledge for tasks such as capability modelling. Prior work further notes that training LLMs to a specific organisational context is often impractical, as business operations evolve over time, necessitating continuous retraining [15]. While fine-tuning LLMs using labelled data can be done for process modelling benefiting from the availability of process models held by enterprise system vendors [15], such approaches rely solely on correlations of data. As a result, they fail to capture the conceptual knowledge and semantics internalised by domain experts, which are essential for modelling more abstract BA constructs such as business capabilities [26]. These gaps highlight the need for an approach that effectively uncover implicitly expressed knowledge found in unstructured organisational documents, enabling the modelling of abstract concepts such as business capabilities.

2.2 Context Adaptation for Domain-Specific LLM Applications

LLM based agents and domain-specific reasoning increasingly rely on context adaptation or context engineering and this involves improving model behavior by constructing inputs to an LLM, instead of modifying its weights [32]. State-of-the-art literature in context adaptation incorporate natural language feedback [1,24]. This includes techniques such as failure reflection [24], optimisation through gradient-like textual feedback [31], refinement based on execution traces [1] and external memory construction [27]. However, in these methods, context is typically derived from execution traces and conversational history, and the feedback used to adapt. In contrast, ur goal is to provide rich context grounded on unstructured documents and expert feedback. In-context learning (ICL) editing and memory based editing are also techniques that update the inherent knowledge in LLMs [9]. Notable limitations of these techniques include their inability to capture rules valid only under specific conditions and their dependence on large data volumes [9].

2.3 Semantics and Structured Knowledge for LLMs

Limitations of existing context adaptation methods discussed in Sect. 2.2 motivate us to explore the use of structured external knowledge such as Knowledge Graphs (KGs). KGs supply stable domain semantics and context beyond what is captured in raw data or iterative feedback for conceptual modelling [4,20]. KGs use the triple structure (Subject-Predicate-Object) to explicitly capture relationships between different entities. Literature successfully uses triples extracted from KGs to train components of the LLM model's architecture and to enrich prompts with textual descriptions of entities in the KG for tasks such as entity disambiguation and question answering [8,20]. In contrast, the proposed approach uses triples to explicitly guide the LLM to infer accurate concepts from related concepts and to validate generated outputs against the intended semantics. Although several BA domain ontologies and semantic models exist at varying levels of expressivity [5,16,21], their integration with LLM-based modeling approaches has not been systematically investigated. A key barrier for the wide adoption of semnatic models with LLMs is that the development and maintenance of comprehensive domain ontologies (e.g. organisation specific KG) are known to be effort-intensive [19]. In contrast, we propose incremental knowledge acquisition from domain experts as a more practical alternative, reducing need for comprehensive domain knowledge representation from the beginning of the modelling process [7].

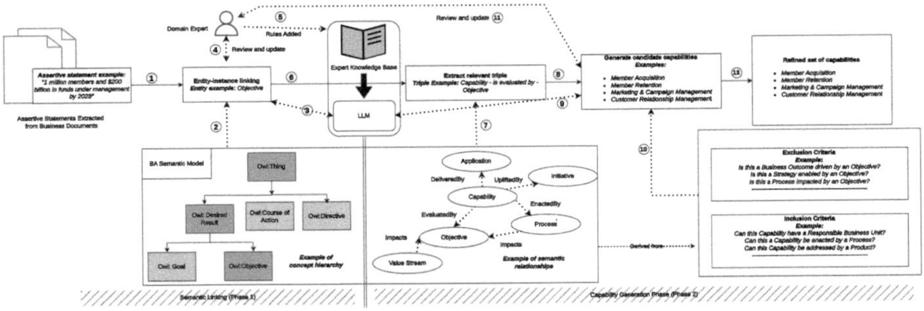

Fig. 1. LLM pipeline with numbers indicating the interaction flow. An end-to-end example of capability generation is included in italic text.

3 Methodology

This paper proposes a novel LLM pipeline that effectively incorporates domain knowledge by leveraging semantics and incremental expert knowledge acquisition. Its components are illustrated with an end-to-end example in Fig. 1. It contains two phases: 1) semantic linking and 2) capability generation as detailed in following sections.

3.1 Phase 1 - Semantic Linking

The pipeline initiates with unstructured text from strategic organisational documents such as strategy booklets and corporate plans. Our approach assumes the existence of a BA ontology or a semantic model that formally represents BA concepts and their semantic relationships, as many contemporary BA modelling tools are grounded in semantic models [6]. The first phase aims to identify and link the information in the document texts to key BA concepts (i.e. class or type) in the semantic model such as goals and objectives. The LLM is supplied with both the input text and a semantic model capturing the hierarchical class structure. It is guided to navigate the hierarchy, incrementally pruning candidate entities until the most semantically aligned entity is identified, along with an explicit justification for the selection. We adapt the hierarchical pruning approach [20] here for its ability to resolve disambiguation in semantically dense domains. For instance, as illustrated in Fig. 1, the semantic hierarchy maps the input text to the *Objective* class via *Desired Result*, while pruning semantically related alternatives such as the *Tactic* class under *Course of Action*. As discussed in C1, organisational documents do not typically follow structures that explicitly represent BA concepts. By leveraging semantic guidance, the proposed approach directs the LLM to identify information that semantically corresponds to BA concepts, thereby filtering out irrelevant content and overcoming this challenge.

While previous studies point out that the LLM performance directly depends on the semantic expressivity of the ontology [20], constructing such a comprehensive ontology is an effort intensive task. To address these limitations of prior methods we incorporate a method to incrementally acquire domain knowledge experts during the process. Experts review the entity-instance links given by the LLM, identify cases that are incorrect, and add simple IF-THEN rules to rectify them. This helps the LLM in handling ambiguous cases where the required knowledge does not exist within the boundaries of the semantic model.

To overcome the need for comprehensive ontologies with high semantic expressivity, we introduce an incremental knowledge acquisition approach that integrates domain expertise into the pipeline. Domain experts review the entity-instance mappings produced by the LLM, identify incorrect associations, and encode corrective knowledge using lightweight IF-THEN rules, providing necessary domain knowledge for the LLM to resolve ambiguous cases not explicitly covered within the semantic model. An expert rule consists of 1) contextual metadata such as unique identifiers and the corresponding classes in the semantic model to which the rule relates to and 2) expert rules that capture reusable strategies, domain concepts or common failure modes in IF-THEN format. In addition to the rules, the cases that prompted the rules are also stored in this expert knowledge base, so that the LLM can leverage these as examples in subsequent iterations. An example rule in Phase 1 is: 'IF a statement outlines the ethical boundaries guiding a course of action, THEN it represents a Directive.' These rules are organised as a structured expert knowledge base rather than a single monolithic prompt so the LLM can consume the rules relevant to respective entities in future tasks.

This design addresses C3 by providing an efficient way to capture case level domain knowledge from experts.

3.2 Phase 2 - Business Capability Generation

This phase uses the entity-instance pairs from Phase 1 and retrieve associated relationships between the identified entity and a capability in the semantic model. For instance, *Capability* is the desired concept and *Objective* is linked to *Capability* through *evaluatedBy* relationship (i.e. capability → evaluatedBy → objective). The LLM leverages these relationships to derive semantically valid capabilities from the input text.

This semantic triple-based generation addresses C2 and C3 by guiding the LLM to understand how other higher-level strategic concepts are relate to other BA constructs.

To address the C4, the need for semantic validity, we introduce validation through a set of inclusion and exclusion criteria derived from the relationships defined in the semantic model. Criteria are generated automatically using the same LLM using predefined templates available at our GitHub repository [1]. The exclusion criteria aim to rule out LLM misclassifications and ensure generated artefact is a business capability and DOES NOT represent different closely related class. For example, the exclusion criteria in Fig. 1 are derived from triples related to *Objective* (e.g.: *Objective*$-$ > *isimpactedby*$-$ > *Process*) and test if the LLM has misinterpreted other concepts related to *Objectives* as business capabilities. Similarly, inclusion criteria are derived from associated outgoing relationships of the Capability concept to verify that the candidate instance adheres to its class characteristics.

Similar to Phase 1, expert review is conducted and the expert knowledge is iteratively accumulated as rules to enhance the capability generation process. An example of expert rule in this phase is; 'IF a concept is not grounded in a business or information object, THEN it cannot be classified as a capability.'

4 Demonstration and Evaluation

4.1 Experiment Setup

For demonstration and evaluation of the proposed pipeline, we implemented it using the CAPSICUM semantic model [21] and the associated commercial BA modelling tool [6]. The reason for choosing CAPSICUM semantic model over other existing ontologies (e.g. Creon [5]) is that CAPSICUM provides explicit semantic specification of how business capabilities relate to other core BA concepts within a particular organisation (e.g., processes, services/products, applications etc.). This suits well for capability modelling task at hand, as the available organisational documents predominantly present capability-relevant evidence through their descriptions of adjacent BA concepts.

We used three language models: GPT-4, Mistral-Large, and Llama 3.3-70B in our experiments. We used GPT-4 to initiate our experiments and populate the knowledge base with expert rules. Then a follow up evaluation of the complete pipeline with and without the rules generated with GPT-4 was conducted on Mistral-Large and Llama 3.3-70B. Five business architects with 7 to 10 years of experience participated in all our experiments as domain experts, to validate LLM outcomes and add new rules to the expert knowledge base (Ethics approval: iRECS7566).

4.2 Experiment Data

As our experiment dataset, we used a set of 495 statements annotated by five experts (with inter-annotator agreement) that focus on specific BA concepts extracted from 16 business (D1–D16) documents. These documents were used in the same order across all our experiments as determined by BA experts in decreasing complexity and ambiguity. The documents used in our experiments and results have been provided in Github.

[1] https://github.com/LLM-drivenBAModelling/BA-Modelling-Pipeline.git.

4.3 Evaluation of Phase 1 Performance

Table 1 presents our experiment results for all 495 statements grouped under the 16 documents (D1-D16). Second raw indicates the number of statements sourced from each document. For experiment purposes, we only considered statements that map to a specific BA concept. The first row in both phases reports the number of rules added after GPT-4 processed each document. The performance of Phase 1 is recorded on the top half of Table 1 and measured as the error rate of the semantic linking across all models compared against the manual expert annotations.

When reviewing the first 284 statements (D1-D8) and the associated GPT-4 outputs, experts reviewed and added 30 rules to the expert knowledge base, and no further rules were added for the subsequent 211 statements. As summarised in Fig. 2, the number of rules added as well as the semantic link error rate decreased rapidly over 320 statements and the error rate was near-zero for the last 131 statements.

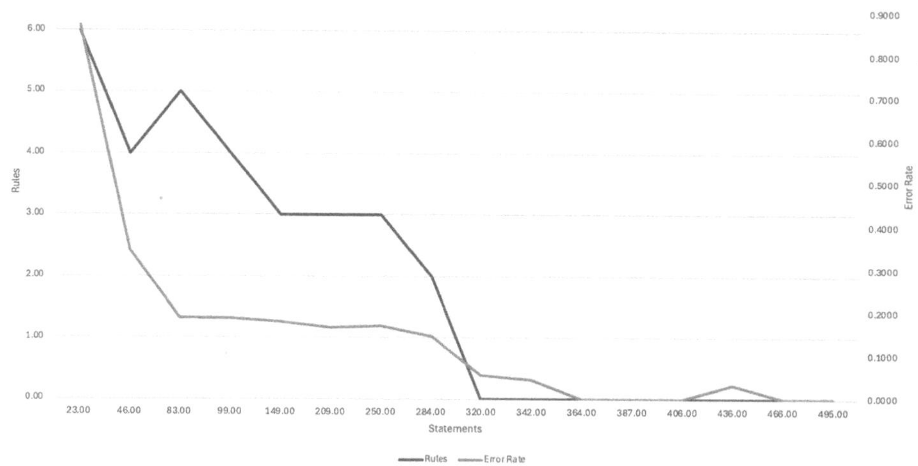

Fig. 2. Decreasing number of rules included and the error rate against the cumulative number of statements processed by GPT-4 during the knowledge base construction process.

The error rate reported for both Llama and Mistral without and with the GPT-4 generated expert knowledge base (Table 1 Phase 1 last 4 rows) also indicates a significant improvement of performance from D1 when expert defined rules are incorporated into semantic linking, compared to their baseline "wihtout rules" in Table 1 where only hierarchical pruning was used. For example, without rules, all Phase 1 outputs for D2 and D11 were incorrect, and by introducing rules to the LLM pipeline the error rates were drastically reduced to 26% and 19% respectively. The lowest error rate reported (best performance) without rules in Llama is 50% in D4, and with rules the error rate reduced to 18.7%. Similarly with Mistral, no correct output was generated for D2, D9, D10, and D12 (100% error rate). By introducing rules, it was reduced to 26%, 14%, 35% and 16% respectively. The best performance (lowest error rate) reported without rules for Mistral is 47.83% for D1, and was reduced to 13% with rules.

Another interesting observation was that the error rates reported by Mistral and Llama are significantly lower than GPT during its rule building phase (D1-D8). This indicates that a smaller open-source model when supplemented with context acquired incrementally from expert as rules, can outperform a commercial model trained on a significantly larger corpus of data.

4.4 Evaluation of Phase 2 Performance

To quantify the acceptance rate of the LLM generated business capabilities in Phase 2, we conducted semi-structured interviews with experts. The performance of this phase is reported as acceptance rate (precision): number of LLM generated capabilities accepted by the experts over the total number of LLM generated capabilities. The comparison baseline for Phase 2 is the current state-of-art: directly prompting the LLM to extract capabilities from documents, providing the standard definition of business capability and few-shot examples and reported in "direct" column in Table 1 bottom half.

As shown in bottom half of the Table 1 under Phase 2, five rules were added to the expert knowledge base while processing the first five documents. D6 reported the best results for GPT-4 with a 94.44% acceptance rate. We then compared the performance of the pipeline using Llama and Mistral, evaluating their effectiveness when supported by both semantic grounding and the GPT-generated rules against baseline performance. The baseline method resulted in eight documents with 0% acceptance and a maximum acceptance rate of only 50% for Llama, while proposed approach had much higher acceptance rates ranging from 75% to 100%. Similarly, for Mistral, the baseline approach produced six documents with 0% acceptance and the maximum reported acceptance rate was 36.36%, while the proposed approach with rules achieved acceptance rates between 66.67% and 100%. Furthermore, the smaller open-source models have shown competitive performance with commercial model GPT in Phase 2.

4.5 Ablation Study

To further understand the effectiveness of the proposed design, we individually assessed the effectiveness of incrementally acquired expert knowledge base and semantic model guided generation and validation. We limited this evaluation to two representative LLMs; one commercial (GPT-4) and one open-source (Mistral-Large) to keep the expert effort manageable. Table 2 presents the results of our ablation study. The average performance improvement achieved through semantic model is 35% and 36% for GPT and Mistral respectively against their corresponding base model. Incrementally acquired expert rules have resulted

Table 1. Evaluation results across three LLMs organised for 16 documents: Error rate of Phase 1 is reported in top and acceptance rate for Phase 2 reported at the bottom. ('–' indicates there was no Phase 2 results for corresponding document as no entities associated to capability were identified in Phase 1)

	D1	D2	D3	D4	D5	D6	D7	D8	D9	D10	D11	D12	D13	D14	D15	D16
Number of statements	23	23	37	16	50	60	41	34	36	22	22	23	19	30	30	29
Phase 1 – Semantic Linking																
Rules added - GPT	6	4	5	4	3	3	3	2	0	0	0	0	0	0	0	0
Error rate - GPT (during rule building)	0.8696	0.7391	0.5405	0.5000	0.2800	0.2000	0.1951	0.1471	0.0555	0.0454	0.0000	0.0000	0.0000	0.0333	0.0000	0.0000
Llama error (with rules)	0.2173	0.2609	0.2703	0.1875	0.1400	0.1167	0.1111	0.4091	0.0454	0.1739	0.1951	0.2105	0.2647	0.2667	0.1667	0.2414
Llama error (without rules)	0.7826	1.0000	0.9189	0.5000	0.8000	0.9000	0.8056	0.7272	0.8636	0.8696	1.0000	0.8947	0.9118	0.7000	0.7333	0.8966
Mistral error (with rules)	0.1304	0.2608	0.2703	0.1250	0.2600	0.2000	0.1944	0.2272	0.1364	0.3478	0.2439	0.1579	0.3235	0.2000	0.1000	0.1724
Mistral error (without rules)	0.4783	1.0000	0.8919	0.6875	0.8000	0.9667	0.9167	0.8636	1.0000	1.0000	0.9024	1.0000	0.7647	0.7333	0.7000	0.7241
Phase 2 – Capability Generation																
Rules added - GPT	3	0	1	0	1	0	0	0	0	0	0	0	0	0	0	0
Acceptance – GPT (during rule building)	0.1667	0.5000	0.7000	–	0.8333	0.9444	0.9000	0.8333	–	0.8889	0.7778	0.0000	0.9000	0.5294	0.6667	0.6667
Acceptance – Llama (proposed)	0.9167	–	0.9231	–	–	0.7500	–	0.7500	0.7500	–	0.9090	0.7500	–	1.0000	0.7778	0.6250
Acceptance – Llama (direct)	0.0667	0.0000	0.5000	0.0000	0.1000	0.2000	0.0000	0.0000	0.0000	0.0000	0.2000	0.1250	0.0000	0.0000	0.1667	0.2000
Acceptance – Mistral (proposed)	1.0000	0.7647	0.7500	–	–	0.9444	–	0.8333	–	–	0.8333	0.8571	0.8000	0.6667	0.6667	–
Acceptance – Mistral (direct)	0.0000	0.0000	0.2000	0.0000	0.0000	0.3636	0.0000	0.1428	0.0833	0.0000	0.0909	0.1538	0.2000	0.6667	0.3000	0.1000

Table 2. Ablation Study: How semantic grounding and incremental acquisition of expert knowledge contribute to improving the expert acceptance rate

	D1	D2	D3	D4	D5	D6	D7	D8	D9	D10	D11	D12	D13	D14	D15	D16
GPT-Base	0.1667	0.2000	0.1000	0.0000	0.0000	0.0869	0.0000	0.1111	0.1579	0.0000	0.2000	0.1500	0.1111	0.1667	0.1429	0.0667
GPT - Semantics only	0.7778 +0.61	0.5556 +0.36	0.5833 +0.48	—	—	0.4444 +0.36	—	0.3333 +0.22	0.5833 +0.43	—	0.6000 +0.40	0.5833 +0.43	0.4444 +0.33	0.4167 +0.25	0.6667 +0.52	0.3333 +0.27
GPT - Rules only	0.2500 +0.08	0.4444 +0.24	0.5000 +0.40	—	—	0.5000 +0.41	—	0.1667 +0.05	0.4167 +0.26	—	0.6000 +0.40	0.3125 +0.60	0.5000 +0.39	0.4444 +0.28	0.4444 +0.30	0.2500 +0.18
Mistral - Base	0.0000	0.0000	0.2000	0.0000	0.0000	0.3636	0.0000	0.1428	0.0833	0.0000	0.0909	0.1538	0.2000	0.1667	0.3000	0.0000
Mistral - Semantics only	0.8750 +0.88	0.1935 +0.19	0.5833 +0.38	—	—	0.5833 +0.22	—	0.5000 +0.36	—	—	0.6667 +0.58	0.4375 +0.28	0.6667 +0.47	0.3333 +0.17	0.4000 +0.10	—
Mistral - Rules only	0.6667 +0.67	0.2857 +0.29	0.3750 +0.18	—	—	0.8750 +0.51	—	0.1667 +0.02	—	—	0.5556 +0.46	0.5333 +0.38	0.5000 +0.30	0.2222 +0.05	0.4000 +0.10	—

in an average performance enhancement of 28% and 29% for GPT and Mistral respectively. The results of this study indicate that both design choices contribute to the substantial performance gains achieved through the proposed pipeline.

4.6 Evaluation of Generalisability Across Other Document Types

To ensure general applicability of the proposed pipeline and expert knowledge base across different document types, we also evaluated its utility on more granular, structured organisational artefacts using a set of 400 entries extracted from five structured catalogues (CT1-CT5). These include application catalogues imported directly from Configuration Management Databases (CMDBs), and entries drawn from project and service catalogues. Table 3 presents the accep-

Table 3. Acceptance rates across structured catalogues for Phase 2

	CT1	CT2	CT3	CT4	C5
GPT-Base	0.3000	0.3157	0.3333	0.1250	0.1000
GPT-Proposed	0.6000	0.4857	0.4925	0.7500	0.6600
Mistral-Base	0.3333	0.2778	0.2727	0.4231	0.0769
Mistral-Proposed	0.5555	0.3809	0.5454	0.8182	0.6429

tance rate results are reported on Phase 2 evaluation across one commercial (GPT-4) and one open-source (Mistral-Large) LLM. Mistral-Large was used as it reported the highest average acceptance rates for previous experiment in Table 1. Phase 1 is not needed for this experiment as entries in each catalogue is linked to a specific known entity (e.g. application catalog contains Application instances).

As presented in Table 3 the proposed approach significantly improve the base model performance for both GPT and Mistral. It is noteworthy that pipeline with Mistral outperforms GPT base model despite being a much smaller LLM. This results indicate that the approach and acquired rule set can be successfully generalised for more granular organisational documents that include lower level business concepts such as applications and services.

4.7 Qualitative Study

We complement our quantitative evaluation with a qualitative assessment, gathering expert insights on how effectively the proposed technique addresses the challenges identified in Sect. 1 (C1–C4). They are the same group of experts who were involved in the process of formulating rules.

All five experts recognised the success of the semantic linking phase in addressing *C1*, with one of them stating that "many organisations struggle with precisely identifying elements such as objectives and initiatives" compared to outcome of the proposed pipeline. They also highlighted the pipeline's capacity to successfully identify enduring business objects despite the information noise in the document (e.g. "Patients" in healthcare context) resulting in capabilities such as "Patient Engagement Management" that are more sustainable over time (*C4*).

In certain instances, the pipeline exceeded their expectations by producing higher-order capabilities by navigating the granularity in text, confirming the pipeline's ability to address (*C3*). For example, the pipeline abstracted the capability "sustainable logistics management" from a statement referring to "vessels". They confirmed the practical value in pipeline's ability to define case specific rules.

All five experts agreed with our observation about the lack of semantic validity *(C4)* in the baseline LLM extraction approach compared to the proposed pipeline. They noted that some instances derived in the direct approach such as 'organisational units and products should never appear in a capability map'.

However, they positively acknowledged the clear conceptual boundaries observed in the capabilities produced by the proposed pipeline, clearly separating them from adjacent constructs like business processes.

Experts also commented on the improved overall efficiency offered by the proposed pipeline for capability modelling. One expert estimated that the manual effort to create a similar BA model to be approximately 8–10 person days (2–3 days with approximately 3 people), followed by 2–3 stakeholder review cycles. One of the experts also highlighted the potential of the proposed approach to address a social dimension to traditional capability modelling practices where differing perspectives among business architects or the dominance of individuals with prior domain experience can hinder innovation and consensus. He stated that "if you put multiple business architects in a room, they never agree", emphasising pipeline's ability to provide an objective starting point to the process compared to existing predefined industry reference models or templates.

However, experts acknowledged that even with a mature expert knowledge base, the proposed pipeline will require certain degree of manual quality checks and refinements. They viewed the proposed approach as an "effective way for the practitioner to influence the automation in a controlled manner", particularly as our evaluation indicate that a large data volume is not necessary for the success of the proposed approach.

5 Discussion

5.1 Key Findings and Implications

Our experiments demonstrate that combining semantic triples with incrementally acquired expert knowledge preserved in an expert knowledge base, provides effective contextual grounding for LLMs to generate semantically valid business capabilities from commonly available organisational documents. Our Phase 1 results discussed under Sect. 4.3 show that supplementing semantic structure based hierarchical pruning with context captured in an evolving expert knowledge base contributes to enhance the semantic linking accuracy by systematically capturing domain-specific heuristics and tactics. This is particularly beneficial for knowledge-intensive domains such as BA modelling where the domain knowledge is tacit and experiential, residing in the minds of practitioners.

Based on our Phase 2 results, semantic triples can be effectively leveraged to enhance and LLM's conceptual understanding of BA constructs. Using LLM to generate explicit inclusion and exclusion criteria based on semantic model and using them to evaluate LLM generated capabilities, allows the LLM to verify the semantic validity of its original outputs. Literature states that experts internalise numerous conceptual relationships [26]. This semantic grounding coupled with incrementally acquired expert knowledge captured in an evolving rule-book facilitates a way to capture this mapping thereby, addressing C4 - a key challenge of ensuring semantic validity in LLM-assisted BA modelling.

The proposed pipeline allows rapid integration of new knowledge at run time while also incrementally constructing artefact; expert knowledge base, that can be shared across models or modules in a compound system. The performance of proposed approach using smaller open-source LLMs surpasses the performance of larger commercial LLMs, indicating its effectiveness over other costly knowledge ingestion techniques such as large-scale pre-training and fine-tuning.

Our findings also suggest that the pipeline does not require a comprehensively defined, fully complete ontology to deliver meaningful improvements. Instead, it can operate with a limited semantic model that specifies core entities and relationships, acquiring the remaining context as incrementally captured expert rules. Since this knowledge is captured as case-specific rules that can be reused across subsequent documents, the level of expert intervention is minimal after early iterations. The performance improvements observed in both Phase 1 (semantic linking) and Phase 2 (capability generation), upon introducing only a small number of rules generated by GPT-4 over few documents, provide empirical evidence that a lightweight semantic representation and incremental rule acquisition are sufficient to support the approach in practice, demonstrating its technical feasibility.

5.2 Limitations and Future Work

While we acknowledge the limited dataset size, the scope of our experiments is still substantial given the expert effort required for evaluation. We also ensure that the selected organisational artefacts for our experiments are representative of documents commonly available in practice, spanning strategic and operational contexts and covering both unstructured narrative sources and structured catalogues.

Our approach relies on expert inputs and cause some burden for the experts. Yet, the results indicate that new rules are introduced at a steadily decreasing rate, approaching zero after only a small number of statements, reducing required expert involvements drastically over time. By having an expert in the loop, our approach aligns with recommendations in the literature that emphasise the need for carefully controlled model generation and expert validation [26]. However, it is necessary to assess the long-term scalability and maintainability of the resulting expert knowledge base within an enterprise. Our approach assumes the existence of a BA semantic model to inject conceptual relationships. While well-founded capability ontologies (e.g. [5]) are also available, our experiments required an explicit semantic specification of how business capabilities relate to other core BA concepts within a particular organisation as presented in CAPSICUM model used. However, unlike building a formal ontology, our results show that the methodology can yield satisfactory results with simple natural language triples representing semantic relationships. This lowers the entry barrier as practitioners can begin with a lightweight set of triples and incrementally enrich it through expert knowledge acquisition. Creating open-sourced and sufficiently detailed semantic frameworks is an open research avenue for future researchers. The chosen processing order of documents may have shaped the results, as expert rules acquired from earlier documents persist and influence how later ones are interpreted. Starting with the most complex and ambiguous documents as per expert opinion likely front-loaded rule acquisition, allowing subsequent inputs to benefit from a more mature knowledge base. An investigation of how different orderings affect pipeline behaviour is left for future work. While our assessment focuses on the validity and the quality of the generated capabilities through expert acceptance rate, the completeness of the generated set has not been assessed due to the lack of a validated ground truth capability maps for the orgnisations under study. Establishing benchmark datasets with expert-validated capability maps is an important direction for future work. It is noteworthy that we have used expert reviews on a single LLM (GPT-4) output to construct our expert knowledge base, and used the generated rules in subsequent experiments with Mistral and Llama. How other LLMs will perform in this process and contribute to reducing error rate over time is an open question. While our ablation study spans a range of document types, a controlled study isolating the impact from the quality of the document on performance remains future work. As the expert rules are accumulated incrementally, formulating and explicit mechanism conflict detection and resolution for the expert knowledge base is also an important direction for future work.

In our future work, we also plan to explore automated rule-generation techniques to further reduce reliance on domain experts. While the focus of this paper is on identifing individual capabilities from strategic text, future work will extend to more operational documents to build full capability maps, and will broaden evaluation across more industries, document types, and expert groups.

6 Conclusion

This study presents semantically-guided LLM pipeline for BA modelling, utilising unstructured strategic documents and incremental expert knowledge acquisition. Our approach yields substantially better results compared to directly prompting an LLM on the documents alone and successfully addresses four key challenges persistent in BA modelling automation: documents lacking BA focus (C1) and their varying granularity (C2), abstracting expert knowledge expressed at case level (C3) and the limited semantic validity (C4). Our design preserves detailed, domain-specific knowledge and yields contexts that are comprehensive and semantically grounded without the need for pre-training or fine-tuning.

Acknowledgment. We are grateful to Orbus software, Dr. Terry Roach, Dr. Rouzbeh Meymandpour, for supporting and sponsoring the research that led to this article.

References

1. Agrawal, L.A., et al.: Gepa: reflective prompt evolution can outperform reinforcement learning. arXiv preprint arXiv:2507.19457 (2025)

2. Bellan, P., Dragoni, M., Ghidini, C.: Extracting business process entities and relations from text using pre-trained language models and in-context learning. In: International Conference on Enterprise Design, Operations, and Computing, pp. 182–199. Springer, Heidelberg (2022). https://doi.org/10.1007/978-3-031-17604-3_11
3. Bondel, G., Faber, A., Matthes, F.: Reporting from the implementation of a business capability map as business-it alignment tool. In: IEEE 22nd International Enterprise Distributed Object Computing Workshop (EDOCW), pp. 125–134. IEEE (2018)
4. Buchmann, R., et al.: Large language models: expectations for semantics-driven systems engineering. Data Knowl. Eng. **152**, 102324 (2024)
5. Calhau, R.F., Almeida, J.P.A., Hoffmann, R., Guizzardi, G.: Creon: a capability reference ontology for the analysis and design of systems capabilities. In: International Conference on Conceptual Modeling, pp. 85–105. Springer, Heidelberg (2025). https://doi.org/10.1007/978-3-032-08623-5_5
6. Capsifi: Capsifi Official Website (2024). https://www.orbussoftware.com/solutions/business-architecture. Accessed 27 July 2024
7. Compton, P.: Experience with discovering knowledge by acquiring it. In: Proceedings of the 18th ACM SIGKDD International Conference on Knowledge Discovery and Data Mining, pp. 814–814 (2012)
8. Ding, Y., Zeng, Q., Weninger, T.: Chatel: entity linking with chatbots. In: Proceedings of the 2024 Joint International Conference on Computational Linguistics, Language Resources and Evaluation (LREC-COLING 2024), pp. 3086–3097 (2024)
9. Dong, Z., Shen, X., Yang, Z., Xia, R.: Chainedit: propagating ripple effects in llm knowledge editing through logical rule-guided chains. In: Proceedings of the 63rd Annual Meeting of the Association for Computational Linguistics, vol. 1: Long Papers, pp. 13558–13571 (2025)
10. Dumas, M., et al.: Process discovery. In: Fundamentals of Business Process Management pp. 159–212 (2018)
11. Dumas, M., et al.: Process identification. In: Fundamentals of Business Process Management, pp. 35–73 (2018)
12. Fill, H.G., Fettke, P., Köpke, J.: Conceptual modeling and large language models: impressions from first experiments with chatgpt. Enterp. Model. Inf. Syst. Arch. (EMISAJ) **18**, 1–15 (2023)
13. Görgen, L., Müller, E., Triller, M., Nast, B., Sandkuhl, K.: Large language models in enterprise modeling: case study and experiences. In: MODELSWARD, pp. 74–85 (2024)
14. Grohs, M., Abb, L., Elsayed, N., Rehse, J.R.: Large language models can accomplish business process management tasks. In: International Conference on Business Process Management, pp. 453–465. Springer, Heidelberg (2023). https://doi.org/10.1007/978-3-031-50974-2_34
15. Kampik, T., et al.: Large process models: a vision for business process management in the age of generative AI. In: KI-Künstliche Intelligenz, pp. 1–15 (2024)
16. Kotusev, S., Alwadain, A.: Modeling business capabilities in enterprise architecture practice: the case of business capability models. Inf. Syst. Manag. **41**(2), 201–223 (2024)
17. Kourani, H., Berti, A., Schuster, D., van der Aalst, W.M.: Process modeling with large language models. In: International Conference on Business Process Modeling, Development and Support, pp. 229–244. Springer, Heidelberg (2024). https://doi.org/10.1007/978-3-031-61007-3_18

18. Kourani, H., Berti, A., Schuster, D., van der Aalst, W.M.: Promoai: process modeling with generative AI. arXiv preprint arXiv:2403.04327 (2024)
19. Noy, N.F., McGuinness, D.L., et al.: Ontology development 101: a guide to creating your first ontology (2001)
20. Pons, G., Bilalli, B., Queralt, A.: Knowledge graphs for enhancing large language models in entity disambiguation. In: International Semantic Web Conference, pp. 162–179. Springer, Heidelberg (2024). https://doi.org/10.1007/978-3-031-77844-5_9
21. Roach, T.: CAPSICUM-a semantic framework for strategically aligned business architecture, UNSW. Ph.D. thesis (2011)
22. Samarasekara, I., Bandara, M., Rabhi, F., Benatallah, B.: Benchmarking llms for business architecture modelling with hierarchical capability maps. In: 37th International Conference on Advanced Information Systems Engineering (2025)
23. Samarasekara, I., Bandara, M., Rabhi, F., Benatallah, B., Meymandpour, R.: Llm driven approach for capability modelling with context-enriched prompt engineering (2024)
24. Shinn, N., Cassano, F., Gopinath, A., Narasimhan, K., Yao, S.: Reflexion: language agents with verbal reinforcement learning. Adv. Neural. Inf. Process. Syst. **36**, 8634–8652 (2023)
25. Simon, C., Haag, S., Zakfeld, L.: Experiments on gpt-3 assisted process model development. In: 37th ECMS 2023 (2023)
26. Storey, V.C., et al.: Large language models for conceptual modeling: assessment and application potential. Data Knowl. Eng. **160**, 102480 (2025)
27. Suzgun, M., Yuksekgonul, M., Bianchi, F., Jurafsky, D., Zou, J.: Dynamic cheatsheet: test-time learning with adaptive memory. arXiv preprint arXiv:2504.07952 (2025)
28. Ulrich, W., Rosen, M.: The business capability map: the rosetta "stone" of business/it alignment. Cutter Consortium, Enterp. Arch. **24**(4) (2011)
29. Van Riel, J., Poels, G.: A method for developing generic capability maps: a design science study in the professional sport industry. Bus. Inf. Syst. Eng. **65**(4), 403–424 (2023)
30. Voelter, M., Hadian, R., Kampik, T., Breitmayer, M., Reichert, M.: Leveraging generative AI for extracting process models from multimodal documents. arXiv preprint arXiv:2406.04959 (2024)
31. Yuksekgonul, M., et al.: Textgrad: Automatic "differentiation" via text. arXiv preprint arXiv:2406.07496 (2024)
32. Zhang, Q., et al.: Agentic context engineering: evolving contexts for self-improving language models. arXiv preprint arXiv:2510.04618 (2025)
33. Ziche, C., Apruzzese, G.: Llm4pm: a case study on using large language models for process modeling in enterprise organizations. In: International Conference on Business Process Management, pp. 472–483. Springer, Heidelberg (2024). https://doi.org/10.1007/978-3-031-70445-1_35

Bridging Accuracy and Interpretability in Business Process Simulation via Knowledge Distillation

Francesco Vinci(✉) and Massimiliano de Leoni

University of Padua, Padua, Italy
{francesco.vinci,massimiliano.deleoni}@unipd.it

Artifact available

Abstract. Business Process Simulation (BPS) is widely used to analyze and optimize processes in a controlled environment. Recent advances integrate AI models to improve the simulation accuracy of process perspectives such as control-flow and time. However, BPS-based improvement requires testing alternative *what-if* analyses. This can only be achieved through white-box models, where analysts can inspect and modify the underlying decision logic. Purely black-box models, although often highly accurate, lack transparency and limit configurability. However, traditional white-box models (e.g. decision trees) provide explicit and interpretable rules but may struggle to capture complex dependencies, leading to lower predictive accuracy, hampering the BPS realism. This paper proposes a distillation-based approach to bridge the gap between predictive accuracy and interpretability in BPS. We first train black-box predictive models (referred to as *teacher* models) to learn complex temporal and behavioral patterns from event logs. We then extract the knowledge captured by the *teacher* models and transfer it into probabilistic decision trees, obtaining interpretable surrogate white-box models (referred to as *student* models). The distilled models approximate the learned behavior of the *teacher* while offering a white-box representation and maintaining competitive predictive performance. Experimental results in four case studies, using four alternative *teacher* models, demonstrate the effectiveness of our approach. The findings show that distillation is able to improve the accuracy of simulation models, compared to standard decision trees without distillation, while maintaining interpretability and supporting reliable *what-if* analyses.

Keywords: Business Process Simulation · Realism · Explainability · Machine and Deep Learning · Knowledge Distillation

1 Introduction

Business Process Simulation (BPS) is one of the most established techniques for analyzing and improving business processes. BPS incorporates key aspects of processes, including control-flow, task durations and resource allocation, to evaluate alternative scenarios and optimize processes in a risk-free environment [7].

F. M. Maggi et al. (Eds.): BPM 2026, LNBIP 599, pp. 290–306, 2027.
https://doi.org/10.1007/978-3-032-37877-4_17

The growing availability of event data has led to the development of automated techniques for discovering simulation models directly from historical process executions [3,17]. Several techniques have been proposed to discover BPS models that build on deep learning techniques (cf. Sect. 2). These techniques have the advantages to increase the simulation realism, but they produce black-box models, hiding the internal decision logic governing the simulations. This hinders the analyst's trust in the model, and also makes it impossible to conduct *what-if* analyses because the simulation logic cannot be varied and alternatives cannot be evaluated [4]. To increase transparency in the definition of the simulation logic, white-box models have been proposed (cf. Sect. 2). However, these white-box models are relatively simpler: they thus can be sensitive to outliers and may struggle to generalize in the presence of highly complex data dependencies [16], potentially limiting their accuracy compared to more sophisticated models.

This paper aims to bridge predictive accuracy of black-box models and interpretability of white-box models through a knowledge distillation approach [8]. Knowledge distillation is a model compression technique in which the knowledge encoded by a complex model (referred to as the *teacher*) is transferred into a simpler and interpretable model (referred to as the *student*). Through distillation, the original values of response variable in the training set are replaced by those returned by the chosen *teacher* model. In this way, the number of outliers in the training set is expected to be substantially reduced and the values are somewhat more standardized.

In our setting, we train black-box *teacher* models to capture complex temporal and control-flow behaviors from event logs, and then distill their knowledge into probabilistic decision trees, which serve as white-box *student* models. The resulting distilled models retain the interpretability and configurability required for *what-if* analyses, while often improving accuracy of directly trained decision trees. The choice of an appropriate *teacher* model is particularly important, as it directly determines the quality of the distilled *student*. To address this, we propose a model selection pipeline that evaluates multiple *teacher* models and selects what performs best.

We evaluated our approach on four real-life and synthetic case studies, using four alternative *teacher* model classes. The experimental results show that distillation noticeably improves simulation accuracy over the non-distilled baseline across multiple process perspectives. The findings confirm that knowledge distillation is an effective strategy for bridging accuracy and configurability in business process simulation.

The paper is structured as follows. Section 2 reports on the related works. Section 3 introduces the preliminary concepts useful for the discussion of the methodology. Section 4 presents our distillation methodology and model selection pipeline. Experimental results are reported in Sect. 5. Finally, Sect. 6 concludes the paper.

2 Related Works

Traditionally, the construction of process simulation models has relied on manual effort from domain experts, making it time-consuming and error-prone due to subjective interpretation and incomplete documentation. To overcome these limitations, Rozinat et al. [17] pioneered the use of process mining for the (semi-)automatic discovery of simulation models from event logs. Their approach reconstructs control-flow and extracts empirical information about timing and resources, enabling realistic and evidence-based simulation models with limited human intervention.

After the seminal work by Rozinat et al. [17], automated discovery of BPS models has gained increasing attention. Recent approaches infer complete simulation parameters directly from event logs, including control-flow structures, task duration distributions, arrival rates, resource assignments, and decision probabilities [3]. By relying on empirical data rather than human intuition, these techniques reduce modeling effort and improve scalability across processes and domains.

With the availability of large-scale event data, deep learning techniques have been adopted for business process simulation [4]. Neural architectures learn complex temporal and behavioral dependencies directly from logs and have been shown to generate more accurate and realistic event logs than traditional approaches [4], capturing intricate correlations among activities, resources, and time-related attributes that are difficult to encode explicitly in classical process models. To mitigate the black-box nature of purely neural approaches, hybrid simulation models have been proposed [2,15]. These approaches combine an interpretable process model (e.g., Petri nets or BPMN) for the control-flow with deep learning predictors for estimating the temporal perspective, thereby integrating learned behavioral patterns into an explicit structure. While hybrid methods enhance realism, a trade-off persists between predictive performance and interpretability, as the inclusion of neural components still limits the analyst's ability to inspect and manipulate simulation parameters explicitly, making it difficult to perform *what-if* analyses [4].

Some white-box approaches have also been proposed. *Simod* [6] automatically discovers simulation models with probabilistic parameters from event logs, but relies on relatively simple distributions and does not explicitly model complex data-dependent dependencies. Vinci et al. [21] introduced the use of probabilistic decision trees for modeling data-dependent control-flow and temporal dependencies in white-box simulation models. The interpretability and configurability of this representation have been validated in [20] through a user study with BPM practitioners interacting with the ProSiT tool [22], achieving high task-completion accuracy. These techniques based on decision trees are known to be unstable and sensitive to outliers, and they may underperform in the presence of highly complex or high-dimensional dependencies, potentially limiting their generalization capability in some scenarios [16]. *AgentSimulator* [10] adopts an agent-based perspective centered on resources rather than control-flow, improving interpretability and adaptability.

Knowledge distillation has seldom been applied in Business Process Management and Mining. In a recent work, Käppel et al. [9] used knowledge distillation in the context of predictive process monitoring to enhance interpretability. Specifically, they distilled black-box models into interpretable decision trees to enable human experts to identify and mitigate unfair decision rules, thereby supporting fairness-aware model refinement. However, their approach specifically focuses on fairness in predictive monitoring and does not consider the construction or enhancement of business process simulation models.

3 Preliminaries

This section introduces the preliminary concepts required to present our methodology. We first provide the basic definitions of events, traces, and event logs. Next, we recall the fundamental notions of Petri nets used to model process control-flow. We finally introduce business process simulation models and their main components.

3.1 Process Data

Process data are typically collected as sequences of events, defining traces and event logs [1]. In particular, an event log can be seen as a set of traces, where each trace records a single process execution as a temporally ordered sequence of events [1]. We first formalize the notion of an event.

Definition 1 (Event). *Let $\mathcal{I}$, $\mathcal{A}$, and $\mathcal{R}$ denote the sets of all possible case identifiers, activities, and resources, respectively. An event e is a record with the following projections: $case(e) \in \mathcal{I}$, the case identifier to which e belongs; $act(e) \in \mathcal{A}$, the activity executed in e; $res(e) \in \mathcal{R}$, the resource involved in e; $time(e) \in \mathbb{N}$, the timestamp of e, inducing the order of events within a trace; $life(e) \in \{start, complete\}$, indicating whether e marks the start or completion of the activity,*

Building on the definition of events, we can now define traces and event logs.

Definition 2 (Traces & Event Logs). *Let $\mathcal{E}$ denote the universe of all possible events, with $\mathcal{I}$ the set of all possible case identifiers. A trace $\sigma = \langle e_1, e_2, \ldots, e_n \rangle \in \mathcal{E}^*$ is a finite sequence of events that belong to the same case, i.e., $case(e_1) = \cdots = case(e_n) \in \mathcal{I}$, and are ordered by their timestamps, i.e., $time(e_i) \leq time(e_{i+1})$ $\forall i = 1, \ldots, n-1$. An event log $\mathcal{L}$ is a finite set of traces, where each trace corresponds to a unique case.*

3.2 Process Models

In this paper, we use Petri nets to model the control-flow perspective [1]. We opted for Petri nets because they have a simple yet formal syntax and semantics for defining the control-flow stochasticity. However, the discussion remains valid even using other process model notations, such as process trees or BPMN models.

Definition 3 (Petri Net). *A Petri net is a tuple (P, T, F, λ), where P is a finite set of places, T is a finite set of transitions, and $F \subseteq (P \times T) \cup (T \times P)$ is a set of arcs, and $\lambda : T \nrightarrow \mathcal{A}$ is a labelling function where $\mathcal{A}$ is a set of activities.*

The state of a Petri net is identified by its current marking: a marking m of a Petri net (P, T, F) is a function $m : P \rightarrow \mathbb{N}$ which assigns for each place $p \in P$ the number $m(p)$ of tokens at this place. A transition $t \in T$ is enabled in a marking m if each input place p of t, i.e. $\exists (p, t) \in F$, contains at least one token, i.e., $m(p) \geq 1$. An enabled transition may fire, consuming one token from each input place p of t and producing one token in each output place p' of t, i.e. $\exists (t, p') \in F$, thus yielding a new marking m'.

In Petri nets, when multiple transitions are enabled at a certain marking, there is no determinism or preference on which is going to fire. In reality, not all runs are equally probable, and some are more likely than others. To model these probabilities, stochastic Petri nets have been proposed where each transition $t \in T$ is associated with a weight $w(t) \in \mathbb{R}^+$ [14]. We employ the concept of stochastic data Petri net, where transition weights are functions dependent on data states $x \in \mathcal{X} \subset \mathbb{R}^n$, that is an assignment to numeric variables [13].

Definition 4 (Stochastic Data Petri Net). *Let $\mathcal{X} \subset \mathbb{R}^n$ be the set of all possible data states. A stochastic data Petri net (P, T, F, λ, w) is a tuple that consists of a Petri net (P, T, F, λ), and $w \colon T \rightarrow (\mathcal{X} \rightarrow \mathbb{R}^+)$ is a weight function.*

Hereafter, we use the term stochastic Petri net to indicate a stochastic data Petri net, without causing ambiguities.

In a stochastic data Petri net, given a marking m and data state $x \in \mathcal{X}$, the probability of firing an enabled transition $t \in E(m)$ is defined as:

$$P(t \mid m, x) = \frac{w(t)(x)}{\sum_{t' \in E(m)} w(t')(x)} \tag{1}$$

where $E(m) \subseteq T$ denotes the set of transitions enabled at marking m. Note that $\sum_{t \in E(m)} P(t \mid m, x) = 1$.

3.3 Business Process Simulation

Business Process Simulation (BPS) is a well established technique for quantitatively analyzing, evaluating, and improving business processes [7]. By combining information on control-flow, task durations, and resource allocation, BPS provides a probabilistic representation of process behavior.

In this paper, we formally define a business process simulation model as follows.

Definition 5 (Business Process Simulation Model). *A business process simulation model is a tuple (N, D, K), where:*

- *$N = (P, T, F, \lambda, w)$ is a stochastic Petri net with $\lambda \colon T \nrightarrow \mathcal{A}$, $w \colon T \rightarrow (\mathcal{X} \rightarrow \mathbb{R}^+)$, and $\mathcal{X} \subset \mathbb{R}^n$, $n \geq 1$, the set of possible data states.*

- $D = (\mathcal{R}, d^{\mathcal{R}}, C^{\mathcal{R}}, MT^{\mathcal{R}}, d^{\mathcal{V}})$ *represents the descriptive parameters of the model, consisting of a set* $\mathcal{R}$ *of resources;* $d^{\mathcal{R}}\colon \mathcal{A} \to (\mathcal{R} \to [0,1])$ *mapping each activity to a function that assigns the probability of using a corresponging resource; a function* $C^{\mathcal{R}}\colon \mathcal{R} \to \{0,1\}^{7\times 24}$ *mapping each resource to its weekly work schedule, represented as a* 7×24 *binary matrix, where* $C^{\mathcal{R}}_{w,h} = 1$ *indicates that the resource is available at hour* h *on weekday* w*; a set* $MT^{\mathcal{R}} \subseteq \mathcal{R}$ *denoting the subset of resources that are allowed to perform multiple activities concurrently; and a function* $d^{\mathcal{V}}\colon \mathcal{V} \to \mathcal{P}$ *mapping each trace attribute to a parametric distribution* $d \in \mathcal{P}$.
- $K = (p^{ET}, p^{WT}, p^{AT})$ *defines the predictive parameters of the model. The function* $p^{ET}\colon \mathcal{A} \to (\mathcal{X} \to \mathcal{P})$ *assigns to each activity a data statedependent parametric distribution modeling its execution time. The function* $p^{WT}\colon \mathcal{R} \to (\mathcal{X} \to \mathcal{P})$ *assigns to each resource a data statedependent parametric distribution predicting waiting times for events performed by that resource. Finally,* $p^{AT} \in (\mathcal{X} \to \mathcal{P})$ *is a function that returns a data statedependent parametric distribution modeling the inter-arrival times of new process instances.*

A BPS model (N, D, K) enables the simulation of the control-flow and behavioral dynamics of process instances. The stochastic Petri net N governs the execution logic of activities, while the descriptive parameters D define the static characteristics of the process environment, such as resource availability and working calendars. The predictive parameters K determine the temporal stochastic and data-driven aspects of the simulation, specifying how timing and case arrivals evolve throughout simulated executions. This definition allows the integration of these components for the runtime generation of events following the approach in [15]. Notably, both the predictive components in K and the weight function w in the stochastic Petri net are data state-dependent, meaning that their values are functions of the current data states in $\mathcal{X} \subset \mathbb{R}^n$, $n \geq 1$. This dependency allows the simulation model to dynamically adjust its behavior based on contextual information available during process execution, such as trace attributes, resource allocations, or temporal features.

In this paper, the predictive component K and the transition's weights w are represented via probabilistic decision trees [21]. Differently from traditional decision trees, leaves are associated with distributions, rather than single values. These distributions coincide with the best-fitting parameter distribution from a given set of potential distributions, where the best fitness is obtained from the distribution minimizing the Wasserstein distance. This combination of decision trees and parametric distributions ensures high interpretability and configurability: the decision rules are expressed analytically, and thus process analysts can directly inspect and modify them to define *what-if* scenarios. The usability of this representation for BPM practitioners has been independently assessed in [20]. Since the present work preserves the same structural representation and only changes how the trees are trained, these usability properties are inherited by the distilled models.

4 Methodology

Several components of the business process simulation model defined in Definition 5 rely on predictive models that estimate probabilities or parametric distributions based on the current data state. These models are typically learned from historical event logs and may adopt complex machine learning techniques.

Section 2 already mentioned that predictive models used in simulation pipelines are often implemented using complex machine learning algorithms such as neural networks. While these models may achieve high predictive performance, they are typically treated as black-box models, meaning that the relationship between input features and predictions is difficult to interpret or analyze directly. On the other hand, white-box models can be directly inspected and configured to perform *what-if* analyses, at the potential cost of a reduced predictive accuracy.

To try to exploit the predictive power of black-box models on white-box models, we employ **knowledge distillation** [8]. Knowledge distillation transfers the predictive behavior of a complex model, referred to as the ***teacher***, into a simpler and more interpretable model, referred to as the ***student***.

Formally, let $\mathcal{X}$ and $\mathcal{Y}$ denote the input feature spaces and the output space, respectively, and let $\mathcal{D} = \{(\boldsymbol{x}_i, y_i)\}_{i=1}^{n}$, with $\boldsymbol{x}_i \in \mathcal{X}$ and $y_i \in \mathcal{Y}$, be a training set. The *teacher* model $f_{teach} : \mathcal{X} \rightarrow \mathcal{Y}$ is first trained on $\mathcal{D}$, mapping input features to predictions with typically high accuracy but limited interpretability. The *student* model $f_{stud} : \mathcal{X} \rightarrow \mathcal{Y}$ is then trained not on the original labels $\mathcal{Y}$, but on the soft outputs produced by the *teacher*, i.e., on $\mathcal{D}' = \{(\boldsymbol{x}_i, f_{teach}(\boldsymbol{x}_i))\}_{i=1}^{n}$. In doing so, the *student* learns to mimic the predictive behavior of the *teacher* while maintaining a significantly simpler and more transparent structure [12]. This allows the *student* model to potentially preserve the predictive behavior captured by the *teacher* while providing a representation that is easier to inspect and analyze.

In the following sections, we first describe how distilled decision tree models can be discovered from event log data (cf. Sect. 4.1), and then introduce a pipeline for incorporating them into simulation models to support accurate and interpretable process simulations (cf. Sect. 4.2).

4.1 Distillation Discovery

Given a business process simulation model (N, D, K), this section describes how knowledge distillation is applied to learn the transition weights $w(t)$, $\forall t \in T$, and the temporal predictive parameters $K = (p^{ET}, p^{WT}, p^{AT})$. In both cases, training sets are constructed from the event log following the feature engineering approach of [21]: transition weight models are conditioned on the history of executed activities and trace attributes; execution time models incorporate the assigned resource, activity history, and trace attributes; waiting time models include resource workload, activity history, and trace attributes; and inter-arrival time models are conditioned on the weekday and hour of the day.

Transition Weights. For each transition $t \in T$, a training set is built from the event log following [21], where each sample corresponds to a process execution reaching a marking that enables t, and the label indicates whether t was fired. A black-box classifier f_{teach} is then trained independently for each transition to predict the firing probability $f_{teach}(\boldsymbol{x}) \in [0, 1]$ given a data state $\boldsymbol{x} \in \mathcal{X}$.

Rather than distilling the hard binary labels directly, we apply *temperature scaling* [8] to soften the *teacher* outputs before distillation. Hard binary labels only indicate which transition was fired, discarding any information about how confidently the *teacher* distinguishes between enabled transitions. Soft targets instead preserve this uncertainty, providing a richer training signal that better captures the *teacher*'s predictive behavior. Specifically, the soft target provided to the *student* is $\sigma(\text{logit}(f_{teach}(\boldsymbol{x}))/\tau)$, where σ is the sigmoid function, $\text{logit}(p) = \log(p/(1-p))$, and $\tau \geq 1$ is the temperature parameter. For $\tau = 1$, the soft target recovers the original *teacher* output; increasing τ flattens the distribution, reducing the *teacher*'s confidence in the most likely transition and redistributing probability mass across the other enabled transitions. The *student* f_{stud} is then trained as a regression tree on the distilled training set $\mathcal{D}'_t = \{(\boldsymbol{x}_i, \sigma(\text{logit}(f_{teach}(\boldsymbol{x}_i))/\tau))\}_{i=1}^{n_t}$, where n_t is the number of training instances for transition t, yielding an interpretable surrogate of the *teacher* predictive behavior.

Temporal Predictive Parameters. For each component of K, dedicated training sets are constructed from the event log [21]: one per activity for execution times, one per resource for waiting times, and a single one for inter-arrival times. For each training set, a black-box regression model f_{teach} is fitted to predict the corresponding duration, capturing complex behavioral dependencies that may be difficult to model directly. The *student* f_{stud} is then trained as a regression tree on the distilled training set $\mathcal{D}' = \{(\boldsymbol{x}_i, f_{teach}(\boldsymbol{x}_i))\}_{i=1}^{n}$. Each leaf of the resulting tree is then enriched with a parametric distribution fitted to the residual values of the training instances assigned to it, preserving the stochastic nature of the simulation while maintaining an interpretable and configurable structure.

4.2 Selection of the Best Teacher Model

There is no unique model that is always preferable as *teacher* model. In fact, a *teacher* model may overfit the training data, thus distilling noise instead of valid data points, or underfit, failing to capture the complex dependencies and causing no benefits in using distillation. It follows that, for each event log, we need to assess alternative *teacher* models, and measure their ability to accurately estimate the weights of the transitions and the predictive parameters of the business process simulation models.

The accuracy is measured through standard metrics widely employed in Business Process Simulation [5]. Given that the transition weights and the predictive parameters focus on the process perspective on control-flow and time, our choice fell into the 3-Gram Distance (3GD), Cycle Time Distribution distance (CTD)

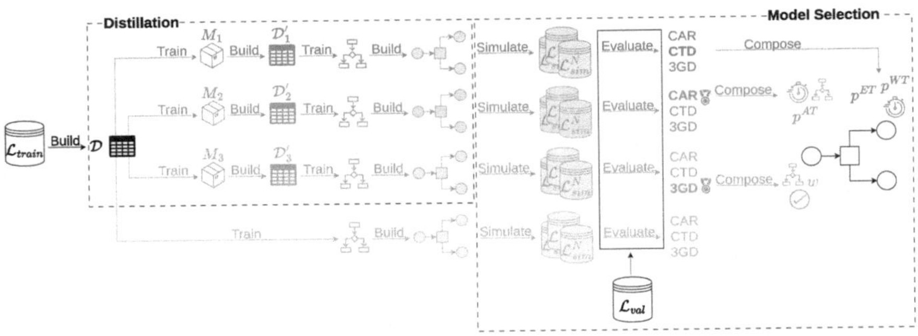

Fig. 1. Overview of the proposed approach. The training log $\mathcal{L}_{train}$ is used to build feature datasets $\mathcal{D}$. For each candidate *teacher* family M_1, M_2, M_3, distilled datasets $\mathcal{D}'_i$ are constructed by replacing the original labels with the black-box teachers' soft outputs, and probabilistic decision trees (*students*) are trained on them. We also include the non-distilled baseline, where the *students* are trained directly on $\mathcal{D}$. Each simulation model is evaluated against $\mathcal{L}_{val}$ using CAR, CTD and 3GD, and the best-performing models are selected independently per predictive component to compose the final simulation model.

and Case Arrival Rate distance (CAR). These metrics are based on the statistical comparison of a validation event log, which records real process executions, and an event log that is generated via simulation. The generated event log contains as many traces as the validation event log for a reliable assessment.

The procedure to select the best *teacher* models requires the provision of a set $\mathcal{M}$ of candidate *teacher* model families, where each family groups models sharing the same algorithm but potentially differing in hyperparameters. We always ensure that the *student* model is in the candidate list (namely, no distillation is actually performed) because it may potentially be that distillation is not beneficial.

Figure 1 depicts the proposed pipeline. The event log $\mathcal{L}$ is temporally split into a training log $\mathcal{L}_{train}$ and a validation log $\mathcal{L}_{val}$. For each candidate family $M \in \mathcal{M}$, $\mathcal{L}_{train}$ is used to train the correspodent models and build the simulation model, via distillation as described in Sect. 4.1, or directly when M is the *student* model. A set of N event logs is then generated and compared against $\mathcal{L}_{val}$ using the three metrics above.

In principle, each predictive component influences the simulation quality differently: the transition weights w primarily affect the 3GD, since they govern the choice of the next activity; the temporal parameters p^{ET} and p^{WT} drive task execution and waiting times, thus jointly affecting the CTD; and the inter-arrival time parameter p^{AT} governs case arrival dynamics, reflected in the CAR metric. Exhaustively searching for the best *teacher* family across all components jointly would require evaluating $|\mathcal{M}|^4$ configurations, which is computationally prohibitive. Instead, our method adopts a heuristic decomposition: the best family is selected independently for each predictive component by optimizing its corresponding metric. Specifically, the family for w is chosen by minimizing the 3GD; the family for p^{AT} by minimizing the CAR; and a single family is selected for both p^{ET} and p^{WT} by minimizing the CTD. We deliberately share the family between p^{ET} and p^{WT} because the CTD metric reflects the joint contribution of execution and waiting times to cycle time, and cannot be attributed to either component in isolation. The final configuration tuple $(M_w, M_{p^{ET}}, M_{p^{WT}}, M_{p^{AT}}) \in \mathcal{M}^4$ is then composed by selecting the best-performing family for each component accordingly.

The simulation model is finally composed by inserting the selected (distilled) models into the corresponding components, resulting in a BPS model (N, D, K): the (distilled) decision trees for w define the stochastic weights of the Petri net N, while the (distilled) probabilistic decision trees for p^{ET}, p^{WT}, and p^{AT} instantiate the predictive parameters in K. The heuristic decomposition is then consistent with the architecture of the BPS model and does not introduce inconsistencies in the composed simulation model.

5 Experiments

This section presents the experiments conducted to evaluate the performance of our distillation approach compared with the baseline approach, namely the approach in [21] which uses directly trained decision trees. The goal is to assess the ability of knowledge distillation to produce accurate and realistic simulations across various process perspectives. Accuracy evaluation focuses on computing distances between the distributions observed in the simulated event logs and the real ones [5]. Moreover, we analyze the performance achieved by the diverse *teacher* models with respect to the characteristics of the process and the available data, to investigate the trade-off between *teacher* complexity and distillation effectiveness discussed in Sect. 4.2.

5.1 Experimental Setup

The approach is implemented in Python and is built upon the `prosit` library from [21].[1]

We evaluate our approach on four real-life and synthetic event logs widely used in the business process simulation literature:

- **BPIC17**, which refers to the same sub-process as BPC12W, but related to the executions in 2016, when the process underwent some significant changes.[2]
- **Academic Credentials Recognition (ACR)**, which records the executions of a credentials recognition process at the University of Los Andes in Colombia.[3]

[1] https://doi.org/10.5281/zenodo.20375680.
[2] https://doi.org/10.4121/uuid:5f3067df-f10b-45da-b98b-86ae4c7a310b.
[3] https://zenodo.org/records/4699983.

Table 1. Hyperparameter search spaces for the *teacher* models. *Max depth* controls the maximum depth of each decision tree; *N. estimators* is the number of trees in the ensemble; *Min samples leaf* is the minimum number of training samples required at a leaf node; *Hidden layer sizes* defines the MLP architecture; α is the L2 regularization penalty.

Model	Hyperparameter	Values
RF/XGB/Cat	Max depth	2, 3, 5, 7
	N. estimators	25, 50, 100, 200, 300
	Min samples leaf	5, 10, 20, 30, 50, 100
MLP	Hidden layer sizes	[32], [64], [128], [128, 64]
	Regularization α	$10^{-4}, 10^{-3}$

- **Synthetic Loan (SynLoan)**, a synthetic event log simulating a loan application process.[3]
- **Purchase to Pay (P2P)**, a synthetic event log registering executions of a purchasing process.[4]

For each event log, the control-flow model is discovered using the Inductive Miner [11], which produces a sound Petri net representation of the process. The discovered Petri net is then used as the structural backbone of the simulation model, as described in Sect. 4.

Each event log is temporally partitioned into training (60%), validation (20%), and test (20%) splits. Training and validation logs are used to discover and select the simulation model as described in Sect. 4.2, while the test log is reserved for the final evaluation of the composed simulation model.

We consider four families of black-box *teacher* models: *Random Forest* (RF), *XGBoost* (XGB), *CatBoost* (Cat), and *Multi-Layer Perceptron* (MLP). For each family, we explore the hyperparameter search spaces reported in Table 1, covering a range of model complexities from shallow trees with few estimators or small networks, which are less prone to overfitting, to deeper trees with more estimators or larger networks, which can capture more intricate behavioral patterns at the cost of a higher risk of overfitting. In these configurations we set the temperature parameter τ to 1.5, which showed a good trade-off between *teacher* confidence and *student* generalization. The best hyperparameter setting for each configuration is selected via 5-fold cross-validation on the training log.

Student decision trees are trained with a maximum depth ranging from 1 to 5 and a minimum number of samples per leaf varying in the set $\{20, 50, 100\}$. The best configuration is selected via 5-fold cross-validation on the training set.

5.2 Experimental Results

We evaluate simulation quality by comparing the simulated logs with the real event log with various metrics. Specifically, for each simulation model we gen-

[4] https://fluxicon.com/academic/material.

Table 2. Validation results per event log and *teacher* model, reported as mean and 95% confidence interval over 10 simulations. Δ% denotes the relative improvement over the baseline, where higher values indicate better performance. Best values per metric and dataset are highlighted in bold.

Dataset	Model	CTD			CAR			3GD		
		Mean	±CI	Δ%	Mean	±CI	Δ%	Mean	±CI	Δ%
BPI17	RF	156.61	±4.37	+25.2%	**250.09**	**±69.42**	**+40.3%**	0.3595	±0.0026	+4.2%
	XGB	180.45	±2.85	+13.8%	284.83	±80.80	+32.0%	0.3585	±0.0027	+4.5%
	Cat	149.31	±3.99	+28.7%	378.39	±176.85	+9.6%	0.3621	±0.0048	+3.5%
	MLP	**127.23**	**±3.23**	**+39.2%**	254.81	±33.48	+39.2%	**0.3459**	**±0.0019**	**+7.8%**
	Baseline	209.34	±4.93	–	418.75	±150.09	–	0.3752	±0.0042	–
ACR	RF	49.28	±12.60	+14.2%	43.08	±11.89	+24.0%	0.3318	±0.0077	-3.2%
	XGB	**36.88**	**±3.54**	**+35.8%**	40.86	±9.76	+27.9%	0.3201	±0.0106	+0.5%
	Cat	55.72	±16.71	+3.0%	44.32	±10.77	+21.8%	0.3214	±0.0080	+0.0%
	MLP	41.38	±10.39	+27.9%	**33.51**	**±10.98**	**+40.9%**	**0.3123**	**±0.0122**	**+2.9%**
	Baseline	57.43	±15.54	–	56.66	±10.94	–	0.3216	±0.0160	–
SynLoan	RF	246.30	±17.04	–15.1%	4.81	±1.53	+15.1%	**0.1267**	**±0.0036**	**+1.2%**
	XGB	230.69	±11.43	-7.8%	**2.52**	**±0.77**	**+55.6%**	0.1341	±0.0040	-4.5%
	Cat	**155.21**	**±7.00**	**+27.4%**	3.68	±1.11	+35.1%	0.1324	±0.0039	–3.1%
	MLP	182.07	±22.24	+14.9%	4.17	±1.40	+26.3%	0.1312	±0.0041	-2.2%
	Baseline	213.91	±19.40	–	5.66	±1.48	–	0.1283	±0.0055	–
P2P	RF	529.69	±15.55	–2.8%	250.08	±26.43	+12.3%	0.2374	±0.0100	–12.7%
	XGB	**507.07**	**±18.82**	**+1.6%**	**209.74**	**±35.87**	**+26.5%**	0.2135	±0.0066	–1.3%
	Cat	525.46	±9.69	–2.0%	252.41	±24.08	+11.5%	0.2210	±0.0119	–4.9%
	MLP	522.41	±13.92	–1.4%	254.99	±17.30	+10.6%	**0.2047**	**±0.0118**	**+2.8%**
	Baseline	515.40	±17.27	–	285.20	±31.32	–	0.2107	±0.0135	–

erate 10 event logs, and compare them with the real event log using the Cycle Time Distribution (CTD) distance, the Case Arrival Rate (CAR) distance and the 3-Gram Distance (3GD) selected from [5]. These metrics are chosen as they directly reflect the accuracy of the three perspectives modeled by our distilled components. For all three metrics, lower values indicate higher simulation quality. Results are reported as mean and 95% confidence interval over the 10 simulation runs.

We first present the validation results, analyzing the performance of individual *teacher* families and its relationship with process characteristics. These results guide the composition of the final simulation model. We then report the test results obtained with the composed simulation model, assembled via the selection pipeline described in Sect. 4.2.

Validation Results. Table 2 reports, for each event log and *teacher* family, the best distillation result obtained on the validation set across all hyperparameter configurations, together within its 95% confidence interval and the relative

Table 3. Ranking of teacher models and baseline per event log and metric (lower is better). The last row reports the average rank across all datasets. Best average ranks per metric are highlighted in bold.

Dataset	CTD					CAR					3GD				
	RF	XGB	Cat	MLP	BL	RF	XGB	Cat	MLP	BL	RF	XGB	Cat	MLP	BL
BPI17	3	4	2	1	5	1	3	4	2	5	3	2	4	1	5
ACR	3	1	4	2	5	3	2	4	1	5	5	2	3	1	4
SynLoan	5	4	1	2	3	4	1	2	3	5	1	5	4	3	2
P2P	5	1	4	3	2	2	1	3	4	5	5	3	4	1	2
Avg.	4.00	2.50	2.75	**2.00**	3.75	2.50	**1.75**	3.25	2.50	5.00	3.50	3.00	3.75	**1.50**	3.25

improvement Δ% over the baseline. Best values per metric and dataset are highlighted in bold.

Overall, distillation consistently improves over the baseline across the majority of datasets and perspectives. Substantial gains are observed on the CTD metric: on BPI17 and ACR all *teacher* families yield improvements with relative gains up to +39.2% (MLP on BPI17) and +35.8% (XGB on ACR). The CAR metric also benefits considerably from distillation, with notable improvements on all four datasets; for instance, XGB achieves a +55.6% gain on SynLoan and RF a +40.3% gain on BPI17. The control-flow perspective (3GD) shows more moderate but consistent improvements across most datasets, with MLP achieving the best results on three out of four logs. Altought not all *teacher* families outperform the baseline in every setting (e.g. CTD on SynLoan and P2P, and 3GD on ACR, SynLoan and P2P), at least one distilled model consistently improves over the baseline in each metric.

To better understand the relative effectiveness of the different *teacher* families across datasets and metrics, we rank each model (including the baseline) based on its mean validation performance. Table 3 reports the ranking of each model per dataset and metric. No single *teacher* family dominates across all settings, confirming the motivation for the model selection pipeline introduced in Sect. 4.2. MLP achieves the best average rank on both CTD (2.00) and 3GD (1.50), while XGB leads on CAR (1.75). Notably, the baseline consistently ranks last on CAR (average rank 5.00), confirming that distillation provides a systematic benefit for the arrival perspective. On 3GD, the baseline ranks 3.25 on average, confirming that distillation still provides a competitive advantage for the control-flow perspective, while also validating the inclusion of the non-distilled baseline as a candidate in the selection pipeline (cf. Sect. 4.2).

To better understand when distillation is most beneficial, we analyze how its effectiveness relates to properties of the process and the available data. Specifically, we compare the improvements with the simplicity of the process model [19], and the number of traces in the event log. Figure 2 shows these relationships. Simplicity shows a negative correlation with all three metrics, indicating that distillation is more beneficial for complex processes where simple

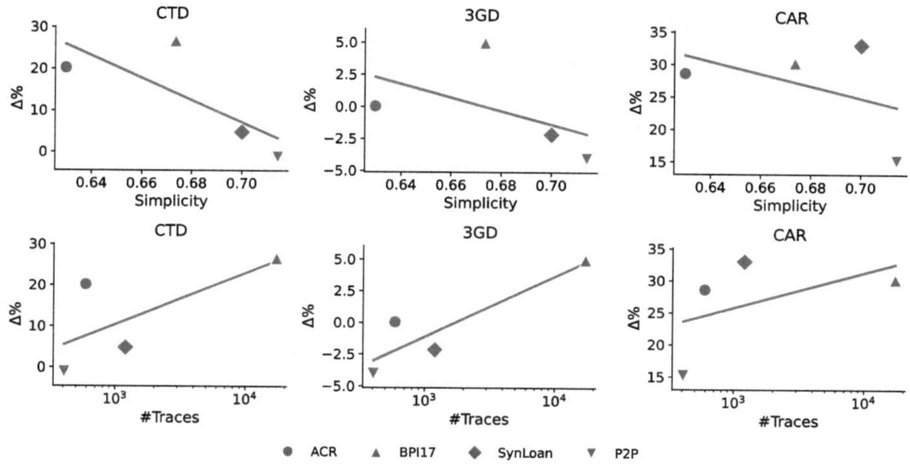

Fig. 2. Distillation effectiveness (Δ% over baseline) and process characteristics for each event log. Lines show linear regression fits.

Table 4. Test results of the composed simulation model against the baseline, reported as mean and 95% confidence interval over 10 simulations. Cohen's d effect size and qualitative size follow [18]: very small $|d| \geq 0.01$, small ≥ 0.2, medium ≥ 0.5, large ≥ 0.8, very large ≥ 1.2, huge ≥ 2.0. Statistical significance is assessed via one-sided Welch t-test ($p < 0.05$): bold indicates the composed model is significantly better than the baseline.

Dataset	Model	CTD			CAR			3GD		
		Mean	±CI	Size	Mean	±CI	Size	Mean	±CI	Size
BPI17	Composed	**122.22**	**±8.65**	**huge**	452.47	±138.63	v. small	**0.3481**	**±0.0035**	**huge**
	Baseline	180.84	±7.15	–	458.67	±178.95	–	0.3785	±0.0060	–
ACR	Composed	**154.47**	**±17.98**	**v.large**	**238.62**	**±11.60**	**large**	0.4105	±0.0184	v. small
	Baseline	199.19	±20.99	–	259.13	±14.52	–	0.4162	±0.0230	–
SynLoan	Composed	**103.04**	**±14.19**	**huge**	**2.87**	**±0.63**	**huge**	**0.1205**	**±0.0075**	**large**
	Baseline	173.71	±11.62	–	7.77	±1.57	–	0.1286	±0.0034	–
P2P	Composed	**502.49**	**±11.38**	**large**	692.50	±14.35	v. small	0.3016	±0.0175	v. small
	Baseline	518.32	±12.88	–	698.06	±29.00	–	0.3053	±0.0139	–

decision trees struggle to capture intricate dependencies. Conversely, the number of traces shows positive correlations, suggesting that larger event logs allow *teacher* models to learn more robust behavioral patterns. These findings provide practical guidance: distillation is expected to yield the largest gains when applied to complex processes with sufficiently large event logs.

Test Results. Based on the validation results, the model selection pipeline described in Sect. 4.2 composes the final simulation model by selecting, for each predictive component, the *teacher* family that yields the best validation performance on the corresponding metric. Table 4 reports the test results of the composed model against the baseline for each dataset and metric, including Cohen's d effect size and its qualitative interpretation following [18]. To assess statistical significance, we apply one-sided Welch t-test ($p < 0.05$), determining whether the composed distilled model significantly outperforms the baseline.

The composed model achieves statistically significant improvements in 8 out of 12 dataset-metric combinations. The CTD metric shows the most consistent gains: the composed model significantly outperforms the baseline on all four datasets, with huge effect sizes on BPI17 (from 180.84 to 122.22) and SynLoan (from 173.71 to 103.04), a very large effect on ACR (199.19 to 154.47), and a large effect on P2P (518.32 to 502.49). The CAR metric yields significant improvements on SynLoan (huge effect, from 7.77 to 2.87) and ACR (large effect, from 259.13 to 238.62), while on BPI17 and P2P the differences are negligible. For 3GD, significant improvements are observed on BPI17 (huge effect, from 0.3785 to 0.3481) and SynLoan (large effect, from 0.1286 to 0.1205), while ACR and P2P show negligible differences. Importantly, no degradation is observed on any metric, confirming that the selection pipeline successfully avoids adopting distilled models when they do not improve over the baseline.

Overall, the results confirm the effectiveness of the proposed distillation approach. The composed model achieves significant improvements in the majority of the cases, with effect sizes ranging from large to huge, and no degradation is observed in any setting. The benefits are most pronounced for processes with higher complexity and larger event logs. Moreover, the variability in which *teacher* family performs best across datasets and metrics (cf. Table 3) validates the design of the selection pipeline, which effectively exploits the complementary strengths of different *teacher* architectures to compose a robust final model.

6 Conclusion

This paper proposes a knowledge distillation approach to bridge the gap between simulation accuracy and interpretability in business process simulation. Given a set of black-box *teacher* models trained to capture complex temporal and control-flow dependencies from event logs, we distill their knowledge into probabilistic decision trees, which serve as interpretable *student* models. The resulting simulation models retain the transparency and configurability required for *what-if* analyses, while improving the accuracy of decision trees, if compared with the trees that are directly trained on the original event logs. To support the selection of the most appropriate *teacher* for each process perspective, we introduce a model selection pipeline that systematically evaluates multiple *teacher* architectures and composes the final simulation model from the best-performing distilled components.

The experimental evaluation on four real-life and synthetic event logs, using four alternative *teacher* families of varying complexity, demonstrated the effectiveness of the proposed approach. Distillation notably improves simulation accuracy over the non-distilled baseline across the majority of datasets and perspectives, with large to huge effect sizes in most cases.

As future work, we aim to extend the distillation approach to additional process perspectives, such as resource allocation, and also to extend the repertoire of potential *teacher* models. Currently, we have opted some shallow and deep models that are often used in the literature as *teacher* models (cf. Sect. 2). Nonetheless, there may be additional models that can even improve further the distillation benefits beyond the already very positive results.

References

1. van der Aalst, W.M.P.: Process Mining - Data Science in Action, 2nd edn. Springer, Heidelberg (2016). https://doi.org/10.1007/978-3-662-49851-4
2. Camargo, M., Báron, D., Dumas, M., Rojas, O.G.: Learning business process simulation models: a hybrid process mining and deep learning approach. Inf. Syst. **117**, 102248 (2023). https://doi.org/10.1016/J.IS.2023.102248
3. Camargo, M., Dumas, M., González, O.: Automated discovery of business process simulation models from event logs. Decis. Support Syst. **134**, 113284 (2020). https://doi.org/10.1016/J.DSS.2020.113284
4. Camargo, M., Dumas, M., Rojas, O.G.: Discovering generative models from event logs: data-driven simulation vs deep learning. PeerJ Comput. Sci. **7**, e577 (2021). https://doi.org/10.7717/PEERJ-CS.577
5. Chapela-Campa, D., Benchekroun, I., Baron, O., Dumas, M., Krass, D., Senderovich, A.: A framework for measuring the quality of business process simulation models. Inf. Syst. **127**, 102447 (2025). https://doi.org/10.1016/J.IS.2024.102447
6. Chapela-Campa, D., López-Pintado, O., Suvorau, I., Dumas, M.: SIMOD: automated discovery of business process simulation models. SoftwareX **30**, 102157 (2025). https://doi.org/10.1016/J.SOFTX.2025.102157
7. Dumas, M., Rosa, M.L., Mendling, J., Reijers, H.A.: Fundamentals of Business Process Management, 2nd edn. Springer, Heidelberg (2018). https://doi.org/10.1007/978-3-662-56509-4
8. Hinton, G.E., Vinyals, O., Dean, J.: Distilling the knowledge in a neural network. CoRR arxiv:1503.02531 (2015)
9. Käppel, M., Neuberger, J., Möhrlein, F., Weinzierl, S., Matzner, M., Jablonski, S.: A human-in-the-loop approach for improving fairness in predictive business process monitoring. In: Business Process Management - 23rd International Conference, BPM 2025, Seville, Spain, 31 August–5 September 2025, Proceedings, vol. 16044, pp. 343–360. Springer, Heidelberg (2025).https://doi.org/10.1007/978-3-032-02867-9_21
10. Kirchdorfer, L., Blümel, R., Kampik, T., van der Aa, H., Stuckenschmidt, H.: Discovering multi-agent systems for resource-centric business process simulation. Process Sci. **2**(1) (2025). https://doi.org/10.1007/S44311-025-00009-5
11. Leemans, S.J.J., Fahland, D., van der Aalst, W.M.P.: Discovering block-structured process models from event logs - a constructive approach. In: Colom, J.-M., Desel, J. (eds.) PETRI NETS 2013. LNCS, vol. 7927, pp. 311–329. Springer, Heidelberg (2013). https://doi.org/10.1007/978-3-642-38697-8_17

12. Liu, X., Wang, X., Matwin, S.: Improving the interpretability of deep neural networks with knowledge distillation. In: 2018 IEEE International Conference on Data Mining Workshops, ICDM Workshops, Singapore, Singapore, 17–20 November 2018, pp. 905–912. IEEE (2018). https://doi.org/10.1109/ICDMW.2018.00132
13. Mannhardt, F., Leemans, S.J.J., Schwanen, C.T., de Leoni, M.: Modelling data-aware stochastic processes - discovery and conformance checking. In: Application and Theory of Petri Nets and Concurrency - 44th International Conference, PETRI NETS 2023, Lisbon, Portugal, 25–30 June 2023, Proceedings, vol. 13929, pp. 77–98. Springer, Heidelberg (2023). https://doi.org/10.1007/978-3-031-33620-1_5
14. Marsan, M.A., Balbo, G., Conte, G., Donatelli, S., Franceschinis, G.: Modelling with generalized stochastic petri nets. SIGMETRICS Perf. Eval. Rev. **26**(2), 2 (1998). https://doi.org/10.1145/288197.581193
15. Meneghello, F., Francescomarino, C.D., Ghidini, C., Ronzani, M.: Runtime integration of machine learning and simulation for business processes: time and decision mining predictions. Inf. Syst. **128**, 102472 (2025). https://doi.org/10.1016/J.IS.2024.102472
16. Molnar, C.: Interpretable Machine Learning, 3 edn. (2025). https://christophm.github.io/interpretable-ml-book
17. Rozinat, A., Mans, R.S., Song, M., van der Aalst, W.M.P.: Discovering simulation models. Inf. Syst. **34**(3), 305–327 (2009). https://doi.org/10.1016/J.IS.2008.09.002
18. Sawilowsky, S.S.: New effect size rules of thumb. J. Mod. Appl. Stat. Methods **8**(2), 26 (2009). https://doi.org/10.22237/jmasm/1257035100
19. Vázquez-Barreiros, B., Mucientes, M., Lama, M.: Prodigen: mining complete, precise and minimal structure process models with a genetic algorithm. Inf. Sci. **294**, 315–333 (2015). https://doi.org/10.1016/J.INS.2014.09.057
20. Vinci, F.: Reliable and Explainable Process Simulation Models: Discovery, Refinement and Applications, Padua, Italy. Ph.D. thesis (2026)
21. Vinci, F., Park, G., van der Aalst, W.M.P., de Leoni, M.: Reliable and configurable process simulations via probabilistic white-box models. In: Service-Oriented Computing - 23rd International Conference, ICSOC 2025, Shenzhen, China, 1–4 December 2025, Proceedings, Part II. vol. 16321, pp. 337–352. Springer, Heidelberg (2025). https://doi.org/10.1007/978-981-95-5015-9_24
22. Vinci, F., Park, G., van der Aalst, W.M.P., de Leoni, M.: ProSiT: a tool for interactive and transparent process simulations. In: Proceedings of the International Conference on Service Oriented Computing (ICSOC): Demonstrations and Resources. LNCS. Springer, Heidelberg (2026)

WARP: Worktime-Aware Representation for Predictive Process Monitoring

Maximilian Vouk[1](✉), Kate Revoredo[2](✉), and Jan Mendling[1,2](✉)

[1] Vienna University of Economics and Business, Vienna, Austria
maximilian.alexander.vouk@s.wu.ac.at
[2] Humboldt-Universität zu Berlin, Berlin, Germany
{kate.revoredo,jan.mendling}@hu-berlin.de

Abstract. Predictive Process Monitoring leverages historical event logs to generate data-driven forecasts about the future development of ongoing process instances, such as the time until the completion of the next event. Recent research has focused on enhancing predictive performance through advanced deep-learning architectures and feature engineering. In this context, existing approaches typically model time as a continuous calendar-based feature. This design choice overlooks an important characteristic of real-world business processes: process execution is often constrained by organizational working patterns, including business hours, shift schedules, and weekends. In this paper, we investigate the integration of organizational working constraints into the predictive pipeline through a worktime-based transformation of the temporal axis. We propose a methodology that maps event timestamps from a conventional clocktime axis to a worktime-based axis, effectively excluding periods of scheduled resource unavailability. This transformation allows predictive models to learn temporal patterns based on active process execution rather than elapsed calendar time. We evaluate the approach using both synthetic and real-world event log data. The results show that the worktime-based transformation improves prediction accuracy, particularly in scenarios characterized by temporal gaps between discrete working intervals.

Keywords: Predictive Process Monitoring · Time Prediction · Temporal Feature Engineering · Worktime

1 Introduction

Predictive Process Monitoring (PPM) [1] focuses on generating data-driven forecasts for the future states and outcomes of business process instances. A key objective within this domain is the prediction of time-related attributes, such as the time remaining until the next event completion [12]. Such predictions support improved resource allocation or provide more accurate information to customers.

F. M. Maggi et al. (Eds.): BPM 2026, LNBIP 599, pp. 307–323, 2027.
https://doi.org/10.1007/978-3-032-37877-4_18

In practical business applications, work on processes is rarely continuous, but is fundamentally constrained by organization-specific working schedules. Factors such as business hours, shift schedules, and statutory holidays dictate the discrete intervals during which work can progress. Previous works approach this problem primarily by integrating temporal features that capture the relative temporal position of events within calendar cycles (e.g., within the day or the week) [14,19]. If a predictive model fails to account for these boundaries, it systematically introduces biases [18]. In such cases, tasks started immediately before off-hours accumulate elapsed time that is purely an artifact of the schedule rather than a reflection of the work's inherent nature. So far, there are no prediction approaches available that explicitly model the constraints by organizational working schedules.

In this paper, we address this research problem. We propose a *transformation of the time axis that is based on dynamically identified working times.* Specifically, we extract active working intervals within a cyclical period (e.g., a week) and eliminate non-working periods from the time axis. This results in a continuous *worktime* axis, which serves as an alternative to standard *clocktime*. We evaluate this transformation by training Long Short-Term Memory (LSTM) models using worktime-based features and targets, comparing it against a clocktime-based baseline. This evaluation is conducted on synthetic event logs to isolate scenarios where the transformation is beneficial, and on public, real-world event logs to demonstrate practical applicability. Our results show that the worktime-based approach improves prediction accuracy, enforces adherence to working hours in predictions, and clarifies actual workload.

The paper is structured as follows: Sect. 2 presents the background on preliminary definitions, related work and problem definition. Section 3 presents our proposed approach for transforming the time axis. Section 4 presents the methodology for its evaluation and the results. In Sect. 5 we present and discuss the results. Section 6 concludes with a summary of the main findings.

2 Background

In this section, we first introduce the preliminary definitions for our work and working and non-working times, review existing work, before stating the research problem.

2.1 Preliminary Definitions

The aim of PPM is to predict the future unfolding of ongoing business process cases [15]. To achieve this, PPM methodologies translate fundamental process mining constructs into machine-readable formats. Specifically, the hierarchical elements of process execution (events, prefixes, full traces, and aggregate event logs) are transformed through feature engineering into the vectorized inputs required by machine learning models [14,20].

Definition 1 (Event). *Let $\mathcal{E}$ be the universe of events. For an event $e \in \mathcal{E}$, $\pi_{\text{case}}(e) \in \mathcal{U}_{\text{case}}$ denotes its case, $\pi_{\text{act}}(e) \in \mathcal{U}_{\text{act}}$ its activity, and $\pi_{\text{time}}(e) \in \mathcal{U}_{\text{time}}$ its timestamp.*

Definition 2 (Trace, Prefix & Event Log). *A trace $\sigma = \langle e_1, \ldots, e_n \rangle \in \mathcal{E}^*$ is a time-ordered sequence of events sharing the same case. A prefix of length k ($0 \leq k \leq n$) is $\text{hd}_k(\sigma) = \langle e_1, \ldots, e_k \rangle$. An event log L is a multiset of traces.*

Definition 3 (Feature). *A feature is a function $f : \mathcal{E}^* \rightarrow \mathcal{U}_{val}$ where $\mathcal{U}_{val}$ is the universe of feature values. The feature value $f(\sigma^k)$ abstracts the event data of a prefix into a structured feature representation.*

2.2 Working and Non-working Time Definitions

An analysis of widely used public event logs (see Sect. 4.1) reveals that process execution often exhibits distinct periodic intervals of resource availability, driven by organizational working schedules. For example, in the BPI 2012O log, 97.5% of events occur between Monday and Friday (09:00-21:00) and on Saturdays (10:00-17:00). To account for the temporal patterns induced by these schedules, we adapt availability constraints [6] to generalize active and inactive periods across the log.

Definition 4 (Period Projection & Partitioning). *Let $p \in \mathcal{U}_{\text{dur}}$ be the duration of a recurring temporal period from the universe of durations, and $t_0 \in \mathcal{U}_{\text{time}}$ a reference start timestamp. The projection $\pi_{\text{period}}(t) = (t - t_0) \mod p$ maps an absolute timestamp to its relative period position. The interval $[0, p)$ is partitioned into a sequence of contiguous, non-overlapping intervals $\mathcal{I}_{\text{period}} = \langle I_1, \ldots, I_q \rangle$.*

Definition 5 (Periodic Working and Non-working Times). *Let E_L be all events in log L, and $\tau \in \mathbb{N}_0$ an event volume threshold. $\mathcal{I}_{period}$ is partitioned into working ($\mathcal{W}$) and non-working ($\mathcal{N}$) intervals based on mapped event volume:*

$$\mathcal{W} = \left\{ I \in \mathcal{I}_{\text{period}} \;\middle|\; \left| \{ e \in E_L \mid \pi_{\text{period}}(\pi_{\text{time}}(e)) \in I \} \right| \geq \tau \right\} \tag{1}$$

$$\mathcal{N} = \mathcal{I}_{\text{period}} \setminus \mathcal{W} \tag{2}$$

2.3 Related Work

Existing PPM literature acknowledges the need to incorporate temporal context to account for schedule-driven patterns [14,19,21]. To achieve calendar-awareness, these methodologies primarily enrich input data with dedicated temporal features that anchor an event within organizational time periods, helping models anticipate non-working intervals. Current approaches focus exclusively on two time horizons—daily and weekly periods—structuring this temporal context into either precise durations or segmented intervals.

Daily Period Semantics: These approaches capture intra-day dynamics to account for daily schedules, shifts, and overnight interruptions by positioning an event within a 24-hour window. To structure this context, methodologies rely on either a precise period progression representing the continuous time elapsed since 00:00 [2,5,11,13,14,16,19], or a segmented period positioning that anchors the event to a specific daily interval by encoding the specific hour [21].

Weekly Period Semantics: These approaches capture intra-week dynamics to account for weekend interruptions and varying weekday workloads by positioning an event within a 7-day schedule. To structure this context, methodologies rely on either a precise period progression representing the continuous time elapsed since the start of the working week (e.g., Monday 00:00) [2,5,11,14,16,19], or a segmented period positioning that anchors the event to a specific weekly interval by identifying the specific day [14,16,21].

Temporal feature-based approaches rely on models to organically learn the periods' impact on the prediction target [19]. However, standard machine learning models generally struggle to infer and enforce hard logical boundaries [3].

2.4 Problem Statement

To ground the limitations of temporal feature-based approaches, consider as an example a bank with the working schedule: Monday to Friday, 09:00 to 17:00, with nights and weekends designated as non-working periods. Consider two consecutive events in an ongoing loan approval process instance: e_1 (Loan Applied, Friday 14:00) and e_2 (Risk Assessed, Monday 10:00). This scenario highlights three distinct problems ($P1$, $P2$, and $P3$):

P1: Increased error at temporal edges: Prediction accuracy drops at working time borders. Let $t_i = \pi_{\text{time}}(e_i)$ be the current timestamp, t_{i+1} the true next timestamp, and $\hat{t}_{i+1}$ the predicted next timestamp. Let $\pi_{\text{period}}(t_i) \in I_k$. The absolute error $|\hat{t}_{i+1} - t_{i+1}|$ magnifies when t_i is in a working interval bounded by a non-working interval:

$$(I_k \in \mathcal{W}) \wedge (I_{k-1} \in \mathcal{N} \vee I_{k+1} \in \mathcal{N}) \tag{3}$$

In our example, for transition $e_1 \rightarrow e_2$, the clocktime elapsed is 68h, though active execution occurred for only 4h. Standard temporal features force the model to capture a massive variance for Friday afternoon tasks. In practice, however, the model often fails to learn this sharp variance successfully, which systematically inflates the edge error for any trace crossing the weekend boundary.

P2: Violation of business schedules: Predictions frequently forecast completions during inactive periods. Let $\pi_{\text{period}}(\hat{t}_{i+1}) \in I_l$. A violation occurs if the destination interval is non-working:

$$I_l \in \mathcal{N} \tag{4}$$

In our example, if a standard model evaluates e_1 and projects e_2's completion time based on an average processing time of 24 clocktime hours, it outputs a completion timestamp of Saturday 14:00. This violates organizational constraints, as no resources are available to execute the process during weekends.

P3: Obfuscation of active workload: Absolute predictions blend actual working time with non-working times, failing to provide actionable workload insights. Let $T_{\mathcal{N}}(t_a, t_b)$ denote the accumulated time in intervals $I \in \mathcal{N}$ between two timestamps. The true active working time $\hat{\Delta}_W$ is obfuscated by this calendar distortion:

$$\hat{\Delta}_W = (\hat{t}_{i+1} - t_i) - T_{\mathcal{N}}(t_i, \hat{t}_{i+1}) \tag{5}$$

In our example, a continuous prediction of 68h for $e_1 \rightarrow e_2$ obscures the fact that the actual workload associated with the risk assessment task requires only 4h of active human resource availability, limiting downstream workforce optimization.

Further illustrations of the three problems are provided in the supplementary material linked below. Despite the significant challenges that organizational working schedules pose to the practical application of PPM, these problems remain explicitly unaddressed in existing literature.

3 WARP Approach

To systematically model the constraints imposed by organizational working schedules, we propose a structural transformation of the underlying time axis to only progress during working intervals: WARP (**W**orktime-**A**ware **R**epresentation for **P**redictive Process Monitoring). The approach is illustrated in Fig. 1.

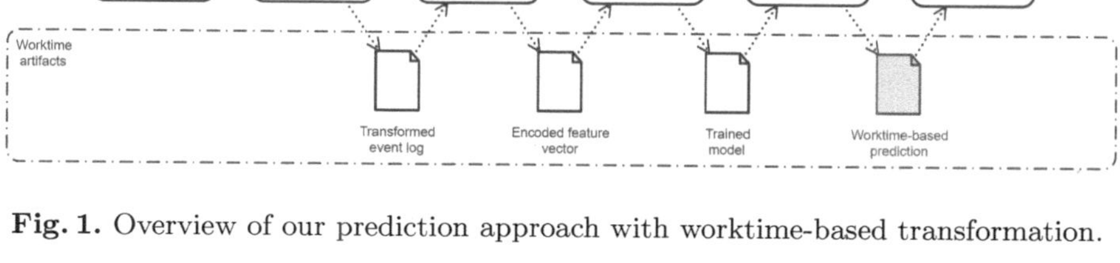

Fig. 1. Overview of our prediction approach with worktime-based transformation.

3.1 Working Times Identification

The initial step is to determine the active working time intervals $\mathcal{W}$, representing the discrete periods during which work usually occurs. These working time intervals can be identified using domain knowledge, such as shift schedules. However, in the absence of domain knowledge, they can be dynamically derived from historical event logs by applying the definition in Sect. 2.2. This dynamic derivation requires suitable choices for the period duration p, period partitioning $\mathcal{I}_{\text{period}}$, and event count threshold τ for the respective domain [6].

3.2 Clocktime-to-Worktime Transformation

With working intervals identified, we establish a mapping between the common *clocktime* and the proposed *worktime*. They are defined as two distinct temporal axes handling absolute timestamps and relative time differences. Let $T_C \in \mathcal{U}_{time}$ represent the clocktime and $T_W \in \mathcal{U}_{time}$ the worktime axis, which is constructed exclusively from the union of all intervals in $\mathcal{W}$. For both axes, an absolute zero point t_0 can be chosen arbitrarily to anchor the mapping.

The *clocktime-to-worktime mapping function* $g : T_C \to T_W$ compresses the clocktime axis onto the worktime axis by accumulating the duration of all elapsed working intervals. We define an indicator function $\mathbf{1}_{\mathcal{W}}(t)$ that evaluates whether a clocktime timestamp t falls within a working interval:

$$\mathbf{1}_{\mathcal{W}}(t) = \begin{cases} 1 & \text{if } \exists I \in \mathcal{W} : \pi_{\text{period}}(t) \in I \\ 0 & \text{otherwise} \end{cases} \tag{6}$$

The equivalent worktime timestamp $t_W \in T_W$ for any clocktime $t_C \in T_C$ is computed as:

$$t_W = g(t_C) = \int_{t_0}^{t_C} \mathbf{1}_{\mathcal{W}}(t)\, dt \tag{7}$$

This integration ensures that any t_C occurring outside of a working interval is effectively snapped to the accumulated working time intervals, maintaining a continuum in T_W.

3.3 Worktime-to-Clocktime Transformation

In the other direction, the *worktime-to-clocktime mapping function* $g^{-1} : T_W \to T_C$ projects the condensed worktime axis back onto the clocktime axis. To resolve this mapping for any $t_W \in T_W$, the corresponding $t_C \in T_C$ is computed by mathematically re-injecting the duration of all traversed inactive gaps back into the timeline.

Because the clocktime-to-worktime transformation $g(t)$ remains constant during non-working intervals, the mapping is not strictly monotonic. Therefore, the inverse function is formally defined as a generalized inverse using the infimum. It identifies the earliest absolute clocktime t at which the accumulated working time reaches the target t_W:

$$t_C = g^{-1}(t_W) = \inf\left\{t_C \geq t_0 \;\middle|\; \int_{t_0}^{t_C} \mathbf{1}_{\mathcal{W}}(t)\,dt \geq t_W\right\} \tag{8}$$

This formulation guarantees schedule compliance for future predictions. By calculating the infimum, the function inherently skips over all periods where $\mathbf{1}_{\mathcal{W}}(u) = 0$ (intervals in $\mathcal{N}$) and places the resulting timestamp t_C precisely within an active working interval $\mathcal{W}$. Figure 2 illustrates this bidirectional transformation framework using the loan application scenario introduced in Sect. 2.4, setting the absolute reference point t_0 at Friday 00:00 of the exemplary week:

- **Event** e_1**:** clocktime timestamp t_{C1} (Friday 10:00) maps directly to the worktime timestamp t_{W1} (1h of accumulated active time since t_0).
- **Event** e_2**:** clocktime timestamp t_{C2} (Friday 14:00) maps to the worktime timestamp t_{W2} (5h). Note that the relative temporal duration between e_1 and e_2 remains constant across both domains ($\Delta t_{C1/2} = \Delta t_{W1/2} = 4\text{h}$).
- **Event** e_3**:** clocktime timestamp t_{C3} (Monday 10:00) maps to the worktime timestamp t_{W3} (9h). Note that the clocktime duration $\Delta t_{C2/3}$ of 68h is effectively compressed to a worktime duration $\Delta t_{W2/3}$ of 4h, by eliminating the 64h weekend gap.
- **Event** e_4**:** clocktime timestamp t_{C4} (Tuesday 03:00) represents an off-schedule anomaly occurring during resource unavailability. The mapping function $g(t_C)$ compresses this absolute timestamp onto the worktime axis at t_{W4} (16h), which sits precisely on the boundary between the Monday and Tuesday working intervals. Therefore, in contrast to the previous examples, t_{W4} cannot be mapped back to its original clocktime timestamp, but is mapped to the end of the preceding working interval at t'_{C4} (Monday 17:00).

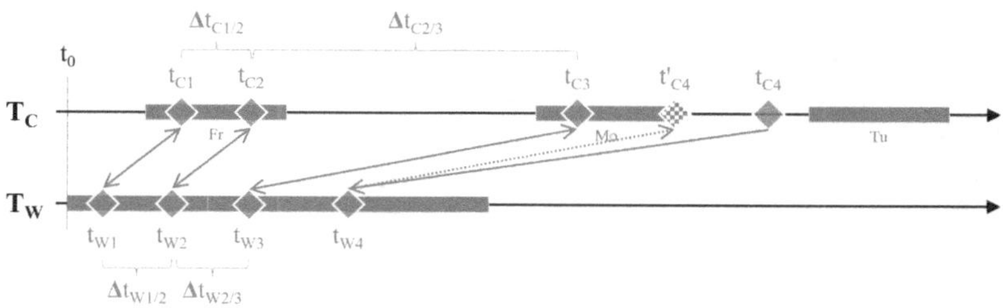

Fig. 2. Illustration of mapping of timestamps and relative time differences.

3.4 Predictive Model Integration

Finally, the time transformations are integrated into the common PPM workflow [21]. By isolating the transformation logic within the pre-processing and post-processing stages, the machine learning architecture is decoupled from the complexities of schedule inference. The integration follows a six-stage sequence:

1. **Identify working intervals:** Determine $\mathcal{W}$ (Sect. 3.1).
2. **Map timestamps:** Apply the mapping function g to all absolute clocktime timestamps t_C within the raw event log, permanently transforming them into worktime timestamps t_W.
3. **Engineer features & encode vector:** Derive all temporal input features and predictive targets entirely within the T_W domain. Relative time differences Δt_W are calculated mathematically as the difference between the previously mapped worktime timestamps.
4. **Train model:** Train the predictive model exclusively using the generated features and targets in the T_W space. With schedule-induced non-working times excised, the algorithm optimizes on a closer approximation of active operational effort.
5. **Generate predictions:** Evaluate the trained model on unseen trace prefixes to output temporal forecasts naturally in T_W.
6. **Map predictions:** Apply the inverse mapping function g^{-1} to the prediction, projecting it back to clocktime.

This transformation framework systematically resolves the limitations outlined in Sect. 2.4:

- **Resolution of P2 (Schedule Violations):** The inverse mapping g^{-1} acts as a rigid constraint. By skipping strict off-times, it structurally guarantees that transformed forecasts land within active business hours.
- **Resolution of P3 (Workload Obfuscation):** The mapping g isolates pure processing periods prior to model training. The resulting targets and predictions more closely reflect net operational effort rather than schedule-conflated durations, yielding actionable metrics for capacity planning.

4 Evaluation Methodology

While the proposed time transformation inherently resolves the violation of business schedules (P2) and the obfuscation of active workload (P3) by design, its impact on predictive performance (P1) requires experimental evaluation. We assess the efficacy of the transformation by comparing the two time axes across distinct feature sets and 21 event logs. Following established benchmarking protocols [14,19,21], our computer experiment is partitioned into an initial *controlled assessment using synthetic event logs*, followed by an *empirical validation using real-world event logs*. This systematic progression generates knowledge about the performance of our proposal under varying environmental uncertainties [10].

4.1 Event Log Data

Our evaluation utilizes a combination of synthetic logs to isolate specific process characteristics and eliminate confounding variables [7,8], and real-world logs to validate practical applicability amid empirical noise.

To classify process traces σ across both domains, we utilize two primary parameters: *trace length* $|\sigma|$ and *boundary crossing density* δ_σ. The tuple $(|\sigma|, \delta_\sigma)$ characterizes structural and temporal complexity independent of specific control flows. We consider a boundary as a temporal gap between non-adjacent working intervals (e.g., overnight periods or weekends). Thus, δ_σ quantifies the relative frequency of schedule-induced boundaries traversed per event transition. Analogous to standard event rate metrics, such as the arrival rate [4], that normalize event volume over a continuous temporal continuum, the boundary crossing density evaluates process discontinuity by quantifying the frequency of schedule-induced boundaries traversed per event transition.

To formalize this density, we define the set of absolute non-working intervals by mapping the relative periodic intervals $\mathcal{N}$ onto the continuous absolute timeline:

$$\mathcal{N}_{\text{abs}} = \left\{ \{t_0 + k \cdot c + t \mid t \in I\} \;\middle|\; I \in \mathcal{N}, k \in \mathbb{Z} \right\} \tag{9}$$

For a trace $\sigma = \langle e_1, \ldots, e_n \rangle$, the total boundary crossing count C_σ is the cardinality of the subset of absolute non-working intervals that intersect with the trace's temporal span:

$$C_\sigma = \left| \left\{ I_{\text{abs}} \in \mathcal{N}_{\text{abs}} \;\middle|\; I_{\text{abs}} \cap (\pi_{\text{time}}(e_1), \pi_{\text{time}}(e_n)) \neq \emptyset \right\} \right| \tag{10}$$

The density metric δ_σ calculates the relative frequency of these interruptions against the total possible transitions between successive events, ensuring the metric is bounded for traces where $|\sigma| > 1$:

$$\delta_\sigma = \begin{cases} \frac{|C_\sigma|}{n-1} & \text{if } |\sigma| > 1 \\ 0 & \text{if } |\sigma| \leq 1 \end{cases} \tag{11}$$

Synthetic Event Log Data. To explore this parameter space, we generate 15 synthetic event logs covering all combinations of three length levels ($|\sigma| \in \{4, 10, 20\}$) and five density levels ($\delta_\sigma \in \{0.0, 0.5, 1.0, 3.0, 9.0\}$). These ranges span from short, uninterrupted sequences to long, highly fragmented traces. The ranges for both the length and the density levels are calibrated to mirror characteristics of real-world event logs (see Fig. 3).

Each log consists of 3,000 traces sharing an invariant, sequential control flow, ensuring that stochastic properties are strictly limited to two independent sources of variability:

- **Structural Variability:** Boundary crossings are randomly interleaved between the given number of events, satisfying the target density parameters.
- **Temporal Variability:** Event timestamps are uniformly sampled at random within the active intervals of a standard business schedule (Monday to Friday, 08:00 to 17:00).

Real-world Event Log Data. To assess the transformation in diffuse, practical settings, we utilize six real-world event logs used in related work [14,19,21]. Figure 3 presents the share of traces in each log that conform to the parameter intervals centered around our synthetic generation ranges.

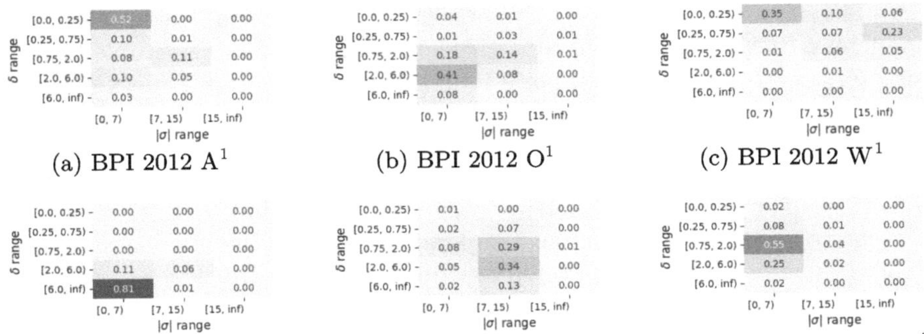

(a) BPI 2012 A[1] (b) BPI 2012 O[1] (c) BPI 2012 W[1]

(d) Helpdesk[2] (e) Prepaid travel cost[3] (f) Request for payment[4]

Fig. 3. Share of traces conforming to parameter tuple per real-world event log.

4.2 Data Processing and Feature Engineering

Each log is temporally partitioned into 80% training and 20% testing data based on the final event's timestamp [19,21]. To emulate ongoing case predictions, we generate all prefixes $\mathrm{hd}_k(\sigma_i)$ with $1 \leq k \leq |\sigma_i|$ for each trace.

Working intervals $\mathcal{W}$ are dynamically identified from the training prefixes following the approach in Sect. 2.2. To align with established temporal features, we utilize a weekly period length c, partitioned into hourly intervals $\mathcal{I}_{\text{period}}$. The event volume threshold τ is set empirically at 20% of the event volume observed in the most active interval.[1]

We engineer four distinct feature sets based on both clocktime and worktime domains. The *calendar-agnostic* set ($F_{agnostic}$) relies exclusively on the elapsed time since the previous event. The *calendar-aware* set (F_{aware}) supplements this with the relative time within the current day and current week, providing an anchoring not only to the relative positions within the trace, but also the absolute position within the weekly calendar.

4.3 Predictive Task, Model and Evaluation

We focus on predicting the *time until the next event's completion* [12]. Capturing shorter time horizons better highlights the isolated impact of singular boundary crossings compared to the more common remaining time prediction.

For predicting the temporal duration until the next event's completion, we utilize LSTM models, as they effectively handle variable-length process traces without the need for padding or lossy aggregation [9,15,21].

We use the Mean Absolute Error (MAE) ($\frac{1}{n}\sum_{i=1}^{n}|y_i - \hat{y}_i|$) as our leading accuracy metric [15,21]. MAE is reported in clocktime across all evaluations to guarantee a consistent baseline, while corresponding worktime MAE is provided as supplementary data.

---

[1] This heuristic was determined via preliminary tuning, balancing off-schedule noise filtration with active window detection across all datasets.

5 Results

In this section, we present the experimental evaluation of the worktime transformation. We first detail a *controlled assessment using synthetic event logs* to identify specific conditions where the transformation adds value, followed by a *validation using real-world event logs* to assess its practical applicability.

5.1 Controlled Assessment on Synthetic Event Log Data

Raw Error. Table 1 presents the MAE for the predictions across the 15 synthetic event logs. The lowest (i.e., best) MAE value measured in T_C per event log is highlighted in blue, and the second-lowest value is highlighted in gray. For the predictions performed in T_W, the MAE measured in T_W is additionally provided in gray font but not used for the comparison.

For 9 out of the 15 event logs, T_W-based predictions yield the best predictive performance. Analyzing the individual MAE values, three distinct extremes emerge regarding the performance differences between the time axes. First, the most pronounced advantage of the T_W-based predictions manifests at $\delta_\sigma = 1.0$. For instance, at $|\sigma| = 4$ (row 3 in Table 1), the calendar-aware T_W configuration achieves an MAE of 1.56 h, substantially outperforming the best T_C equivalent at 6.88 h. Second, the maximum disparity favoring T_C-based predictions occurs at $\delta_\sigma = 0.0$. In this scenario, at $|\sigma| = 4$ (row 1), the calendar-agnostic T_C model yields a minimal MAE of 0.48 h, whereas the best T_W counterpart trails at 1.23 h. Third, the smallest relative performance difference between the time axes is observed at the highest sparsity level ($\delta_\sigma = 9.0$). In these cases, the absolute errors converge, e.g., at $|\sigma| = 20$ (row 15), the optimal T_C configuration (245.63 h) and the optimal T_W configuration (249.52 h) report nearly identical results relative to their overall error magnitude. Across all these configurations, an increase in trace length ($|\sigma|$) consistently drives up the absolute MAE.

Patterns of Error. To highlight the predictive patterns across parameter configurations, Fig. 4 normalizes the MAE per log against the T_C/calendar-agnostic baseline. The impact of the worktime transformation maps precisely to three distinct regimes:

Regime 1 ($\delta_\sigma \in \{0.0\}$): Characterized by highly dense traces where all events occur within a single day, corresponding to rows 1, 6, 11 in Table 1. This regime exhibits substantially increased error for T_W-based predictions, an effect amplified by shorter trace lengths.

Regime 2 ($\delta_\sigma \in \{0.5, 1.0\}$): Characterized by medium-density traces containing occasional boundaries between events, corresponding to rows 2, 3, 7, 8, 12, 13. This regime represents the 'sweet spot' for T_W-based predictions, achieving the lowest error rates across all configurations.

Regime 3 ($\delta_\sigma \in \{3.0, 9.0\}$): Characterized by sparse traces featuring multiple boundaries between events, corresponding to rows 4, 5, 9, 10, 14, 15. This regime shows comparable predictive performance across all four configurations of time axis and feature set.

Table 1. Mean Absolute Error (hours) per synthetic event log and feature set. Lowest error per event log highlighted in blue, second-lowest in gray.

Prediction in:		Worktime T_W				Clocktime T_C	
Feature set:		calendar-agnostic		calendar-aware		c.-agnostic	c.-aware
Metric in:		T_W	T_C	T_W	T_C	T_C	T_C
$\|\sigma\|$	δ_σ						
(1) 4	0.0	0.48	1.23	0.62	5.59	0.48	0.61
(2) 4	0.5	1.58	4.65	1.61	5.81	8.06	8.15
(3) 4	1.0	1.79	2.03	1.56	1.56	8.55	6.88
(4) 4	3.0	6.94	24.92	7.73	28.27	29.34	28.63
(5) 4	9.0	24.78	92.53	25.14	93.90	84.01	89.52
(6) 10	0.0	0.38	2.38	0.38	1.11	0.38	0.42
(7) 10	0.5	1.76	6.08	1.99	6.77	9.52	8.23
(8) 10	1.0	2.94	6.61	2.11	2.11	13.56	5.23
(9) 10	3.0	17.54	64.90	17.84	66.00	66.46	62.77
(10) 10	9.0	55.88	208.19	57.88	216.11	209.04	212.87
(11) 20	0.0	0.17	0.40	0.17	0.38	0.17	0.16
(12) 20	0.5	2.85	9.88	3.03	10.47	12.69	10.42
(13) 20	1.0	3.56	10.11	2.27	2.28	16.64	3.89
(14) 20	3.0	22.71	83.98	20.97	77.48	80.78	82.01
(15) 20	9.0	66.85	249.52	67.40	251.50	249.03	245.63

Sources of Error. To analyze the underlying causes of these patterns, Fig. 5 decomposes the total MAE (for $|\sigma| = 10$) into hours of the week, mimicking a typical calendar page for each subplot. Darker cells indicate higher error contributions. The three regimes are characterized by distinct error generation mechanisms:

Regime 1 ($\delta_\sigma \in \{0.0\}$): T_C-based errors are homogeneously low, as the model easily learns dense, single-day traces. Conversely, T_W-based errors peak towards the end of the day, and particularly on Friday. This stems from the discontinuity in g^{-1} when mapping back to T_C. Small overestimations in T_W near the end of a working interval translate into substantial T_C errors by pushing the prediction into the next day or week.

Regime 2 ($\delta_\sigma \in \{0.5, 1.0\}$): T_C-based errors concentrate heavily around the edges of the working week and days. This represents the model's inability to handle the noise of boundary interval crossings, even with calendar-aware features (see P1 in Sect. 2.4). The worktime transformation delivers its primary benefit here by structurally mitigating this boundary ambiguity, though residual discontinuity errors remain.

Regime 3 ($\delta_\sigma \in \{3.0, 9.0\}$): Errors in both T_C and T_W are distributed uniformly. The high frequency of boundary crossings results in a regression to the mean,

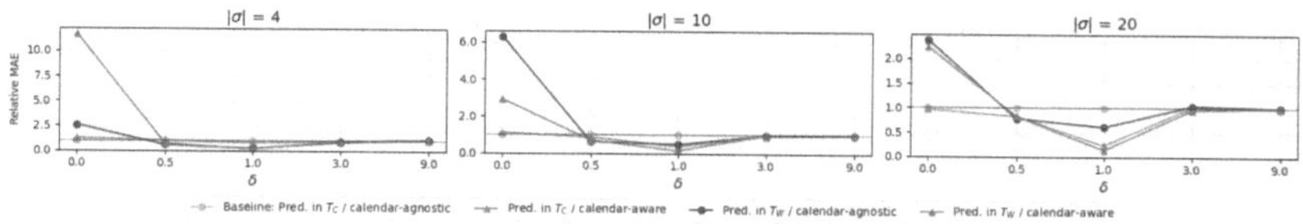

(a) All synthetic event logs.

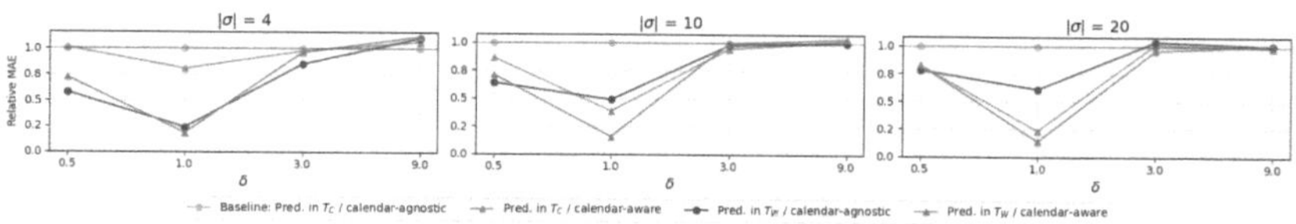

(b) Excluding event logs with $\delta = 0$.

Fig. 4. Normalized Mean Absolute Error per synthetic event log and feature set. Lowest error per event log highlighted in blue, second-lowest in gray. (Color figure online)

neutralizing the impact of individual interruptions. Consequently, the informative value of both calendar-aware features and the time transformation is limited.

In summary, the T_W-based prediction fails in the first regime due to erroneous crossovers into subsequent days or weeks, excels in the second regime by alleviating uncertainty from boundary crossings, and performs on par with T_C-based predictions in the third regime as boundary crossings even out over longer temporal gaps.

5.2 Validation on Real-World Event Log Data

Raw error. Table 2 presents the MAE across the six real-world event logs. For five of the six logs, T_W-based predictions yield the best performance. Compared to the optimal T_C-based baseline, T_W configurations achieve error reductions of 6.0% for Request for Payment, 3.1% for BPI 2012 W and Helpdesk, 2.0% for Prepaid Travel Cost, and 0.1% for BPI 2012 A. For BPI 2012O, the error increases marginally by less than 0.1%. Overall, improvements in predictive accuracy are achieved using both calendar-agnostic and calendar-aware feature sets, depending on the specific event log.

While T_W predictions in general improve predictive performance, the relative reductions are smaller than in synthetic data. This discrepancy arises because synthetic logs isolate temporal ambiguity, which our transformation targets directly, whereas real-world logs introduce multi-dimensional contextual ambiguity (complex control flows, internal actors, external behaviors) [17] that remains unaffected by time axis modifications.

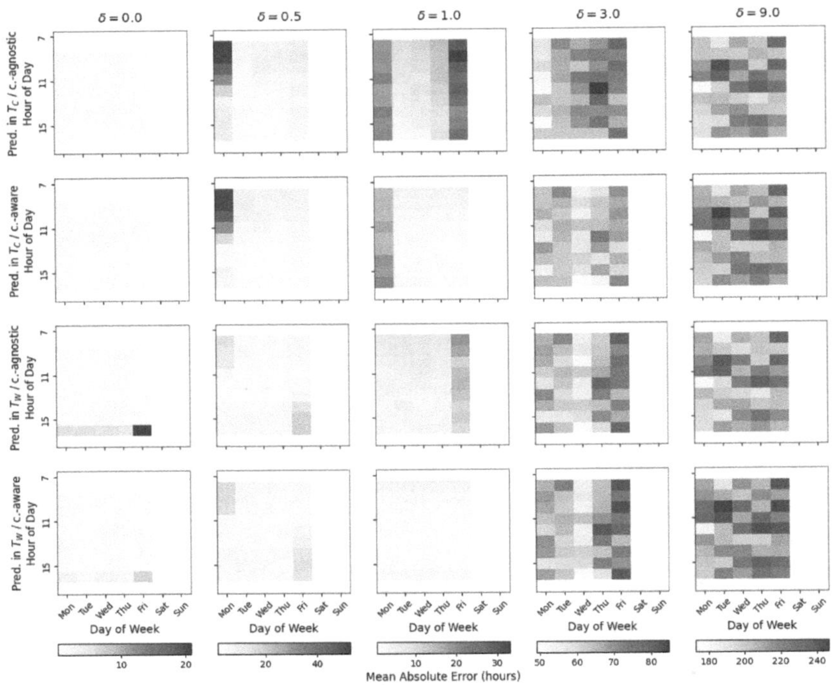

Fig. 5. Heatmap of Mean Absolute Error per δ (all with $|\sigma| = 10$), time axis, feature set and hour in the week (of true value).

Table 2. Mean Absolute Error (hours) per event log and feature set.

Prediction in:	Worktime T_W				Clocktime T_C	
Feature set:	calendar-agnostic		calendar-aware		c.-agnostic	c.-aware
Metric in:	T_W	T_C	T_W	T_C	T_C	T_C
Event log						
(1) BPI 2012 A	26.54	44.99	26.54	45.00	45.94	45.29
(2) BPI 2012 O	23.64	59.54	23.67	59.56	59.87	59.52
(3) BPI 2012 W	6.62	16.16	6.69	16.32	16.67	16.71
(4) Helpdesk	36.67	107.89	34.94	103.12	110.97	106.38
(5) Prepaid travel cost	34.95	107.73	34.84	106.25	109.32	108.37
(6) Request for payment	17.57	55.77	17.21	52.74	57.58	56.10

Sources of Error. Decomposing the MAE into weekly hours reveals why these differences occur (Fig. 6). In four of the six logs, T_C-based predictions exhibit elevated error rates around boundary intervals, following two main patterns:

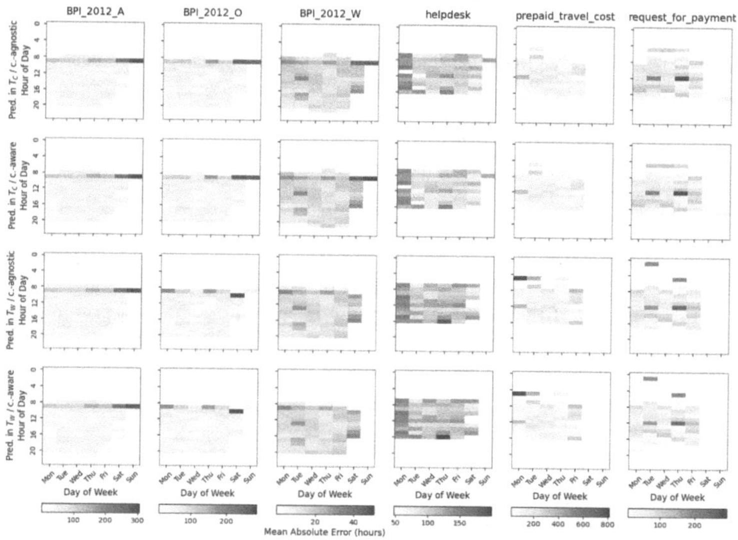

Fig. 6. Heatmap of Mean Absolute Error per real-world event log, time axis, feature set and hour in the week (of true value).

- **Daily Boundaries:** In BPI 2012 A, O, and W, errors peak early in the working day and progressively amplify toward the end of the week.
- **Weekly Boundaries:** In the Helpdesk log, errors concentrate predominantly at the start of the week following the weekend gap.

Applying the worktime transformation demonstrably alleviates some of these error spikes, particularly noticeable in the BPI 2012O and W logs.

6 Conclusion

In this paper, we introduced a worktime-based time representation to address the systematic biases of working schedules inherent in Predictive Process Monitoring. By transforming the standard clocktime axis into a worktime axis, we effectively eliminate non-working periods where process progress is stagnant. Our evaluation across synthetic and real-world event logs demonstrates that this transformation mitigates three critical limitations. Specifically, it reduces the increased error at temporal edges, it prevents the violation of business schedules, and it resolves the obfuscation of active workload.

The findings open several avenues for future research. First, future work could investigate the application of the transformation to further prediction tasks, such as remaining time or next activity prediction. Second, future work could assess the impact of the transformation on further prediction models, such as tree-based models, transformer-based architectures or graph neural networks. Third, future work could develop ensemble prediction methods that utilize the identified, specific strengths of the transformation, such as combinations of clocktime and worktime models.

Reproducibility. The source code required to reproduce our experimental analysis is available under https://github.com/maalevo/WARP (https://doi.org/10.5281/zenodo.20408504).

Acknowledgments. The research was supported by the Einstein Foundation Berlin under grant EPP-2019-524, by the German Federal Ministry of Research, Technology and Space under grant 16DII1433, and by Deutsche Forschungsgemeinschaft under grant 496119880. Kate Revoredo is funded by the Berliner Chancengleichheitsprogramm (BCP) as part of the DiGiTal Graduate Program.

References

1. Ceravolo, P., Comuzzi, M., De Weerdt, J., Di Francescomarino, C., Maggi, F.M.: Predictive process monitoring: concepts, challenges, and future research directions. Process Sci. **1**(1), 2 (2024). https://doi.org/10.1007/s44311-024-00002-4
2. Di Francescomarino, C., Ghidini, C., Maggi, F.M., Petrucci, G., Yeshchenko, A.: An eye into the future: Leveraging a-priori knowledge in predictive business process monitoring. In: Business Process Management, vol. 10445, pp. 252–268 (2017). https://doi.org/10.1007/978-3-319-65000-5_15
3. Donti, P.L., Rolnick, D., Kolter, J.Z.: DC3: A learning method for optimization with hard constraints (Apr 2021). https://doi.org/10.48550/arXiv.2104.12225
4. Dumas, M., La Rosa, M., Mendling, J., Reijers, H.A.: Fundamentals of Business Process Management, Berlin, Heidelberg (2018). https://doi.org/10.1007/978-3-662-56509-4
5. Duong, L.T., Travé-Massuyès, L., Subias, A., Merle, C.: Remaining cycle time prediction with graph neural networks for predictive process monitoring. In: Proceedings of the 2023 8th International Conference on Machine Learning Technologies, pp. 95–101. (2023). https://doi.org/10.1145/3589883.3589897
6. Estrada-Torres, B., Camargo, M., Dumas, M., García-Bañuelos, L., Mahdy, I., Yerokhin, M.: Discovering business process simulation models in the presence of multitasking and availability constraints. Data Knowl. Eng. **134**, 101897 (2021). https://doi.org/10.1016/j.datak.2021.101897
7. Klinkmüller, C., Van Beest, N.R.T.P., Weber, I.: Towards reliable predictive process monitoring. In: Information Systems in the Big Data Era, vol. 317, pp. 163–181 (2018). https://doi.org/10.1007/978-3-319-92901-9_15
8. Maldonado, A., Frey, C.M.M., Tavares, G.M., Rehwald, N., Seidl, T.: GEDI: Generating event data with intentional features for benchmarking process mining. In: Business Process Management. vol. 14940, pp. 221–237 (2024). https://doi.org/10.1007/978-3-031-70396-6_13
9. Marquez-Chamorro, A.E., Resinas, M., Ruiz-Cortes, A.: Predictive monitoring of business processes: a survey. IEEE Trans. Serv. Comput. **11**(6), 962–977 (2018). https://doi.org/10.1109/TSC.2017.2772256
10. Mendling, J., Leopold, H., Meyerhenke, H., Depaire, B.: Methodology of algorithm engineering. ACM Comput. Surv. **58**(4), 1–38 (2026). https://doi.org/10.1145/3769071

11. Navarin, N., Vincenzi, B., Polato, M., Sperduti, A.: LSTM networks for data-aware remaining time prediction of business process instances. In: 2017 IEEE Symposium Series on Computational Intelligence (SSCI), pp. 1–7. (2017). https://doi.org/10.1109/SSCI.2017.8285184
12. Neu, D.A., Lahann, J., Fettke, P.: A systematic literature review on state-of-the-art deep learning methods for process prediction. Artif. Intell. Rev., 1–27 (2021). https://doi.org/10.1007/s10462-021-09960-8
13. Nguyen, A., Chatterjee, S., Weinzierl, S., Schwinn, L., Matzner, M., Eskofier, B.: Time matters: Time-aware LSTMs for predictive business process monitoring. In: Process Mining Workshops, pp. 112–123 (2021). https://doi.org/10.1007/978-3-030-72693-5_9
14. Oyamada, R.S., Tavares, G.M., Junior, S.B., Ceravolo, P.: Enhancing predictive process monitoring with time-related feature engineering. In: Advanced Information Systems Engineering. vol. 14663, pp. 71–86. https://doi.org/10.1007/978-3-031-61057-8_5
15. Rama-Maneiro, E., Vidal, J., Lama, M.: Deep learning for predictive business process monitoring: Review and benchmark. IEEE Trans. Serv. Comput. 1–1. https://doi.org/10.1109/TSC.2021.3139807
16. Roider, J., Nguyen, A., Zanca, D., Eskofier, B.M.: Assessing the performance of remaining time prediction methods for business processes. IEEE Access **12**, 130583–130601. https://doi.org/10.1109/ACCESS.2024.3459648
17. Rosemann, M., Recker, J., Flender, C.: Contextualisation of business processes. Int. J. Bus. Process. Integr. Manag. **3**(1), 47 (2008). https://doi.org/10.1504/IJBPIM.2008.019347
18. Suriadi, S., Ouyang, C., Van Der Aalst, W.M., Ter Hofstede, A.H.: Event interval analysis: why do processes take time? Decis. Support Syst. **79**, 77–98 (2025). https://doi.org/10.1016/j.dss.2015.07.007
19. Tax, N., Verenich, I., La Rosa, M., Dumas, M.: Predictive business process monitoring with LSTM neural networks. In: Advanced Information Systems Engineering, vol. 10253, pp. 477–492 (2017). https://doi.org/10.1007/978-3-319-59536-8_30
20. Van Der Aalst, W.: Process Mining, Berlin, Heidelberg (2016). https://doi.org/10.1007/978-3-662-49851-4
21. Verenich, I., Dumas, M., Rosa, M.L., Maggi, F.M., Teinemaa, I.: Survey and cross-benchmark comparison of remaining time prediction methods in business process monitoring. ACM Trans. Intell. Syst. Technol. **10**(4), 1–34 (2019). https://doi.org/10.1145/3331449

David vs. Goliath in Next Activity Prediction: Argmax vs. LSTM, Transformer, and LLM

Hans Weytjens[1(✉)] and Ingo Weber[1,2]

[1] Technical University of Munich, School of CIT, Munich, Germany
{hans.weytjens,ingo.weber}@tum.de
[2] Fraunhofer Gesellschaft, Munich, Germany

Abstract. Next activity prediction (NAP) is a cornerstone of predictive process monitoring (PPM), enabling organizations to move from retrospective analysis to proactive process steering. The PPM field has progressed from classical machine learning through deep learning architectures such as LSTMs and Transformers to large language models (LLMs). Despite growing model complexity, no benchmark jointly compares LLMs, Transformers, LSTMs, and simple baselines in a direct sequence modeling setting for NAP. In this paper, we fill this gap with a systematic benchmark. We compare vocabulary-adapted LLMs, Transformers trained from scratch, LLM-distilled Transformers, and LSTMs against a simple counting-based argmax baseline across seven real-life event logs. Our results tell a David vs. Goliath story: pretraining confers no consistent improvement over training from scratch, model size shows little effect on performance, and on most datasets the argmax baseline matches or approaches the performance of billion-parameter LLMs.

Keywords: Predictive Process Monitoring · Next Activity Prediction · LLM Fine-tuning · Distillation · Transformer

1 Introduction

Next activity prediction (NAP) constitutes one of predictive process monitoring (PPM)'s cornerstones. It enables organizations to move from retrospective analysis to proactive process steering.

Initially, researchers turned to classical machine learning (ML) techniques such as decision trees, support vector machines, etc. for NAP tasks [21]. As deep learning (DL) models became increasingly available, the field experimented with some multi-layer perceptrons (MLPs), before pivoting to convolutional neural networks (CNNs), and more importantly, recurrent neural networks (RNNs) with an emphasis on Long Short-Term Memory Networks (LSTMs) [25]. Later, some researchers followed up with Transformers [4,12,13,28]. The breakthrough of general-purpose large language models (LLMs) opened a new NAP research

F. M. Maggi et al. (Eds.): BPM 2026, LNBIP 599, pp. 324–341, 2027.
https://doi.org/10.1007/978-3-032-37877-4_19

direction with text-serialization approaches, using in-context learning (ICL) [15, 20], fine-tuning [5,17,20], Retrieval-Augmented Generation (RAG) [5], etc.

Oyamada et al. [14] applied a vocabulary-adapted direct sequence modeling approach (as used in classical ML, DL, and Transformers) to pretrained LLMs: They adapted the LLM's vocabulary by replacing the LLMs' input and output layers by custom embedding and linear layers to account for the NAP dataset's smaller vocabulary and different tokenization. These customized LLMs were then fine-tuned using the NAP datasets. The goal was to "evaluate the LLM's intrinsic capability to interpret sequential process information [...] bypassing the need for natural language conversion" [14].

In our work, we wish to test this hypothesis and establish to what extent the favorable comparison of the vocabulary-adapted, fine-tuned LLMs to LSTMs can be attributed to their pretraining. We also investigate how smaller pretrained LLMs compare with legacy GPT-2 and more recent Qwen-3 models. For lack of pretrained small LLMs, we use distillation to derive such models from larger teacher models. To relate to earlier research, we compare the Transformer-based models to LSTMs, and additionally, a simple count-based argmax model. The latter comparison led us to pose the provocative question of whether these advanced DL models are worth the effort.

Within the context of direct sequence modeling approaches for NAP, this paper makes the following contributions:

- A systematic evaluation of the effect of generic LLM pretraining versus training from scratch.
- A distillation-based method to compress fine-tuned LLMs into smaller model architectures for which no pretrained versions exist.
- A comprehensive benchmark, comparing a spectrum of modern sequence modeling approaches– from LSTMs through Transformers to LLMs of varying scale and vintage.
- A sobering reality check: by including a count-based argmax baseline absent from prior work, we reveal the degree to which reported progress in NAP may have been overstated.

In the following section, we provide background about NAP, LSTMs, Transformers, and LLMs and derive our research questions from it. Section 3 explains the methodology and setup for our experiments. Next, we present the experimental results and answers to the research questions in Sect. 4. We interpret these results in Sect. 5, before concluding the paper in Sect. 6. The code for our work is available in our repository at https://github.com/hansweytjens/DavidGoliath.

2 Background and Research Questions

2.1 Predictive Process Monitoring (PPM) and Next Activity Prediction (NAP)

PPM concerns predictions about the future of ongoing business process executions from their prefixes. In their systematic literature review of PPM methods, Di Francescomarino et al. [6] identify three broad prediction categories: (1)

numeric, such as time and cost predictions, (2) categorical, such as risk and categorical outcome predictions, and (3) next activities (plural) predictions. Within the latter, next activity (singular) prediction can be viewed as a specific case.

A process execution or case $c \in \mathcal{C}$ is represented by a trace $t_c = \langle e_1, e_2, \ldots, e_n \rangle$, where e_i denotes the i-th event and $\mathcal{C}$ the set of all case identifiers. Each event e_i consists of an activity label a_i, a timestamp τ_i, and, optionally, further event attributes. Formally, $e_i = (a_i, \tau_i, \mathcal{A}_i)$ where $\mathcal{A}_i$ is the (possibly empty) set of additional attributes. An event log $\mathcal{L}$ is a multiset of traces, i.e., $\mathcal{L} = \{t_c \mid c \in \mathcal{C}\}$, with traces $t_c = \langle e_1, \ldots, e_n \rangle$, for which $\tau_i \leq \tau_{i+1}$. A trace's prefix of length k, with $1 \leq k \leq n$, is defined as $\text{pref}_k(t_c) = \langle e_1, \ldots, e_k \rangle$. The set of all prefixes is therefore $\mathcal{P} = \{\text{pref}_k(t_c) \mid t_c \in \mathcal{L},\ 1 \leq k \leq |t_c|\}$. Given a running case with an observed prefix $\sigma = \text{pref}_m(t_c)$, with $1 \leq m < |t_c|$, next activity prediction aims to predict the activity label a_{m+1} of the next event e_{m+1}.

2.2 Long Short-Term Memory Networks (LSTMs) for NAP

In 1997, Hochreiter et al. [10] proposed the LSTM, an enhancement of the RNN, particularly suited to capture long-term dependencies in sequential data. Two decades later, Evermann et al. [7] and Tax et al. [22] applied LSTMs to NAP tasks. LSTMs emerged as a strong and widely adopted technique for NAP, often outperforming classical machine learning approaches [22], even though they are not universally superior to all alternative neural architectures [19]. However, the sequential nature of LSTMs limits parallelization and requires information to be propagated across many time steps, making the modeling of very long-range dependencies in event log sequences challenging. These limitations motivated the adoption of attention-based architectures such as the Transformer.

2.3 Transformers and Large Language Models

Twenty years after the LSTM, Vaswani et al. [23] proposed the Transformer, originally formulated as an encoderdecoder architecture based on attention mechanisms. The following year, Radford et al. [18] adopted a decoder-only Transformer and demonstrated that large-scale self-supervised pretraining with a language modeling objective, followed by task-specific fine-tuning, yields strong performance across diverse downstream tasks. This model family, termed the Generative Pre-trained Transformer (GPT), forms the basis of many modern LLMs – a term that was adopted as model capacity and training data scale increased.

In practice, when applying Transformer architectures and LLMs to dataset-driven tasks (rather than conversational settings), several modes of use can be distinguished. First, a Transformer model can be trained from scratch, i.e., initialized with random weights and optimized directly on the target dataset, analogous to conventional DL models. Second, with in-context learning (ICL) [3], a pretrained LLM is used without modifying its parameters by providing a prompt that may contain task instructions, contextual information, and data ranging from a few representative examples to an entire dataset in structured form. Third, the model can be fine-tuned [18], where the pretrained weights are further adapted to a specific domain or task using supervised or self-supervised objectives on task-specific data. Finally, knowledge distillation [9] can be applied, where a smaller "student" model is trained on task-specific data using supervision derived from a larger pretrained "teacher" model (e.g., soft labels or logits), enabling more efficient deployment while retaining much of the teacher's performance.

2.4 Transformers and LLMs in NAP

We distinguish two main approaches to using Transformers and LLMs for NAP: namely text-serialization approaches, which convert event data to natural language, and direct sequence modeling approaches, which preserve the structured nature of event data by representing event attributes directly as learned embeddings, without natural language conversion. Our work tackles the latter category.

1) Text-serialization approaches aim to leverage the contextual reasoning capabilities of LLMs by transforming structured event log data into textual encodings that can be processed by LLMs.

Rebmann et al. [20] found NAP too challenging for in-context learning (ICL) with off-the-shelf LLMs but reported improved performance after task-specific fine-tuning. As an alternative to both ICL and fine-tuning, Casciani et al. [5] rely on Retrieval-Augmented Generation (RAG) to provide selected event data as additional context to the model. Rather than fine-tuning on a single task such as NAP, Pyrih et al. [17] adopt instruction-tuning across multiple downstream tasks, achieving mixed results with modest improvements for NAP compared to earlier findings [20]. Further variations include fine-tuning LLMs directly on serialized XES logs [2], and training encoder-only models for suffix prediction [15]. The text-serialization approach permits fine-tuning LLMs on multiple event log datasets from different domains [17,20], effectively realizing transfer training. However, since some publicly available datasets may have been included in LLM pretraining corpora, data contamination poses a potential threat to the validity of empirical evaluations.

2) Direct sequence modeling has two variants:

Training Transformers from scratch: Wuyts et al. [28] trained a randomly-initialized encoder-decoder Transformer for suffix prediction. Later, Bukhsh et al. [4] opted for a decoder-only variant, claiming superior NAP performance compared to earlier methods including LSTM. Higher accuracy rates were claimed by adding a mixture-of-experts (MoE) mechanism [13] and transfer learning [12]. Berti et al. [1] also use an MoE and train an "in-context foundation model" using several synthetic datasets at once.

Vocabulary Adaptation: is a transfer learning strategy in which a pretrained LLM backbone is reused while a corpus-specific vocabulary and corresponding tokenization scheme is defined a priori for the target dataset. This induces a new input embedding matrix and output projection that are trained to map the new tokens into and out of the pretrained model's latent space, whereas the core Transformer backbone remains fixed or only minimally adapted. De Vries et al. [24] explored this idea in the context of cross-lingual transfer. Oyamada et al. [14] extended the approach to the PPM domain for NAP and remaining time predictions, applying various parameter-efficient fine-tuning (PEFT) techniques.

2.5 Research Questions

Against this backdrop, we are now ready to formulate our research questions. In this paper, we investigate the added value of generic next-token pretraining in LLMs for NAP using vocabulary adaptation and systematically comparing pretrained LLM backbones to Transformers trained from scratch as well as to smaller distilled models. To structure this investigation, we pose the following research questions:

RQ1: Does generic LLM pretraining improve NAP performance?
RQ2: Can knowledge distillation from fine-tuned teacher models improve the performance of smaller architectures?
RQ3: Do larger models outperform smaller ones?
RQ4: Do more recent LLM architectures (e.g., Qwen-3 from 2025) outperform earlier ones (e.g., GPT-2 from 2019)?
RQ5: Does the high complexity of (fine-tuned) LLMs and Transformers offer performance gains over LSTMs and an argmax baseline?

3 Methodology and Experimental Setup

3.1 Models and Baseline

Table 1 lists the models[1] used in our experiments. We selected three models from OpenAI's open source GPT-2[2][3] LLM family enabling controlled scaling comparisons across model sizes under a shared architecture, training corpus (OpenWebText, ≈ 40 GB), vocabulary, and training procedure. To this, we added GPT-2-mini[4], a similar but smaller model trained on a subset of OpenWebText. The Qwen-3 LLM series[5] adds more modern and larger GPT-style models to our selection, whilst retaining some capacity overlap with the larger GPT-2 models. Our choice for GPT-2 was also motivated by Andrej Karpathy's nanoGPT,

1 Model sizes differ from the official numbers because we modified the vocabulary, which changes the input and output layers. Since each dataset has a different vocabulary, model sizes vary slightly across datasets; the reported values correspond to BPI12. Architectural differences between GPT-2 and nanoGPT also lead to different parameter counts.

2 Hugging Face GPT-2 checkpoints: `openai-community/gpt2`, `gpt2-medium`, and `gpt2-large`, (https://huggingface.co/openai-community, accessed 2026-03-10).

3 To avoid confusion, we refer to "GPT-2" as "GPT-2-small".

4 https://huggingface.co/erwanf/gpt2-mini, accessed 2026-03-10.

5 https://huggingface.co/collections/Qwen/qwen3, https://qwen.ai/blog?id=qwen3, accessed 2026-03-10.

Table 1. Models. Sizes refer to values for BPI12.

Model	Total Params	Hidden (embedding) size	Layers	Att. heads
GPT-2				
GPT-2-large	774 M	1,280	36	20
GPT-2-medium	354 M	1,024	24	16
GPT-2-small	124 M	768	12	12
GPT-2-mini (not OpenAI)	39 M	512	4	8
Qwen-3				(Q/KV)
Qwen-3-4B	4 B	2,560	36	32/8
Qwen-3-1.7B	1.7 B	2,048	28	16/8
Qwen-3-0.6B	596 M	1,024	28	16/8
nanoGPT				
Nano-768-12-12	85 M	768	12	12
Nano-512-12-8	38 M	512	12	8
Nano-256-12-4	9.5 M	256	12	4
Nano-128-12-2	2.4 M	128	12	2
Nano-768-4-12	28 M	768	4	12
Nano-512-4-8	13 M	512	4	8
Nano-256-4-4	3.2 M	256	4	4
Nano-128-4-2	0.8 M	128	4	2
Nano-64-4-1	0.2 M	64	4	1
Nano-64-2-1	0.1 M	64	2	1
LSTM				
LSTM-256-2	1.1 M	256	2	–
LSTM-128-2	0.3 M	128	2	–
LSTM-64-2	0.07 M	64	2	–
LSTM-64-1	0.04 M	64	1	–
Baseline				
argmax	–	–	–	–

a minimal GPT-2style Transformer implementation[6] which we instantiate in different sizes, reflected in their names in Table 1. We use nanoGPT for from-scratch training and distillation. To compare with a classical ML model, we also included four LSTM models of different sizes.

Finally, we compare these models with an **argmax** baseline. Our David is a first-order empirical transition model that predicts the next activity as $a_{k+1} = \arg\max_a P(a \mid a_k)$. $P(a \mid a_k)$ is estimated from the relative transition frequencies

[6] https://github.com/karpathy/nanochat, accessed 2026-03-10.

in the training set. It conditions only on the current activity and requires no optimization.

3.2 Datasets and Preprocessing

We use the BPI challenge event logs[7] BPI12, BPI15, BPI17, BPI19, Prepaid Travel Costs (BPI20_PTC), Travel Permit Data (BPI20_TPD), and Request for payment (BPI20_RFP)[8]. Table 2 demonstrates how these datasets vary along multiple dimensions affecting NAP difficulty. A high share of top-5 and top-10 prefixes (BPI19: 0.42/0.59, BPI20_RFP: 0.58/0.78) means that a small number of distinct prefixes occur very frequently, indicating strong concentration in the prefix distribution. A low weighted conditional entropy[9] (BPI15: 0.25, BPI17: 0.32) indicates that, given a prefix, the next activity is highly predictable. Datasets combining frequent prefixes with few classes and low conditional entropy (BPI20_RFP, BPI20_PTC) are expected to favor the argmax baseline. BPI15 – with 168K unique prefixes out of 201K and 324 classes – presents the opposite case, despite its low entropy. The performance of more sophisticated models also depends on the availability of explanatory variables in the dataset (unknown) and the number (known) and quality (unknown) of samples to learn from.

We reject traces with fewer than three events. For greater relevance, we incorporate lifecycle information by augmenting activity labels where available. We attach an "EOS" token to each case because case completion is itself a relevant next-activity prediction outcome. We do not cap the number of classes to observe the detrimental effect of high numbers (see BPI15). To avoid data

Table 2. Datasets after preprocessing (as per Sect. 3.2).

Dataset	No. of cases	No. of classes	No. of prefixes	No. of unique prefixes	Top 5 prefixes (%)	Top 10 prefixes (%)	Weighted conditional entropy
BPI12	9,487	37	173,498	37.060	16.93	24.36	0.40
BPI15	4,411	324	201,243	168,344	3.70	4.65	0.25
BPI17	29,306	63	1,100,691	246,844	9.22	13.98	0.32
BPI19	156,072	41	914,988	40,058	41.67	59.03	0.75
BPI20_PTC	1,791	30	15,730	940	30.03	44.69	0.47
BPI20_TPD	6,847	52	83,620	13,134	20.76	28.04	0.58
BPI20_RFP	5,701	18	30,153	245	57.57	78.39	0.49

[7] https://data.4tu.nl/, accessed 2026-03-10.

[8] These are the same datasets as Oyamada et al. [14], with the exception of BPI15. All other cited papers using real-life datasets [2,4,5,7,12,13,15,19,22,28] relied at least partially on some of these as well.

[9] $H = -\sum_{\sigma \in \mathcal{P}} P(\sigma) \sum_{a \in \mathcal{A}} P(a \mid \sigma) \log_2 P(a \mid \sigma)$.

leakage and test set bias, we build train and test sets (20%) according to Weytjens et al. [26]. A validation set is then extracted from the training set. We calculate the "accumulated_time" (since the case's first event's timestamp) attribute for each event.

3.3 Training

Data Representation and Model Adaptation: Preliminary test runs using the validation set (see our code repository) led to the selection of the following attributes[10] for training: "activity" (all datasets), "resource" (all), "amount" (BPI12, BPI17), and "accumulated_time" (all). The numerical features are standardized. The categorical feature "activity_name" is concatenated with the "lifecycle" information (e.g., "start", "complete") when available. Categorical values are mapped from strings to integer IDs using the training dataset vocabulary (with special PAD/UNK/EOS tokens). Each ID is mapped via a learned embedding table to an E-dimensional vector (with E equal to the model's embedding/hidden size). In parallel, the numerical feature is projected to the same E-dimensional space through a linear layer. The categorical and numerical representations are then combined by element-wise addition ("sum" strategy) before being passed to the backbone. Prefixes are variable-length event sequences and are right-padded with PAD tokens to the longest trace in each mini-batch; an attention mask is applied to ensure that padded positions are ignored by the attention mechanism. The effective maximum sequence length is bounded by the backbone context window.

General: Following the same procedure [26] as for the test set, we carve validation sets (10%) out of the training sets. We minimize the cross-entropy loss with the Adam optimizer. Batch sizes are 64. We do not regularize the models, but enforce early stopping (patience = 10) to combat overfitting. We conducted preliminary pilot experiments to select appropriate learning rates. Based on these experiments, a learning rate of 0.005 was used for nanoGPT and smaller GPT-2 models (up to GPT-2-medium), while 0.0005 was used for larger models. Learning-rate tuning results are available in our code repository.

Fine-Tuning: We replaced the input and output layers of the pretrained LLMs to map embeddings and logits to our dataset-specific vocabulary, applying weight-tying [16] between the input embedding and output projection matrices, while keeping the Transformer backbone architecture intact. We used parameter-efficient fine-tuning (PEFT [8]), exploring selective layer freezing strategies [11]. Given our vocabulary adaptation of both input and output layers, we considered unfreezing both bottom and top layers. Our experiments showed that freezing all blocks except the first and last, with LayerNorm kept trainable, consistently matched or outperformed other configurations including full fine-tuning. Ablation results are available in our code repository.

[10] The attribute names are standardized; original names differ across datasets.

Distillation: We use logits-only knowledge distillation, where the student is trained to match the teacher's output distribution directly. This design is architecture-agnostic and requires no intermediate feature alignment. The training objective $\mathcal{L} = (1 - \alpha)\mathcal{L}_{CE} + \alpha\mathcal{L}_{KL}$ combines hard-label cross-entropy and KL divergence to teacher logits, with α increased from 0.1 to 0.9 over training. For each dataset, the fine-tuned model (GPT-2-x or Qwen-3-x) with the lowest validation loss assumes the role of teacher, with all Nano-x-x-x models its students.

Hardware and Framework: We used NVIDIA A100 GPUs and a NVIDIA GB10 Grace-Blackwell superchip, depending on availability. All LLMs were sourced from Hugging Face and run through the Transformers library.

3.4 Evaluation Metrics

We evaluate model performance using two complementary metrics: accuracy and macro-F1, both commonly used in the next-activity prediction literature.

Accuracy measures the proportion of prefixes for which the correct next activity is predicted. However, the datasets exhibit substantial class imbalance, as indicated by the prefix concentration in Table 2. In such settings, predicting only the dominant activity may yield high accuracy while failing on rare activities. We therefore additionally report macro-F1, the harmonic mean of recall and precision, which assigns equal weight to each activity class regardless of frequency. Metrics are computed at the prefix level, where each prefix corresponds to one prediction instance in the test set.

Rather than reporting the full set of statistical significance tests inline, we provide Wilcoxon signed-rank test statistics and p-values in the code repository. Figures 4, 5 and 6 include standard deviations to support visual interpretation.

4 Results

Figures 1, 2 and 3 present the complete results; the figures in the remainder of this section highlight cross-sections corresponding to each research question. Preliminary hyperparameter search results (attribute selection, layer freezing, learning rates) are available in our code repository (see Sect. 1).

4.1 Pretraining Effect: Direct Measurement (RQ1)

In Fig. 4, we compare the fine-tuned GPT-2-mini and GPT-2-small to their corresponding nanoGPT versions (Nano-512-4-8 and Nano-768-12-12), trained from scratch. No consistent pattern across datasets or model sizes emerges. Performance differences vary in magnitude and direction, indicating that pretraining yields no systematic advantage.

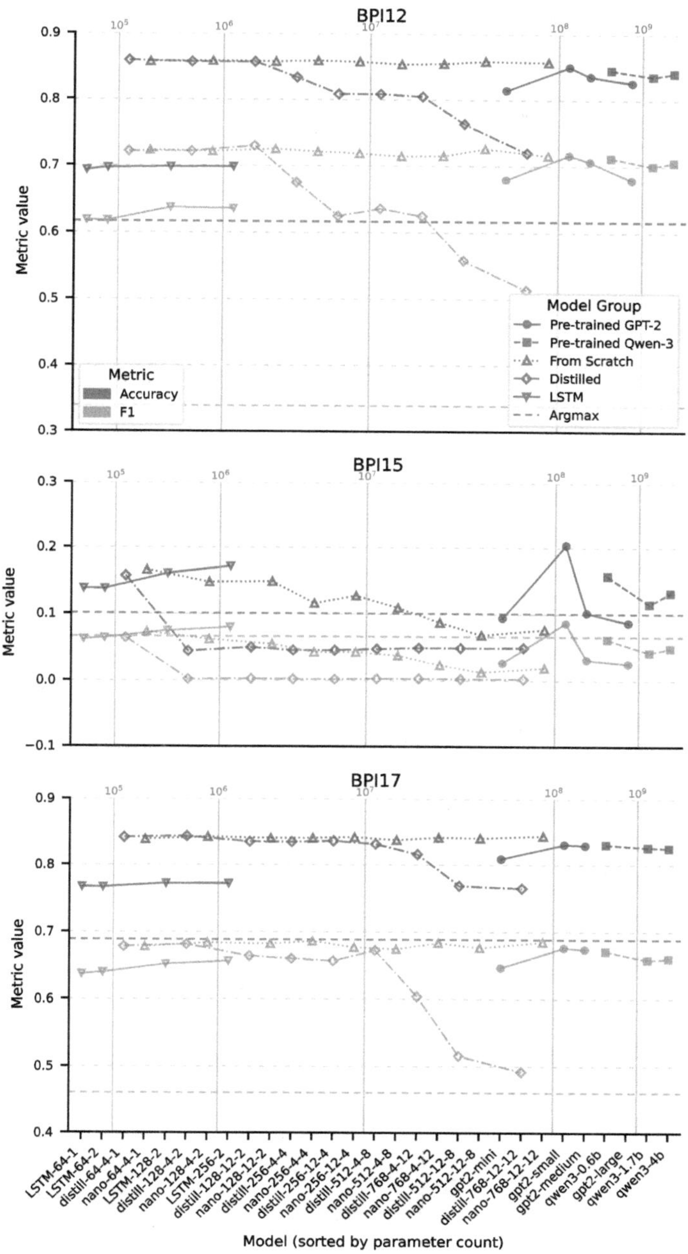

Fig. 1. Accuracy and macro-F1 as a function of total number of parameters. All models are displayed at equidistance, with vertical grid lines showing approximate size on a log scale. The vertical axes differ per dataset, but are at the same scale for better comparison. Standard deviations are omitted for clarity but are reported in tables in the code repository (see Sect. 1).

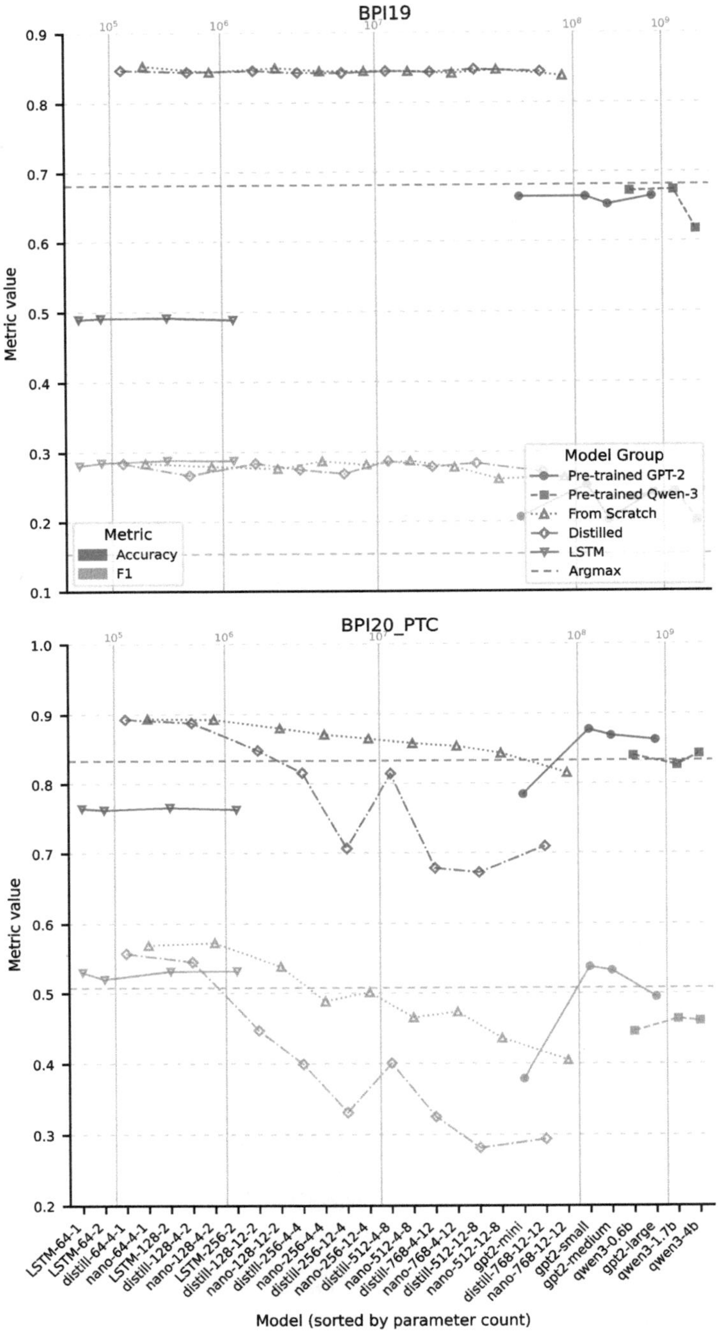

Fig. 2. Accuracy and macro-F1 as a function of total number of trainable parameters. Continuation of graphs on previous page.

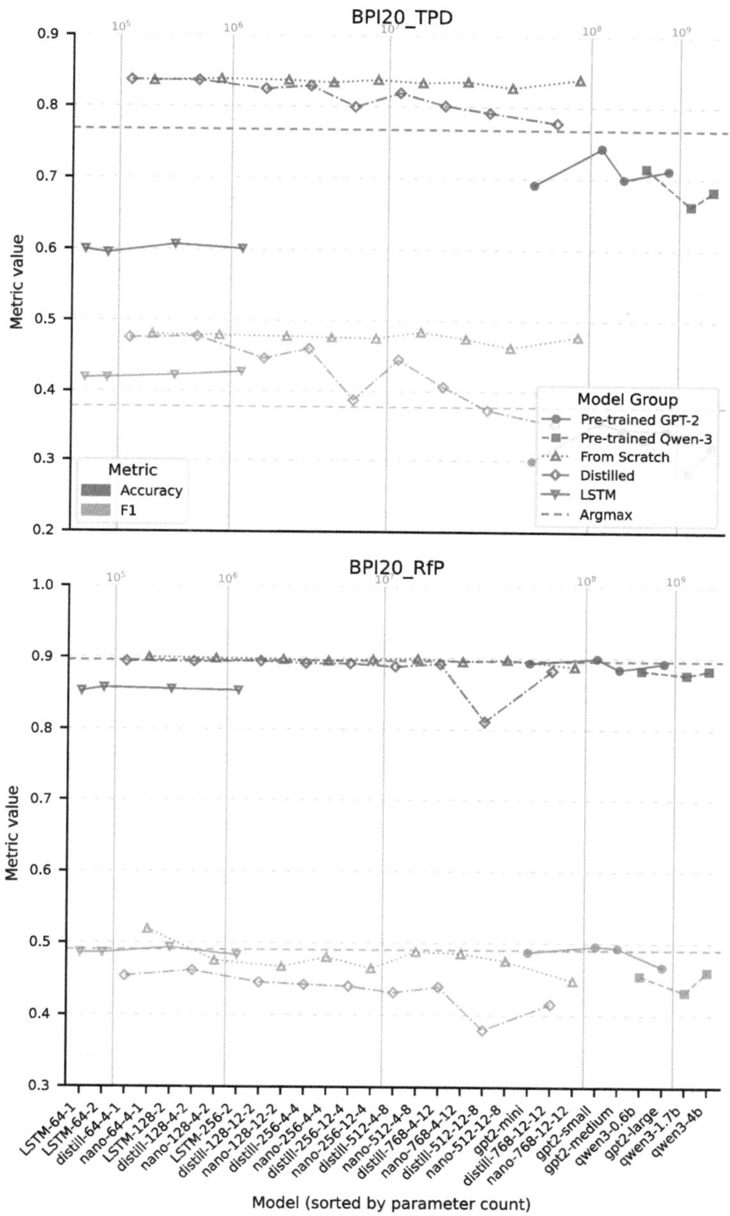

Fig. 3. Accuracy and macro-F1 as a function of total number of trainable parameters. Continuation of graphs on two previous pages.

RQ1 Result: We observe no consistent added value of pretraining.

4.2 Pretraining Effect: Measurement with Distillation (RQ2)

To investigate whether the observed phenomenon holds for smaller model architectures for which we have no pretrained version, we use distillation to "create" smaller fine-tuned models: the distilled student model learns concurrently from both the training data and the task-specific output distribution of its pretrained and fine-tuned teacher (see Sect. 2.3). The student thus inherits the teacher's task-specific distribution, itself a product of its pretraining, carrying richer supervision than hard labels alone [9]. We train both from scratch and distill all remaining nanoGPT models. Across model sizes in Fig. 5, average performance differences between distilled models and their corresponding models trained from scratch are mostly negative despite variation across datasets, suggesting that distillation provides no systematic benefit.

Closer inspection of Figs. 1, 2 and 3 reveals that the smallest models trained from scratch score higher on both metrics than the much larger fine-tuned GPT-2 and Qwen-3 models, further undermining the case for using pretrained models.

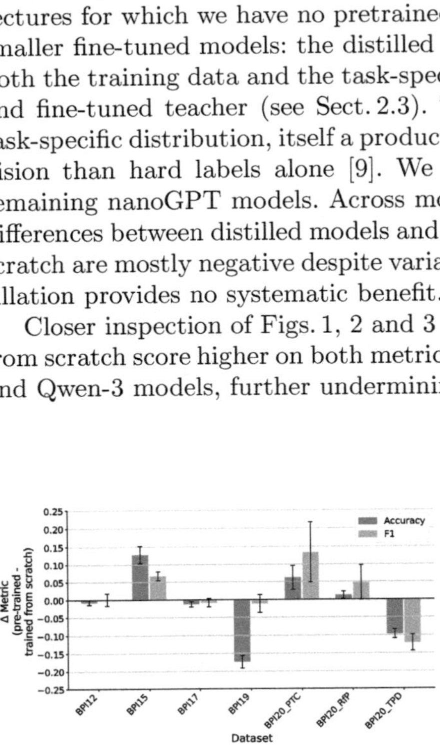

(a) GPT-2-mini vs. Nano_512_4_8.

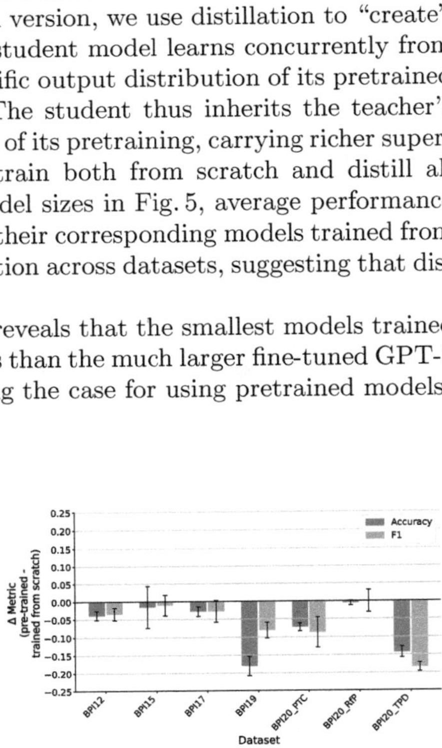

(b) GPT-2-small vs. Nano_768_12_12.

Fig. 4. Net pretraining effect for two model sizes.

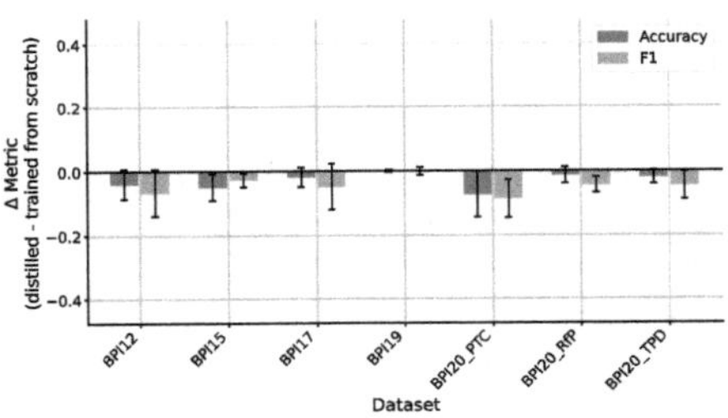

(a) Per dataset, averaged over models.

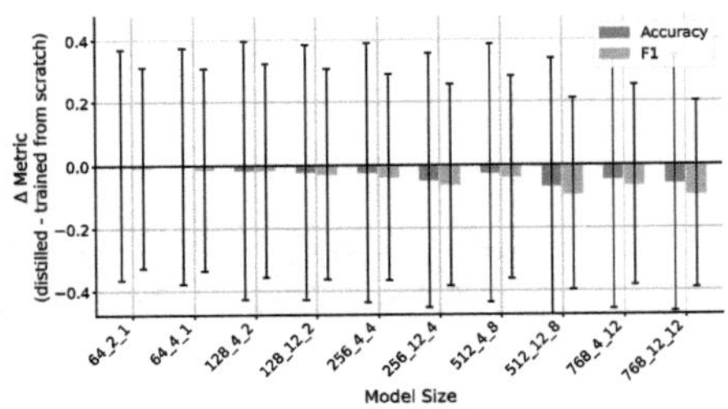

(b) Per model, averaged over datasets.

Fig. 5. Performance difference between distilled models and their corresponding models trained from scratch. Negative values indicate an advantage for training from scratch.

RQ2 Result: Logits-only distillation from fine-tuned pretrained models does not improve smaller architectures over training them directly from scratch.

4.3 Model Capacity and Modernity (RQ3 and RQ4)

Figures 1, 2 and 3 show that for most model families (nanoGPT, GPT-2, Qwen-3) results are very similar across sizes within the respective families. The early stopping mechanism effectively prevents overfitting during training. The family of distilled models is the exception: performance *degrades* with model size.

RQ3 Result: Within the range tested, increasing model size confers no consistent performance benefit.

For similar sizes, the newer Qwen-3 models outperform the GPT-2 models slightly for BPI12 and BPI15, but lag behind for the BPI20 datasets. The picture is unclear for BPI17 and BPI19. In summary, our limited sample suggests no advantage for newer models.

RQ4 Result: The more recent Qwen-3 architecture shows no consistent advantage over GPT-2.

4.4 LSTMs and Baseline (RQ5)

Overall, the LLMs (fine-tuned GPT-2 and Qwen-3) reach higher accuracy than the LSTMs, as shown in the left panel of Fig. 6. Interestingly, the reverse is true for the macro-F1, where the LLMs only score significantly better on the BPI12 dataset. The picture for the LLMs further darkens when comparing them against the argmax baseline (right panel): only for the BPI12 and BPI17 datasets, the LLMs outperform argmax.

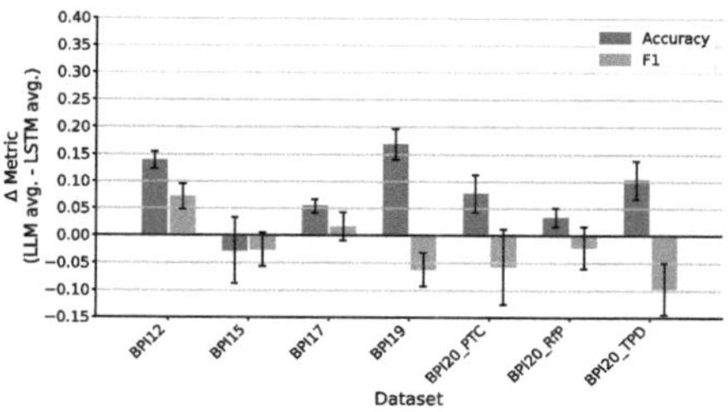

(a) LLM vs. LSTM

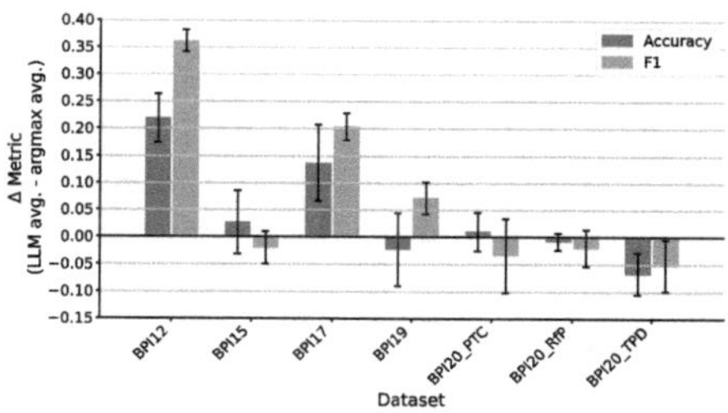

(b) LLM vs. baseline

Fig. 6. LLMs vs. LSTM and baseline per log. Averaged over all models within the family (fine-tuned GPT-2 and Qwen-3).

RQ5 Result: LSTMs come close to the LLMs' performance when considering both metrics. But for datasets with few training samples, high prefix variance, insufficient explanatory variables, and/or many different activities, defaulting to an argmax model is the practitioner's best option.

4.5 Wall-Clock Time

As modern GPUs are fast and relatively affordable, wall-clock time is of secondary importance compared to accuracy and F1. On our hardware, times range from 1–2 s for argmax, 0.5–11 min for LSTM, 0.6–15 min for nanoGPT models, 0.4–38 min for the distilled models, 2–56 min for GPT-2, and 4–281 min for Qwen-3. More details can be found in our code repository (see Sect. 1).

5 Discussion

Overall, the results indicate limited benefits of using pretraining and model scaling in our setting. A simple frequency-based baseline can be a competitive option, although comparatively small Transformer models trained from scratch achieve the best predictive performance on several datasets.

Datasets. Across datasets, the ranking of models by accuracy correlates highly with rankings by F1 (except for LSTM, see RQ5), validating our conclusions across both metrics. Only BPI12 and BPI17 yield meaningful improvements over the argmax baseline. The F1 gains exceed the accuracy gains, suggesting that learned models handle minority activity classes better than the argmax baseline. BPI15 proves intractable: 324 activity classes overwhelm all models. The high accuracy but very low F1 for BPI19 reflects the dataset's class imbalance: models learn the dominant classes while ignoring rare ones. The high argmax baseline scores for the BPI20 datasets are explained by their relatively low number of activity classes and highly skewed prefix distributions (see Table 2). This leaves little room for learned models to improve upon our simple argmax model.

Why Goliath stumbled: Our research suggests that LLMs' pretrained priors are not effectively leveraged for NAP, as fine-tuning alone does not resolve the vocabulary mismatch. Moreover, the relatively low business process complexity (see Table 2) does not appear to warrant larger and more complex architectures. Instead, comparatively small models trained from scratch can fit the target distribution directly, unencumbered by mismatched priors.

Implications: Our research has three major implications. First, researchers and practitioners alike should first consider a simple and explainable baseline model like our argmax. None of the NAP papers we cited included such a simple baseline model, creating uncertainty about the relevance of the proposed techniques, models, and methods. Second, when the baseline proves insufficient and a more powerful model is warranted, priority should be given to training small Transformer models from scratch, as our results suggest that this is more beneficial than increasing model size or relying on pretrained or distilled architectures. Third, for datasets with high activity diversity or limited training data, uncertainty quantification [27] can improve practical utility by trading coverage for accuracy.

Limitations: The validity and generalizibility of our results is limited by our selection of pretrained models from the GPT-2 and Qwen-3 families only, none larger than 4B. Training from scratch and distillation was only performed on one architecture (nanoGPT). Only part of the hyperparameter space was explored. Underperformance of distilled models may partly reflect limitations of logits-only distillation rather than the absence of transferable pretraining knowledge. On the data side, only 7 datasets were used, all drawn from the BPI challenge logs commonly used in the PPM literature; broader coverage across domains and process types would improve the generalizability of our findings. The datasets may not contain all relevant explanatory variables. Process interdependencies were ignored.

Future Research: We suggest extending this work to other PPM tasks, potentially in a multitask setting, and considering process interdependencies. Transfer learning [1] with alternative generic representations for different datasets could improve results. Enhancing our sequence modeling approach with symbolic AI techniques or combining it with classical process mining techniques present further research directions. Stratifying by prefix length could reveal whether learned models gain an advantage over argmax as context grows.

6 Conclusion

David held his own. On most datasets, a count-based argmax model – conditioning only on the last activity – performs close to fine-tuned LLMs up to eight orders of magnitude larger, despite the latter having access to the complete prefix. However, comparatively small Transformer models trained from scratch often outperform both the argmax baseline and fine-tuned LLMs. Only on BPI12 and BPI17 do Transformer models clearly justify the additional complexity. The broader message for researchers and practitioners is clear: do not bring a giant to a fight where a sling will do. The PPM field's tendency to benchmark against other DL models, rather than against simple baselines, has obscured how much of the apparent progress is of practical relevance. Furthermore, smaller Transformer models trained from scratch consistently match or outperform fine-tuned LLMs, suggesting that generic language priors learned during pretraining do not transfer to NAP via vocabulary-adapted direct sequence modeling.

Rather than discouraging further research, these findings establish a clear starting point for the community. Together with dataset benchmarks and standardized preprocessing for NAP [26], the findings provide the infrastructure against which genuine progress can be measured. With the bar now clearly set, future work can move beyond architectural scaling and focus on what actually drives performance—whether through richer process-aware representations, transfer learning, multi-task objectives, or entirely new modeling paradigms.

Note on AI Use: We used AI agents, based on GPT-5.2 (later 5.4) and Sonnet-4.6 for coding, visualization, and language editing assistance. *Credits:* We thank Daniel Emrani for contributing to earlier implementation work.

References

1. Berti, A., Van Der Aalst, W.M.: An in-context foundation model for predictive process monitoring on event logs. IEEE Access (2026)
2. Brennig, K., Kaltenpoth, S., Müller, O.: Straight outta logs: can large language models overcome preprocessing in next event prediction? In: Gdowska, K., Gómez-López, M.T., Rehse, JR. (eds.) BPM Workshops, pp. 197–208. Springer Nature Switzerland, Cham (2025). https://doi.org/10.1007/978-3-031-78666-2_15
3. Brown, T., et al.: Language models are few-shot learners. In: Advances in Neural Information Processing Systems, vol. 33, pp. 1877–1901 (2020)
4. Bukhsh, Z.A., Saeed, A., Dijkman, R.M.: ProcessTransformer: predictive business process monitoring with transformer network (2021). https://arxiv.org/abs/2104.00721
5. Casciani, A., Bernardi, M.L., Cimitile, M., Marrella, A.: Enhancing next activity prediction in process mining with retrieval-augmented generation. Inf. Syst. **137**, 102642 (2026). https://doi.org/10.1016/j.is.2025.102642
6. Di Francescomarino, C., Ghidini, C., Maggi, F.M., Milani, F.: Predictive process monitoring methods: which one suits me best? In: Weske, M., Montali, M., Weber, I., vom Brocke, J. (eds.) BPM 2018. LNCS, vol. 11080, pp. 462–479. Springer, Cham (2018). https://doi.org/10.1007/978-3-319-98648-7_27
7. Evermann, J., Rehse, J.R., Fettke, P.: Predicting process behaviour using deep learning. Decis. Support Syst. **100**, 129–140 (2017). https://doi.org/10.1016/j.dss.2017.04.003
8. Han, Z., Gao, C., Liu, J., Zhang, J., Zhang, S.Q.: Parameter-efficient fine-tuning for large models: a comprehensive survey. arXiv preprint arXiv:2403.14608 (2024)
9. Hinton, G., Vinyals, O., Dean, J.: Distilling the knowledge in a neural network. arXiv preprint arXiv:1503.02531 (2015)
10. Hochreiter, S., Schmidhuber, J.: Long Short-Term Memory. Neural Comput. **9**(8), 1735–1780 (1997). https://doi.org/10.1162/neco.1997.9.8.1735
11. Hwang, T., Seo, H., Jung, J., Jung, S.: Exploring selective layer freezing strategies in transformer fine-tuning: nli classifiers with sub-3b parameter models. Appl. Sci. **15**(19) (2025). https://doi.org/10.3390/app151910434, https://www.mdpi.com/2076-3417/15/19/10434
12. van Luijken, M., Ketykó, I., Mannhardt, F.: An experiment on transfer learning for suffix prediction on event logs. In: De Weerdt, J., Pufahl, L. (eds.) International Conference on BPM, pp. 31–43. Springer, Cham (2023). https://doi.org/10.1007/978-3-031-50974-2_3
13. Nguyen, T.H., Pham, D.L., Kim, K.P.: Next process-activity prediction using switch-transformer: approach, visualization, and performance evaluation. Knowl. Inf. Syst. **67**(12), 11827–11854 (2025)
14. Oyamada, R.S., Peeperkorn, J., Weerdt, J.D., Smedt, J.D.: Domain adaptation of LLMs for process data (2025). https://arxiv.org/abs/2509.03161

15. Pasquadibisceglie, V., Appice, A., Malerba, D.: Lupin: a LLM approach for activity suffix prediction in business process event logs. In: 2024 6th ICPM, pp. 1–8. IEEE (2024)
16. Press, O., Wolf, L.: Using the output embedding to improve language models. In: Proceedings of the 15th Conference of the European Chapter of the Association for Computational Linguistics: Volume 2, Short Papers, pp. 157–163 (2017)
17. Pyrih, V., Rebmann, A., van der Aa, H.: LLMs that understand processes: instruction-tuning for semantics-aware process mining. In: 2025 7th ICPM, pp. 1–8. IEEE (2025)
18. Radford, A., Narasimhan, K., Salimans, T., Sutskever, I.: Improving language understanding by generative pre-training (2018)
19. Rama-Maneiro, E., Vidal, J.C., Lama, M.: Deep learning for predictive business process monitoring: review and benchmark. IEEE Trans. Serv. Comput. **16**(1), 739–756 (2021)
20. Rebmann, A., Schmidt, F.D., Glavaš, G., van Der Aa, H.: Evaluating the ability of LLMs to solve semantics-aware process mining tasks. In: 2024 6th ICPM, pp. 9–16. IEEE (2024)
21. Tama, B.A., Comuzzi, M.: An empirical comparison of classification techniques for next event prediction using business process event logs. Expert Syst. Appl. **129**, 233–245 (2019). https://doi.org/10.1016/j.eswa.2019.04.016
22. Tax, N., Verenich, I., La Rosa, M., Dumas, M.: Predictive business process monitoring with LSTM Neural Networks. In: Dubois, E., Pohl, K. (eds.) CAiSE 2017. LNCS, vol. 10253, pp. 477–492. Springer, Cham (2017). https://doi.org/10.1007/978-3-319-59536-8_30
23. Vaswani, A., et al.: Attention is all you need. In: Advances in Neural Information Processing Systems, vol. 30 (2017)
24. de Vries, W., Nissim, M.: As Good as new. How to successfully recycle English GPT-2 to make models for other languages. In: Findings of the Association for Computational Linguistics: ACL-IJCNLP 2021, pp. 836–846 (2021)
25. Weinzierl, S., et al.: An empirical comparison of deep-neural-network architectures for next activity prediction using context-enriched process event logs (2020). https://arxiv.org/abs/2005.01194
26. Weytjens, H., De Weerdt, J.: Creating unbiased public benchmark datasets with data leakage prevention for predictive process monitoring. In: Marrella, A., Weber, B. (eds.) BPM Workshops, pp. 18–29. Springer International Publishing, Cham (2022). https://doi.org/10.1007/978-3-030-94343-1_2
27. Weytjens, H., De Weerdt, J.: Learning uncertainty with artificial neural networks for predictive process monitoring. Appl. Soft Comput. **125**, 109134 (2022)
28. Wuyts, B., Vanden Broucke, S., De Weerdt, J.: Sutran: an encoder-decoder transformer for full-context-aware suffix prediction of business processes. In: 2024 6th ICPM, pp. 17–24. IEEE (2024)

Track III: Management

Systematic and Automated Diagnosis of Process Models Using Reference Frameworks

Simone Agostinelli[1], Francesca De Luzi[2(✉)], Sara Giovannetti[2], Andrea Marrella[2], and Fabio Nonino[2]

[1] Universitas Mercatorum, Rome, Italy
simone.agostinelli@unimercatorum.it
[2] Sapienza Universitá di Roma, Rome, Italy
{deluzi,giovannetti,marrella,nonino}@diag.uniroma1.it

Abstract. Reference frameworks in Business Process Management (BPM) are generic best-practice templates used to design, analyze, benchmark, and improve business processes (BPs). They promote consistency in BP structure and terminology and support BP model diagnosis by revealing gaps and weaknesses through comparison with a reference structure. Despite extensive research on the development and adaptation of reference frameworks, the BPM literature offers limited methodological support for their systematic application to existing BP models, leaving much of the analysis to manual work by domain experts. However, this analysis typically demands significant time, human, and financial resources and does not scale well to enterprise-wide BP sets, limiting its practical use for BP diagnosis. In this paper, following a Design Science Research (DSR) methodology, we analyze existing approaches that leverage reference frameworks for BP diagnosis to identify their key characteristics, and propose a maturity model to systematically guide their application to existing BP models. To demonstrate feasibility, we use a Large Language Model (LLM) to automate the application of the maturity model, generating suggestions across all diagnostic levels.

Keywords: Reference framework · Systematic literature review · Maturity model · Automated diagnosis of process models · DSR methodology

1 Introduction

Business process standardization (PS) is the practice of establishing uniform procedures across an organization to ensure business processes (BPs) are performed in a consistent way [11]. PS is recognized as a key component of business process management (BPM) principles [29] and has been shown to positively influence organizational performance [24]. By reducing variability in the specification of BP activities, standardized BPs provide a stable foundation for initiatives such as

F. M. Maggi et al. (Eds.): BPM 2026, LNBIP 599, pp. 345–361, 2027.
https://doi.org/10.1007/978-3-032-37877-4_20

BP identification [7], automation [8], and outsourcing [28]. Additionally, adopting standardized BPs enables organizations to effectively identify and address issues and inefficiencies within their BPs [5].

Among the PS strategies documented in the literature [11], several rely on the use of *reference frameworks* [4]. Reference frameworks consist of reusable BP representations that capture industry best practices, such as the Information Technology Infrastructure Library (ITIL[1]) and the APQC Process Classification Framework (PCF[2]). According to Dumas et al. [7], reference frameworks are essential tools for identifying major organizational process areas, structuring BPs into a hierarchical classification that captures their relationships, and defining performance metrics (KPIs). In addition, they support BP redesign through completeness checks and a common labeling vocabulary for BP activities.

Although reference frameworks are widely adopted by leading companies worldwide [7], their effective application remains challenging, as it requires in-depth knowledge of BPs and careful alignment with organizational practices [15]. As a result, PS initiatives often involve complex, organization-wide implementations and demand significant investments in time, human resources, and financial effort [1]. Surprisingly, despite extensive research on the development and use of reference frameworks [11], the BPM literature offers limited methodological support for their systematic application to existing BPs. Consequently, analyses based on reference frameworks largely rely on manual expert effort, which does not scale to enterprise-wide BP sets, limiting their practical use for BP diagnosis.

In this paper, we tackle this issue by proposing a *maturity model* for the *systematic diagnosis* of BPs modeled in BPMN (ISO/IEC 19510:2013) using reference frameworks. Assuming in input a syntactically correct BPMN model along with its process profile, which outlines its purpose, scope, inputs, outputs, resources and KPIs, the maturity model is structured into five levels, namely *labeling*, *mapping*, *granularity*, *scoping*, and *targeting*. These levels, derived from a systematic literature review (SLR) of approaches that use reference frameworks for BP diagnosis, guide the stepwise alignment of a BP with best practices in reference frameworks. They support tasks ranging from refining activity names and adjusting BP granularity to identifying scope issues and verifying whether the BP effectively contributes to the organization's KPIs. To demonstrate feasibility, we implemented the maturity model as a collaborative tool powered by a Large Language Model (LLM), which automates suggestions across all diagnostic levels and reduces manual effort in BP diagnosis.

This paper follows the Design Science Research (DSR) methodology outlined in [16]. After introducing the problem and motivating its relevance through a review of related work in Sect. 2 (`Step 1: Problem explication`), we conduct an SLR to identify the key characteristics of reference frameworks in Sect. 3, which inform the requirements of our approach (`Step 2: Requirements specification`). We then present our maturity model and illustrate its application to a real-world BP model in Sect. 4 (`Step 3: Design &`

[1] axelos.com/certifications/itil-service-management.

[2] apqc.org/pcf.

`Development`). Next, in Sect. 5, we demonstrate how the analysis enabled by the maturity model can be automated with an LLM (`Step 4: Demonstration`). Finally, Sect. 6 discusses limitations and outlines directions for future work.

2 Related Work

The adoption of reference frameworks to support the standardization of BPs has been extensively investigated in the BPM literature. A first research stream analyzes how organizations can benefit from reusable BP knowledge embedded in reference frameworks. For example, in [11], the authors explore 12 types of PS strategies (some including the use of reference frameworks) and apply them to 21 real-world BPs, with the aim of supporting practitioners in selecting the most suitable PS strategy for a given application domain. This study shows that no reference framework can be universally applied to any context. Nevertheless, some works explore how to adapt and extend organizational reference models through questionnaire-driven configuration [3,20], even if their effectiveness is limited to internal best practices rather than generic reference frameworks [19].

A second line of research focuses on configurable reference models [27,33], where variability is explicitly embedded into BP models to support adaptation to different organizational contexts. While these approaches enable reuse through configuration points and process variants, they assume prior integration of the reference model into the organization and do not address how existing BPs are mapped and aligned to the reference models. Other contributions investigate the role of reference frameworks in PS and redesign. For instance, the works [24, 29] discuss the impact of PS on organizational performance and outline factors influencing successful adoption. Similarly, the work [7] highlights how reference frameworks can support BP identification and redesign. However, these studies provide general recommendations rather than step-by-step methods.

Overall, even if the literature demonstrates the value of reference frameworks for PS, it largely lacks methodological support for their systematic application to existing BPs, leaving the actual alignment process dependent on manual, expert-driven analysis, which is often impractical. Indeed, as reported in [30], the complexity and number of BPs within an organization directly affect PS activities, influencing whether standardization is feasible, the extent to which it can be applied, and the effort required to implement it. This gap motivates the *need for approaches that provide systematic and automated guidance for applying reference frameworks to perform BP diagnosis*, as addressed in this paper.

3 Systematic Literature Review

In this section, we summarize the planning and result phases of the SLR conducted to identify the characteristics of state-of-the-art reference frameworks.
Search Protocol. The SLR was driven by the research question (RQ): *What are the key characteristics of reference frameworks to consider when applied to existing BPs?* The search string for the SLR was developed based on the authors'

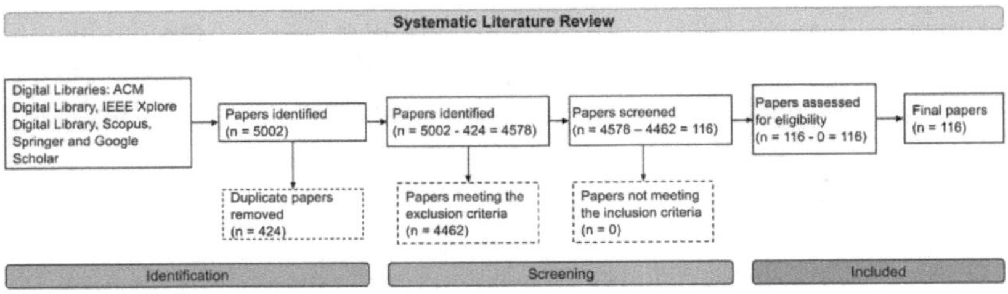

Fig. 1. Overview of the SLR phases

domain knowledge. Starting from the main term "*reference framework*" derived from the research question, we expanded the query with relevant synonyms (such as "*taxonomy*", "*reference model*", "*classification framework*", etc.) to capture different term variations and ensure comprehensive coverage of relevant studies. The search terms were connected with AND, while synonyms and alternative spellings were combined using OR. The resulting search string is shown below:

(business process ***OR*** *workflow* ***OR*** *business processes* ***OR*** *workflows* ***OR*** *supply chain* ***OR*** *BPM* ***OR*** *business process management)* ***AND*** *(reference framework* ***OR*** *classification framework* ***OR*** *taxonomy* ***OR*** *reference model* ***OR*** *process standardization* ***OR*** *process architecture)*

The mentioned search string was applied on five search engines, namely Google Scholar, Scopus, SpringerLink, IEEE Xplore Digital Library and ACM Digital Library, retrieving studies based on the occurrence of the string in the title, the keywords or the abstract of a paper. To ensure a rigorous SLR, we defined inclusion and exclusion criteria to retain only relevant studies. The exclusion criteria were formulated to exclude: (E1) non-English studies; (E2) non-peer-reviewed papers; (E3) studies that are not electronically accessible or require paid access; (E4) papers deemed irrelevant or out of scope. Consequently, we applied a single (implicit) inclusion criterion (I1) to retain only the studies not explicitly filtered out by the exclusion criteria.

Results Analysis. The search string applied to the search engines retrieved 5,002 papers. After removing 424 duplicates and applying the exclusion criteria to the remaining 4,578, 116 papers were selected for analysis, as shown in Fig. 1. In the following, for the sake of space, we present a summary of the analysis conducted to answer the research question. Interested readers can explore the details of the SLR in: https://github.com/bpm-diag/AutoReF.

A first noteworthy result is that the literature on reference frameworks provides many ad-hoc guidelines, taxonomies, and best practices for constructing syntactically correct BP models from scratch. Several studies investigate the quality of BP model syntax by proposing constraints and techniques to ensure the correctness of workflow specifications [17,26]. The Seven Process Modeling Guidelines (7PMG) consolidate the best practices for improving the representational quality of BP models, including specific attention to the textual style used for syntactic clarity [22]. In particular, the *verb-object style* is widely recommended in the literature as the least ambiguous and most effective approach, based on empirical findings and experimental evidence [22,23]. Moreover, label refactoring approaches using parsing algorithms highlight the importance of *consistent activity labeling*, though they have limited generalizability [21].

Other identified studies from the SLR examine the features of well-known reference frameworks, such as APQC Process Classification Framework (PCF), Information Technology Infrastructure Library (ITIL), and Supply Chain Operations Reference (SCOR) [14,31,32]. These works highlight the importance of *granularity*, stressing the need to decompose organizational functions to an appropriate level of detail to enable process linkage [31]. This principle is further supported by the 7PMG, which strongly recommend BP decomposition [22]. Furthermore, considering the structure of the aforementioned reference frameworks, BPs are typically grouped into areas or process groups based on their *scope*, which refers to the boundaries that define what is included and what is excluded within the framework. More specifically, scope determines which BPs are covered (e.g., core, supporting, or management BPs) and which part of the organization is addressed (e.g., entire enterprise or a specific department) [14].

Finally, several works highlight a sustained focus on BP improvement, with approaches based on semantic similarity measures aimed at refining existing BPs to support the achievement of organizational KPIs [10,18]. In this regard, reference frameworks such as ITIL and APQC PCF support benchmarking and enable organizations to assess and improve performance [14].

Based on the SLR results, we identified the following key characteristics of reference frameworks to consider when diagnosing and standardizing BPs:

- *Labeling clarity*: Ensures consistent, unambiguous naming of BP elements.
- *Levels of granularity*: Determine the appropriate degree of refinement at which BP elements should be represented.
- *Scope classification*: Categorizes BP activities based on their shared characteristics or functions, defining the boundaries within which they apply.
- *Goal orientation*: Focuses on associating BP activities with specific KPIs.

4 Maturity Model

In this section, based on the characteristics of reference frameworks identified through the SLR, we propose a *maturity model* designed to systematically enhance the BP's compliance with a selected reference framework. As shown in Fig. 2, the proposed maturity model consists of five levels. The final outcome of the maturity model is a Process Diagnostic Report (PDR), which tracks the BP model's evolution through the maturity model's stages, similar to how a clinical record follows a patient's journey.

We selected the APQC Process Classification Framework (PCF) as the sample reference framework. We chose PCF because it is universally applicable,

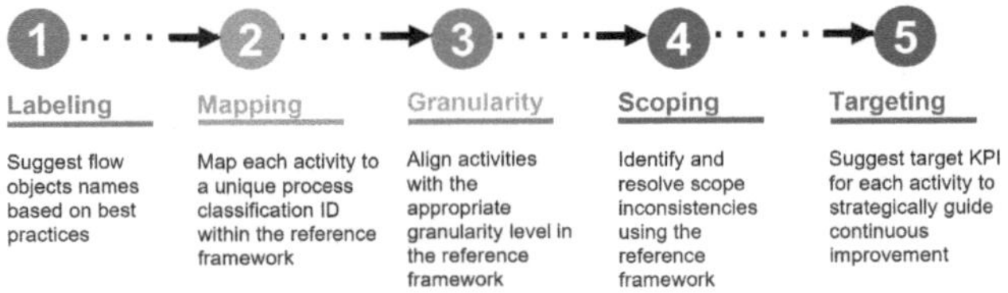

Fig. 2. Overview of the five levels of the maturity model.

regardless of industry type, organization size, or geographic location. Additionally, PCF embodies all the key characteristics identified in the previous section, making it a suitable choice for practical application within a BP model. Specifically, the PCF offers a structured approach to categorizing BPs across an organization, ensuring that all BPs are grouped into one of its 13 high-level categories (e.g., manufacturing and production, finance and accounting, etc.). Each BP or activity within the framework is assigned a unique identifier (ID). The PCF organizes BPs hierarchically in 5 levels (L1-Category, L2-Process Group, L3-Process, L4-Activity, L5-Task), with broad categories at the top (e.g., core, support, and management BPs) and more detailed sub-processes at a lower level. Furthermore, the framework ensures alignment between BP activities and business objectives, guaranteeing that the BP under analysis may contribute directly to the organization's KPIs.

4.1 Running Example

As a running example, we use a BP model that describes a real-world BP for managing a request to create a customized (ad-personam) product by a manufacturer, a scenario common in the luxury manufacturing sector. The BP begins when the manufacturer accepts the customer's request, cf. Fig. 3:

- First, the request is documented, initiating a preliminary analysis. The After Sales employee then requests an internal assessment to evaluate feasibility and determine the manufacturing requirements.
- Next, the Cost and Quality employee conducts the internal analysis, represented as a sub-process. Once completed, the results are integrated into the main process and used to develop the customer proposal. The analysis consists of three activities: (i) feasibility control, where the details of the ad-personam request are examined for technical and production feasibility; (ii) time schedule verification, which evaluates the required time for creating the customized product based on resource availability, production planning, and impact on existing schedules; and (iii) cost analysis, which estimates costs considering materials, production time, and production adaptations.
- After two working days, the procedure's status is checked. If not completed, an apology is sent to the customer, and the BP waits an additional three working days before proceeding. Once the results are available, they are retrieved by the main process and used to prepare the proposal.

- Based on the analysis results, a commercial proposal is created, detailing the product's manufacturing conditions and price. The proposal is then sent to the Financial Controller for evaluation. If the proposal is deemed inadequate, the price is revised by adjusting the economic conditions until a satisfactory solution is reached. If approved, the BP moves forward by updating the price in the company's management system (SAP).
- Finally, two notifications are sent: one to the customer, informing them of the proposal's outcome and final conditions, and another to the Boutique Retailer, who must be updated for commercial and operational purposes.

In this context, a process profile is defined as a descriptive input to the BP model, specifying elements such as triggers (e.g., customer request), outputs (e.g., commercial proposal), KPIs (e.g., customer satisfaction, operational costs, etc.), vision (efficient handling of customized orders), resources (After Sales Employee, Financial Controller, etc.), and process ownership.

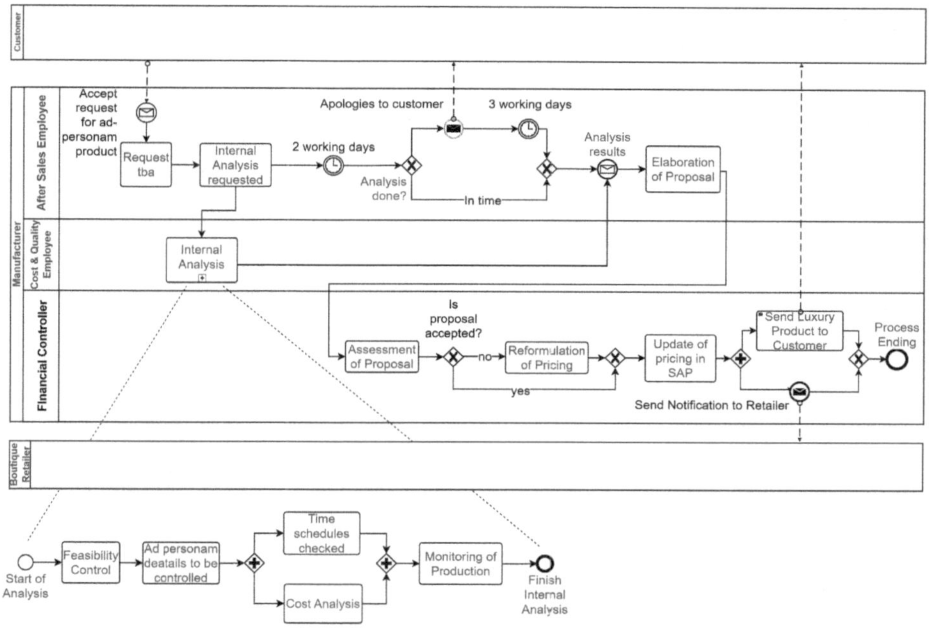

Fig. 3. BP model of the running example.

4.2 Level 1 Labeling

At this level, we assume the input BP model is syntactically correct in terms of structural rules governing its graph-based structure (i.e., free of deadlocks, isolated nodes, or connectivity errors). According to the literature [7], quality assurance begins with verification (syntactic quality). However, we extend this definition to include control over meaningful labels as an additional requirement [23]. Although labeling is often linked to certification (pragmatic quality), it is a key bottleneck for automated BP diagnosis. This level serves as the foundation for the maturity model: a BP that is not syntactically correct or uses ambiguous labeling can not be reliably mapped to reference frameworks. Therefore, Level 1 focuses on the grammatical refinement of the name attribute of all flow objects. In this direction, activities and other flow objects must use labels that adhere to specific naming conventions to guarantee the pragmatic quality of the BP model [7]. By adopting recognized best practices, such as the Camunda Naming Guidelines[3], Level 1 compliance is achieved by identifying attribute names that do not conform to the verb-object style and transforming them into correct labels. This convention has been shown to reduce ambiguity and improve users' perceived usefulness of BP models [23].

Level-1 compliant model

The first step takes a syntactically correct BPMN model as *input*, refactors non-compliant flow objects' labeling issues, and produces a labeling-compliant model as *output*, populating the first section of the PDR.

Consider the BP model of the running example. In its initial state, several elements violate Level 1 requirements despite structural correctness. The labels follow inconsistent styles and lack a common glossary, leading to ambiguity and reduced understandability. For instance, the activity "Elaboration of Proposal" and the sub-process "Internal Analysis" use action-noun styles, which should be refactored to verb-object form (e.g., "Elaborate proposal" and "Perform internal analysis"). The activity "Internal Analysis requested", which describes a state rather than an action, can be revised to "Request internal analysis" to restore its active nature. Similarly, "Time schedules checked" is corrected to "Check time schedules" for consistency. For events and gateways, the start event "Accept request for ad-personam product" is changed to "Ad-personam product request received" to better reflect the process trigger. The XOR-split label "Analysis done?" is reformulated as "Is analysis done?" to clarify the decision logic.

As a result, the output of Level 1 is a BP model where flow object labels adhere to established naming best practices. This correction is captured in the first section of the PDR, which records naming violations and the rationale for each optimization.

4.3 Level 2 Mapping

Level 2 of the maturity model transitions from a specialized BP view to a structured process architecture. While Level 1 focuses on labeling corrections, Level 2 bridges the gap between the specific BP model nomenclature and standardized business language defined by reference frameworks. This phase is essential

[3] https://docs.camunda.io/docs/components/best-practices/modeling/naming-bpmn-elements/.

for higher levels of the maturity model. Without successful mapping at Level 2, granularity assessment (Level 3) and scope analysis (Level 4) cannot be performed. Compliance at Level 2 is achieved when each activity in the renamed BP model (from Level 1) is cross-referenced and mapped to the appropriate BP in the chosen reference framework. While the model is agnostic to the framework used, our example demonstrates this mapping using the PCF for luxury manufacturing operations.

Level-2 compliant model

The second step takes the labeling-compliant model and process profile as *input* to map each activity to a unique process classification ID within the selected reference framework. The *output* is a set of activity-process ID associations, which populate the mapping section of the PDR.

In our example, the activity "Request TBA" is mapped to PCF ID 11781 ("Receive Request For Quote"), marking its role in demand management. Similarly, "Update Pricing in SAP" is mapped to PCF ID 11497 ("Communicate and implement price changes"). Level 2 also clarifies activities with governance or strategic importance, like "Assess Proposal", mapped to PCF ID 20016 ("Manage internal reviews"), highlighting its role as a decision point requiring financial approval before proceeding. Achieving Level 2 does not represent static alignment; it enables assessment at the next maturity level to ensure activities, such as "Monitor Production" (mapped to PCF ID 11417), are managed at the appropriate hierarchical level within the BP.

4.4 Level 3 Granularity

In a mature BP architecture, activities must align at a consistent level of granularity to provide an actionable value chain model. A BP model mixing activities related with strategic goals with technical operations is unsuitable for resource estimation or performance analysis. Level 3 assumes activities have been mapped to the reference framework and uses its hierarchical metadata (e.g., the five-level of the PCF) to identify granularity discrepancies. Misalignments, such as strategic activities requiring more resources than simpler tasks, are fixed by suggesting alternative activities that may differ in level or category within the framework.

Level-3 compliant model

The third step uses the activity-process ID mapping, the process profile, and the reference framework as *input* to identify and correct granularity variance, realigning tasks across inconsistent hierarchical levels. The *output* is a structurally homogeneous model and the granularity evaluation section of the PDR.

In our example, applying diagnostics of the Level 3 of the maturity model uncovers architectural imbalances in the BP model. For instance, the sub-process "Internal Analysis" (renamed "Perform Internal Analysis") was initially mapped to PCF ID 10019 ("Assess the internal environment"), corresponding to L3-Process of the PCF. This incorrectly treats a high-level organizational function the same as execution events. The maturity model should recommend correcting it to PCF ID 19948 (activity "Analyze internal operations"), which aligns it with the rest of the activities in the BP model, which lie at L4-Activity of the PCF. Similar discrepancies are found with "Check Time Schedules" (mapped to L3-Process, PCF ID 10224) and "Monitor Production" (mapped to L5-Task, PCF ID 11417). The application of the maturity model suggests correcting these mappings to their L4-Activity counterparts of the PCF, such as PCF ID 20024 (activity "Create master production schedule") and PCF ID 10100 (activity "Install and validate production/service delivery process"). Once the Level 3 of the maturity model is considered as achieved, the PDR is updated with the corresponding PCF ID, the PCF hierarchical level for each BP activity, the granularity status (aligned/misaligned), and the needed corrections to achieve homogeneity.

4.5 Level 4 Scoping

Level 4 evaluates the relevance of BP activities within the organization's boundaries. Inspired by process landscape models [7], it helps balance broad and narrow BPs to optimize scope [6]. Specifically, Level 4 identifies scope discrepancies, ensuring the BP forms a coherent sequence aligned with the reference framework's strategic domains. This reflects Davenport's [6] idea that, in a mature process architecture, each activity must contribute to its parent category's end-to-end objective.

Therefore, to achieve Level 4 compliance, the maturity model verifies the L1-Categories of the PCF for all mapped activities. It then suggests corrections by identifying alternative activities within the appropriate functional category that meet the same operational need, bringing the BP back within its proper scope. For example, consider the "Employee Onboarding" process, which belongs to the Human Resources domain (PCF Category 7.0: "Develop and Manage Human Capital"). Its primary goal is to integrate a new hire into the corporate culture and administrative systems. However, including a technical task like "Configure Network Firewall for New Workstation" (from Category 8.0: "Manage Information Technology") creates a scope discrepancy. While necessary for the employee to begin work, this task disrupts the process's functional cohesion within human resource management.

Level-4 compliant model

The fourth step takes the hierarchically balanced BP model, process profile, and reference framework as *input* to perform scoping, ensuring functional balance by realigning activities that do not contribute to their parent category's goal. The *output* is a cohesive model and the populated scoping section of the PDR.

The running example falls within the sales and customer relations domain. However, Level 4 analysis reveals functional fragmentation. Although the process is primarily mapped to Category 3.0 ("Market and Sell Products and Services"), several activities are linked to other categories, such as "Vision and Strategy" (1.0), "Product Management" (2.0), or "Supply Chain" (4.0). For instance, the "Check Time Schedules" activity, initially mapped at PCF L4-Activity, is now associated with PCF ID 20024 (activity "Create master production schedule") in Category 4.0 ("Manage Supply Chain for Physical Products"). While functional, this mapping causes a scope inconsistency in a sales-driven pricing process. Level 4 of the maturity model resolves this by reassigning the activity to PCF ID 11788 (activity "Develop pricing and scheduling estimates") in the "Sales" category (3.0), maintaining the BP within its proper scope. At Level 4, the process is also updated in the PDR, which documents the scope, highlights functional deviations, and provides a detailed log of out-of-scope activities with their realignment rationale.

4.6 Level 5 Targeting

The fifth and final level of the maturity model represents the strategic outcome of the diagnostic task. It progresses from improved labeling (Level 1) to activity mapping (Level 2), hierarchical homogeneity (Level 3), and functional cohesion (Level 4), ultimately ensuring a measurable value chain. The goal is to enable process performance targeting, or KPI assignment, by linking specific KPIs to each activity based on the reference framework and strategic intent in the process profile. This level moves beyond structural analysis to assess BP effectiveness in terms of efficiency, quality, and cost, aligning operational activities with high-level strategic outcomes like customer satisfaction. To achieve Level 5 compliance, the model uses reference framework coordinates (PCF IDs) and process profile data to identify the most relevant KPIs for each activity.

Level-5 compliant model

The final step uses the functionally cohesive model, process profile, and reference framework as *input* to assign KPIs (efficiency, quality, or cost) to each activity based on its process ID. The *output* is a BP model enriched with context-aligned KPIs and the populated strategic alignment section of the PDR.

In the "Ad Personam" pricing process, Level 5 is achieved by assigning specific metrics to activities from the previous levels (renamed, homogenized, and re-scoped). For example, "Control Feasibility" (mapped to PCF 11784 at L4-Activity) is monitored through the KPI "Strategic Alignment Rate", tracking the percentage of custom requests meeting feasibility criteria without redesign. Performance alignment is crucial in later phases, such as with "Analyze Costs" (PCF 11788), linked to the KPI "Cost Estimation Accuracy", which tracks the variance between estimated and actual manufacturing costs, directly impacting profitability. Similarly, the final activity, "Send Product" (PCF 11790), is tied to the KPI "Proposal Win Rate", validating the tailored pricing strategy.

5 Demonstration

To demonstrate feasibility, we used Gemini 3 Flash [13] LLM to automate the application of the maturity model, generating suggestions across all diagnostic levels. Gemini 3 Flash [13] provides high-speed inference, advanced multimodal capabilities, and native support for agentic reasoning [12]. The maturity model was implemented using five LLM agents, one for each level of the model. In line with design choices, the APQC PCF was used as the target reference framework.

Prompt structuring followed a multi-layered methodology combining OpenAI's prompt engineering best practices [25], and Chain-of-Thought (CoT) prompting [34] to maximize clarity and reduce AI hallucinations. Below, we present the general structure of each prompt used to implement the AI agents corresponding to each level of the maturity model.

Level 1 - Labeling. Prompt 1 guides the LLM to act as an expert BPM Analyst and Camunda Certified Consultant to review a BPMN 2.0 diagram. The task involves identifying all activities (and sub-processes), events, and gateways in the provided BPMN diagram, evaluating their current labels against Camunda's naming conventions, and proposing updated names that strictly follow best practices. The expected output is a Markdown table listing each BPMN element's type, original name, updated name, and a rationale based on Camunda guidelines. This ensures the BPMN diagram is standardized and ready to be analyzed by the subsequent stage of the maturity model.

Prompt 1: Labeling

Role Assignment: Expert BPM Analyst and Camunda Certified Consultant specializing in BPMN 2.0.

Task Definition: Analyze the BPMN diagram.

Reference Inputs: BPMN diagram and Camunda's official naming convention documentation.

Step-by-Step Instructions: 1. Identify BPMN elements (activities, sub-processes, events, and gateways); 2. Evaluate labels; 3. Propose updates.

Specific Output Format: Markdown table: BPMN Element Type, Original Name, Updated Name, and Rationale.

Level 2 - Mapping. The primary goal of Prompt 2 is to map BP activities to the APQC PCF. This establishes a standardized foundation for BP diagnosis and performance measurement by assigning a precise five-digit PCF ID and a rationale to each activity or sub-process within the BPMN diagram. Ultimately, cross-referencing a BPMN diagram with a strategic process profile anchors the mapping in its real business context.

Prompt 2: Mapping

Role Assignment: Senior Process Architect and APQC Certified Specialist.

Task Definition: Mapping between a BPMN diagram and the APQC PCF.

Reference Inputs: BPMN Diagram (Flow), Process Profile (Context), and APQC PCF (Reference).

Step-by-Step Instructions: 1. Content Extraction; 2. Taxonomy Mapping; 3. PCF ID Identification.

Specific Output Format: Markdown table: BPMN Activity Name, Mapped APQC PCF Activity/Sub-Process, PCF ID, and Rationale.

Level 3 - Granularity. Prompt 3 ensures hierarchical consistency within a BPMN diagram by normalizing the depth of its APQC PCF mappings. By identifying the specific level (L1L5) of each assigned PCF ID, the prompt resolves granularity issues where activities may be mapped inconsistently (for example, mixing L3, L4, and L5 PCF IDs). This "Granularity Correction" process maintains the integrity of the original PCF ID categories (e.g., "3.0 Market and Sell Products and Services", "4.0 Manage Supply Chain for Physical Products", etc.), which is essential for the next stage of the maturity model.

Prompt 3: Granularity

Role Assignment: Senior Business Architect and Process Governance Specialist (Expert in APQC PCF hierarchies).

Task Definition: Perform a "Granularity Correction" to align the PCF ID levels of BPMN activities.

Reference Inputs: Markdown table as the output of Stage 2, Process Profile (Context), and APQC PCF (Reference).

Step-by-Step Instructions: 1. PCF ID Levels Extraction; 2. Alignment Check; 3. Granularity Correction.

Specific Output Format: Markdown table: BPMN Activity Name, Mapped APQC PCF Activity/Sub-Process, PCF ID, PCF ID Level, Granularity Status (Aligned/Misaligned), and PCF ID Correction Suggestion (if applicable).

Level 4 - Scoping. Prompt 4 ensures all BPMN activities are assigned to the correct PCF ID category. It flags "Scope Discrepancies" when activities are mapped to unrelated categories (e.g., a "3.0 Market and Sell Products and Services" activity mistakenly assigned to "4.0 Manage Supply Chain for Physical Products"). This step realigns out-of-scope IDs with appropriate alternatives, while preserving the granularities established in Stage 3.

Prompt 4: Scoping

Role Assignment: Senior Business Architect and Process Governance Specialist (Expert in APQC PCF hierarchies).

Task Definition: Perform a "Scope Correction" to ensure PCF IDs associated with BPMN activities are bounded to the same PCF ID Category.

Reference Inputs: Markdown table as the output of Stage 3, Process Profile (Context), and APQC PCF (Reference).

Step-by-Step Instructions: 1. PCF ID Category Extraction; 2. Alignment Check; 3. Scope Correction.

Specific Output Format: Markdown table: BPMN Activity Name, Mapped APQC PCF Activity/Sub-Process, PCF ID as the output of Stage 3, PCF ID Category, Scope Status (in-Scope/Discrepancy), and PCF ID Correction Suggestion (if applicable).

Level 5 - Targeting. By integrating the APQC PCF with the Balanced Scorecard methodology, Prompt 5 identifies specific KPIs for efficiency (e.g., cycle time), quality (e.g., error rate), and cost (e.g., cost per proposal) that are tailored to the process profile context. In this way, the outcomes are strategically aligned with the organization's mission.

Prompt 5: Targeting

Role Assignment: Senior Business Architect and Performance Measurement Specialist (Expert in the APQC PCF and the Balanced Scorecard).

Task Definition: Perform KPI Targeting for Performance Measurement.

Reference Inputs: Markdown table as the output of Stage 4, Process Profile (Context), and APQC KPI (Reference).

Step-by-Step Instructions: 1. Activity-Level KPI Targeting; 2. End-to-End KPI Synthesis; 3. Strategic Alignment.

Specific Output Format: Markdown table: BPMN Activity Name, Mapped APQC PCF Activity/Sub-Process, PCF ID as the output of Stage 4, Activity-Level KPI, and Metric Definition.

All prompts, the automatically produced PDR, and the source materials, can be inspected at: https://github.com/bpm-diag/AutoReF, where the prompts were instantiated using the running example from Sect. 4.

6 Conclusion

This paper addresses a relevant challenge in BPM initiatives, namely the manual and time-consuming process of diagnosing existing BPs against reference frameworks. Using a DSR methodology, the study introduces a novel five-level maturity model covering *labeling*, *mapping*, *granularity*, *scoping*, and *targeting* levels.

The model offers a systematic, domain-agnostic approach for analyzing BPs using reference frameworks (e.g., ITIL, APQC PCF, SCOR, etc.). To demonstrate the maturity model's feasibility, we have implemented each stage as an LLM agent, specifically utilizing Gemini 3 Flash for automated analysis. This approach reduces the need for extensive domain expertise while maintaining a collaborative loop with human modelers. By using specialized LLM agents with CoT prompting techniques, we demonstrated that standardized reference frameworks, like APQC PCF, can be applied automatically to support BP diagnosis. The results from the running example highlight how the maturity model can refactor BPMN labels, align activities with hierarchical standards, identify scope inconsistencies often missed in manual inspections, and assist in performance measurement.

A potential threat to the validity of this work arises from the inherent creativity of LLMs. When prompts are run multiple times, the PCF-ID assigned to BPMN activities may differ, as can the results from the granularity, scoping, and targeting stages. However, the primary limitations of this study are the application of the maturity model to a single real-world use case, which restricts the generalizability of our findings, and the lack of end-user evaluation to investigate the effectiveness of our approach to automated BP diagnosis compared to manual analysis. Therefore, future work will focus on conducting extensive evaluations across various PCF ID categories to enhance the applicability of the findings, as well as engaging with user experts to validate the effectiveness of the maturity model and the quality of the suggestions provided. Additionally, expanding the model to support a broader range of reference frameworks, such as ITIL and SCOR, will be explored.

This research ultimately lays the foundation for AI-augmented BPM systems [9], where the maturity model's output (e.g., the PDR of a given BP) can be used by intelligent tools for automated KPI computation [2]. Currently, the output of Level 5 consists of KPI definitions without computed values. It would be valuable to connect these definitions to an intelligent tool that, given a KPI definition and the BP's event log, computes the value based on the definition.

Acknowledgments. The work of F. De Luzi has been supported by MUR-Sapienza Tuscia CUP B87G23000100001, accounting project "000055_23_AP_ANGELONI_Polo Universitario Rieti". The work of A. Marrella has been supported by the Sapienza projects CAI4PPI and FOND-AIBPM, and the EU Horizon Europe Research and Innovation Programme under Grant 101299128 (euroFMX). The work of F. Nonino has been supported by the project "Human-centric adaptive automation for sustainable manufacturing: dynamic task allocation and techno-economic integration" CALL-0745-B83C26000880005, and received funding from the Sapienza University of Rome.

References

1. Afflerbach, P., Bolsinger, M., Röglinger, M.: An economic decision model for determining the appropriate level of business process standardization. Bus. Res. **9**(2), 335–375 (2016). https://doi.org/10.1007/s40685-016-0035-6
2. Agostinelli, S.: Automating performance insights: suggesting and computing process performance indicators from event logs. In: CAiSE, pp. 221–237 (2025)
3. Balko, S., et al.: Business process extensibility. EMISAJ **5**(3), 4–23 (2010)
4. vom Brocke, J.: Design principles for reference modelling. In: Innovations in Information Systems Modeling: Methods and Best Practices. IGI Global (2009)
5. Carmona, J., van Dongen, B., Solti, A., Weidlich, M.: Conformance Checking: Relating Processes and Models. Springer (2018)
6. Davenport, T.H.: Process innovation: reengineering work through information technology. Harvard business press (1993)
7. Dumas, M., La Rosa, M., Mendling, J., Reijers, H.A.: Fundamentals of Business Process Management. Springer, Heidelberg (2018). https://doi.org/10.1007/978-3-662-56509-4_9
8. Dumas, M.: Robotic Process Mining, pp. 468–491. Springer (2022)
9. Dumas, M., et al.: AI-augmented business process management systems: a research manifesto. ACM Trans. Manage. Inf. Syst. **14**(1), 1–19 (2023)
10. Ehrig, M., Koschmider, A., Oberweis, A.: Measuring similarity between semantic business process models. In: APCCM, pp. 71–80 (2007)
11. Goel, K., Bandara, W., Gable, G.: A typology of business process standardization strategies. In: Goel, K., et al. (eds.) BISE 63(6), pp. 621–635 (2021)
12. Google AI: Gemini api documentation (2025). https://ai.google.dev/gemini-api/docs/gemini-3. Accessed 2026-03-18
13. Google DeepMind: Gemini 3 flash (2025). https://deepmind.google/models/gemini/flash/. Accessed 2026-03-18
14. Houy, C., Fettke, P., Loos, P.: Business Process Frameworks, pp. 153–175 (2015)
15. Houy, C., Fettke, P., Loos, P., van der Aalst, W.M., Krogstie, J.: Business process management in the large. BISE **3**(6), 385–388 (2011)
16. Johannesson, P., Perjons, E.: An Introduction to Design Science. Springer (2014)
17. Kamath, M., Ramamritham, K.: Correctness issues in workflow management. Distrib. Sys. Eng. **3**(4), 213–221 (1996)
18. Khalaj, M.: A semantic framework for business process modeling based on architecture styles. In: IEEE/ICIS, pp. 513–520 (2012)
19. Koschmider, A., Oberweis, A.: Recommendation-Based Business Processes Design, pp. 323–336. Springer (2015)
20. La Rosa, M., et al.: Questionnaire-driven configuration of reference process models. In: CAISE, pp. 424–438. Springer (2007)
21. Leopold, H., Smirnov, S., Mendling, J.: On the refactoring of activity labels in business process models. Inf. Syst. **37**(5), 443–459 (2012)
22. Mendling, J., Reijers, H., van der Aalst, W.: Seven process modeling guidelines (7PMG). Inf. Softw. Technol. **52**(2), 127–136 (2010)
23. Mendling, J., Reijers, H.A., Recker, J.: Activity labeling in process modeling: empirical insights and recommendations. Inf. Syst. **35**(4), 467–482 (2010)
24. Münstermann, B.: Business Process Standardization: A Multi-Methodological Analysis of Drivers and Consequences. IGI Global (2014)
25. OpenAI: Prompt engineering guide (2025). https://developers.openai.com/api/docs/guides/prompt-engineering. Accessed 2026-03-18

26. Orlowska, W.S.M.E.: On correctness issues in conceptual modeling of workflows. In: ECIS (1997)
27. Rosemann, M., Van der Aalst, W.M.: A configurable reference modelling language. Inf. Sys. **32**(1), 1–23 (2007)
28. Rosenkranz, C., et al.: Towards a framework for business process standardization. In: BPM, pp. 53–63. Springer (2009)
29. Schafermeyer, M.: Factors influencing business process standardization: a multiple case study. In: HICSS, pp. 1–10. IEEE (2010)
30. Schäfermeyer, M., et al.: The impact of business process complexity on business process standardization: an empirical study. BISE **4**(5), 261–270 (2012)
31. Suárez, M., Sánchez, M., Villalobos, J.: Towards business process architecture analysis. In: 2015 CCC, pp. 535–542 (2015)
32. Tregear, R.: Business Process Standardization, pp. 421–441. Springer (2015)
33. van der Aalst, W.M.P., et al.: Configurable process models as a basis for reference modeling. In: BPM Workshops, pp. 512–518. Springer (2006)
34. Wei, J., et al.: Chain-of-thought prompting elicits reasoning in large language models. Adv. Neural. Inf. Process. Syst. **35**, 24824–24837 (2022)

How Organizations Use Process Mining to Enable Automations: Evidence from Qualitative Interviews in Canada

Najah Mary El-Gharib(✉), Daniel Amyot, and Danielle Cruise

University of Ottawa, Ottawa, Canada
{nelgh031,damyot,dcrui022}@uottawa.ca

Abstract. Robotic Process Automation (RPA) and Agentic AI (AAI) are increasingly positioned as central enablers of digital transformation, supporting organizations in automating repetitive, rule-based tasks as well as more dynamic and decision-intensive processes. However, organizations often face challenges in systematically identifying, prioritizing, and planning automation initiatives that align with strategic goals and deliver measurable value. This study investigates how industry practitioners plan and prioritize automation initiatives and how Process Mining (PM) can be leveraged to identify suitable candidates for RPA or AAI deployment. The research examines practitioners' approaches to selecting automation opportunities, identifying technical feasibility with business impact, and measuring the automation impact on the improved business processes. The study's qualitative research design involves semi-structured interviews with fifteen industry experts in Canada, including process analysts, automation leads, consultants, and technology specialists. Using Nvivo and concept coding, the analysis identified recurring patterns for each of the interview themes (6) and subthemes (19). The findings indicate that organizations strongly recognize the value of PM in accelerating the identification of RPA/AAI automation opportunities. However, maximizing its adoption and impact requires addressing several challenges, including data availability, budget constraints, limited technology awareness, and insufficient collaboration among teams.

Keywords: Process Mining · Robotic Process Automation · Agentic AI · Technology adoption · UTAUT · Qualitative study

1 Introduction

In the context of digital transformation, many organizations are automating their manual processes to improve performance, save costs, and minimize errors. Innovation is essential for organizations to be effective at carrying out their mission. Yet, there is currently a large amount of guess work in assessing the processes (or parts thereof) that can be automated, and in monitoring their actual improvement, if any, once automated.

F. M. Maggi et al. (Eds.): BPM 2026, LNBIP 599, pp. 362–379, 2027.
https://doi.org/10.1007/978-3-032-37877-4_21

Over the past decade, there has been a steep increase in the use of *Robotic Process Automation* (RPA) in organizations, together with related tool support. RPA, which uses software robots to automate human tasks, has often been applied in the areas of administration and finance. RPA is frequently implemented in cases where high volumes of data are processed through repetitive tasks that can be automated [23]. The market of RPA solutions includes over 60 vendors who develop tools that provide different functionalities to automate office tasks in an intelligent way [10].

The rise of *Agentic Artificial Intelligence* (AAI) [6] recently marked a turning point in the evolution of automation, reshaping the role of RPA. While RPA automates rule-based, repetitive tasks by mimicking human interactions with digital systems, it offers limited flexibility [23]. In contrast, AAI systems can set goals, support interactions, make autonomous decisions, and adapt to dynamic environments, enabling more intelligent and flexible automation [19,22]. Organizations are increasingly integrating AAI into business processes, where AI agents act as intelligent assistants [13]. Industry reports indicate strong interest in this technology. For example, UiPath's 2025 Agentic AI report [6] states that 93% of U.S. IT executives are interested in adopting AAI, with 45% planning investments and 37% already deploying such technology.

Process Mining (PM) is a technology that constructs a complete visual model of an existing process based on digital event logs available in information systems [17]. Such evidence-based model reflects how the process is truly being carried out, as opposed to theoretically or as perceived by stakeholders, and enables different types of performance and conformance analyses. PM can enhance the implementation of RPA and AAI by providing a data-driven understanding of how processes are executed. PM enables organizations to analyze process behavior, identify inefficiencies, and discover suitable automation opportunities. By revealing high-volume and repetitive tasks, PM supports informed automation design and helps evaluate the impact of implementations [1,15].

Despite the growing interest in integrating PM with RPA and AAI, the existing literature focuses on technical capabilities, conceptual integrations, or case-based implementations [8]. There is limited empirical understanding of how practitioners in real organizational settings identify automation candidates using PM insights and how they evaluate automation performance post-implementation. This gap highlights the need for practice-oriented evidence that explains how these technologies are operationalized and assessed in industry contexts.

The main objectives of this paper are to determine how PM and RPA/AAI practitioners in Canadian organizations actually use these technologies together to decide which processes to automate, and then monitor their performance. Three research questions (RQs) are identified:

- **RQ1:** How do organizations leverage PM to identify RPA/AAI implementation options?
- **RQ2:** How do practitioners assess and monitor the performance of RPA and AAI automations after implementation?
- **RQ3:** What challenges do practitioners face when using PM to identify and assess RPA and AAI automation opportunities?

In order to go beyond academic literature reviews (e.g. [2,8]) and answer the RQs with a more industrial perspective, we conducted semi-structured interviews to collect information from 15 PM and RPA/AAI specialists from the private and public sectors in Canada, conducted qualitative coding on the interview transcripts, and used thematic analysis to structure our answers.

In rest of the paper, Sect. 2 presents the background on PM and RPA/AAI in practice, and then Sect. 3 highlights the research method. Section 4 presents the findings of the study, while Sect. 5 discusses these findings and the study limitations. Finally, Sect. 6 summarizes the RQ answers and concludes.

2 Background

2.1 Enhancing RPA and Agentic AI with Process Mining

Recent systematic literature reviews [2,8] synthesize the growing body of research at the intersection of PM and RPA, highlighting the increasing integration of data-driven process discovery with automation initiatives. By leveraging PM, organizations can identify automation candidates based on objective criteria such as execution time, rework rates, and resources used, rather than by relying solely on expert judgment or workshops. This data-driven alignment helps mitigate common RPA challenges such as automating low-value tasks, overlooking process dependencies, or failing to capture end-to-end performance impacts.

El-Gharib and Amyot [8] consolidated empirical and conceptual contributions in this domain and identified key themes related to automation candidate selection, performance improvement, and value realization through process intelligence. More recently, Alonso-Rocha et al. [2] explored categories of process/routine optimization issues in a PM/RPA context, including data handling, routine consistency, UI adaptation, and data cleaning & filtering.

Recent research exploring the integration between PM and AAI highlights a growing interest in combining data-driven process insights with autonomous agents [15]. The relationship between PM and AAI is both complementary and transformative in the field of intelligent automation. PM reliably uncovers how business processes truly operate, whereas AAI builds on these insights by executing tasks, making decisions, and adapting workflows. Recent studies, e.g., by Berti et al. [3], have begun to explore ways AAI can benefit from PM. PM improves decision-making, continuous learning and optimization, root cause analysis, and exception handling. Schmidt et al. [15] also investigated AAI readiness for process automation, with evidence-based insights provided by PM.

2.2 Using Process Mining and RPA/AAI in Practice

Following the increasing PM adoption in industry, researchers have investigated the use and adoption of PM in practice [21]. For example, Zerbato et al. [25] investigated how PM is actually practiced in real organizations by analyzing evidence from semi-structured interviews with 37 experienced practitioners and researchers. They focused specifically on the analysis stage of PM projects. A key result is that analysts heavily rely on interactions with stakeholders and domain knowledge to interpret models and metrics, and that data issues, process complexity, and tool limitations strongly influence which strategies are applied. Syed et al. [16] conducted an interview study with 9 stakeholders in the context of a Dutch pension fund, identifying challenges and enablers of PM adoption. Their findings touch upon different levels, going from tensions between teams to user challenges with learning tools. Grisold et al. [11] conducted a focus group study with 22 process managers to investigate the benefits and challenges of PM adoption in organizations. The challenges are related to: 1) planning and business case, 2) process selection, 3) implementation, and 4) use and operationalization.

Regarding the adoption of RPA in organizations, Vollenberg et al. [20] investigated key factors that drive adoption and usage behavior of RPA in public sector organizations. Their study examines what motivates institutions to accept and use RPA technologies, considering both organizational and individual perspectives. Through empirical investigation, the paper identifies drivers of acceptance–such as perceived usefulness, organizational support, and process characteristics–that influence whether employees and managers are willing to use RPA in their work. The results highlight that acceptance in the public sector is shaped by a combination of technological beliefs, organizational context, and managerial support, and offers insights into how public institutions can better promote RPA usage and overcome barriers typical of bureaucratic environments. Similarly, Hohenadl and Axmann [12] explored the evolving role of RPA within the broader automation and digital transformation platforms. Their work highlights how RPA, initially designed to automate repetitive user-interface tasks, is increasingly integrated with other new emerging technologies such as low-code/no-code platforms, intelligent automation, and orchestration layers that support end-to-end process automation. The study also discusses architectural and organizational trends driving this evolution, including the convergence of RPA with business process management and AI-assisted automation design.

Despite these contributions, existing practitioner-related studies rarely examined the relationship between PM and RPA/AAI automations. Also, limited attention was paid to how these technologies are applied within specific national contexts. Addressing this gap, this paper investigates how PM supports the identification and evaluation of automation opportunities in the Canadian industry.

3 Research Method

This section presents the methodological design, execution, and analytical procedures of the qualitative study, which is based on semi-structured interviews [5].

Interview Study Design. This is the first time a study targets the Canadian market on the use of PM to identify automation opportunities and monitor their implementation. Given the exploratory nature of this research and the limited theoretical grounding at the intersection of PM, RPA, and Agentic AI, this study adopts a qualitative research design to uncover the underlying factors, practices, and decision-making mechanisms shaping automation initiatives. For this reason, the well-known *Unified Theory of Acceptance and Use of Technology* (UTAUT) [4] was selected as the primary theoretical framework for our research. This theory proposes four core constructs of intention and use: performance expectancy, effort expectancy, social influence, and facilitating conditions [18]. In this study, PM serves as the mechanism for identifying potential RPA/AAI implementation opportunities, while UTAUT provides a structured lens for understanding how organizations evaluate the perceived usefulness, ease of use, social influence (e.g., support from senior management), and facilitating conditions (e.g., organizational support requirements) associated with these opportunities. More specifically, UTAUT is used to analyze the adoption of process mining for identifying RPA and AAI automation opportunities by examining how these four core constructs influence stakeholders' intention to use and adopt such tools.

Semi-structured interviews were conducted to ensure consistency across participants while retaining sufficient flexibility to explore individual experiences in PM and automations in depth, with examples. The interview guide consisted of open-ended questions (available online [9]) organized around the UTAUT model's key constructs stated above. The participants were asked to answer those questions and explain their experience. Demographic information was also collected as participants held diverse positions and responsibilities relevant to PM and automation initiatives in their organizations. The study was approved by the Research Ethics Board of the University of Ottawa, certificate H-10-25-12117.

Recruitment of Participants. We recruited practitioners from Canadian organizations (public or private) with experience in i) PM, and ii) automation with RPA and/or AAI. The interviews were in English. There were no other restrictions on participation. Fifteen participants were interviewed, and recruitment was concluded once data saturation was reached (i.e., the point at which additional interviews no longer yielded new themes or additional insights). Table 1 describes the participants (P1–P15). Unsurprisingly, experience with PM and RPA is higher and more frequent than with more recent AAI technologies.

Data Collection. The interviews were conducted virtually via Microsoft Teams. Each interview lasted approximately 60 min and was recorded with the participants' consent. Transcripts were automatically generated using Microsoft Teams and then reviewed manually for accuracy prior to analysis.

Data Analysis and Validation. The interview transcripts were analyzed using thematic analysis following a structured, multi-step procedure consistent with established qualitative guidelines [5]. Two co-authors first familiarized themselves with the transcripts by reading them and recording analytic memos to capture initial impressions and potential patterns. Second, an initial coding process was conducted in which meaningful segments of text were systematically identified and assigned inductive codes relevant to the RQs. This stage involved the use of both NVivo (i.e., participant's own words and expressions) and descriptive codes (i.e., a word/phrase summarizing the idea/topic presented in the data excerpt). Following this, a deductive coding approach was conducted in which the inductive codes were mapped onto the core UTAUT constructs. Specifically, this involved comparing the meaning and content of each inductive code with the conceptual definitions of the UTAUT constructs. For example, codes reflecting training or IT support were mapped to facilitating conditions. Coding decisions were made based on the extent to which participants' reflected perceptions of performance expectancy/perceived usefulness, effort expectancy/ease of

Table 1. Overview of the Interviewed Participants.

ID	Organization Type	Role	Experience (Years)		
			PM	RPA	AAI
P1	Public	Technical Advisor	3	2 - 3	0
P2	Public	Senior Analyst	1.5	< 1	0
P3	Public	Senior Analyst, Process Engineering	1	1 - 2	0
P4	Private Consulting	Partner and Process Intelligence Lead	6	1	1
P5	Public	Business Analyst and Process Engineer	3	4	1
P6	Public	Technical Advisor	> 5	10	< 1
P7	Crown Corporation	Process Automation Program Lead	3 - 4	6	< 1
P8	Crown Corporation	Intelligent Automation Lead	8	8	1 - 2
P9	Private	Chief Technology Officer (CTO)	5	1 - 2	3 -4
P10	Private	PMO Manager - Process Intelligence Function	6	7	0
P11	Private Consulting	Senior Manager in AI and Data Practice	7	9	2
P12	Private	Team Lead for Process Mining and Digital Adoption Platform	3	6	0
P13	Public	Senior Manager, Business Process Optimization	2	2	< 1
P14	Private	Director of Digital Transformation and Automation	8	8	3
P15	Private Consulting	Senior Business Analyst	3	3	< 1
		Average	≈ 4.3	≈ 4.7	≈ 1.1

use, social influence, and facilitating conditions. The level of agreement between the two coders was 83.25% (high). Discrepancies were discussed and resolved through consensus.

The 19 subthemes were reviewed to ensure there were no duplications. During this iterative process (where the NVivo tool was used), some codes were merged, split, or discarded. Candidate subthemes were then refined to ensure internal coherence and clear distinctions between them by revisiting the original transcripts for consistency and completeness. Finally, the themes were clearly defined and reported (with a resulting codebook of 429 codes [9]), supported by representative interview excerpts to enhance transparency and analytical rigor.

4 Findings

This section summarizes the results of the qualitative thematic analysis (see Fig. 1 for an excerpt) and reports on our findings from the expert interviews along the six themes identified.

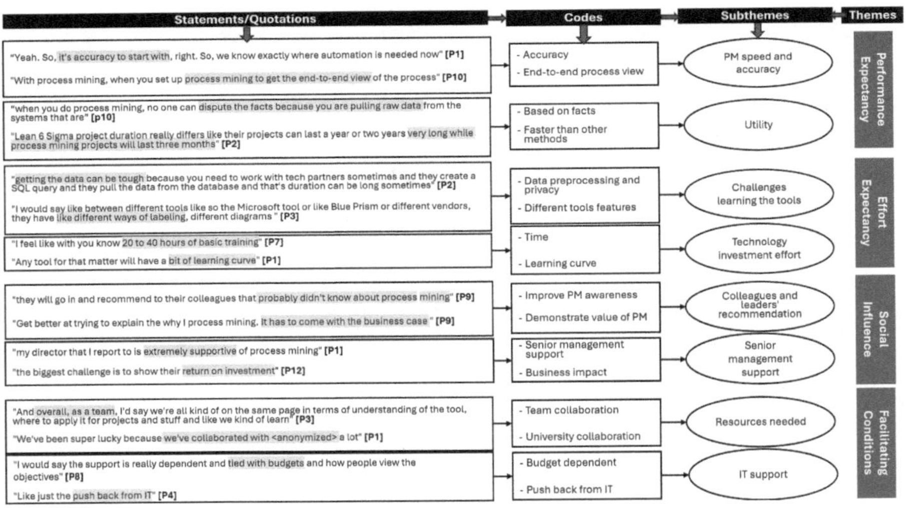

Fig. 1. Summary of coding results (extract) aligned to UTAUT themes.

4.1 General Context

Automation Goals. To understand how organizations prioritize automation initiatives, participants were asked about the goals they typically consider when automating a process or part of a process. The interview responses were analyzed to identify the primary objectives guiding the selection of automation candidates using RPA or Agentic AI. Through iterative coding, goal-related statements were grouped into higher-level categories representing organizational decision criteria.

The findings indicate that automation decisions are predominantly driven by operational performance objectives. The most frequently mentioned goals include manual work reduction (11 participants), cost reduction (9), cycle time reduction (9), automation of high-volume processes (8), and automation of repetitive tasks (6). These criteria reflect a strong focus on efficiency gains and workload optimization when identifying opportunities.

In addition, six of the fifteen participants emphasized process standardization as an explicit goal. These participants highlighted that automation initiatives are often used as a mechanism to enforce uniform execution, reduce process variability, and increase governance across organizational units.

Process Discovery Techniques. The findings indicate that automation opportunity identification remains largely exploratory and human-driven. Ten out of fifteen participants emphasized conducting stakeholder interviews to uncover pain points, inefficiencies, and repetitive tasks. Seven participants highlighted process mapping and modelling to visualize workflows and identify rule-based or high-volume activities. Six participants mentioned workshops as collaborative mechanisms to explore automation potential, while four referred to job shadowing to observe tasks in real time. Two additional participants relied on existing process documentation to assess automation feasibility. Moreover, three participants indicated that structured forms are sometimes distributed to operations or RPA teams to collect automation ideas.

Although traditional qualitative approaches are still widely used to identify automation opportunities, *eleven out of fifteen participants explicitly noted that PM significantly enhances automation discovery*. They emphasized that PM provides a data-driven and objective view of the end-to-end process, reduces reliance on subjective opinions, and substantially saves time, particularly in consulting engagements where rapid assessment is required. Participants noted that PM enables faster identification of high-volume, repetitive, and bottleneck-prone activities that are strong candidates for RPA implementations.

Identification of Automation Opportunities. The analysis reveals that the identification of automation opportunities is influenced by several interrelated factors. First, *automation suitability* emerged as an important consideration. Participants emphasized that tasks characterized by repetitive executions (7 participants), structured rule-based (6), and high volume (5), and are more suitable candidates for automation, while activities requiring significant human judgment (6) or complex cognitive processing are less likely to be automated. Second, *practitioner knowledge* and expertise play a key role in identifying automation opportunities. Participants highlighted the importance of domain knowledge (6), experience, and consultation with subject matter experts when exploring potential automation candidates. Additionally, *stakeholder collaboration* (5 participants) and *organizational priorities* (4) and also shape automation decisions.

Factors such as business goals, departmental priorities, and stakeholder expectations play an important role in determining which automation opportunities are pursued. Two participants mentioned that a process is at times automated because other departments are automating it as well, whereas two participants stated that in some cases it is the operations team's decision. System integration challenges were mentioned by four participants, who indicated that automation is often considered when multiple systems require manual data transfer or coordination. Three participants identified cost as a main factor in automation decisions. Automation initiatives were often justified based on expected cost reduction, return on investment (ROI), or resource reallocation potential.

Overall, the results suggest that automation opportunity identification is primarily guided by operational characteristics – such as volume, repetitiveness, manual effort, and rule-based logic – while also considering the level of complexity and knowledge involved in the task.

Automation Monitoring Opportunities. Dashboard-based monitoring was the most frequently mentioned approach. Six participants reported using it to track bot performance, execution status, and operational metrics. Three also specified that these dashboards are embedded within automation platforms (e.g., native RPA tool dashboards), enabling real-time visibility into bot activities.

Beyond visualization tools, five participants highlighted the use of automated alerts to detect failures and exception rates, and six mentioned benchmarking practices to compare bot performance over time or against predefined baselines. Additionally, five participants emphasized the use of predefined Key Performance Indicators (KPIs) to assess automation success, while six participants referred to periodic reporting mechanisms to communicate performance outcomes to stakeholders.

The results suggest that post-deployment monitoring relies on tool-supported dashboards complemented by KPI tracking, benchmarking, alert systems, and structured reporting practices. This indicates a predominantly operational focus on performance visibility, reliability, and value realization after deployment.

4.2 Performance Expectancy

Process Mining Effectiveness for Automation. Overall, the perceptions were largely positive regarding the effectiveness of PM in identifying potential areas for RPA/AAI-based activity automation. Nine participants explicitly stated that PM is effective in identifying automation opportunities. A stronger theme emerged around visibility and transparency, with eleven participants emphasizing that PM enables detailed, data-driven insight into actual process execution. Several participants noted that PM helps identify the "right" automation points (5 participants) and detect process bottlenecks (5), thereby supporting more targeted and impactful automation decisions. One participant highlighted that PM facilitates the identification of larger-scale automation opportunities that might otherwise be overlooked."*It has made a huge impact, honestly, in my department. Before process mining, I know it has only been just a stakeholder meetings and SME meetings and trying to understand their process, right? And many times I've seen some biased responses would come in from them and we would make some wrong automation.*" [P1]

In addition, four participants suggested that PM can partially replace traditional workshop-based discovery approaches, reducing reliance on subjective assessments. However, limitations were also acknowledged. Five participants pointed out constraints related to data availability and process maturity. Specifically, one participant noted that PM is less effective when processes are poorly defined, while another emphasized that non-digitalized processes limit PM applicability. Furthermore, five participants argued that combining PM with Task Mining (TM) provides a more comprehensive view, particularly for capturing granular, user-level activities relevant for RPA or AAI implementation.

Process Mining Speed and Accuracy. A majority of participants emphasized this as a positive contribution. Nine participants highlighted that PM significantly reduces the time required to discover and understand processes, thereby accelerating the automation assessment phase. Seven also indicated that PM increases the accuracy of automation opportunity identification by relying on data-driven insights rather than subjective assessments. In addition, five participants noted that PM provides an end-to-end view of processes, enabling a more comprehensive evaluation of automation potential across the entire workflow. Finally, two participants mentioned cost-related benefits, suggesting that the improved transparency and prioritization enabled by PM can contribute to cost reduction during automation initiatives.

Utility of Process Mining. Overall, PM was perceived as offering distinct advantages over traditional approaches such as workshops or interviews. Participants (7) emphasized that PM is fact-based, relying on objective event log data rather than subjective perceptions, which enhances the credibility of automation decisions. PM is faster (7) than manual discovery methods, as it reduces the time required to analyze processes and identify candidates for automation. Notably, PM provides an end-to-end view of processes (12), enabling organizations to understand the full workflow across systems and departments. Together, these responses suggest that PM is perceived as a more data-driven, efficient, and holistic method for identifying RPA/AAI automation opportunities.

4.3 Effort Expectancy

Learning Process Mining and RPA Tools. When participants were asked about the ease of learning PM tools, responses revealed a generally positive but nuanced perspective. Nine of the 15 participants described the tools as relatively easy to learn, particularly once foundational concepts were understood. However, many (8) emphasized that the level of difficulty largely depends on the specific tool and its features, noting variations in usability, interface design, and required technical knowledge. Despite the overall positive responses, creating event logs (5) remains one of the most challenging aspects of learning and applying PM. One participant explained that it typically takes 2–3 projects to become comfortable with event log creation (depending on background knowledge), while two participants estimated that achieving proficiency with the tools may require approximately 3–4 months of project hands-on experience. Additionally, two participants mentioned that vendor-provided support and online training resources significantly facilitate the learning process. One participant further noted that AI-based assistants, such as Copilot, are increasingly making it easier to learn and use PM tools by supporting query generation, data preparation, and interpretation tasks.

Challenges. The challenges that were pointed out by the participants can be grouped into four categories (with frequencies), as presented in Table 2.

Table 2. Overview of the Challenges.

Challenges	Sample Collective Codes	# of Instances
Data Related	– Working with large datasets – Getting the data – Creating a data model and building the event logs	37
Technical and Tool Related	– Tool procurement – Multiple systems – Learning how the applications work	18
Organizational and Stakeholder Challenge	– Buy-in from business owner – Aligning with stakeholder expectations – Business and technical clash	15
Skill-set and Knowledge Gap	– Understanding the domain – Reading documents – Understanding full capability	25

Consistent with prior research [26], participants frequently highlighted data-related challenges such as incomplete or inconsistent event logs. Participants also noted limitations in tool capabilities and integration with other automation technologies, which aligns with earlier findings regarding difficulties in configuring PM tools and interpreting complex models [26]. Organizational challenges such as stakeholder resistance and lack of process ownership were also identified, supporting previous research highlighting the importance of collaboration between business and IT stakeholders during PM initiatives.

Investment Efforts in Learning the Technology. Regarding RPA learning investment, responses varied considerably. Four participants indicated that achieving a functional level of proficiency may require 3–4 months as one needs to learn how to write the program for the RPA bot, whereas three participants suggested that 4–5 weeks could be sufficient, depending on prior technical background and the complexity of the automation use cases. Interestingly, several participants noted that learning RPA tools is generally more difficult than learning PM tools, particularly due to scripting requirements, exception handling, and integrations with systems challenges.

Seven participants emphasized that PM tools still require a significant amount of learning, especially in understanding data structures, event logs, building the data model, and process concepts. There is a learning curve on how to build event logs from the different data tables. Five participants further noted that the learning curve depends heavily on the maturity of the team; teams with established data engineering practices and analytical capabilities tend to adopt PM tools more quickly. Estimates of the time required to learn PM tools varied widely, ranging from short-term training of one week (or 40 h) for basic proficiency (7 participants) to 6–12 months for full proficiency in conducting multiple projects (2 participant). These findings suggest that while both RPA and PM tools require dedicated learning effort, the perceived difficulty and time investment depend on organizational maturity, prior technical expertise, and the scope of intended use.

4.4 Social Influence

Colleagues and Leaders Recommendation. To understand how frequently PM is recommended for identifying RPA/AAI automation opportunities, participants were asked about advocacy and adoption practices within their organizations. Ten participants reported that recommendations from colleagues to adopt PM were little to none, mainly because of limited awareness of PM within organizations. Seven participants emphasized that increasing organizational awareness of PM remains necessary, suggesting that PM often requires internal promotion and education. Six participants highlighted that demonstrating the tangible value of PM–such as data-driven insights and improved decision accuracy–is essential for gaining leadership support and broader adoption. Though three participants indicated that they personally recommended PM to colleagues when identifying automation opportunities. Additionally, five participants noted that the cost of PM platforms can act as a barrier to frequent recommendation, particularly in organizations with budget constraints or limited digital maturity. Other challenges that were mentioned once each include *fear of highlighting people processes*, *time and resource investment*, and *change management*.

Senior Management Support. Most participants (14) indicated that senior management actively supports the use of new technologies, including PM, particularly when these technologies contribute to measurable business improvements. However, four participants highlighted budget constraints as a recurring concern, noting that financial approval often depends on a clear business case. In this regard, five participants emphasized that demonstrating ROI and tangible business impact—such as cost reduction, efficiency gains, or risk mitigation–is critical to securing and maintaining executive support. Interestingly, two participant noted that senior management is generally tool-agnostic, stating that leadership does not specifically prefer PM over other technologies, but rather prioritizes solutions that deliver business value regardless of the tool used.

4.5 Facilitating Conditions

Access to Resources. When asked about the access to resources required to effectively use PM and RPA tools, the responses revealed that while tool access is generally available (7 participants), successful adoption is strongly influenced by complementary organizational capabilities. Enabling factors include internal team collaboration and skill availability (3), partnerships with process owners (3), engagement of external consultants (2), vendor-provided support (2), and academic collaboration and student hiring (1). These findings suggest that beyond technological availability, human expertise and cross-functional collaboration play a critical role in realizing value from PM-driven automation initiatives.

IT Support. When participants were asked how well they are supported by IT or other departments when using these tools, the answers varied. Two fifths (6 out of 15) mentioned that IT is very supportive, while four more got some support. Three participants stated that they get push back from IT, while for consultants such support differs based on clients. Additionally, data access (a common challenge [24,26]) and licenses (7) impact directly their project costs and timeline. We noted that support from tool vendors is also available (5).

Tools Training. Participants were asked whether they had received formal training or guidance on how to use PM, RPA, or AAI tools. Fifteen participants reported receiving vendor-provided training, including online documentation, structured academies, and professional certifications. In addition to formal training, eleven participants indicated that they primarily developed their expertise through on-the-job experience and self-directed learning. Two participants noted that their organization invested in building internal expertise and developed in-house specialists. Conversely, another participant highlighted that "*a lot of companies are not big enough to set up their own center of excellence, right? Or whatever you would call it, their own PM center. They don't have the foresight, they don't have the knowledge, they don't have the resources, either one of the above*" [P9].

Regarding RPA/AAI tool training, two participant highlighted the existence of a significant learning curve, estimating that it can take about two months to become sufficiently comfortable and productive with automation platforms.

4.6 Behavioral Intention

The findings indicate a strong behavioral intention to continue using PM for automation opportunity identification among current users, primarily driven by its perceived business value and its role as a data-driven "source of truth" within their teams. However, despite this positive intention, budget constraints emerged as the primary factor that could limit sustained or expanded use. Table 3 summarizes the drivers and barriers to the adoption of the PM technology for discovering automation (RPA/AAI) opportunities.

Table 3. Drivers and Barriers Influencing PM Continuation and Adoption

Subtheme	Code	# Participants
Continuation Drivers	Mandate and organizational embedding	9
	Perceived business value	9
Continuation Barriers	Budget constraints	7
	Demonstrating value of PM	5
Adoption Intent (Non-Users)	Likely future adoption	4
	Lean Six Sigma alignment	3
	Vendor collaboration	2
	PM + Task Mining (TM) integration	2
Adoption Resistance	Perceived low value for RPA discovery	2

5 Discussion and Limitations

From interviews with 15 practitioners in Canada, we have derived 6 themes and 19 subthemes on how they identify automation opportunities and monitor post-implementation performance. Our findings show that automation is still not technical and data-driven but rather shaped by organizational awareness, data availability, and stakeholder involvement: *"how do we evaluate and prioritize the RPA project is we ask like team lead how much time it is going to take for your employee to do it"* [P12]. Participants noted that PM alone may not capture detailed user interactions or decision logic. Therefore, practitioners suggest the need to combine PM with task mining and human expertise to obtain a more comprehensive understanding of process behavior *"I think personally I do believe that RPA teams would be interested more in task mining rather than process mining in terms of reducing the workload that they need to do within their team"* [P2]. Additionally, practitioners reported that automation opportunities are often identified during PM projects (among other improvement opportunities) and then communicated to automation teams, rather than initiating PM with the explicit goal of automating a process. This suggests that the level of collaboration between PM and automation teams, including who takes the initiative and the leadership, plays a critical role in identifying, prioritizing, and implementing automation solutions based on PM-driven analysis.

These findings are consistent with recent industry reports; e.g., a recent survey [7] highlights the strong connection between PM and automation initiatives. When participants were asked which measures were introduced as a result of PM insights, process automation was the most frequent response (57%). In addition, when asked about third-party technologies used alongside PM tools, RPA ranked second (31%). Our findings indicate that while key UTAUT constructs remain relevant to PM adoption, their manifestation is shaped by PM-specific organizational and socio-technical contexts. Beyond traditional facilitating conditions such as training and IT support, participants emphasized institutional capacity and system interoperability as critical enablers. Similarly, effort expectancy extended beyond system usability to include challenges related to data access and availability [18]. These findings suggest that PM adoption is influenced by broader organizational dynamics not fully captured by the generic application of UTAUT, highlighting the importance of contextualizing UTAUT constructs within PM implementation environments.

Limitations. Our findings are subject to validity threats typical of interview studies [14]. First, this study involves interviews with 15 practitioners in Canada, which may limit the generalization of the findings. In addition, the results rely on participants' perceptions and experiences and reflect organizations already exposed to PM or automation technologies. Second, the participants' answers could have been biased, e.g., by the study setting. Third, the data collection and analysis could have been biased by the researchers' personal interpretations. Future research could complement these insights with larger samples and quantitative analyses evaluating the measurable impact of PM-driven automation.

6 Conclusion

This study investigated how 15 practitioners in Canada identify, evaluate, and experience challenges when using PM for RPA/AAI initiatives. This paper highlights practical considerations, challenges, and opportunities for integrating PM with RPA/AAI to support data-driven automation decision-making. Thematic analysis of the interview transcripts led to many findings, structured along six themes (Sect. 4). Guided by three RQs (Sect. 1), these findings provide several important insights.

> **RQ1 Answer:** Traditional qualitative approaches are still widely used to identify process activities suitable for automations using RPA. Those approaches include stakeholder interviews, process mapping workshops, and job shadowing. Eleven (of 15) participants indicated that they complement traditional approaches with PM techniques when identifying automation opportunities. In these cases, automation candidates were often discovered after conducting a PM analysis, which revealed inefficiencies, repetitive tasks, and potential areas for improvement.

RQ2 Answer: The performance of RPA automations is mainly monitored using dashboards available in automation platforms or external monitoring and business intelligence tools. Monitoring typically focuses on individual bot performance and failure rates rather than on evaluating the overall end-to-end process performance. The findings show that participants currently make very limited use of PM as a technique for monitoring the performance of implemented automations.

RQ3 Answer: Several challenges were identified when using PM to support the discovery of automation (RPA/AAI) opportunities, including data-related issues such as limited availability and poor quality of event logs, technical limitations of PM tools, and difficulties integrating them with existing systems. Additionally, organizational barriers—including budget constraints, limited expertise, and stakeholder resistance—can delay the effective adoption and use of PM for automation initiatives.

This research contributes to the growing body of knowledge by providing empirical evidence of the practical connection between PM and automations (RPA/AAI) in real Canadian organizational contexts. Overall, this study demonstrates that while PM is increasingly recognized as a valuable technology to identify automation opportunities, its adoption remains dependent on organizational context, PM awareness, human expertise, and data readiness.

Future research should focus on developing structured methodologies that integrate PM throughout the automation lifecycle, from opportunity identification and prioritization to implementation and continuous monitoring. Such approaches would help organizations maximize the benefits of process transparency and data-driven decision-making. The findings emphasize the need for more integrated and human-centric approaches that combine data-driven techniques with stakeholder collaboration. The findings also highlight the need for increased education, training, and awareness initiatives in Canada to improve understanding of PM capabilities and support wider organizational adoption.

Acknowledgments. We thank our participants for taking the time to share their valuable experience. This work was supported by the Natural Sciences and Engineering Research Council of Canada (Discovery Grant) and the University of Ottawa.

References

1. van der Aalst, W.M.P.: On the Pareto principle in process mining, task mining, and robotic process automation. In: Proceedings of the 9th International Conference on Data Science, Technology and Applications - DATA, pp. 5–12. INSTICC, SciTePress (2020). https://doi.org/10.5220/0009979200050012
2. Alonso-Rocha, J.L., Martínez-Rojas, A., Jiménez-Ramírez, A., Enríquez, J.G.: Research challenges in routine optimization for synthesizing software robots. In: Research Challenges in Information Science, pp. 98–108. Springer, Cham (2025). https://doi.org/10.1007/978-3-031-92471-2_7
3. Berti, A., Maatallah, M., Jessen, U., et al.: Re-thinking process mining in the AI-based agents era (2024). https://arxiv.org/abs/2408.07720

4. Chang, A.: UTAUT and UTAUT 2: A review and agenda for future research. The Winners **13**(2), 106–114 (2012). https://doi.org/10.21512/tw.v13i2.656
5. Clarke, V., Braun, V.: Thematic analysis. J. Posit. Psychol. **12**(3), 297–298 (2017). https://doi.org/10.1080/17439760.2016.1262613
6. Davenport, T., UiPath: UiPath 2025 agentic AI research report: Preparing for the agentic era. Tech. rep., UiPath (2025). https://www.uipath.com/resources/automation-analyst-reports/agentic-ai-research-report
7. Deloitte: Global process mining survey 2025 (2025). https://www.deloitte.com/de/de/services/consulting-financial/research/global-process-mining-survey.html. Accessed 10 Mar 2026
8. El-Gharib, N.M., Amyot, D.: Robotic process automation using process mining - a systematic literature review. Data Knowl. Eng. **148**, 102229 (2023). https://doi.org/10.1016/j.datak.2023.102229
9. El-Gharib, N.M., Amyot, D., Cruise, D.: How organizations use process mining to enable automations: Questionnaire and Codebook (2026). https://doi.org/10.5281/zenodo.19140105
10. Gartner Inc.: Robotic process automation reviews and ratings. Gartner Peer Insights (2026). https://www.gartner.com/reviews/market/robotic-process-automation. Accessed 10 Mar 2026
11. Grisold, T., Mendling, J., Otto, M., vom Brocke, J.: Adoption, use and management of process mining in practice. Bus. Process. Manag. J. **27**(2), 369–387 (2021). https://doi.org/10.1108/BPMJ-03-2020-0112
12. Hohenadl, T., Axmann, B.: The future of RPA in the context of automation platforms. In: Business Process Management (BPM 2024). LNBIP, vol. 489, pp. 263–278. Springer, Cham (2025). https://doi.org/10.1007/978-3-032-02936-2_18
13. Krishnan, N.: AI agents: Evolution, architecture, and real-world applications (2025). https://arxiv.org/abs/2503.12687
14. Padgett, D.K.: Qualitative methods in social work research, 3rd edition. Sage Publications (2016). https://bit.ly/Pargett2016
15. Schmidt, R., Alt, R., Zimmermann, A.: Agentic AI readiness: A process-oriented assessment framework. In: Proceedings of the 59th Hawaii International Conference on System Sciences (2026). https://hdl.handle.net/10125/111878
16. Syed, R., Leemans, S.J.J., Eden, R., Buijs, J.A.C.M.: Process mining adoption. In: Business Process Management Forum, pp. 229–245. Springer, Cham (2020). https://doi.org/10.1007/978-3-030-58638-6_14
17. van der Aalst, W.M.P.: Process Mining: Data Science in Action. Springer, Cham (2016). https://doi.org/10.1007/978-3-662-49851-4
18. Venkatesh, V., Morris, M.G., Davis, G.B., Davis, F.D.: User acceptance of information technology: Toward a unified view. MIS Q. **27**(3), 425–478 (2003). https://doi.org/10.2307/30036540
19. Vepachedu, N.: The future of RPA: How agentic frameworks and LLMs are transforming intelligent automation. IJRCAIT **8**(1), 1114–1128 (2025). https://doi.org/10.34218/IJRCAIT_08_01_083
20. Vollenberg, C., Hackl, J., Matthies, B., Coners, A.: Acceptance of RPA in public sector institutions. J. Organ. Comput. Electron. Commer. **34**(4), 400–418 (2024). https://doi.org/10.1080/10919392.2024.2365446
21. vom Brocke, J., Jans, M., Mendling, J., Reijers, H.A.: A five-level framework for research on process mining. Bus. Inf. Syst. Eng. **63**(5), 483–490 (2021). https://doi.org/10.1007/s12599-021-00718-8

22. Vu, H., Klievtsova, N., et al.: Agentic business process management: Practitioner perspectives on agent governance in business processes. In: Business Process Management: Responsible BPM Forum, Process Technology Forum, Educators Forum, pp. 29–43. Springer, Cham (2026). https://doi.org/10.1007/978-3-032-02936-2_3
23. Wewerka, J., Reichert, M.: Robotic process automation - a systematic mapping study and classification framework. Enterp. Inf. Syst. **17**(2), 1986862 (2023). https://doi.org/10.1080/17517575.2021.1986862
24. Wurm, B., Mendling, J., Pentland, B.T.: Using Process Mining to Investigate Organizational Processes. In: Wurm, B., Mendling, J. (eds.) Digital Trace Data Research in Information Systems pp. 175–197. Springer, Cham (2026). https://doi.org/10.1007/978-3-032-05497-5_8
25. Zerbato, F., Soffer, P., Weber, B.: Process mining practices: Evidence from interviews. In: Di Ciccio, C., Dijkman, R., del Río Ortega, A., Rinderle-Ma, S. (eds.) Business Process Management (BPM 2022), LNCS, vol. 13420, pp. 268–285. Springer, Cham (2022). https://doi.org/10.1007/978-3-031-16103-2_19
26. Zimmermann, L., Zerbato, F., Weber, B.: Process mining challenges perceived by analysts: An interview study. In: Augusto, A., Gill, A., Bork, D., Nurcan, S., Reinhartz-Berger, I., Schmidt, R. (eds.) Enterprise, Business-Process and Information Systems Modeling, pp. 3–17. Springer, Cham (2022). https://doi.org/10.1007/978-3-031-07475-2_1

Advancing AI Integration: Exploring Business Process Management Capabilities for AI-Driven Processes

Henriette Grashoff(✉) and Christian Bartelheimer(✉)

University of Göttingen, 37073 Göttingen, Germany
{henriette.grashoff,christian.bartelheimer}@uni-goettingen.de

Abstract. The increased integration of artificial intelligence (AI) into business processes challenges established business process management (BPM) assumptions. AI-driven processes are shaped by autonomy, opacity, and high data dependency. This paper examines the BPM capabilities organizations need to establish when integrating AI into business processes. Drawing on an exploratory qualitative study based on 18 semi-structured interviews, we reveal that existing BPM capabilities must be enhanced to account for autonomy shifts toward AI actors that are characterized by non-deterministic behavior, opacity, and dependence on reliable data. Building on the interview data, we identify AI-specific capability demands and derive three cross-cutting capability shifts that organizations need to establish to successfully and sustainably integrate AI into their business processes. We contribute by showing that effective AI integration requires coordinated attention to process selection, data readiness, governance mechanisms, role development, and organizational conditions enabling effective oversight and continuous adaptation of processes.

Keywords: BPM Capabilities · Artificial Intelligence · AI-Driven Business Processes · Business Process Management

1 Introduction

Business process management (BPM) is a boundary-spanning management approach for designing, executing, and continuously improving business processes [1, 2]. To implement BPM effectively, organizations must master diverse methods, techniques, and tools and develop a set of capabilities [1–3]. BPM capabilities represent the organizational areas most critical for establishing efficient and effective process management and for becoming a process-centered organization [2, 4]. The BPM capabilities required for leveraging digital technologies comprise 30 elements organized into five dimensions, comprising strategic alignment, governance, methods/information technology, people, and culture [3]. Recent research has further explored how advanced, often data-driven and algorithmic technologies augment the BPM lifecycle [5, 6]. For instance, process mining provides end-to-end transparency and diagnostics, predictive monitoring supports proactive interventions, and robotic process automation reconfigures task structures and

F. M. Maggi et al. (Eds.): BPM 2026, LNBIP 599, pp. 380–397, 2027.
https://doi.org/10.1007/978-3-032-37877-4_22

human-machine collaboration [7–9]. However, BPM capability requirements have not yet been systematically revisited in light of recent advances in artificial intelligence (AI), despite prior research highlighting the significant implications of implementing such technologies for organizational capabilities [3, 6].

Rapid advances in AI, including machine learning, generative AI, and intelligent autonomous systems, are reshaping how work is organized. Many organizations aim to integrate AI systems to realize benefits such as increased efficiency, improved service quality, reduced costs, and faster decision cycles [10–13]. Yet, AI's specific characteristics such as data dependence, learning ability, autonomy, and inscrutability introduce distinctive oversight and control demands that go beyond traditional information technology systems and BPM practices [10, 14]. These characteristics create new opportunities for process innovation and transformation while simultaneously exposing limitations of established BPM capability thinking [6, 15]. AI affects not only the implementation and execution of business processes, but also BPM as a management discipline, indicating that existing BPM capability frameworks may require reinterpretation and extension. Hence, to explore the evolving BPM capability demands, we address the following research question: *What BPM capabilities do organizations require when integrating AI into business processes?*

To address this question, we conducted an exploratory qualitative study based on semi-structured interviews with experts from academia and practice. Drawing on qualitative analysis, we examined the capability demands that arise when organizations integrate AI into business processes. We identify capability demands across established BPM capability dimensions and point to three cross-cutting shifts: the increasing need for dynamic management of business processes across dimensions, the role of data as an operational prerequisite of AI-driven processes, and the growing need for new governance and management requirements for emerging process forms.

The remainder of this paper is structured as follows. Section 2 introduces BPM capability literature and addresses AI in the context of BPM. Section 3 outlines the qualitative research design. Section 4 presents the study's findings including BPM capability demands. Section 0 discusses the results in relation to existing literature and outlines the paper's contributions, and Sect. 6 concludes the paper.

2 Theoretical Background

2.1 Business Process Management Capabilities

BPM evolved into an established body of knowledge encompassing methods and tools for the identification, discovery, analysis, redesign, implementation, and monitoring of business processes [1, 2]. It represents a comprehensive management discipline aimed at enhancing organizational performance and value creation [1, 2]. However, BPM remains a complex practice that many organizations struggle to adopt and advance effectively [16]. To address these challenges, BPM research has increasingly conceptualized and investigated BPM from a capability perspective [2]. Capabilities can be understood as an organization's repeatable action patterns that enable it to systematically leverage competencies, such as technical and managerial skills, to achieve intended process outcomes [17]. Accordingly, they function as organizational resources capable of generating sustained competitive advantage [17, 18].

Capability frameworks are central to the successful and sustainable establishment of BPM and support the prioritization of BPM-related investments [1, 3]. The framework proposed by de Bruin and Rosemann [16] is widely recognized, comprising 30 capability areas structured around the core elements of BPM, including *strategic alignment, governance, methods, information technology, people, and culture* [2]. *Strategic alignment* refers to the consistent linkage between an organization's overarching strategic priorities and its business processes. In this context, BPM is designed and managed in accordance with the overall strategy, ensuring that organizational activities are synchronized in a way that enables continuous and effective improvement of business performance. The core element *governance* includes the establishment of transparent and relevant responsibilities, roles, reward processes, and decision-making to guide process-related actions. *Methods and information technology* in the context of BPM are defined as the set of tools, techniques, and information-technology-based solutions that facilitate and support process activities. The core BPM element *people* encompass the human capital of an organization, meaning individuals or groups who continuously adapt and use their process knowledge and skills to improve business performance. In the context of BPM, *culture* refers to the shared values within an organization that shape a supportive environment for process initiative and significantly influence their success [2]. However, BPM frameworks are not static but must adapt to dynamic environments that drive changes in organizational requirements. Current research emphasizes that such frameworks need to evolve in response to changing organizational contexts and environmental dynamics [19–21]. In line with this, Kerpedzhiev et al. [3] examined which capability areas remain relevant in the context of digitalization and proposed an updated version of the de Bruin and Rosemann [16] framework, which highlights the ongoing need to reassess and adapt BPM capability models to emerging technologies [3].

2.2 Artificial Intelligence and BPM

One of the most transformative technological developments currently reshaping organizations is the rapid advancement of AI. AI is an umbrella term for the evolving frontier of computational systems that reference human intelligence to solve increasingly complex decision problems [10, 22]. These systems rely on advanced, data-driven algorithms (e.g. machine learning and neural networks) to learn from large-scale data, act with varying degrees of autonomy, and continually extend the scope of digital computation [10, 22]. Subsequently, we focus on AI technologies as autonomous intelligent systems rather than rule-based systems. With the rapid development of AI technologies, organizations increasingly aim to integrate them to improve operations, reduce costs, enhance decision-making, and gain a competitive advantage [10–13].

However, successfully embedding AI systems into business processes remains a major challenge [23]. Understanding how AI can be efficiently integrated into organizational processes is therefore of growing importance. AI integration can have different effects on organizations. First, AI transforms business processes by altering process activities. This in turn, influences process performance outcomes, such as process efficiency and process quality, which contribute to the overall organizational performance [24, 25]. Second, AI affects BPM as a discipline by challenging existing BPM methods and tools required to manage AI-driven processes. For instance, research underscored the importance of capabilities for managing operations with AI [26]. As a result of these developments, a vibrant research field has emerged at the intersection of BPM and AI [27]. For example, it has been discussed how large process models can transform BPM by integrating large language models with process data and organizational knowledge to enable more intelligent, context-aware analysis, automation, and decision support within BPM systems [28]. Additionally, Dumas et al. [5] introduced AI-augmented business process management systems.

Current literature increasingly addresses how AI can be integrated into business processes to realize its operational benefits and how it reshapes the nature and practice of BPM. The literature suggests that AI requires different approaches, tools, and competencies that differ from those associated with traditional information technology [10, 14]. Consequently, it becomes essential to understand how the BPM discipline needs to evolve concerning AI integration. Developing the organizational ability to manage and govern AI-driven processes effectively represents a critical new dimension of BPM maturity and a key prerequisite for realizing AI's full potential.

3 Research Method

To explore what BPM capabilities are required when organizations integrate AI into their processes, we adopt an inductive qualitative approach. We organized our semi-structured interviews with the core BPM elements outlined by Kerpedzhiev et al. (2021) [3]. Given the exploratory nature of our research and the challenges associated with AI integration, we sought to collect rich and detailed empirical data to derive theoretical insights from both practice and academia [29]. A qualitative approach is particularly appropriate for research seeking to generate deep, contextualized, and nuanced insights into complex phenomena [30, 31]. Accordingly, we conducted 18 semi-structured interviews with practitioners and researchers experienced in BPM and AI, allowing for the exploration of diverse perspectives. As no substantial differences were observed between the responses of academics and practitioners, this distinction is not discussed further.

To strengthen the validity and reliability of the study, we carefully selected participants guided by two sampling criteria: (1) The expert must hold a position related to BPM and AI, either as a core research focus or as part of their professional responsibilities, (2) the expert must have held their position for at least six month to possess sufficient knowledge. The experts were therefore well positioned to evaluate the implications of AI integration for BPM. To ensure a holistic perspective, the sample included experts from both academia and industry. Initial contact was established via direct messaging after a detailed review of their profiles. Potential participants received an information sheet outlining the purpose of the study and the interview procedures. In addition, we applied snowball sampling to identify further qualified participants through referrals [33]. We conducted 18 interviews until we reached the point of saturation, at which interviewees did not report any new information anymore. The interviewees had between three and more than 30 years of professional experience in their respective fields. A complete overview of the interviews is provided in Table 1.

All interviews were conducted online between August and October 2025 and lasted 39 min on average. The interviews were held in German, recorded, translated into English by both authors, and subsequently transcribed for analysis. We followed a semi-structured interview guide with open-ended questions, which ensured that all essential topics were addressed consistently while still allowing flexibility to explore emerging ideas, experiences, and anecdotes raised by the interviewees [32, 34]. The interviews were structured alongside the BPM elements, however leaving space to depart from the guide freely and not forcing interviewee statements or opinions [2]. In line with the iterative nature of qualitative research, we refined the interview guide based on initial insights [32]. Hence, interviewees were invited to discuss the challenges, opportunities and skills that they believed to be most relevant to BPM in the context of AI integration.

We coded and analyzed the obtained data with MAXQDA. First, we conducted open coding by reading the interview transcripts and identifying initial codes guided by the research question [35], while remaining close to the participants' original wording. Second, by using the approach of axial coding, we identified relationships between similar codes and grouped these first-order codes into higher-order themes [35]. Third, building on these themes, we then derived aggregated dimensions and structured the findings based on the BPM dimensions.

Table 1. Overview of the interviewees and interview duration

#	Position	Professional Experience (years)	Interview Duration (minutes)
IP1	Professor of Business Information Systems and Organization	>20	34
IP2	Lecturer and Process Manager	>20	38
IP3	IT Consultant (AI specialization)	>30	27
IP4	Chief Executive Officer	>20	52
IP5	AI Integration Consultant	>20	37
IP6	Chief Executive Officer of a Consulting Firm	>20	30
IP7	Partner at a Consulting Firm	>20	35
IP8	Managing Director and Partner	11	32
IP9	Professor of Digital Business and Industry Practitioner	25	43
IP10	IT Auditor and Process Manager	20	58

(continued)

Table 1. (*continued*)

#	Position	Professional Experience (years)	Interview Duration (minutes)
IP11	IT Auditor	3	58
IP12	IT Consultant	15	24
IP13	IT Consultant	14	43
IP14	IT Project Manager	25	39
IP15	Partner at a Consulting Firm	>20	45
IP16	Chief Financial and Transformation Officer	>30	30
IP17	Senior Associate	4	37
IP18	Lecturer and AI Consultant	>20	34

4 Results

By analyzing the data obtained from the interviews, we identified 17 BPM capability demands for the management of AI-driven processes. The results reveal that the integration of AI into business processes introduces distinct capability demands that organizations must address. The following section presents the interviewees' perspectives on these demands across the BPM dimensions, as summarized in Table 2.

Considering *strategic alignment* when focusing on AI integration, interviewees described the need for BPM to define which types of processes and AI systems are appropriate for integration, as different AI technologies carry different requirements and outcomes. Interviewees consistently noted that not all AI technologies are universally applicable, making systematic use case identification and prioritization at the right level of integration a strategic necessity. This aspect is of particular importance, given that organizations often lack a common understanding of what AI technologies entail. Additionally, interviewees stressed the need for future-oriented process design, including the organization's consideration of their intended business journey and proactively align processes accordingly. Furthermore, interviewees stated that the level of AI integration raises new questions around stakeholders' perception and corporate image. For example, if an organization integrates autonomous AI agents into customer-facing processes such as service-centered processes to communicate with their customers, the interaction of those may impact the customers' perception of the organization. Hence, this dimension requires explicit strategic consideration.

With regard to the dimension *governance*, participants highlighted that governance topics significantly gained importance due to the increasing use of AI. Interviewees stated that with the development of AI also new regulations such as the EU-AI act appear that introduce legal challenges that must be addressed when deploying AI systems. Beyond regulatory compliance, AI introduces new ethical challenges that many organizations have not yet addressed. Hence, interviewees stressed the need to define and govern the ethical, moral, and societal dimensions of AI-driven processes. Moreover, interviewees highlighted that autonomous AI systems extract, collect, change, and create data as part of process execution, requiring BPM to define and enforce standards for how data is captured, stored, linked, and accessed within AI-driven processes. This ensures that all process data used by AI comply with security and privacy requirements and can be adapted when processes and AI use cases evolve.

Interviewees also described new challenges regarding roles and responsibilities when integrating AI systems. Organizations must define accountability for AI-driven outcomes and decision rights, including how responsibility is transferred between humans, AI agents, and external partners. Interviewees additionally questioned whether existing BPM roles remain adequate. For instance, the role of process owners should evolve beyond existing responsibilities to encompass a multidisciplinary function, incorporating business, technical, and societal aspects. In this regard, the question of whether they should transform into a more advisory role was discussed. Finally, the importance of establishing a dynamic process governance capable of rapidly adjusting to regulatory and technological changes was strongly emphasized across interviews.

One recurring aspect in relation to the dimension *methods and information technology* concerns the pace of process management. Interviewees stressed that managing AI-driven processes requires continuous, agile, and dynamic improvements, allowing AI-driven processes to be rapidly (re)designed, tested, and adapted using flexible methods and tools. For instance, one interviewee described the need for tools "similar to agile tools like DevOps, but no release anymore". Try-and-error was described as a legitimate and necessary approach, supported by low−/no-code tools and prompt-based guidance for process managers. Furthermore, interviewees described a fundamental shift in how processes should be analyzed and selected for AI integration. Rather than documenting sequential activity flows and relying on traditional performance metrics alone, AI integration requires systematically evaluating processes according to their AI-readiness, identifying weak points, and deciding whether processes should be retained, improved, or eliminated. Some interviewees went further, arguing that processes only realize genuine efficiency gains through AI when they are fundamentally rethought rather than incrementally optimized.

Furthermore, interviewees emphasized the importance of end-to-end viewing and management of processes when integrating AI. As AI systems access information across departmental boundaries and may autonomously improve, a holistic process understanding becomes essential, focusing not only on individual process chains but on the entire process landscape. This requires designing and maintaining an integrated, enterprise-wide process architecture that connects end-to-end processes, defines and manages interfaces between AI-driven and legacy process components, and uses a robust, standardized technology stack and orchestration tools to enable seamless data and activity flows. In this context, interviewees also pointed out that many organizations acquire process management tools that serve merely as graphical documentation rather than active management instruments. When integrating AI into processes, however, BPM tools must evolve into operational management platforms capable of governing AI-driven process execution, otherwise AI becomes a disconnected element rather than an integrated part of the process landscape. Closely connected, data readiness was consistently highlighted as a critical and largely overlooked challenge. Traditional BPM focuses on the sequential logic of process steps but does not systematically assess whether the necessary data prerequisites for AI deployment exist. AI systems require access to high-quality, machine-readable data pools, yet organizations often combine AI systems with legacy infrastructures, resulting in the same data existing in different contexts and quality levels. As one interviewee noted: "I need to have my data under control to use AI efficiently." Interviewees stressed the importance of systematically managing data flows, defining input and output parameters, and clarifying what data AI systems are allowed to access and use. Finally, autonomous AI systems introduce new monitoring and controlling demands. Unlike traditional process participants, AI systems act independently and are often difficult to supervise due to their black box characteristics. This raises questions that go beyond traditional process management, such as the determination of when an autonomous, AI-driven process should be terminated and the individual with the authority to make that decision. Interviewees stressed the need for new AI-specific performance indicators, and several interviewees suggested BPM must develop contingency mechanisms including backup processes to ensure reliable and controllable outcomes for scenarios when AI systems do not perform as expected.

Concerning the dimension *people*, interviewees described the necessity of fundamentally extending existing knowledge and skill profiles in BPM in order to facilitate AI integration. Interviewees consistently emphasized the need for a conceptual and practical understanding of AI systems. This includes not only their functionalities, limitations and basic interaction mechanisms, such as prompting and working with AI-agents, but also differences from traditional IT-systems, and regulatory constraints. It also includes digital and technical knowledge that enables employees to understand the risks and benefits of AI, leading effective integration and collaboration with AI-based applications. This is especially important because AI led to fears across employees, including BPM professionals, which results in not efficiently realizing AI's value. Specific AI knowledge was seen as primary means of addressing this. Beyond AI literacy, respondents indicated the need for a different thinking framework, which includes analytical, problem-solving, interdisciplinary, critical, and logical thinking to evaluate and understand AI outputs, recommendations, and to use it for decision-making. This also requires the knowledge and the understanding of data flows, data storage, data quality and how data can be responsibly used to design, train, and operate AI systems in business processes. Interviewees frequently highlighted the lack of such skills in organizations. Another often observed deficit relates to BPM and process knowledge. As one interviewee explained, AI integration requires a "completely different logic, a switch from static processes with if else logic to self-regulating systems". Organizations currently lacking transparency about their own processes were seen as fundamentally unable to leverage AI efficiently. Furthermore, as AI systems work autonomously and may change or improve processes without explicit human instruction, process understanding becomes critical to prevent information loss and maintain organizational control. Finally, interviewees emphasized the importance of dynamic skills, understanding the ability to continuously adapt to technological and market changes. This was considered particularly important given AI's rapid development and the constantly shifting environment surrounding it.

The findings of the interviews addressed several *cultural* aspects. In all previous discussed dimensions, the aspect of developing a dynamic BPM has surfaced. Interviewees strongly stressed that managing AI-driven processes requires a cultural shift toward a dynamic process culture including rapid, responsive adjustments of AI-driven processes while maintaining a transparent, end-to-end view on process flows. In general, a willingness to change and to take risks was described as essential in order to disrupt existing business processes and to rebuild them. This was particularly emphasized because interviewees underlined that the full potential of AI can only be realized if processes are completely rethought, as opposed to merely implementing optimization using AI. Interviewees also emphasized the need of commitment to design and improve AI-enabled processes jointly across functions, without silos or strict hierarchies. AI systems access information across departmental boundaries, making a fragmented or hierarchical approach to process management fundamentally incompatible with effective AI integration. Moreover, interviewees pointed out that realizing AI's potential in processes requires a cultural shift toward experimentation and continuous learning. Decisions about AI integration should be grounded in evidence such as data, facts, and KPIs, rather than driven by hype or avoided out of fear. Interviewees emphasized that AI should be approached as a powerful tool, requiring a mindset of curiosity and openness to failure rather than anxiety about the technology. Table 2 outlines the identified capability demands exemplified by illustrative interview quotes. Additionally, we classified the demands as traditional BPM capabilities, capabilities that become increasingly important due to AI integration, and new capabilities demanded through AI integration

Table 2. Overview of the demanded capabilities supported by interview quotes pre-existing in BPM; intensified by AI-specific characteristics; novel capability demand

Dimension	Demanded Capability	Illustrative Quotations
Strategic Alignment	Ability to systematically identify and prioritize AI use cases, determine the appropriate level of AI integration (e.g., process, task, specific AI technology), and define the intended purpose and value contribution of AI systems within the organizational process landscape.	"What kind of processes do I need in the future" (IP2), "On what level do we integrate AI?" (IP3), "Align business and IT to identify use cases" (IP15), "Bring together AI know-how and process know-how" (IP12), "Focus where the business journey is headed and align processes, accordingly, be future-oriented" (IP2)
Strategic Alignment	Ability to align AI-driven process deployment with the organization's corporate image and its stakeholder-facing modes of interaction.	"How is the company presented, how do we use AI?" (IP9), "How to act with customers and AI?" (IP15), "Different point of view, benefits, different departments" (IP15)
Governance	Ability to ensure that every AI-driven process is designed, executed, and monitored in line with applicable external and internal regulatory policies, and to continuously adapt these to evolving regulatory and technological conditions.	"Need for central AI process governance to ensure that all processes comply with regulations, e.g., EU-AI act" (IP18), "Regulate the allowed usage of AI" (IP8), "Verify whether AI is legally entitled to make the specific decision" (IP13), "Adapt regulations, be flexible" (IP12)
Governance	Ability to define and govern the ethical, moral, societal, and responsible dimensions of AI-driven processes.	"AI ethics as one part of regulations" (IP3), "Ethics as a new discipline whether AI acts within ethical, moral, responsible areas" (IP14), "How can I act ethically correctly?" (IP14)
Governance	Ability to define and enforce standards for how data are captured, stored, linked, and accessed within AI-driven processes, ensuring that all process data used by AI systems comply with security and privacy requirements and can be adapted as processes and AI use cases evolve.	"Data Governance to ensure the source, security, reliability of these data" (IP3), "Data privacy is more important considering AI-driven processes" (IP2), "Information security" (IP14), "How to handle which information?" (IP15), "Data ownership" (IP3), "Determine where to process data" (IP8)

(continued)

Table 2. (*continued*)

Governance	Ability to define standards and guidelines regarding responsibilities for AI-driven outcomes and decision rights, and how accountability is transferred and governed between teams, humans, AI agents, and external partners.	"Determine who is responsible for AI's decisions" (IP18), "Clear regulations regarding responsibilities" (IP11), "Different roles, e.g., what does the process owner?" (IP1), "Governance between agents" (IP6)
Methods and Information Technology	Ability to manage AI-driven processes rapidly and flexibly by using agile methods, dynamic tools, and dynamic low-/no-code programming, supported by prompt-based guidance for process managers and employees.	"Adjust processes more agile and quicker" (IP8), "Act fast, be flexible" (IP7), "Try and error model" (IP12), "Integrate prompts to narrow down employees' use" (IP5), "Low-/no-code for process modeling" (IP9)
Methods and Information Technology	Ability to employ methods for systematically analyzing whole processes and process chains with AI, identifying weak points, deciding whether processes should be retained, eliminated, or radically quickly reengineered with AI to keep up with the speed at which changes occur.	"Process manager should be allowed to interrogate, change the whole process" (IP2), "People who carry out the processes need to keep up with the speed at which things change" (IP1), "Analyze processes and then decide whether processes still exist and can be improved with AI" (IP16), "Integration of AI at weak points" (IP9)
Methods and Information Technology	Ability to design and maintain an integrated, end-to-end process architecture that defines and manages interfaces between AI-driven and legacy process components, and enables seamless data and activity flows across the overall process landscape through robust, standardized technology stacks and orchestration tools.	"Use of one main platform, not several 'island solutions'" (IP6), "Connect processes end-to-end and not only one process" (IP6), "Enable architecture, a semantic network of information as basic for AI-systems" (IP4), "To integrate AI you need to look end-to-end, have a holistic view over the whole processes" (IP2)
Methods and Information Technology	Ability to provide secure, compliant and high-quality data and analytics infrastructure (e.g., data lakes, data pipelines, transformation, and analytics methods) so that AI systems in processes can access machine-readable, brand-safe data for training and optimization of process chains.	"If the fundamentals are not right, then nothing sensible will come out of it" (IP4), "I need to have my data under control to use AI efficiently" (IP14), "Before you manage data, know your data" (IP10), "Data storage via data lake or other data system that makes the data available centrally" (IP11), "Make your data machine readable" (IP12), "The network of information needs to be mastered to train the AI on processes" (IP4)

(*continued*)

Table 2. (*continued*)

Methods and Information Technology	Ability to manage, monitor, and control AI-driven processes, including AI-generated outputs, external-partner activities, and the criteria for ending autonomous AI operations, while tracking key performance indicators and mitigating output risk.	"Methods to end an AI-driven autonomous process" (IP8), "Closely control, monitor outputs" (IP8), "KPIs for monitoring AI-driven processes" (IP10), "Control and make ultimate ratio-decisions in the process chains" (IP15), "Control what external partners are doing" (IP11)
People	Ability to understand AI systems conceptually and practically, including their core functionalities, limitations, interaction mechanisms, differences from traditional IT systems, regulatory constraints, and relevant digital and technical knowledge for effective AI-driven process management.	"Many people don't even know what it's all about" (IP3), "It is important to understand how the systems works" (IP5), "Understand AI's possibilities" (IP2), "What is an AI based application and what is traditional IT?" (IP1), "What is it, what does AI, what tools are available?" (IP10), "Programming skills" (IP9), "Software development" (IP9)
People	Ability to apply analytical, problem-solving, interdisciplinary, critical, and logical thinking to assess AI outputs and recommendations to make informed decisions.	"Critical thinking is needed" (IP10), "Using AI to generate value, I need analytic skills, interdisciplinary thinking" (IP14), "Understanding what data flows where, in and out, and how I can use them" (IP10), "We need problem-solving methods" (IP10)
People	Ability to understand process structures, interdependencies, and the shift from static rule-based processes to more adaptive and self-regulating process logics.	"Completely different logic: switch from static processes, if else, to self-regulating systems" (IP18), "Knowledge about processes and holistic view on dependences" (IP8)
Culture	Commitment to prioritize rapid, responsive adjustments of AI-driven processes while maintaining a transparent, end-to-end view of process flows and gaps.	"Overall view of processes because it is not transparent what happens next" (IP5), "Respond extremely quickly to changes" (IP1), "Cultural aspect whether companies move slowly or quickly" (IP3), "there are gaps in the processes that no one wants to admit" (IP15)
Culture	Commitment to proactively disrupt and rebuild existing business processes with AI, including the willingness to change and to take increased risks.	"Intrinsic motivation to change something" (IP11), "Be curious and open" (IP10), "Courage to disrupt existing business processes, tear them down and rebuild them" (IP2)
Culture	Commitment to design and improve AI-enabled processes jointly across functions, without silos or strict hierarchies, based on teamwork and knowledge sharing.	"Culture to develop processes together, not top down" (IP18), "No silo mindset with AI" (IP15), "Not based on division of labour, hierarchical, fragmented" (IP7), "Ability to be open and share knowledge" (IP14)

5 Discussion

The findings of this study indicate that integrating AI into business processes creates capability demands across all established BPM dimensions. At the same time, the results show that although BPM has long addressed change, process flexibility, and knowledge-intensive processes, established capabilities do not fully capture the nature of AI and its impact on managing business processes [3, 10, 16, 36]. Across the interviews, three cross-cutting shifts emerged when we synthesized the 17 capability demands across the BPM dimensions and identified patterns that recurred in more than one dimension: (1) the intensification of dynamic BPM across dimensions, (2) the elevation of data from a supporting resource to an operational prerequisite, and (3) the need to govern and manage novel, partially autonomous process forms.

First, the findings indicate that AI increases the need for a more dynamic understanding of BPM. While prior BPM research has already emphasized adaptation, change, and innovation in process management [3, 37, 38], the interview data indicate that dynamic behavior becomes a pervasive requirement across all capability areas rather than a localized concern. For instance, AI use cases must be reassessed continuously (strategic alignment), governance arrangements must remain adaptable under changing technological and regulatory conditions (governance), methods and tools must support rapid redesign and experimentation (methods/information technology), employees must continuously update relevant competencies and dynamic skills (people), and the overall BPM culture needs to be dynamically driven, prioritizing rapid, responsive process adjustments (culture). The findings therefore do not propose that AI introduces dynamism into BPM for the first time. Instead, they indicate that AI intensifies and broadens the need for dynamic BPM across the full set of capability dimensions.

Second, the findings show that AI changes the role of data in BPM. In contrast to prior work, where data primarily supports process analysis and improvement [3, 39], data becomes a foundational condition for process execution, adaptation, and control considering AI-driven processes. AI systems depend on accessible, machine-readable, and sufficiently reliable data to operate within process environments, while at the same time generating and transforming data during process execution. As a result, data-related requirements cannot be reduced to a technical issue within methods and information technology alone. Instead, they need to be addressed across all dimensions. This shift is grounded in the data-related capability demands in Table 2, for instance, the definition and enforcement of data standards (governance), the provision of a secure and high-quality data and analytics infrastructure (methods and information technology), and data-related knowledge and thinking (people). The findings therefore go beyond the established observation that data quality matters. They underline that the ability to provide, govern, and interpret data becomes a core condition for integrating AI into business processes.

Third, the results highlight a shift in BPM thinking from static, rule-based processes to emergent, self-regulating process forms that lead to new challenges in the BPM discipline. Additionally, traditional BPM has largely been developed for settings in which business processes are specified in advance and enacted on the basis of explicit process models and rules [40]. The interview data indicate that AI challenges this assumption. As

AI systems act with varying degrees of autonomy, produce outputs that are less deterministic as classic software solutions but are more likely to lead to unintended or erroneous results and often remain partly opaque in their operation, BPM must address issues that extend beyond conventional process execution and control [10]. In the interviews and derived capability demands, this was reflected in a stronger emphasis on accountability for AI-driven outcomes and decision rights (governance), criteria for stopping autonomous process behavior as well as monitoring and intervention mechanisms (methods and information technology), redesigned roles and responsibilities (governance), the understanding of self-regulating process logics (people), and the maintenance of organizational control despite increasing autonomy and limited transparency. Our findings indicate that AI reshapes the conditions for enacting BPM, pointing to a more dynamic, data-driven, and extended conceptualization of BPM capabilities in AI-driven process contexts.

Our study contributes to the existing literature in multiple ways. First, we respond to calls to reconsider existing BPM assumptions considering advanced technologies [3, 19–21] by exploring what BPM capability demands occur when integrating AI into business processes. In doing so, we highlight AI-specific relevant BPM capability demands as well as discuss those in context with existing literature and thus, support the core BPM literature on BPM capabilities. Second, the study contributes to the emerging literature at the intersection of AI and BPM research by exploring BPM capability demands when integrating AI into business processes and identifying three cross-cutting shifts in BPM associated with this transformation. These shifts provide a more focused conceptual lens for explaining how AI changes BPM beyond conventional digitalization and automation discussions. Third, the study contributes to the still limited connection between BPM and dynamic capability research. The findings suggest that AI integration increases the need for BPM to support the continuous reassessment of use cases, the adaptation of governance arrangements, and the renewal of organizational competencies. In this way, the study clarifies how BPM can support ongoing organizational adaptation in AI-driven process environments. From a practical perspective, our findings have several significant implications, especially for organizations aiming to realize AI's potential for innovation and competitive advantage. We support organizations' understanding of BPM in a more dynamic way and highlight what aspects are especially important when integrating AI effectively into business processes in highly dynamic environments. In particular, our findings indicate that organizations should focus on strengthening and adapting their BPM capabilities. Realizing AI-driven change requires going beyond purely technological focus and instead managing the dynamic interplay of people, governance, business, processes, and technology. The three cross-cutting shifts identified in this study further support organizations to interpret capability demands for AI-integrated business processes. These shifts point to broader areas in which organizations need to develop and direct their BPM capabilities: the need for dynamic BPM, the centrality of data as a foundational layer, and the emergence of new governance and management requirements for autonomous process forms. In doing so, we provide a starting point for operationalizing BPM capability development in AI-driven process environments. Organizations may use these demands and shifts as assessment criteria to diagnose existing capability gaps, as input for maturity models that distinguish different levels of AI-enabled BPM readiness, or for structuring governance, roles, methods, and competencies around AI-driven processes.

6 Conclusion, Limitations, and Outlook

The increasing integration of AI into business processes challenges organizations to reassess how they manage and govern their processes. In this light, this study explored BPM capability demands for integrating AI into business processes. Drawing on 18 semi-structured interviews with BPM and AI experts, the findings indicate three cross-cutting shifts in BPM: (1) the need for dynamic BPM across all dimensions, (2) the centrality of data as a foundational layer, and (3) the emergence of new governance and management requirements for autonomous process forms. Overall, the study demonstrates that AI does not simply augment BPM but fundamentally reshapes its capability requirements, calling for a reconceptualization of BPM as a dynamic, data-driven, and continuously adaptive discipline.

Naturally, our study is subject to limitations, that at the same time, provide opportunities for future research. As a qualitative interview study, the findings are based on the perceptions and experiences of a limited number of experts from academia and industry in Germany. This allows for an in-depth understanding of current challenges and perspectives but limits the extent to which the findings can be generalized beyond the studied context. The EU AI Act, for instance, features as a driver of the identified governance-related capability demands. In jurisdictions with different regulatory regimes, the relative emphasis among the demands may differ. In addition, the field itself is still evolving, as many organizations are only beginning to integrate AI into business processes and are doing so under rapidly changing technological and regulatory conditions. The capability demands identified in this study should therefore be understood as reflecting the current state of expert understanding rather than final capability categories. Furthermore, the study deliberately treated AI as a broad umbrella term in order to identify overarching implications for BPM. While this approach was useful for capturing cross-cutting shifts, it also means that differences between specific AI types, levels of autonomy, and concrete process contexts remain underexplored. As a result, the findings remain at a relatively general level. Finally, as with qualitative interpretive research more broadly, the analysis is shaped by the researchers' coding and interpretation despite careful efforts to ensure rigor and transparency. The study should therefore not be understood as proposing a validated extension of existing BPM capability frameworks, but as providing an empirically grounded basis for their further refinement in AI-driven process contexts.

Building on our findings and limitations, several avenues for future research emerge. Future research should examine BPM capability demands in more specific AI and process contexts, as different AI types and levels of autonomy are likely to create different requirements for process management. In addition, future research could further operationalize the identified capability demands and shifts by translating them into more concrete organizational support mechanisms, such as assessment instruments, maturity models, capability roadmaps, or design principles for BPM considering AI-driven processes. Such work would move the field from identifying capability demands toward instruments that organizations can apply directly. Moreover, longitudinal research could examine how BPM roles, governance arrangements, and competencies evolve over time as organizations gain more experience with AI-driven processes.

Finally, we acknowledge that although BPM research has long recognized dynamic, knowledge-intensive, and difficult-to-standardize process environments, we suggest that AI extends and intensifies BPM capability dimensions. Since AI requires approaches, tools, and competencies that differ from those associated with traditional information technology [10, 14], it becomes essential to understand how BPM capabilities may need to evolve in light of AI and provide a starting point for future research.

Disclosure of Interests.. The authors have no competing interests to declare that are relevant to the content of this article.

References

1. Dumas, M., La Rosa, M., Mendling, J., Reijers, H.A.: Fundamentals of Business Process Management. Springer, Berlin Heidelberg, Berlin, Heidelberg (2018). https://doi.org/10.1007/978-3-662-56509-4
2. Vom Brocke, J., Rosemann, M. (eds.): Handbook on Business Process Management 1: Introduction, Methods, and Information Systems. Springer, Berlin Heidelberg (2015). https://doi.org/10.1007/978-3-642-45100-3
3. Kerpedzhiev, G.D., König, U.M., Röglinger, M., Rosemann, M.: An exploration into future business process management capabilities in view of digitalization: results from a Delphi study. Bus. Inf. Syst. Eng. **63**, 83–96 (2021). https://doi.org/10.1007/s12599-020-00637-0
4. Poeppelbuss, J., Plattfaut, R., Niehaves, B.: How do we Progress? An exploration of alternate explanations for BPM capability development. CAIS. **36** (2015). https://doi.org/10.17705/1CAIS.03601
5. Dumas, M., et al.: AI-augmented business process management systems: a research manifesto. ACM Trans. Manage. Inf. Syst. **14**, 1–19 (2023). https://doi.org/10.1145/3576047
6. Weinzierl, S., Zilker, S., Dunzer, S., Matzner, M.: Machine learning in business process management: a systematic literature review. Expert Syst. Appl. **253**, 124181 (2024). https://doi.org/10.1016/j.eswa.2024.124181
7. Ceravolo, P., Comuzzi, M., De Weerdt, J., Di Francescomarino, C., Maggi, F.M.: Predictive process monitoring: concepts, challenges, and future research directions. Proc. Sci. **1**, 2 (2024). https://doi.org/10.1007/s44311-024-00002-4
8. Van Der Aalst, W.: Process mining: overview and opportunities. ACM Trans. Manag. Inf. Syst. **3**, 1–17 (2012). https://doi.org/10.1145/2229156.2229157
9. van der Aalst, W.M.P., Bichler, M., Heinzl, A.: Robotic process automation. Bus. Inf. Syst. Eng. **60**, 269–272 (2018). https://doi.org/10.1007/s12599-018-0542-4
10. Berente, N., Gu, B., Recker, J., Santhanam, R.: Managing artificial intelligence. MIS Q. **45**, 1433–1450 (2021). https://doi.org/10.25300/MISQ/2021/16274
11. Koehler, J.: Business process innovation with artificial intelligence: levering benefits and controlling operational risks. EBM. **4**, 55 (2018). https://doi.org/10.11648/j.ebm.20180402.12
12. Mayer, A.-S., Strich, F., Fiedler, M.: Unintended consequences of introducing AI Systems for Decision Making. MIS Q. Exec. **239–257** (2020). https://doi.org/10.17705/2msqe.00036
13. Meierhöfer, S., Dorner, C.-M.: Through the lens of cognitive functions and contextual factors-a study of the action potentials of artificial intelligence in business process management. In: Beverungen, D., Lehrer, C., Trier, M. (eds.) Solutions and Technologies for Responsible Digitalization, pp. 203–218. Springer Nature, Switzerland, Cham (2025). https://doi.org/10.1007/978-3-031-80122-8_13

14. Weber, M., Engert, M., Schaffer, N., Weking, J., Krcmar, H.: Organizational capabilities for AI implementation-coping with inscrutability and data dependency in AI. Inf. Syst. Front. **25**, 1549–1569 (2023). https://doi.org/10.1007/s10796-022-10297-y
15. Mendling, J., Decker, G., Hull, R., Reijers, H.A., Weber, I.: How do machine learning, robotic process automation, and Blockchains affect the human factor in business process management? CAIS. **297–320** (2018) https://doi.org/10.17705/1CAIS.04319
16. de Bruin, T., Rosemann, M.: Using the Delphi Technique to Identify BPM Capability Areas. In, ACIS Proceedings (2007)
17. Wade, M., Hulland, J.: Review: the resource-based view and information systems research: Review, extension, and suggestions for future research. MIS Q. **28**, 107–142 (2004). https://doi.org/10.2307/25148626
18. Barney, J.: Firm resources and sustained competitive advantage. J. Manag. **17**, 99–120 (1991)
19. Alter, S.: Extending BPM by treating processes as components of work systems that produce product/services valued by their customers. In: ICIS 2024 Proceedings (2024)
20. Klun, M., Trkman, P.: Business process management – at the crossroads. Bus. Process. Manag. J. **24**, 786–813 (2018). https://doi.org/10.1108/BPMJ-11-2016-0226
21. Recker, J.: Suggestions for the next wave of BPM research: strengthening the theoretical core and exploring the protective belt. J. Inf. Technol. **15**(2), 5–20 (2014)
22. Collins, C., Dennehy, D., Conboy, K., Mikalef, P.: Artificial intelligence in information systems research: a systematic literature review and research agenda. Int. J. Inf. Manag. **60**, 102383 (2021). https://doi.org/10.1016/j.ijinfomgt.2021.102383
23. Mishra, A.N., Pani, A.K.: Business value appropriation roadmap for artificial intelligence. VJIKMS. **51**, 353–368 (2021). https://doi.org/10.1108/VJIKMS-07-2019-0107
24. Kunz, P.C., Jussupow, E., Spohrer, K., Heinzl, A.: How the application of machine learning systems changes business processes: a multiple case study. In: ECIS Proceedings (2022)
25. Melville, N., Kraemer, K., Gurbaxani, V.: Review: Information technology and organizational performance: an integrative model of IT business value. MIS Q. **28**(2), 283–322 (2004). https://doi.org/10.2307/25148636
26. Kirchmer, M., Havaligi, S.: Realizing the full potential of AI applications through business process management. In: Shishkov, B. (ed.) Business Modeling and Software Design, pp. 108–125. Springer Nature, Switzerland, Cham (2026). https://doi.org/10.1007/978-3-031-98033-6_7
27. Fettke, P., Di Francescomarino, C.: Business process management and artificial intelligence: literature survey and future research. Künstl. Intell. **39**, 67–79 (2025). https://doi.org/10.1007/s13218-025-00891-y
28. Kampik, T., et al.: Large process models: a vision for business process Management in the age of generative AI. Künstl. Intell. **39**, 81–95 (2025). https://doi.org/10.1007/s13218-024-00863-8
29. Van Maanen, J., Sørensen, J.B., Mitchell, T.R.: Introduction to special topic forum: the interplay between theory and method. Acad. Manag. Rev. **32**, 1145–1154 (2007). https://doi.org/10.5465/amr.2007.26586080
30. Draper, J.: The relationship between research question and research design. In: Research into Practice: Essential Skills for Reading and Applying Research in Nursing and Health Care, pp. 69–84. Bailliere Tindall (2004)
31. Schultze, U., Avital, M.: Designing interviews to generate rich data for information systems research. Inf. Organ. **21**, 1–16 (2011). https://doi.org/10.1016/j.infoandorg.2010.11.001
32. Gioia, D.A., Corley, K.G., Hamilton, A.L.: Seeking qualitative rigor in inductive research: notes on the Gioia methodology. Organ. Res. Methods. **16**, 15–31 (2013). https://doi.org/10.1177/1094428112452151
33. Lincoln, Y.S., Guba, E.G.: Naturalistic Inquiry. SAGE. (1985).

34. Myers, M.D., Newman, M.: The qualitative interview in IS research: examining the craft. Inf. Organ. **17**, 2–26 (2007). https://doi.org/10.1016/j.infoandorg.2006.11.001
35. Strauss, A.L., Corbin, J.M.: Basics of Qualitative Research: Techniques and Procedures for Developing Grounded Theory. Sage Publications, Inc., Thousand Oaks (1998).
36. Rosemann, M., Brocke, J.V., Van Looy, A., Santoro, F.: Business process management in the age of AI – three essential drifts. IseB. **22**, 415–429 (2024). https://doi.org/10.1007/s10257-024-00689-9
37. Grisold, T., Janiesch, C., Röglinger, M., Wynn, M.T.: Call for papers, issue 5/2024: managing the dynamics of business processes. Bus. Inf. Syst. Eng. **64**, 841–843 (2022). https://doi.org/10.1007/s12599-022-00777-5
38. Grisold, T., Janiesch, C., Röglinger, M., Wynn, M.T.: Managing dynamics in and around business processes. Bus. Inf. Syst. Eng. **66**, 533–540 (2024). https://doi.org/10.1007/s12599-024-00895-2
39. van der Aalst, W.M.P.: Business process management: a comprehensive survey. Int. Sch. Res. Not. 507984 (2013). https://doi.org/10.1155/2013/507984
40. Weske, M.: Introduction. In: Weske, M. (ed.) Business Process Management: Concepts, Languages, Architectures, pp. 3–22. Springer, Berlin, Heidelberg (2019). https://doi.org/10.1007/978-3-662-59432-2_1

Automatic AI-Based Extraction of Processes from Administrative Texts: The Case of Waste Management

Alessia Membrini[1], Maria Teresa Rossi[2(✉)], Leonardo Mariani[1], and Daniela Micucci[1]

[1] University of Milano-Bicocca, Milan, Italy
{alessia.membrini,leonardo.mariani,daniela.micucci}@unimib.it
[2] Gran Sasso Science Institute, L'Aquila, Italy
mariateresa.rossi@gssi.it

Abstract. Public administrations produce large volumes of unstructured administrative documents that regulate operational procedures, responsibilities, and regulatory constraints. These documents are characterized by high information density and fragmented procedural knowledge, making both human interpretation and automated extraction challenging. This issue is particularly critical in the waste management domain, where regulatory complexity and local variability significantly impact operational processes. This paper presents a hybrid pipeline for the automatic extraction and structuring of operational procedures from unstructured administrative texts. The proposed approach combines LLMs with RAG to ensure faithful, traceable extraction constrained to the source document. The pipeline identifies procedural sections, extracts relevant information from each section, and generates visual workflows. The system is evaluated on real-world administrative documents for waste management. Results show that the integration of generative AI within a structured pipeline improves the accuracy and reliability of procedural knowledge extraction, while mitigating hallucination risks.

Keywords: Procedural Knowledge Extraction · Automated Workflow Generation · Retrieval-Augmented Generation (RAG)

1 Introduction

Public administrations and private organizations continuously produce large volumes of administrative documents aimed at regulating operational activities, assigning responsibilities, and ensuring compliance with evolving regulatory frameworks [11]. Such documents are predominantly written in natural language and are characterized by high informational density, implicit structural hierarchies, and a mixture of normative, descriptive, and procedural content. In domains subject to complex and dynamic regulation, such as waste management, these *documents play a crucial role in defining operational workflows and*

F. M. Maggi et al. (Eds.): BPM 2026, LNBIP 599, pp. 398–414, 2027.
https://doi.org/10.1007/978-3-032-37877-4_23

accountability chains. However, their discursive and often fragmented structure makes systematic interpretation and reuse challenging, particularly when procedures are distributed across multiple sections and interwoven with references to definitions and legal provisions. The need to *bridge the gap between textual prescriptions and structured process representations* has thus become increasingly relevant [12].

The problem addressed in this work concerns the *automatic identification, extraction, and structuring of operational procedures from unstructured administrative documents* with a specific focus on the waste management domain. More precisely, the objective is to reconstruct coherent and verifiable procedural workflows from documents in which actions, responsible agents, temporal constraints, and conditional branches are not explicitly formalized but embedded in heterogeneous textual segments. This task requires not only detecting procedural sections but also resolving implicit references, integrating distributed definitions, and preserving links between extracted elements and their textual evidence.

The relevance of this problem is twofold. From an operational perspective, the manual interpretation of lengthy administrative documents is time-consuming and prone to inconsistencies, especially when multiple actors are involved and responsibilities depend on context-specific definitions. From a computational perspective, the extraction of structured processes from natural language represents a challenging instance of information extraction and semantic reconstruction. In regulatory domains, such as waste management, incorrect interpretation may lead to compliance issues or procedural delays. Therefore, any automated solution must ensure *fidelity* to the source text, *traceability* of extracted information, and *robustness* to linguistic variability and structural fragmentation.

State-of-the-art approaches can be broadly categorized into pattern-based methods [1], rule- or ontology-driven pipelines [6], and end-to-end Large Language Model (LLM) solutions [2]. Pattern-based approaches rely on recurring statistical regularities and are effective in homogeneous datasets, but they struggle with long, heterogeneous, and unique administrative documents. Rule-based and ontology-driven methods provide stronger formal guarantees but require substantial domain modeling and are less scalable when regulations vary across institutions. End-to-end LLM-based systems demonstrate strong linguistic generalization capabilities, but they are prone to hallucination and limited traceability, particularly when processing long documents. Retrieval-Augmented Generation (RAG) [8] mitigates some of these limitations by grounding generation in retrieved evidence; however, naive configurations may introduce irrelevant context and reduce interpretability.

Our approach advances knowledge in the area by investigating a combination of *structured document analysis*, grounded on the characteristics of waste management texts, and *LLM-based text interpretation*. Instead of relying exclusively on generative inference or rule-based heuristics, the proposed approach combines them: document hierarchy reconstruction and section segmentation are performed deterministically using structural heuristics, while LLMs are employed only in the interpretative tasks. Retrieval mechanisms are activated selectively to recover distributed definitions and contextual evidence, reducing the risk of unsupported inferences. LLMs are used to extract actions, agents, conditions, and dependencies, producing a JSON representation that is transformed into graphical workflows.

We evaluated the approach on three real administrative documents from the waste management domain. We considered two alternative configurations of the proposed approach, one using a RAG populated with similar procedures and one not using a RAG, to assess the impact of the knowledge of similar procedures on the accuracy of the results. We also considered a pure LLM-based configuration as a baseline. The results indicate that the RAG-based configuration achieves a more balanced trade-off between precision and recall, particularly in the identification of procedural sections and responsible agents. The purely generative configuration tends to maximize recall at the cost of low precision, highlighting the importance of contextual constraints and structural grounding.

2 Processes in Waste Management Documents

Waste management represents a particularly challenging context for automatic process extraction due to the complexity and dynamic evolution of its regulatory framework. In fact, waste management procedures are highly context-dependent and often vary across regional and local administrations. As discussed in [5], procedural obligations may differ depending on territorial jurisdiction, organizational structure, and administrative competence. This variability reduces the feasibility of purely template-driven or rule-based solutions.

A procedure can be modeled as a structured entity composed of interdependent elements, including *actions*, *responsible agents*, *temporal constraints*, *conditional branches*, and possible *exceptions*. In unstructured administrative texts, however, this structure is rarely explicitly formalized and procedural information is rarely localized in a single coherent section. Instead, relevant content is unstructured, typically written in discursive natural language and characterized by: *cross-references* (sections that include references to concepts defined in other sections), *implicit dependencies* (dependencies between activities are often left implicit, and must be deduced from the text), *high informational density and length* (documents may span dozens of pages and contain numerous interdependent clauses), *mixed hierarchical organization* (titles and subtitles may exist but do not necessarily follow standardized patterns), *heterogeneous content types* (operational, normative, explanatory, and contextual content coexist within the same structural layer), *semi-formal linguistic style* (imperative, conditional, and narrative formulations are interwoven, increasing ambiguity), *anaphoric references and implicit agents* (entities are often defined once, in dedicated sections, and later recalled in subsequent sections).

In Fig. 1, we report an excerpt of a typical procedure coming from an administrative text. The objective is to reconstruct complete procedural workflows, and not just to summarize textual content. This requires explicitly identifying actions, agents, responsibilities, and constraints, resolving ambiguities, and preserving a verifiable link to the originating textual evidence.

Title: Registration of Enterprises in the Animal By-Products Sector
Article 23 of Regulation (EC) 1069/2009 introduces the possibility of registering enterprises operating in the animal by-products (SOA) sector.

Definitions
Operator: who submits notifications to the competent authority.
N.LAs.: Notification of commencement of sanitary activity submitted to the competent authority.
Competent Authorities: the territorially competent Local Health Authority and the Regional Veterinary Health and Food Safety Service.
SUAP: Single Desk for Productive Activities.

Registration Procedure
1. The Operator submits a notification to the Local Health Authority through the SUAP for each establishment under its control. The notification must include all required information according to the prescribed form.
  – If the SUAP is not operational, the Operator submits the notification directly to the competent Local Health Authority. In all cases, certified electronic mail (PEC) is preferred.

2. The Veterinary Service of the Local Health Authority verifies the documentation and proceeds with the registration using the BDR-SIVRA system, informing the Regional Veterinary Authority.

3. When required, the Regional Authority assigns a registration code through the Ministry of Health's SINTESI system and communicates it to the SUAP and the Local Health Authority.

4. Operators declaring compliance with regulatory requirements are registered by the Local Health Authority. Compliance verification is performed later through periodic inspection activities.

Fig. 1. Excerpt of a typical procedure coming from an administrative text

Due to the structural dispersion of the text (e.g., in Fig. 1 the SUAP (Sportello Unico per le Attività Produttive) is described in the *Definitions* section and the corresponding action appears in a separate section), procedural extraction cannot rely on local sentence-level analysis alone, but requires context-aware integration across structural boundaries. The heterogeneity of expression further increases complexity. Actions may be formulated narratively; agents may be expressed through pronouns or generic labels; conditions may be introduced through subordinate clauses; temporal ordering may be implied rather than explicitly stated. For instance, in the example in Fig. 1 the first step of the *Registration procedure* contains a conditional clause indented differently in the text. The challenge is therefore not limited to determining *where* procedural information appears, but extends to understanding *how* it is semantically expressed and how it relates to other elements.

3 Related Work

The problem of extracting structured process representations from administrative documents has been addressed with pattern-based approaches, rule-based approaches, and LLM-based approaches.

Pattern-Based Approaches. Early attempts to extract process models from text adopt bottom-up strategies based on pattern discovery and sequence mining. These approaches assume that procedural knowledge can be inferred from recurring linguistic patterns or statistically frequent sequences observed across document collections. For instance, techniques inspired by process mining [1] have been applied to textual artifacts such as email logs or structured operational reports, where textual segments are treated as event traces. Systems such as MailOfMine [4] reconstruct workflows from email corpora by first classifying messages and then applying process discovery algorithms. Similarly, Shing *et al.* [10] combine topic modeling and sequence mining to identify activity clusters and reconstruct procedural dependencies. While these methods demonstrate that latent process structures can emerge from corpora, they rely on the availability of multiple similar instances and recurring patterns. In the context of administrative regulations, where each document may describe unique procedures and where textual regularity is limited, such corpus-driven assumptions are difficult to satisfy. Moreover, these approaches do not explicitly address traceability to source text segments, a requirement central to regulatory domains.

Rule-Based Methods. A second line of research adopts top-down approaches based on syntactic parsing, handcrafted rules, and domain ontologies. In these systems, linguistic structures such as SubjectVerbObject triples are extracted and mapped to predefined process constructs. For instance, Friedrich *et al.*[6] propose a pipeline for transforming natural language text into BPMN models using dependency parsing and rule-based transformation. Their method demonstrates that structured process models can be derived from textual descriptions when the input adheres to relatively controlled syntactic patterns. Similar approaches leverage semantic role labeling [3] and domain ontologies [13] to improve actionagent association. Although rule-based methods provide strong formal control, they require extensive engineering effort and domain adaptation. In heterogeneous administrative contexts characterized by evolving regulations and linguistic variability, maintaining rule sets and ontological mappings becomes increasingly costly. Furthermore, these approaches struggle with distributed definitions and long-range dependencies typical of regulatory texts.

LLM-Based Approaches. The recent emergence of LLMs has significantly influenced document understanding tasks. In the context of process modeling, Ajmal *et al.* [2] propose an automated BPMN diagram generation framework in which a fine-tuned generative model directly extracts structured process representations from text. This end-to-end paradigm reduces the need for handcrafted rules and improves robustness to linguistic variation. However, purely generative approaches present well-documented limitations, including hallucination phenomena and limited traceability [7]. In regulatory domains, where faithfulness to the source document is mandatory, unconstrained generation poses reliability risks. Moreover, the fixed context window of LLMs complicates the processing of long administrative documents with hierarchical structures. To mitigate the limitations of purely generative models, hybrid architectures combining retrieval and generation have been proposed. RAG [8] has been successfully applied in question answering and knowledge-intensive NLP tasks, where grounding reduces hallucination and improves factual consistency. However, its application to procedural extraction from administrative documents remains underexplored. In particular, naive RAG configurations may retrieve semantically related but structurally irrelevant fragments, potentially contaminating the generative phase. Recent research emphasizes the importance of structured context selection for reliable information extraction [9]. This work shows that combining semantic retrieval with structural constraints can enhance interpretability and robustness.

4 Proposed Approach

The architecture of the proposed pipeline separates structural analysis from semantic interpretation tasks delegated to LLMs, and integrates RAG only where context is required. This design choice ensures traceability and robustness to the linguistic and structural variability typical of administrative texts.

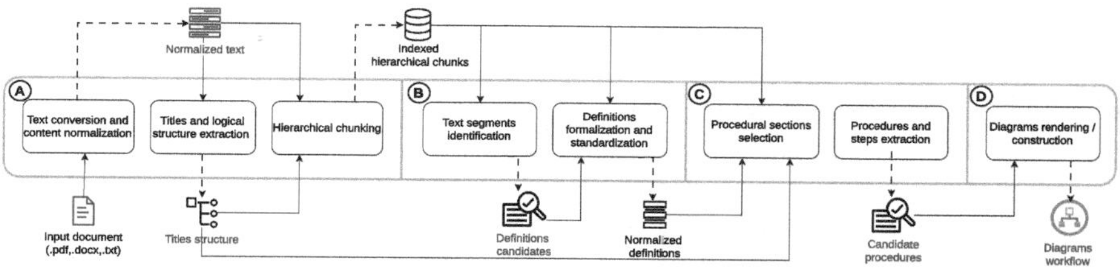

Fig. 2. Automatic process extraction.

Figure 2 illustrates the architecture of the pipeline composed of four main phases: Ⓐ *Knowledge Base Creation*, Ⓑ *Definitions Detection*, Ⓒ *Procedures Detection*, and Ⓓ *Diagrams Generation*. The input is a waste management document in unstructured textual form and the output is a diagrammatic representation of the extracted procedures.

At a glance, the workflow operates as follows. In Ⓐ *Knowledge Base Creation*, the document is normalized by removing consecutive blank lines, repeated headers and page numbers, and segmented into hierarchically organized chunks enriched with metadata. This phase produces an indexed internal representation that serves as a persistent and interrogable knowledge base. In Ⓑ *Definitions Detection*, the definitional segments are identified, normalized, and stored to build a terminological reference (e.g., "*Operator*: the person who submits notifications to the competent authority."). In Ⓒ *Procedures Detection*, candidate procedural sections are analyzed using LLMs, supported by retrieval over the knowledge base if necessary. The output is a structured JSON representation of actions, agents, conditions, and dependencies that shall appear in the extracted workflow. Finally, in Ⓓ *Diagrams Generation*, the structured representation is transformed into graphical workflows using dedicated tools. In the following, we illustrate these steps in detail using the fragment of an administrative regulation in Fig. 1 as a running example.

4.1 Knowledge Base Creation

The first phase of the approach is devoted to constructing a structured internal representation of the input document. The goal is to transform a heterogeneous text into a hierarchically organized and traceable text where titles and text between titles are isolated, to support subsequent retrieval and interpretation.

The process begins with *document acquisition and normalization*, which involves the removal of duplicated headers, page numbers, and redundant blank lines. This normalization reduces structural noise and prevents artificial fragmentation during segmentation.

Next, the system performs *structural analysis and heading detection*, which identifies titles and section headers using structural heuristics based on the following recurring patterns observed in administrative documents: (i) numbering patterns (e.g., "Art. 5", "1.", "1.1"), (ii) typographic conventions such as capitalization and short standalone lines, and (iii) domain-specific lexical markers (e.g., "Definitions", "Procedure", "Title"). These heuristics are used to extract a set of candidate headings, which are passed to an LLM that identifies false positives. This design reflects the observation that heading identification is primarily driven by syntactic and typographic features, while semantic interpretation is only required to disambiguate borderline cases.

In the running example in Fig. 1, the document contains section headers such as TITLE: REGISTRATION OF ENTERPRISES IN THE ANIMAL BY-PRODUCTS SECTOR. This heading is detected through a combination of structural heuristics, including its standalone position, typographic prominence (e.g., capitalization and absence of surrounding text), and lexical cues typical of section headers such as the marker "Title". The candidate is then validated through an LLM-based classification step to confirm its role as a structural heading.

Once the hierarchical structure is reconstructed, the text between titles/headers is segmented into *hierarchical chunks*. Each chunk corresponds to a semantically coherent portion of text within a specific section and is enriched with metadata, including: (i) section identifier, (ii) position within the section, and (iii) document identifier. This segmentation strategy produces chunks of variable length, as they are defined by the document structure rather than by fixed-size windows. Each chunk preserves semantic coherence, since it corresponds to a meaningful unit of text (e.g., a section or sub-section). In the example, the block describing the notification submission by the Operator and the subsequent verification by the Local Health Authority is stored as a chunk associated with the section "Registration of Enterprises in the Animal By-Products Sector". This chunk becomes an indexed unit in the internal knowledge base and can later be retrieved either semantically (via embeddings) or structurally (via metadata filters constrained to the parent section).

4.2 Definitions Detection

Administrative documents frequently introduce definitions that specify the meaning of roles, entities, or procedures. These definitions are often distributed across different sections and may not be globally applicable. The *Definitions Detection* phase extracts, normalizes, and organizes them before procedural interpretation.

The detection process starts from the indexed chunks produced in the previous phase. Candidate definitional segments are identified using pattern-based heuristics, derived from recurring linguistic formulations in administrative texts. In particular, the approach targets structures such as termdefinition patterns (e.g., "X: ...") and explicit definitional expressions (e.g., "For X it is meant ..." or "For the purposes of this regulation, X means ..."). These patterns are used to identify candidate segments likely to introduce definitions. In addition, structural cues are considered, such as sections explicitly labeled as *"Definitions"*, which are treated as high-confidence sources of definitional content.

In the running example, suppose the document contains the statement *"Operator who submits notifications to the competent authority."*. This segment is detected as definitional and extracted together with its structural metadata. Since definitions may vary slightly in formulation across sections, a normalization step is performed. Every time a chunk potentially including a definition is detected, an LLM is prompted to (i) isolate the defined term, (ii) extract its corresponding description through minimal paraphrasing for clarity, (iii) avoid introducing new concepts or external knowledge not present in the text, and (iv) produce a structured output in JSON.

The links between sections and the included definitions are stored for reference. This reference is later used to recognize agents in the procedural steps. For example, in the *Registration Procedure* section of the running example, a sentence such as *"The Operator submits the notification to the ASL through the SUAP."* can be interpreted by linking the term "Operator" to its definition, thereby avoiding any generic or ambiguous interpretation of the terms, ensuring coherence across sections.

4.3 Procedures Detection

The *Procedures Detection* phase constitutes the core interpretative component of the pipeline. Its objective is to identify procedures represented as actions performed by agents under specific constraints, also considering control flows information (the links between actions).

The process begins with a *deterministic pre-filtering* of sections. Rather than performing an inefficient global semantic search for actions, the system analyzes each section returned by the previous step according to terms that mark the possible presence of procedures. These include deontic expressions (e.g., "must", "is responsible for"), temporal markers (e.g., "within", "subsequently"), enumerations, and action verbs. For each fragment that is marked as possibly describing procedures, the approach interprets the text semantically by prompting an LLM to use the knowledge base of the definitions extracted so far for returning a JSON structured representation of the agent doing the action, the recipient of any data produced by the action, the temporal constraints and dependencies from other actions, if any.

Consider the following fragment from the running example: *"The Operator submits the notification to the Local Health Authority through the SUAP. The Veterinary Service of the Local Health Authority verifies the documentation and proceeds with the registration."* From this fragment, the LLM generates a structured representation including two actions. The first action is `submission of notification`, with `Operator` as agent and `Local Health Authority (via SUAP)` as recipient. The second action is `verification and registration`, with `Veterinary Service of the Local Health Authority` as agent.

The model also identifies a sequential dependency, namely that the verification activity occurs after the submission of the notification. If temporal expressions such as "within 7 days" are present, they are extracted as temporal constraints associated with the relevant action.

4.4 Diagrams Generation

The final phase transforms the structured JSON representation into graphical process models. The objective is to produce visual workflows that facilitate human interpretation while preserving alignment with the extracted structure. The transformation process adjusts the data stored in the JSON format according to the target visualization library. For instance, our prototype implementation supports both Graphviz[1], for simple free-form graph-like representations, and ProcessPiper[2] for the generation of BPMN-compliant diagrams.

In particular, the BPMN representation is obtained by applying the following rules: (a) Each action is converted into a BPMN activity node; (b) The agent is mapped into lanes, so that the responsibilities of the actions are visually attributed; and (c) The control-flow relations are translated into directed connections between activities.

5 Empirical Assessment

Our evaluation is driven by the following three research questions:
RQ1-Structure: *How accurate is the reconstruction of the hierarchical document schema?* This question focuses on the ability of the system to correctly identify titles, subtitles, and the hierarchical organization of the document, which constitutes the structural backbone for subsequent extraction steps.
RQ2-Agents: *How accurate is the identification and canonicalization of agents?* This question investigates how accurate the extraction and normalization of the agents is.
RQ3-Procedures: *How accurate is the identification of the procedures?* This question investigates the effectiveness of the proposed pipeline in detecting procedural sections and extracting structured representations of procedures.

[1] https://graphviz.org/.
[2] https://github.com/csgoh/processpiper.

5.1 Experiment Setup

The experimental evaluation was conducted on a corpus of three real administrative documents belonging to the waste management domain coming from different municipalities and regulating the waste disposal for public and private entities. These documents were selected because of their high regulatory complexity, heterogeneous structure, and strong contextual dependency between procedural sections and distributed definitions[3] The documents include regional and local regulations, operational guidelines, and administrative provisions describing notification procedures, authorization workflows, and verification processes. In several cases, definitions of roles such as "Operator", "Portal", or "Local Health Authority" are introduced in dedicated sections and later referenced anaphorically in procedural paragraphs, thus providing a realistic setting to test both structural reconstruction and semantic extraction.

For each document, we manually constructed a ground truth, where we identified titles and subtitles, marked procedural paragraphs, and explicitly specified the extracted procedures and associated agents. The first document is 16 pages long and includes 31 agent definitions and 9 procedures. The second document is 1 page and includes 10 agent definitions and 1 procedure. The third document is 10 pages long and includes 14 agent definitions and 13 procedures.

We used the three documents and the associated ground truth to address RQ1–RQ3. We studied three configurations: *pipeline-w/o RAG*, which corresponds to the structured pipeline described in this paper without activating retrieval; *pipeline with RAG*, which uses the same pipeline but integrates a RAG populated with the hierarchical chunks extracted from the input document (sections, paragraphs, and definitional segments enriched with structural metadata), and *LLM-only*, which serves as a baseline approach relying exclusively on an LLM prompted to extract procedures without structural constraints or retrieval.

In the RAG configuration, retrieval is performed over the indexed chunks of the same document and is used to help the LLM in identifying candidate procedural sections and retrieving relevant definitions. In all experiments, we used the same GPT-based[4] Large Language Model deployed through a cloud infrastructure, ensuring that differences across configurations depend on architectural choices rather than model capacity.

To answer the research questions, we compared the data extracted by the automatic procedures to the ground truth we created, computing precision, recall, and F1-score for the identification of titles and subtitles (RQ1), agents (RQ2), and procedures (RQ3).

For RQ1 and RQ2, retrieved items and ground truth elements are checked using exact matching. In the case of agents, surface variations (e.g., abbreviated forms or minor lexical differences) are tolerated only when they clearly refer to the same organizational role.

For RQ3, a predicted procedure is considered correct if it corresponds to a gold-standard procedure describing the same operational sequence within the

[3] https://github.com/AlessiaMembrini/DocuProcessAI.

[4] https://openai.com/it-IT/index/gpt-4/.

same document section. Minor lexical variations are tolerated, provided that (i) the actions exactly match those annotated in the ground truth, (ii) the responsible agents match the annotated ones, and (iii) the ordering relations between actions are preserved. A predicted procedure that partially overlaps with a gold procedure but omits essential actions or introduces additional unrelated actions is considered an incorrect match. Each gold-standard procedure can be matched to at most one predicted procedure.

The experimental material to reproduce our results is available online[5].

5.2 RQ1 - Structure

RQ1 examines how accurately the structural elements (i.e., titles and subtitles) are identified in real administrative texts, so that the document can be hierarchically partitioned into multiple text segments. In the proposed pipeline, heading detection is based on typographic and syntactic heuristics and LLM is used as a validation step to confirm or reject the candidate headings. In the LLM-only configuration, by contrast, the model is directly prompted to identify titles without the support of structural constraints. Table 1 column RQ1-Structure reports the accuracy of the three considered configurations.

Table 1. Results about the extraction of the structure.

Pipeline	RQ1-Structure			RQ2-Agents			RQ3-Procedures		
	Precision	Recall	F1	Precision	Recall	F1	Precision	Recall	F1
pipeline with RAG	**0.89**	0.76	0.82	**0.69**	**0.30**	**0.40**	0.68	0.60	0.63
pipeline w/o RAG	0.88	0.83	**0.86**	0.67	0.21	0.30	**0.67**	0.68	**0.67**
LLM-Only (baseline)	0.44	**0.98**	0.60	0.63	0.29	0.39	0.32	**0.76**	0.44

The results highlight markedly different behaviors across configurations. The LLM-only approach achieves the highest recall (0.9833), indicating that it tends to identify almost all true headings. However, its precision is significantly lower (0.4435), meaning that it frequently misclassifies non-structural text fragments as titles. In practice, the model confuses descriptive paragraphs or emphasized sentences with actual section headers. This behavior results in a large number of false positives and consequently a low F1 score.

The pipeline without RAG achieves the best F1 score (0.8561), representing the most balanced trade-off between precision and recall. The deterministic structural heuristics effectively limit the search space to plausible headings, while the LLM validation step reduces residual ambiguities. Errors in this configuration mainly occur in cases where headings deviate from expected formatting patterns or when numbering conventions are inconsistent.

[5] https://github.com/AlessiaMembrini/DocuProcessAI.

The pipeline with RAG shows slightly lower recall compared to the non-RAG configuration. This suggests that, in the context of structural detection, retrieval may introduce additional noise by supplying semantically related but structurally irrelevant fragments. Since heading identification is primarily a syntactic problem rather than a semantic one, constraining the model with semantically retrieved context does not necessarily improve performance and may occasionally mislead the validation phase.

Overall, these findings indicate that the combination of rule-based candidate generation and controlled LLM validation provides a robust compromise between completeness and precision in identifying document hierarchy.

5.3 RQ2 - Agents

RQ2 investigates the identification of the agents involved in procedures, which is a task sensitive to distributed definitions, anaphoric references, and variability in role naming. The results are reported in Tables 1 column RQ2 - Agents.

The results confirm that agent identification is the most challenging aspect of the extraction process. Recall values are consistently low across configurations, indicating that many agents are not explicitly detected. This difficulty stems from the characteristics of administrative texts: agents are frequently expressed implicitly, introduced once in a definitional section and later referred to through anaphoric expressions (e.g., "the operator", "the competent authority"), or embedded within complex syntactic constructions. As a result, purely local sentence-level interpretation often fails to recover all responsible roles.

Despite the low recall, precision remains comparatively higher across all configurations. This indicates that when the system identifies an agent, it is often correct. In other words, the approach tends to be conservative: it avoids over-generating agents, but struggles to capture all implicit references.

The pipeline with RAG achieves the highest precision, recall, and F1 score among the three configurations. Although the improvement in recall is moderate, it is consistent. This suggests that retrieval plays a beneficial role in agent identification. In particular, the RAG component enables the LLM to access definitional segments and structurally related fragments within the same document, supporting the disambiguation and canonicalization of roles. When an agent is mentioned indirectly in a procedural section, retrieval of the corresponding definition helps the model associate the expression with the correct role.

The pipeline without RAG shows lower recall, reflecting the difficulty of resolving distributed definitions without explicit contextual retrieval. The LLM-only configuration achieves a recall comparable to the RAG setting but with slightly lower precision, indicating a tendency to occasionally infer agents without sufficient grounding. R1.2These results suggest that the current RAG configuration, which is populated with chunks extracted from the same input document, is only partially effective for agent identification. A possible limitation is that the retrieved context does not always include sufficiently explicit organizational knowledge to resolve implicit or anaphoric references to responsible actors. Improving the content used to populate the retrieval component represents a promising direction for increasing recall without sacrificing precision.

5.4 RQ3 - Procedures

The objective of RQ3 is to assess the accuracy of the identification and extraction of procedural content. Results are shown in Table 1 Column RQ3-Procedures.

The LLM-only configuration achieves the highest recall (0.7578), indicating that it tends to identify a large portion of the procedures present in the documents. However, its precision is substantially low (0.3241), meaning that it frequently extracts spurious or over-segmented procedural fragments. In practice, the model often interprets descriptive or normative passages as independent procedures, resulting in a high number of false positives and consequently a low F1 score.

The pipeline without RAG achieves the highest F1 score (0.6686), representing the most balanced trade-off between completeness and correctness. The deterministic pre-filtering of candidate sections effectively reduces the search space, limiting the extraction to structurally plausible procedural segments. The subsequent LLM interpretation operates within these constrained contexts, reducing over-generation while preserving most of the relevant procedures.

The pipeline with RAG does not outperform the non-RAG configuration in this setting. Although retrieval is intended to provide additional contextual grounding, in the case of procedure detection it may introduce semantically related but procedurally irrelevant fragments from nearby sections. This additional context can blur section boundaries and lead to under-segmentation or misalignment with the ground truth, slightly reducing recall and overall F1.

5.5 Qualitative Analysis of the Generated Diagrams

To complement the quantitative evaluation, we provide a qualitative comparison of the generated diagrams. Figures 3 and 4 illustrate two complementary views of the extracted procedure.

Figure 3 shows the complete multi-boxes diagram generated from the structured JSON representation. The figure corresponds to the "Business Registration" process and displays the activities distributed across organizational actors such as the *Business Owner – Animal By-Products Sector*, the *Local Health Authority*, the *Veterinary Service*, and the *Regional Veterinary and Food Safety Service*.

The diagram can be read starting from the green "Beginning of the procedure" node and ending at the red "Conclusion of the business registration procedure" node. Each box represents a distinct organizational role, and the activities inside each box correspond to the actions extracted for that specific agent. The decision node "Check whether the portal is active" illustrates how conditional logic identified in the text is translated into an explicit branching structure.

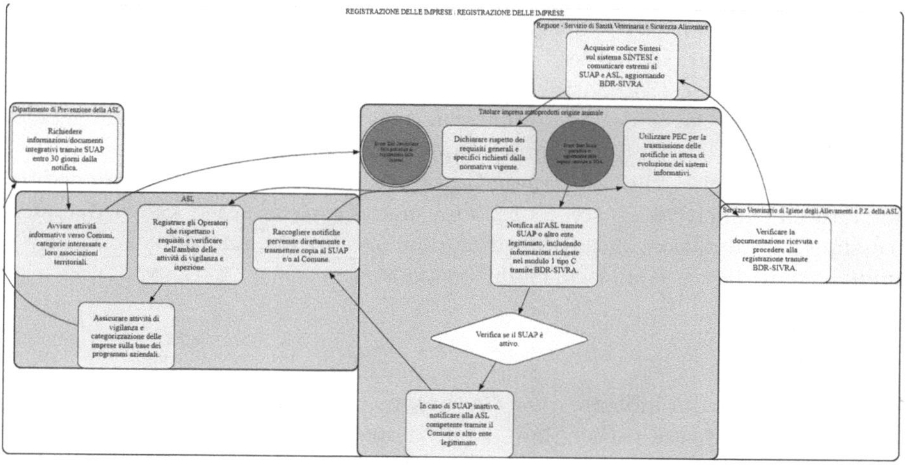

Fig. 3. Multi-lane diagram of the business registration procedure generated from the extracted structured representation.

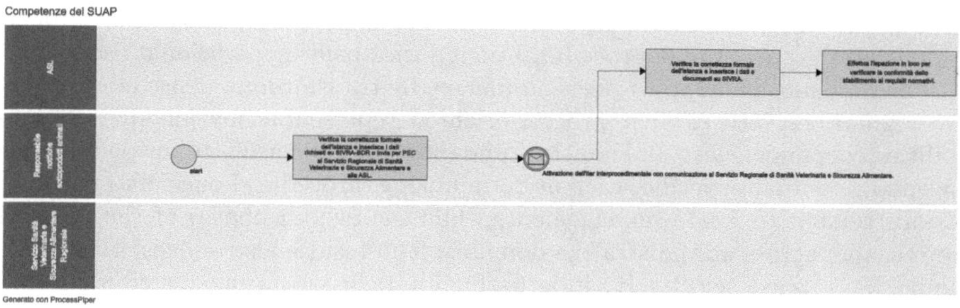

Fig. 4. Excerpt of the obtained BPMN representation of SUAP responsibilities.

Compared to the original textual description, the diagram makes the distribution of responsibilities immediately visible. For instance, the distinction between actions performed by the *Business Owner* (e.g., submitting notifications), those performed by the *Veterinary Service* (verification and registration), and those performed by the *Regional Authority* (SINTESI code assignment) becomes structurally explicit rather than being dispersed across paragraphs.

Figure 4 shows an excerpt of a BPMN-compliant representation generated with ProcessPiper. This diagram focuses on the responsibilities of the SUAP and related authorities. In this BPMN view, activities are placed within lanes corresponding to institutional roles (e.g., Local Health Authority, Responsible for animal by-products registration, Regional veterinary health and food safety service). The start event, task nodes, and sequence flows provide a standardized representation of the procedural flow.

Beyond improved clarity, the graphical representation offers analytical advantages over plain text. It enables immediate inspection of (i) responsibility distribution across actors, (ii) control-flow structure (sequence and branching), (iii) potential bottlenecks or decision points, and (iv) missing or ambiguous agents.

While textual descriptions require sequential reading and mental reconstruction of dependencies, the diagrammatic representation externalizes procedural logic into an explicit visual structure. This makes the extracted knowledge more accessible for validation, discussion with domain experts, and potential reuse in compliance or digital transformation scenarios.

5.6 Discussion

The empirical results indicate that the proposed pipeline should be considered a decision-support tool rather than a fully autonomous replacement for human process modeling. The scores obtained for structure and procedures suggest that the approach can produce useful initial process representations, especially compared to the LLM-only baseline. However, the lower performance on agent identification shows that human validation remains necessary in compliance-sensitive settings. In practice, experts have to validate agents, correct missing or implicit responsibilities, and refine procedural boundaries before operational use.

The pipeline can support decision makers by transforming dense administrative regulations into explicit process-oriented representations, making responsibilities, dependencies, ambiguities, and bottlenecks easier to inspect. Some aspects are domain-specific, such as terminology, procedural cues, and role definitions related to waste management, while the overall phases of the pipeline are reusable across administrative domains. The results also suggest that RAG should be applied selectively: it is useful for tasks requiring distributed contextual knowledge, such as resolving implicit agents or role definitions, but may introduce noise in tasks mainly driven by local structural evidence, such as heading detection. Therefore, administrations adopting similar pipelines should carefully design the knowledge base and retrieval configuration, possibly enriching them with stakeholder lists, organizational structures, and role descriptions.

These findings align with prior work on process extraction and document understanding. Pure LLM-based approaches provide flexibility, but our results confirm their risk of over-generation and limited grounding, as shown by the low precision of the LLM-only baseline for procedures. The proposed hybrid pipeline addresses these trade-offs through structural constraints, controlled LLM interpretation, and selective retrieval.

5.7 Threats to Validity

This empirical assessment is subject to several threats to validity.

Internal Validity. A first threat concerns the construction of the ground truth. Titles, procedures, and agents were manually annotated, and although the annotation process followed consistent guidelines, the identification of procedural boundaries and the canonicalization of agents may still involve subjective judgment, especially when procedures are fragmented across sections or when agents are implicit. This threat was mitigated by adopting one-to-one matching and by evaluating procedures through structural criteria (actions, agents, and ordering) rather than relying on purely textual similarity. This aspect can be strengthened in future studies by involving multiple annotators or domain experts and assessing inter-annotator agreement.

Construct Validity. The evaluation relies on precision, recall, and F1-score computed through exact matching for structure and agents, and through a structural one-to-one correspondence for procedures. These metrics capture correctness and completeness of extraction, but they do not directly measure the usefulness of the extracted artifacts for downstream tasks nor the degree of interpretability of the produced diagrams. In addition, while the procedure matching tolerates minor lexical variation, it still requires alignment on core actions, agents, and ordering, potentially penalizing alternative but equivalent decompositions of the same procedure.

External Validity. The dataset consists of three administrative documents from the waste management domain. Although the documents originate from different municipalities and exhibit heterogeneous structures, the sample size is limited, and the domain is specific. As a consequence, generalization to other administrative domains, document styles, or jurisdictions cannot be claimed without further validation. Yet, our empirical evidence is based on real-world documents. The evaluation relies on a single GPT-based model, compares against an internal LLM-only baseline, and provides an illustrative rather than standardized qualitative assessment. Future work should address these limitations, extending the empirical evidence to additional datasets, baselines, and models.

Conclusion Validity. Results may be sensitive to configuration choices and prompt formulation. To reduce this threat, all configurations used the same GPT-based model and were evaluated on the same annotated corpus, but future studies should replicate the evaluation under different model versions and parameter settings.

6 Conclusion and Future Work

This work addresses the problem of extracting structured and verifiable procedural representations from unstructured administrative documents, specifically considering the case of waste management documents. The proposed approach combines structural analysis of the documents with controlled use of LLMs, and selective use of the RAG to obtain a pipeline that can provide a visual illustration of articulated procedures.

The evaluation demonstrates the effectiveness of the approach, while it shows challenges in extracting agents and fully accurate procedural descriptions.

Future work mainly concerns the definition of strategies to improve the accuracy of agent identification. Further, considering additional documents in other areas of public administration is another relevant research direction.

References

1. Van der Aalst, W.: Process Mining: Data Science in Action, 2nd edn. Springer Publishing Company, Incorporated (2016)
2. Ajmal, F., Wijekoon, P., Dhanamina, H., Ravishan, Y., Nawinna, D., Attanayaka, B.: Automated bpmn diagram generation. In: 2024 6th International Conference on Advancements in Computing (ICAC), pp. 7–12 (2024). https://doi.org/10.1109/ICAC64487.2024.10851120
3. Cui, M., Huang, R., Hu, Z., Xia, F., Xu, X., Qi, L.: Semantic rule-based information extraction for meteorological reports. Int. J. Mach. Learn. Cybern. **15**(1), 177–188 (2024)
4. Di Ciccio, C., Mecella, M., Scannapieco, M., Zardetto, D., Catarci, T.: MailOfMine – Analyzing Mail Messages for Mining Artful Collaborative Processes. In: Aberer, K., Damiani, E., Dillon, T. (eds.) SIMPDA 2011. LNBIP, vol. 116, pp. 55–81. Springer, Heidelberg (2012). https://doi.org/10.1007/978-3-642-34044-4_4
5. Di Salle, A., et al.: Waste management through digital twins and business process modeling. In: Proceedings of the ACM/IEEE 27th International Conference on Model Driven Engineering Languages and Systems Companion, pp. 1–6 (2024)
6. Friedrich, F., Mendling, J., Puhlmann, F.: Process model generation from natural language text. In: Mouratidis, H., Rolland, C. (eds.) CAiSE 2011. LNCS, vol. 6741, pp. 482–496. Springer, Cham (2011). https://doi.org/10.1007/978-3-642-21640-4_36
7. Ji, Z., et al.: Survey of hallucination in natural language generation. ACM Comput. Surv. **55**(12) (2023)
8. Lewis, P., et al.: Retrieval-augmented generation for knowledge-intensive nlp tasks. In: Larochelle, H., Ranzato, M., Hadsell, R., Balcan, M., Lin, H. (eds.) Advances in Neural Information Processing Systems, vol. 33, pp. 9459–9474. Curran Associates, Inc. (2020)
9. Perot, V., et al.: LMDX: Language model-based document information extraction and localization. In: Findings of the Association for Computational Linguistics: ACL 2024 (2024)
10. Shing, L., et al.: Extracting Workflows from Natural Language Documents: A First Step. In: Daniel, F., Sheng, Q.Z., Motahari, H. (eds.) BPM 2018. LNBIP, vol. 342, pp. 294–300. Springer, Cham (2019). https://doi.org/10.1007/978-3-030-11641-5_23
11. Soliman, K.S., Affisco, J.F., Klischewski, R.: Ontologies for e-document management in public administration. Bus. Process. Manag. J. **12**(1), 34–47 (2006). https://doi.org/10.1108/14637150610643742
12. Vesperi, W., Melina, A.M., Ventura, M.: Organizing decision making processes in public administration: The impact of knowledge visualization. In: ECKM 2023 24th European Conference on Knowledge Management Vol 2. Academic Conferences and publishing limited (2023)
13. Zhong, B., Wu, H., Xiang, R., Guo, J.: Automatic information extraction from construction quality inspection regulations: A knowledge pattern-based ontological method. J. Constr. Eng. Manag. **148**(3), 04021207 (2022)

AI-Driven Automation in Knowledge-Intensive License Management: An Empirical Analysis of Enterprise Constraints and Automation Potential

Ume Rubab(✉) and Bernhard Axmann

Technische Hochschule Ingolstadt,
Esplanade 10, 85049 Ingolstadt, Germany
{ume.rubab,bernhard.axmann}@thi.de

BPM Artifact available

Abstract. Software License Management (SLM) is a complex, Knowledge-Intensive Process (KiP) in which fragmented data, heterogeneous IT landscapes, and frequent vendor audits contribute to significant process governance gaps, reducing organizational agility and increasing contractual and financial risk exposure. This study investigates the relationship between operational burden and contractual risk by examining SLM in the context of the Agentic Business Process Management (Agentic BPM) framework and the design–emergence gap. An explanatory sequential mixed-methods design was used, starting with a quantitative survey (N = 60) to establish three novel metrics: the Cost and Risk Exposure Index (CREI), the Operational Burden and Visibility Index (OBVI), and the Automation Readiness and Demand Index (ARDI). These quantitative findings were refined through twelve semi-structured interviews with senior IT and license managers. The findings reveal that traditional rule-based systems fail to address the dynamic nature of license governance, leaving the SLM undermanaged despite growing financial exposure. To bridge the design–emergence gap, this study presents three process governance-oriented design principles for Agentic BPM: lifecycle visibility, cross-functional coordination, and human-in-the-loop oversight. The study concludes with a proposal for a cross-functional process governance framework that shifts the focus from rigid task automation to a lifecycle-oriented agentic oversight model, thus enhancing audit readiness and supporting the development of process-oriented governance capabilities.

Keywords: Software License Management (SLM) · Knowledge-intensive Process (KiP) · Agentic BPM · Design-Emergence Gap · Process-Oriented Capabilities · Process Governance · Mixed-Methods Research

F. M. Maggi et al. (Eds.): BPM 2026, LNBIP 599, pp. 415–431, 2027.
https://doi.org/10.1007/978-3-032-37877-4_24

1 Introduction

Digital transformation has evolved beyond static automation toward the Agentic Business Process Management (Agentic BPM), where autonomous AI agents increasingly handle complex, self-directed decision-making [3,22]. Within BPM, this shift is linked to digital innovation and platform-based structures [13]. However, in governance-intensive domains, automated execution often outpaces the development of corresponding process governance structures and coordination mechanisms. This creates a structural tension between increased execution speed and the capacity for effective process governance oversight: as AI-enabled agents introduce autonomy and self-directed decision-making, responsibility for compliance, traceability, and decision authority may become fragmented across human and technical actors, complicating accountability in compliance-sensitive processes [22]. Software License Management (SLM) represents a governance-critical process domain. Organizations must coordinate the procurement, deployment, and retirement of heterogeneous licensing agreements across distributed IT infrastructures [9,15]. The growing heterogeneity of licensing models and vendor ecosystems further increases coordination and oversight requirements [18]. Prior research highlights persistent inefficiencies in license management [11] and substantial compliance risks resulting from inadequate governance structures [9,10]. Crucially, these dynamics suggest that traditional automation (e.g., RPA) fails to resolve governance vulnerabilities because it lacks the cognitive flexibility to navigate the **"Knowledge-Intensive"** [7] and often ambiguous nature of licensing contracts. While traditional automation is usually not suitable for knowledge-intensive tasks, which typically rely on manual knowledge management rather than rigid computer scripts, modern Agentic AI changes this. Rather than attempting to replace human work entirely, it aims to provide intelligent support for human experts [8]. Aligned with the principles of Agentic BPM [8], many tasks in such domains do not reflect routine work, meaning full automation is often technically unfeasible. Instead, SLM requires "augmented" systems that do not replace humans but rather leverage AI to handle non-routine tasks and adapt to context-sensitive licensing scenarios, thereby supporting process governance rather than merely increasing execution speed. This approach is distinct in its emphasis on agentic architectures, setting it apart from conventional deterministic workflows. By doing so, it successfully tackles the interpretive and cross-functional challenges associated with expert decision-making within the process.

From a BPM perspective, SLM can be conceptualized as a Knowledge-Intensive Process (KiP) [7], characterized by emergent decision-making and expertise distributed across IT, legal, and procurement functions. Such environments resist full formalization, creating a design–emergence gap in which static compliance models and predefined control mechanisms may diverge from the autonomous, goal-directed actions enacted during real-time operation [22]. This gap generates operational burden and increases financial and contractual risk exposure. Although prior research has examined RPA adoption [3] and automation in license management contexts [5,11], limited empirical work systematically connects operational workload with risk exposure in governance-intensive environments. The central challenge appears to lie in the persistent gap between formally designed governance structures and enacted operational practice.

To address this gap, this study investigates the following: **How do operational burden and contractual risk exposure interact in SLM, and how can Agentic automation be institutionally embedded to bridge the design-emergence gap?** Using an explanatory sequential mixed-methods design [4,16], the study combines a survey of SLM practitioners in the **DACH region (Germany, Austria, and Switzerland)** ($N = 60$) with semi-structured expert interviews ($N = 12$). The findings indicate a consistent relationship between manual reconciliation effort and perceived audit risk, leading to governance-oriented automation design principles emphasizing lifecycle transparency, cross-functional coordination, and human-in-the-loop oversight. This study makes three contributions to BPM research. First, it conceptualizes Software License Management as a Knowledge-Intensive Process and reframes automation challenges as a design–emergence misalignment problem. Second, it provides empirical evidence linking operational workload to contractual risk exposure, demonstrating that automation without governance integration may amplify rather than reduce compliance vulnerability. Third, it derives process governance-oriented design principles for Agentic BPM, extending current automation research beyond efficiency logics toward institutional embedding and the development of process-oriented governance capabilities.

2 Related Work

2.1 Software License Management as a Knowledge-Intensive Process

SLM in large enterprise environments constitutes a structurally complex governance activity shaped by heterogeneous vendor ecosystems and evolving contractual models. Modern B2B software portfolios increasingly consist of bundled offerings, embedded third-party components, and subscription- or consumption-based licensing schemes [14]. Organizations often manage thousands of license types across hundreds of vendors, generating millions of distributed data records across servers and client systems [9]. Beyond technical scale, SLM is characterized by interpretative uncertainty. Licensing agreements require contextual legal interpretation rather than purely rule-based enforcement. These features position SLM as a **KiP**, where outcomes depend on expert judgment and cross-functional coordination rather than deterministic workflows [7]. This decentralized complexity aligns with the foundational view of agentic BPM, which conceptualizes business processes as a collection of autonomous, problem-solving entities that interact when interdependencies arise [12]. By viewing organizational groups as relatively autonomous actors that coordinate via Service Level Agreements (SLAs), this framework provides a theoretical basis for the "Agentic" oversight required in SLM. Consequently, SLM reflects two core KiP characteristics: **Process Unpredictability and Constraint-Driven Decision Making:** Governance trajectories evolve dynamically in response to audits, contract changes, infrastructure changes, or usage fluctuations. Managers interpret contractual entitlements under incomplete visibility, relying on expert heuristics and tacit knowledge [6,7,22]. This kind of knowledge is essential because the wording of explicit contracts often leaves product use cases open to interpretation. This means that managers have to rely on historical unwritten agreements and past audit experiences to evaluate actual compliance positions. **Knowledge Fragmentation and Coordination Intensity:** Compliance-relevant information is distributed across IT, procurement, legal and vendors, requiring sustained cross-departmental synchronization [5,13].

2.2 Governance and Automation Gaps: From RPA to Agentic BPM

Despite strategic importance of automation, a significant **automation gap** persists between the formal design of the SLM process and management. This gap is intensified by the rapid expansion of governance and legal technology markets, projected to grow from US$27.6 billion in 2021 to over US$35.6 billion by 2027 [19]. As portfolios grow, exposure to audit and contractual risk increases accordingly. Empirical evidence shows that 61% of senior executives experience supplier inspections within a single year [2]. Inadequate compliance governance can lead to financial penalties and reputational damage [2,10,15]. Current automation approaches frequently fail to address this complexity. While RPA supports repetitive tasks, its rule-based architecture lacks adaptability for interpretative licensing contexts [3,11,22]. According to the AI-Augmented BPM Manifesto, these automation gaps occur because traditional systems cannot handle tasks that do not reflect routine work or require context-sensitivity [8]. Consequently, automated components often reach only 12%–16% of their intended utility [10], leaving substantial manual workload. Consequently, traditional automated components achieve only 12%–16% of their intended utility [5,11], leaving a "manual residue" where managers spend 80%–90% of effort on the validation of low-value data [14,20]. This manual coordination creates transparency deficits, directly increasing compliance vulnerability [9,18].

These limitations motivate a transition to agentic business process management (Agentic BPM) [11]. Rather than embedding static automation in predefined workflows, Agentic BPM coordinates adaptive agents capable of handling routine analytical tasks while escalating high-risk legal decisions. This modern agentic approach realizes the core agent property of "autonomy," where systems perform tasks without direct intervention and exercise control over their internal state to manage complex, distributed resources [12]. However, the inherently unpredictable nature of generative AI poses challenges for governance [22]. Configurable autonomy and structured human-in-the-loop (HITL) oversight are therefore essential [13,22]. In addition, immutable event logs are required to ensure traceable and audit-ready compliance documentation [21]. The grounding of this transition within the design-emergence gap establishes a clear relationship between administrative strains and legal vulnerabilities. When an organization faces an extensive manual workload due to disconnected data silos, operational oversight drops. This transparency deficit makes it difficult to track contractual obligations accurately, directly increasing compliance gaps and unexpected audit penalties. To empirically investigate this relationship, the core hypothesis is formulated as follows:

- H_1: Manual operational burden in SLM is positively correlated with perceived contractual risk exposure.

3 Methodology

3.1 Research Approach

This study employs an explanatory sequential mixed-methods research design to investigate how operational burden and financial/contractual risk exposure interact in SLM, and how automation can be institutionally embedded to bridge the design–emergence gap. This approach was chosen to address the inherent complexity of the research problem, combining the breadth of quantitative data with the depth of qualitative insights to provide a comprehensive answer to the research question [4]. In the quantitative phase, an online survey was conducted among practitioners in the DACH region (N=60) to identify baseline patterns in license management practices and challenges. Survey items focused on operational pain points, responsibility delineation, dispute frequency, Service Level Agreement (SLA) verification, and desired automation features. The qualitative phase followed an explanatory logic, using semi-structured expert interviews (N=12) to elaborate on and contextualize the quantitative findings. Interview participants were purposively sampled, and the data were analyzed using thematic analysis to uncover recurring themes and explanatory mechanisms. Figure 1 illustrates the explanatory sequential mixed-methods research design employed in this study.

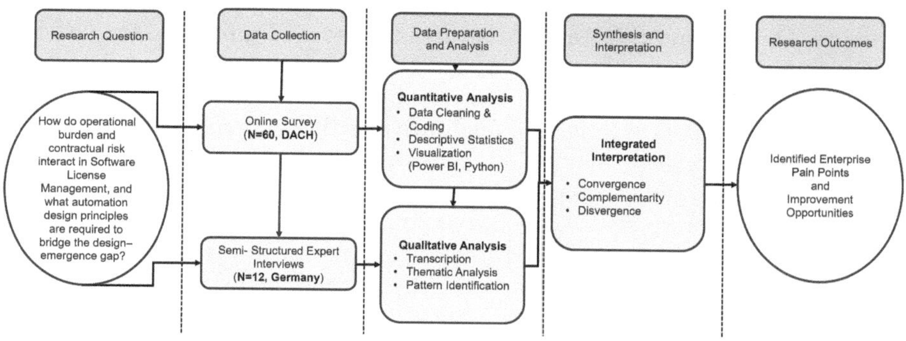

Fig. 1. Explanatory sequential mixed-methods research design.

During the survey and the phase of finding interview partners, it was a significant challenge to find support; feedback very often indicated that Software

License Management has no high priority in the company. Quantitative and qualitative findings were integrated at the interpretation stage through joint analysis, allowing for the identification of convergence and complementarity between data sources. Within the research framework, the mixed-methods approach combines survey and interview strategies to ensure both breadth and depth in this explanatory BPM study.

3.2 Quantitative Survey Design

The quantitative phase adopted a cross-sectional online survey design to map prevailing practices and identify enterprise pain points in the DACH region. A non-probability purposive sampling approach was applied due to the absence of a comprehensive public sampling frame for software license managers, making probability-based sampling impractical in specialized organizational contexts [17]. Participants were recruited via professional networks, targeting roles in procurement, IT asset management, and contract governance. The survey instrument was developed through a multi-stage validation pipeline to ensure structural and content integrity. Initial item pools were derived from a systematic review of established SLM governance literature to ground the constructs in verified empirical themes. To establish face validity, these items were subjected to an internal practitioner review by enterprise senior management, who refined the phrasing to align with corporate workflows. To ensure instrument reliability, a pilot study was initiated by contacting seven industry practitioners. A final panel of three field experts responded and completed the pilot test ($N = 3$), providing targeted improvements to item clarity and resolving technical ambiguities before the deployment of the final dual-language instrument. Data analysis was performed using Pearson correlation and simple linear regression. To facilitate rigorous analysis, individual multi-point survey items were normalized and aggregated into three composite constructs to capture cumulative enterprise friction: the Cost and Risk Exposure Index (CREI), the Operational Burden and Visibility Index (OBVI), and the Automation Readiness and Demand Index (ARDI), $\text{CREI} = \text{Q1} + \text{Q2}$ (ranging from 0 to 4) and $\text{OBVI} = \text{Q3} + \text{Q4} + \text{Q5} + \text{Q6}$ (ranging from 2 to 15), while $\text{ARDI} = \sum \text{Q7}$ tracks the total count of demanded automation features. Calculation logic, and qualitative analysis documentation are fully available as a permanent open-science research artifact [23]. The final survey reached 4,826 potential respondents via targeted digital outreach, resulting in 60 fully completed responses. The low response rate (1.24%) reflects the niche, compliance-sensitive nature of SLM roles, where practitioners often deprioritize participation due to time constraints and the perception that **"if everything is running, all is well."** While the achieved sample remains analytically valid and robust for explanatory modeling in specialized business-to-business domains [4,17].

3.3 Qualitative Expert Interviews

Following the survey, twelve experts were selected via purposive sampling for semi-structured interviews to provide an explanatory sequential depth to the quantitative patterns. This approach allows the **operational burden** of SLM to be interpreted through practitioners lived experiences [4]. The interview guide targeted inter-departmental silos, SLA compliance verification, and requirements for transitioning toward agentic automation [22].

Participants, including senior consultants and contract and license managers, provided 30–45 min accounts which were transcribed and analyzed using Braun and Clarke's reflexive thematic analysis [1]. The coding pipeline followed an inductive approach: raw transcript segments were compiled into an Excel-sheet to sort participant expressions into initial codes and subsequent subthemes. An LLM acted strictly as a technical assistant within this spreadsheet workflow to expedite initial categorization; however, all conceptual abstraction, thematic refinement, and final data interpretations were executed entirely by the researcher. Through inductive coding, broader patterns such as fragmented responsibility and transparency deficits were identified. This methodological triangulation, synthesizing qualitative themes with descriptive statistics, strengthens the study's validity and addresses knowledge gaps regarding how digital innovation emerges *in situ* within complex BPM environments [13]. Table 1 summarizing demographic and professional characteristics of 12 expert participants across company sizes, roles, organizational contexts, and experience levels

Table 1. Expert participant characteristics

ID	Company Size*	Role	Organizational Context	Experience
1	Medium	Senior Consultant	Automation Consulting firm	10–15 years
2	Large	Head of Business Intelligence	Global Logistics/Rental Services	25+ years
3	Medium	Director System Operations	Insurance and Financial Services	20–25 years
4	Large	Academic & Former Manager	Business Services	20+ years
5	Small	IT Operations Coordinator	IT Service Providers	10–15 years
6	Small	Senior Consultant	Digital Transformation Firm	5–10 years
7	Large	Team Coach for Software	Original Equipment Manufacturer	15–20 years
8	Medium	Service Manager	Digital Solution Provider	Not reported
9	Large	General Manager	Original Equipment Manufacturer	25+ years
10	Medium	Contract Manager	IT Infrastructure Services	5–10 years
11	Large	Software Asset Manager	Telecommunication	20+ years
12	Large	License Manager	Service Provider	20+ years

*Definitions per EU Recommendation 2003/361: **Small**: $<$50 employees; **Medium**: $<$250 employees; **Large**: $\geq$250 employees

4 Findings

4.1 Overview of Empirical Data

The results of this empirical study are derived from an explanatory, sequential, mixed-methods approach (see Table 2). The analysis focuses on three key constructs designed to measure the current state of SLM: **Operational Burden and Visibility Index (OBVI)**: This index quantifies the "manual residue" and administrative effort within the license lifecycle. It aggregates scores from survey items related to SLA verification, usage tracking, and cost analysis (Q3–Q6). To facilitate variance analysis, a categorical OBVI-Level (Low, Medium, High) is also derived from this continuous score. **Cost and Risk Exposure Index (CREI)**: This construct captures the perceived financial and legal vulnerabilities. It is calculated based on the frequency of missed contract deadlines and the occurrence of disputes arising from ambiguous clauses (Q1–Q2). **Automation Readiness and Demand Index (ARDI)**: Unlike the perception-based indices, the ARDI is a count-based metric representing the volume of specific technical interventions, such as automated renewal alerts and compliance monitoring demanded by practitioners (Q7). Table 2 provides an overview of these constructions, including their full names and the specific measurement focus each serves in the analysis. The sample comprises a diverse range of industries and organizational sizes. While segmentation variables, including role, experience, and sector, were not used for index construction, they facilitated subgroup analysis and comparative visualizations to validate the findings.

Table 2. Mapping of Research Constructs to Survey Items

Abbreviation	Full Name	Survey Items
CREI	Cost and Risk Exposure Index	Q1, Q2 (Contractual Risk)
OBVI	Operational Burden and Visibility Index	Q3, Q4, Q5, Q6 (Admin Burden)
OBVI-Level	OBVI Intensity Level	Derived from OBVI
ARDI	Automation Readiness and Demand Index	Q7 (Automation Demand)

Note: Full item descriptions and survey link are available in the Data Availability section

4.2 Quantitative Survey Results

Descriptive statistics provide an initial overview of how participating organizations perceive operational burden, contractual risk, and demand for automation, prior to testing the specific research hypotheses. Figure 2 shows the shape of distribution and variability to justify treating the indices as approximately continuous and suitable for correlation/regression.

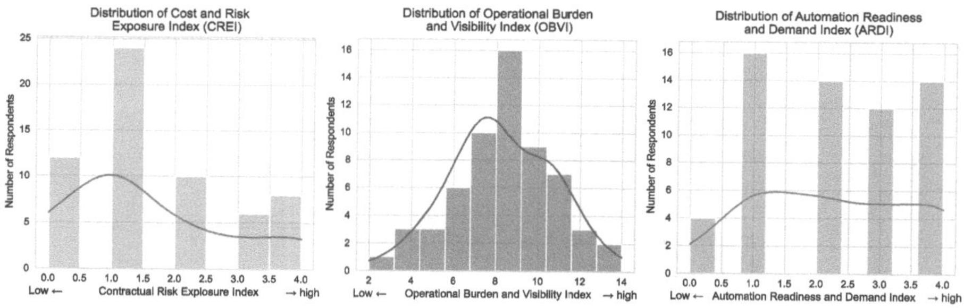

Fig. 2. Distribution of Indices derived from Survey Data.

The descriptive statistics for the three constructs are summarized in Table 3. To facilitate comparison across different scale ranges, the Percentage of Maximum Possible (POMP) score was calculated for each construct using the formula:

$$POMP = \frac{Mean - \min}{\max - \min} \times 100 \quad (1)$$

The approximately bell-shaped distribution of the OBVI (see Fig. 2) justifies the use of parametric analyses. The POMP scores provide a normalized view of the findings. The ARDI (56.75%) shows the highest relative intensity, indicating a strong baseline demand for automation among practitioners. The OBVI (52.25%) suggests that the perceived administrative burden is moderate-to-high across the sample, while the CREI (39.25%) indicates that while contractual risk is present, it is currently perceived as lower than the operational challenges.

Table 3. Descriptive statistics and POMP scores for research constructs

Index	N	Min	Max	Mean	SD	POMP(%)
CREI	60	0.0	4.0	1.57	1.29	39.25
OBVI	60	2.0	14.0	8.27	2.44	52.25
ARDI	60	0.0	4.0	2.27	1.27	56.75

To enhance interpretability, the OBVI scores were categorized into three intensity levels based on the cumulative index range of 1 to 15: Low Burden (1–5), Moderate Burden (6–10), and High Burden (11–15). This equal-interval grouping allows for a clear comparative analysis of how increasing administrative complexity correlates with contractual risk and automation demand. Specifically, only a small minority of participants (11.7%, $n = 7$) reported a Low Burden level. In contrast, the largest segment of the sample (68.3%, $n = 41$) is characterized by a Moderate Burden, while 20.0% ($n = 12$) experience a High Burden. As illustrated in Fig. 3, a clear progression emerges: increasing operational burden

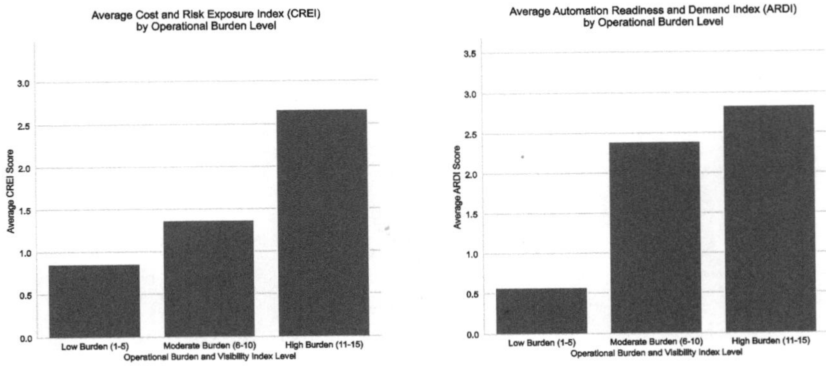

Fig. 3. Distribution of Indices derived from Survey Data.

corresponds with higher contractual risk and automation demand. This reinforces the hypothesis that automation's perceived value is highest where manual governance is unsustainable [9]. To examine the association between operational burden and contractual risk, this study utilizes Pearson's product-moment correlation and a simple linear regression model. The analysis reveals a Pearson correlation of $r = 0.4880$ (Table 4), signifying a moderate positive association. Substantively, organizations reporting higher manual effort in license management consistently report higher risk exposure indicators. To test H_1, a linear regression model was estimated:

$$\mathrm{CREI}_i = \beta_0 + \beta_1 \mathrm{OBVI}_i + \epsilon_i \tag{2}$$

Table 4. Correlation and Regression Results: OBVI as a Predictor of CREI

Statistic	Value
Pearson correlation (r)	0.4880
Regression coefficient (β_1)	0.2590
Intercept (β_0)	−0.5710
Coefficient of Determination (R^2)	0.2380
p-value (for β_1)	0.0001

As detailed in Table 4, the estimated slope ($\beta_1 = 0.2590$, $p = 0.0001$) confirms that a one-unit increase in operational burden leads to a statistically significant rise in risk exposure. With an R^2 of 0.2380, the model explains approximately 23.8% of the variance in CREI through OBVI. The relative demand for specific automation functionalities such as automated usage versus contract checks, renewal reminders, SLA tracking and GDPR related checks were also recorded. The largest share of expressed interest concerns automated comparison of license usage versus contractual entitlements (37.5%), followed by automated renewal reminders (27.9%), while automated SLA tracking (16.2%) and automated GDPR checks (18.4%) are somewhat less prominent, which helps prioritize feature design for potential tooling.

In summary, the empirical data highlights a critical **automation gap** within current SLM practices. The analysis demonstrates that a significant majority of practitioners (88.3%) operate under moderate-to-high operational burdens, which directly correlates ($r = 0.4880, p < 0.001$) with increased contractual and financial risk exposure. Furthermore, the high relative intensity of the ARDI (56.75%) underscores a proactive demand for automated interventions, particularly regarding license usage and renewal tracking. These findings statistically validate the transition from manual governance to an Agentic BPM approach as a necessary evolution for mitigating risk and reducing **manual residue** in complex organizational environments.

4.3 Qualitative Interview Insights

This section presents the qualitative findings derived from twelve semi-structured interviews with key stakeholders, including consultants, IT service providers, contract managers, and license managers. To assess the perceived urgency of SLM governance, expert interviews were mapped against a 5-point priority scale. A high concentration of "Strategic" urgency (Rank 4–5) was observed among 75% of respondents. Notably, participants 4, 7, 10, and 11 categorized SLM as a "Highest Priority" (Rank 5), citing non-negotiable drivers such as audit-related financial risk, tax compliance, and SaaS forecasting. In contrast, only one participant (Expert 5) characterized the process as a **necessary evil**, suggesting a broader industry shift toward professionalized, high-stakes governance.

Thematic analysis revealed six overarching themes that enrich the survey findings: Structural fragmentation, operational burden, automation constraints, financial risk, managerial awareness, and usage transparency. The interviews suggest a perceived sequential pattern of organizational friction, in which fragmentation of data repositories and limited contract visibility are described by participants as foundational structural conditions. According to Experts, these visibility gaps contribute to reactive administrative cycles, as human actors engage in labor-intensive reconciliation processes to align formal license entitlements with actual software consumption. Expert 3 emphasized this operational strain, noting that contract renewals become exceptionally complex because teams must gather extensive, disparate utilization datasets. Rather than representing isolated inefficiencies, participants characterized these activities as recurring "firefighting" patterns that absorb managerial attention and operational resources.

As these manual coordination efforts intensify, experts reported increasing difficulty in achieving strategic oversight of the license estate. In this interpretation, fragmentation and coordination ambiguity are perceived as reinforcing conditions that increase both compliance vulnerability and financial exposure. These themes are visually integrated into an Interdependency Model (Fig. 4),

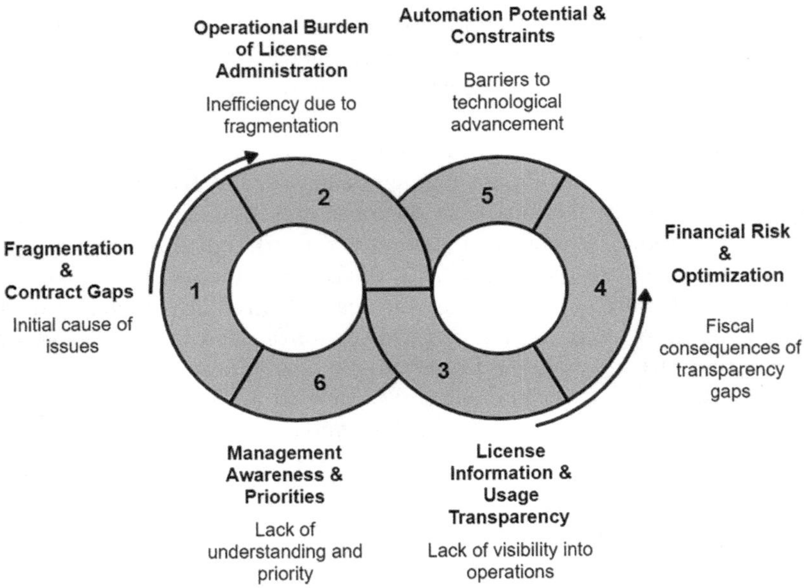

Fig. 4. Interdependency Model of Software License Management Pain Points

which illustrates a sequential increase in company friction. Mapping 208 coded segments against a six-theme framework (see Fig. 5) reveals that friction is concentrated in Fragmentation and Contract Gaps (24%, n=50), Operational Burden (22.1%, n=46), and Automation Constraints (21.6%, n=45), followed by

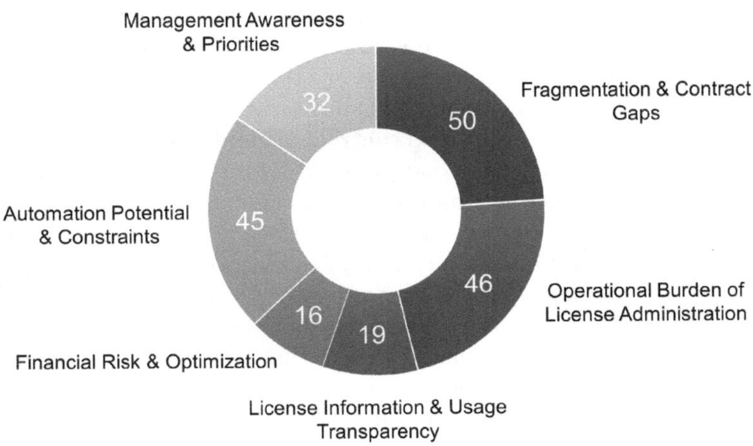

Fig. 5. Distribution of Thematic Codes and Narrative Prominence

Management Awareness (15.3%), Usage Transparency (9.1%), and Financial Risk (7.7%).

Participants identified ambiguous ownership across IT, procurement, and legal as a primary friction driver; Expert 11 highlighted this systemic strain, noting that the centralized license group holds financial accountability while individual departments generate the resource consumption. This structural fragmentation is exacerbated by data silos and a heavy dependency on localized tribal knowledge. Consequently, daily operations are defined by incident-driven "firefighting" and manual tracking, leading to renewal time pressures and a practitioner-estimated "waste premium" of 20% to 40% from defensive over-licensing to mitigate audit risks. Despite these stakes, a cultural gap treats license management as a secondary task; shifting to Agentic BPM remains hindered by technical heterogeneity, with Expert 3 noting that non-standardized vendor APIs block consistent data aggregation.

4.4 Synthesis and Integration of Results

The synthesis of qualitative and quantitative findings provides a triangulated perspective on enterprise governance friction (see Table 5).

Table 5. Triangulation of Qualitative (Interview) and Quantitative (Survey) Findings

Qualitative Theme (Interview)	Quantitative Index (Survey)	Synthesis of Mixed-Methods Discovery
Organizational Fragmentation and Visibility Gaps	CREI (Risk)	Thematic Lead (24%): Diffuse ownership and "Tribal Knowledge" provide the causal context for why 39.25% of respondents feel high risk exposure.
Operational Burden and Manual Residue	OBVI (Burden)	High Prominence (22.1%): Qualitative descriptions of "Firefighting" and reactive cycles validate the 88.3% Moderate-to-High burden measured in the survey.
Automation Potential, Constraints, and Expectations	ARDI (Demand)	Practitioner demand for Agentic BPM and proactive monitoring is reflected in the high ARDI intensity (56.75%).
Financial Risk Exposure and Over-Licensing	CREI (Indirect)	Experts estimate a 20–40% expenditure loss due to defensive over-licensing, explaining the high stakes behind CREI scores.
Managerial Awareness and Priority Gaps	Contextual Variable	The Governance Paradox: While 75% of experts see "Strategic Urgency," institutional blindness often results in SLM being underweighted as a "secondary task.
License Information and Usage Transparency	OBVI / CREI (Drivers)	Structural Friction: Limited visibility into entitlements (9.1% prominence) is identified as the root cause driving both administrative strain and audit vulnerability.

Although quantitative analysis identifies a statistically significant positive association between OBVI and CREI ($r = 0.48$, $R^2 = 0.238$), providing empirical support for H_1-**Manual operational burden in SLM is positively correlated with perceived contractual risk exposure**. Qualitative interviews offer an explanation of the mechanisms underlying this relationship. Participants consistently described how structural fragmentation, diffuse ownership, and limited contract transparency translate into reactive administrative workloads, thereby increasing perceived compliance and financial exposure.

For example, the structural tension over budget ownership highlighted by Expert 11, where centralized license teams shoulder financial accountability while decentralized business units drive consumption without cost sensitivity directly illustrates how administrative fragmentation feeds day-to-day operational strain. During the survey and the recruitment of interview partners, it was a significant challenge to secure organizational support; feedback frequently indicated that Software License Management is not perceived as a high priority within the company. In this sense, the qualitative evidence contextualizes the statistical association by illustrating how governance fragmentation manifests in day-to-day operational strain. This recruitment difficulty reflects a "Governance Paradox": while the data proves SLM carries high financial risk, the lack of organizational priority during the research phase suggests a widespread institutional blindness to these costs.

Furthermore, the alignment between high OBVI levels and ARDI is mirrored in interview narratives emphasizing the need for integrated data visibility, proactive monitoring, and reduced reliance on manual reconciliation. Rather than suggesting deterministic causality, the combined findings indicate that fragmentation, burden, and risk are perceived by practitioners as mutually reinforcing elements of a broader governance challenge. The proposed AI-enabled knowledge and automation framework is therefore positioned not as the sole solution, but as a potential governance intervention aimed at mitigating these interdependencies by enhancing transparency, accountability, and process integration.

5 Practical Implications and Limitations

The findings provide actionable guidance for organizations seeking to reduce enterprise friction and optimize SLM. To address the structural fragmentation identified as a root driver of operational burden and contractual risk, firms should establish permanent cross-functional governance structures that integrate IT, procurement, legal, and finance functions. Standardized handover protocols and centralized contract repositories can reduce information asymmetries and mitigate the siloed decision-making patterns described by interview participants.

Automation investments with Agentic BPM should be strategically prioritized. Survey results indicate that usage-contract reconciliation (37%) and renewal management (28%) represent the most critical intervention areas. To bridge this gap, three design principles were proposed for Agentic BPM: (1) AI agents must connect data across separate IT, legal, and procurement silos, (2) the system must require human approval for high-risk or high-cost decisions, and (3) every AI action must be permanently logged to ensure the company is always ready for a vendor audit. High-burden organizations should prioritize interoperable Software as a Service (SaaS) platforms with API-based metering capabilities to enable real-time entitlement tracking. However, development teams must account for the integration roadblocks raised by Expert 3, who noted that because every software utility leverages distinct, non-standardized APIs, uniform data aggregation remains an explicit barrier. Executive dashboards that visualize the relationship between operational burden and financial exposure can further enhance managerial awareness and reposition SLM from an administrative back-office task to a strategically managed cost-control function capable of reducing the estimated 20–40% waste premium. As highlighted by Expert 11, such visualization tools are critical for injecting cost sensitivity across external departments that often act as if software budgets "get from heaven." Despite these contributions, several limitations must be acknowledged. The survey sample (N=60) and the predominance of medium-sized firms (51–1,000 employees) may limit generalizability to micro-enterprises or large multinational corporations. Moreover, the use of self-reported indices introduces potential common method bias, although qualitative triangulation partially mitigates this concern.

Future research should employ longitudinal designs to clarify causal dynamics within the Burden–Risk Nexus and assess whether fragmentation precedes risk exposure or reflects broader organizational immaturity. As AI-driven governance solutions with Agentic BPM mature, further studies should empirically evaluate the effectiveness of specific automation frameworks in narrowing the execution gap between system-level usage monitoring and contractual compliance. Expanding the empirical base across industries and organizational scales would strengthen validation of the enterprise friction model proposed in this study.

6 Conclusion

The findings of this study confirm H_1: Manual operational burden in SLM is positively correlated with perceived contractual risk exposure ($r = 0.48$), demonstrating that contemporary software governance is structurally constrained by fragmentation. As corroborated by experts, tracking hardware environments is notoriously complex, carrying a permanent risk that unmanaged servers remain disconnected from tracking systems. These decentralized structures and information asymmetries are systemic drivers of compliance exposure. A central contribution of this research is identifying an **institutional blind spot**: while recruitment proved difficult due to a perception that SLM is a low-priority task, 75% of engaged experts categorized it as a "Highest Priority," creating a sharp contrast between cultural de-prioritization and actual financial weight. Qualitative insights revealed a **waste premium**, where organizations deliberately over-license by 20–40% to mitigate audit uncertainty. This waste is compounded by multi-tiered vendor structures where individual products carry five to six specialized user roles, penalizing organizations by both volume and functional access.

Together, these findings show that lacking a transparent data foundation perpetuates reactive governance and limits strategic control, even within firms spending up to 120 million annually. To address this friction, we propose transitioning toward integrated Agentic BPM frameworks. Advancing SLM maturity requires bridging the design-emergence gap by shifting from rigid task-automation to autonomous, goal-directed agents that navigate data silos. By targeting fragmentation and embedding traceable, human-supervised automation, organizations can reposition SLM from a reactive cost-control function to a strategically managed capability that enhances regulatory resilience. While this study establishes the empirical baseline and conceptual principles, the physical technical deployment and orchestration of these software agents constitute the immediate next phase of this research program.

Data Availability

The research instrument, including the survey questionnaire, index operationalization logic (OBVI, CREI, ARDI), and qualitative interview protocols, is available in an anonymized repository [23]. Raw interview transcripts are withheld to protect participant confidentiality, will be provided on demand.

Disclosure of AI Assistance

Authors used a Large Language Model to assist with the grammatical refinement and stylistic editing of the manuscript. The conceptual framework, empirical data analysis, and conclusions represent the original intellectual work of the authors.

Disclosure of Interests. Authors have no competing interests to declare that are relevant to the content.

References

1. Braun, V., Clarke, V.: Using thematic analysis in psychology. Qual. Res. Psychol. **3**(2), 77–101 (2006). https://doi.org/10.1191/1478088706qp063oa
2. CDW: Software License Management White Paper. CDW LLC (2013)
3. Costa, D.A.D.S., Mamede, H.S., Mira Da Silva, M.: Robotic process automation (RPA) adoption: a systematic literature review. Eng. Manag. Prod. Serv. **14**(2), 1–12 (2022). https://doi.org/10.2478/emj-2022-0012
4. Creswell, J.W.: Research Design: Qualitative, Quantitative, and Mixed Methods Approaches, 4th edn. SAGE Publications, Thousand Oaks (2014)
5. Daliparthi, V.S.S.A., et al.: A License Management System for Collaborative AI Engineering. In: Proceedings of AICCC 2024, pp. 77–86. ACM (2024). https://doi.org/10.1145/3719384.3719395
6. De Reuver, M., Sørensen, C., Basole, R.C.: The digital platform: a research agenda. J. Inf. Technol. **33**(2), 124–135 (2018). https://doi.org/10.1057/s41265-016-0033-3
7. Di Ciccio, C., Marrella, A., Russo, A.: Knowledge-intensive processes: characteristics, requirements and analysis. J. Data Semant. **4**(1), 29–57 (2015). https://doi.org/10.1007/s13740-014-0038-4
8. Dumas, M., et al.: AI-augmented business process management systems: a research manifesto. ACM Trans. Softw. Eng. Methodol. **32**(3), 1–20 (2023). https://doi.org/10.1145/3583565
9. Durdu, A., Tutğaç, A.: Examination of risks in management of software licenses. Bil. Tek. Derg. **15**(4), 467–479 (2022). https://doi.org/10.17671/gazibtd.1019869
10. Flexera: State of the Stack: Audit costs and licensing chaos (2025). https://www.flexera.com

11. Janik, D.: Optimization of license management for BPM with RPA. Int. J. Contemp. Manag. **60**(1), 280–289 (2024). https://doi.org/10.2478/ijcm-2024-0017
12. Jennings, N.R., Faratin, P., Johnson, M.J., Norman, T.J., O'Brien, P., Wiegand, M.E.: Agent-based business process management. Int. J. Coop. Inf. Syst. **5**(02n03), 105–130 (1996). https://doi.org/10.1142/S021884309600006X
13. Mendling, J., Pentland, B.T., Recker, J.: Building a complementary agenda for BPM and digital innovation. Eur. J. Inf. Syst. **29**(3), 208–219 (2020). https://doi.org/10.1080/0960085X.2020.1755207
14. Parycek, P., Schmid, V., Novak, A.S.: AI and automation in administrative procedures. J. Knowl. Econ. **15**(2), 8390–8415 (2023). https://doi.org/10.1007/s13132-023-01433-3
15. Pozder, N., Axelsson, J.: End-to-End Software License Management, Master & Thesis. Mälardalen University (2022)
16. Punter, T., et al.: Conducting on-line surveys in software engineering. In: Proceedings of ISESE 2003, pp. 80–88. IEEE (2003). https://doi.org/10.1109/ISESE.2003.1237967
17. Saunders, M., Lewis, P., Thornhill, A.: Research Methods for Business Students, 9th edn. Pearson, Harlow (2023)
18. Statista: Revenue of the Global Software Market. Statista Market Insights (2023)
19. Statista: Global Legal Tech Market Revenue.Statista, pp. 2021–2027 (2024)
20. Van Der Aalst, W.: Process Mining. Springer, Cham (2016). https://doi.org/10.1007/978-3-662-49851-4
21. Van Der Aalst, W.M.P., Carmona, J. (eds.): Process Mining Handbook. LNBIP, vol. 448. Springer (2022). https://doi.org/10.1007/978-3-031-08848-3
22. Vu, H., et al.: Agentic BPM: Practitioner Perspectives. arXiv:2504.03693 (2025). https://doi.org/10.48550/arXiv.2504.03693
23. Ume Rubab., Axmann, Bernhard.: Research Artifacts for: AI-Driven Automation in Knowledge-Intensive License Management. Zenodo Dataset (2026). https://doi.org/10.5281/zenodo.20376810

Author Index

F. M. Maggi et al. (Eds.): BPM 2026, LNBIP 599, pp. 433–434, 2027.
https://doi.org/10.1007/978-3-032-37877-4

FSC
www.fsc.org
MIX
Papier | Fördert gute Waldnutzung
FSC® C083411

Kolibri 360 GmbH
Ferdinand-Jühlke-Straße 7
99095 Erfurt, Deutschland
produktsicherheit@kolibri360.de